Industry 4.0 Solutions for Building Design and Construction

This book provides in-depth results and case studies in innovation from actual work undertaken in collaboration with industry partners in Architecture, Engineering, and Construction (AEC). Scientific advances and innovative technologies in the sector are key to shaping the changes emerging as a result of Industry 4.0. Mainstream Building Information Management (BIM) is seen as a vehicle for addressing issues such as industry fragmentation, value-driven solutions, decision-making, client engagement, and design/process flow; however, advanced simulation, computer vision, Internet of Things (IoT), blockchain, machine learning, deep learning, and linked data all provide immense opportunities for dealing with these challenges and can provide evidence-based innovative solutions not seen before. These technologies are perceived as the "true" enablers of future practice, but only recently has the AEC sector recognised terms such as "golden key" and "golden thread" as part of BIM processes and workflows.

This book builds on the success of a number of initiatives and projects by the authors, which include seminal findings from the literature, research and development, and practice-based solutions produced for industry. It presents these findings through real projects and case studies developed by the authors and reports on how these technologies made a real-world impact.

The chapters and cases in the book are developed around these overarching themes:

- BIM and AEC Design and Optimisation: Application of Artificial Intelligence in Design
- BIM and XR as Advanced Visualisation and Simulation Tools
- Design Informatics and Advancements in BIM Authoring
- Green Building Assessment: Emerging Design Support Tools
- Computer Vision and Image Processing for Expediting Project Management and Operations
- Blockchain, Big Data, and IoT for Facilitated Project Management
- BIM Strategies and Leveraged Solutions

This book is a timely and relevant synthesis of a number of cogent subjects underpinning the paradigm shift needed for the AEC industry and is essential reading for all involved in the sector. It is particularly suited for use in Masters-level programs in Architecture, Engineering, and Construction.

Farzad Pour Rahimian is Professor of Digital Engineering and Manufacturing at Teesside University, UK.

Jack Steven Goulding is Professor of Construction Project Management at the University of Wolverhampton, UK and Director of a specialist BIM consultancy service.

Sepehr Abrishami is a Senior Lecturer and BIM Programme Leader at the University of Portsmouth, UK.

Saleh Seyedzadeh is a Data Scientist at Offshore Renewable Energy Catapult, Scotland, UK.

Faris Elghaish is a Lecturer in Construction Project Management at Queen's University Belfast, Northern Ireland, UK.

Industry 4.0 Solutions for Building Design and Construction

A Paradigm of New Opportunities

Farzad Pour Rahimian, Jack Steven Goulding,
Sepehr Abrishami, Saleh Seyedzadeh,
and Faris Elghaish

Routledge
Taylor & Francis Group

LONDON AND NEW YORK

First published 2022
by Routledge
2 Park Square, Milton Park, Abingdon, Oxon OX14 4RN

and by Routledge
605 Third Avenue, New York, NY 10158

Routledge is an imprint of the Taylor & Francis Group, an informa business

British Library Cataloguing-in-Publication Data
A catalogue record for this book is available from the British Library

Library of Congress Cataloging-in-Publication Data
Names: Rahimian, Farzad Pour, author.
Title: Industry 4.0 solutions for building design and construction : a paradigm of new opportunities / Farzad Pour Rahimian, Jack Goulding, Sepehr Abrishami, Saleh Seyedzadeh, and Faris Elghaish.
Description: Abingdon, Oxon ; New York, NY : Routledge, 2022. | Includes bibliographical references and index.
Identifiers: LCCN 2021030897 (print) | LCCN 2021030898 (ebook) | ISBN 9780367618803 (hbk) | ISBN 9780367618780 (pbk) | ISBN 9781003106944 (ebk)
Subjects: LCSH: Building—Data processing. | Building information modeling. | Industry 4.0.
Classification: LCC TH438.13 .R34 2022 (print) | LCC TH438.13 (ebook) | DDC 690.068—dc23
LC record available at https://lccn.loc.gov/2021030897
LC ebook record available at https://lccn.loc.gov/2021030898

ISBN: 978-0-367-61880-3 (hbk)
ISBN: 978-0-367-61878-0 (pbk)
ISBN: 978-1-003-10694-4 (ebk)

DOI: 10.1201/9781003106944

Typeset in Goudy
by Apex CoVantage, LLC

Contents

1 Industry 4.0 solutions for building design and construction

A paradigm of new opportunities

1.1. Introduction

One of the main reasons this book came to fruition was in part inspired by frustration and in part driven by a collective soliloquy of wanting to present readers with a rich picture of golden opportunities. Frustration in this sense relates to the way through which Architecture, Engineering, and Construction (AEC) has responded to change (particularly over the last 50 years), where, for example, several global reports have repeatedly mentioned that AEC needed to change. Key report recommendations have included several issues, from industry fragmentation through to the need for higher skills, improved quality, enhanced performance and productivity, better value, tangible progress in innovation, improved communication and integration, and the need for more meaningful collaborative relationships. Arguably, this list could be extended almost *ad infinitum*; however, an interesting point to note here is that several of these reports have attempted to compare the performance of the sector against others, such as automotive, aerospace, engineering, healthcare, and manufacturing – all of which seem to have performed significantly better than AEC. The question is why? Whilst this book does not seek to provide solutions to this specific question *per se*, it does open debate in several important areas, with a view of challenging the current perception and status quo (in the hope that this will inspire change). For example, the global AEC geospatial market is expected to reach US$12.26 trillion by 2023 (Narain, 2020). This is not only tangible and significant, but this also offers AEC a unique opportunity to step beyond introspection, to an industry that fervently aspires to continually evolve as new industry leaders and pioneers.

It is acknowledged from the outset that this journey may not be easy. Moreover, it would be rather naïve of the authors to focus on all issues and challenges facing the industry. That being said, we had to start somewhere; collectively, we decided to focus on some of the underpinning themes and challenges relating to design. In this respect, design-related issues have been seen as causal contributors to many of these high-level challenges. These issues include (but are not limited to): communication and information processing, technology adoption, collaboration, integration, automation, interoperability, labour, and skills (Egan, 1998; Peansupap & Walker, 2005; Goulding & Pour Rahimian, 2019; Fruchter et al., 2016; Day, 2019; Pour Rahimian et al., 2019; Elghaish et al., 2020; Leon & Laing, 2021). These factors are not only fundamental and integral (throughout the project lifecycle), but they also have a direct or indirect knock-on effect with many other support services.

To address some of these issues (particularly within the context of design), this book seeks to raise awareness by presenting several practice-based solutions, with the expressed aim of unlocking AEC's digitalisation potential. For example, whilst client organisations are predominantly seen as the core initiators of the design process, they (arguably) often tend

DOI: 10.1201/9781003106944-1

to lack knowledge and awareness needed to inform or shape the professional capability to deliver real value (CLC, 2018). This lack of understanding or wider appreciation of nuance influences the project lifecycle from day one – from conceptual design through to handover, maintenance, and deconstruction. This is a significant challenge to address, especially as projects seem to be increasingly more complex. Several solutions have started to enter the market, from advanced virtual reality–based collaborative technologies (Pour Rahimian et al., 2019) to artificial intelligence–based optimisation (Pilechiha et al., 2020) and data-driven decision support systems (Seyedzadeh et al., 2019). These solutions offer AEC significant opportunities – enabling (or empowering) them to not only unlock their digital potential to improve performance and capability, but also leverage better value throughout the whole process (McKinsey, 2017). In fact, significant markets have now started to leverage success by unlocking this digital potential (Herr & Fischer, 2019; Ahuja et al., 2020).

Part of the journey of unlocking AEC's digital potential involves moving towards Industry 4.0. This may seem a little daunting to most; however, this is seen as the way forward – a real paradigm shift for the sector – a transformative journey which more purposefully engages new ways of thinking, where digital technologies converge to provide significant advantages. Whilst this transition to Industry 4.0 may not be easy, it is encouraging to note that many AEC entities have made significant progress to achieve this goal (Maskuriy et al., 2019; Alaloul et al., 2020). This paradigm shift is not only significant and transformative, but it is starting to open up many new revenue streams and divested services for AEC (Figure 1.1);

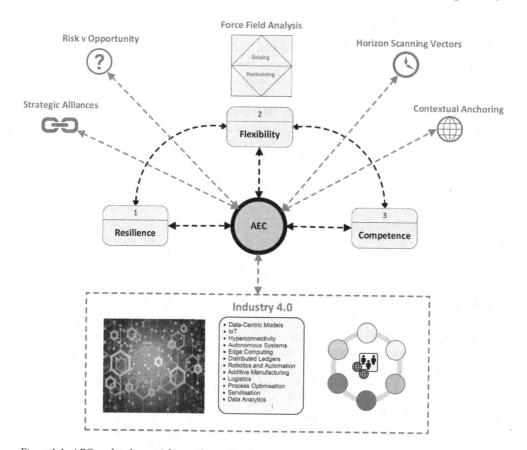

Figure 1.1 AEC and industry 4.0 transformational opportunities

however, a caveat of caution needs to be raised at this juncture. This transition is not free; it comes at a cost. This 'cost' requires conjoined thinking and a willingness (acceptance) to embrace change, not just at the individual or organisational level, but also at the macro level (involving the whole sector and supply chain). In this respect, fragmentation and siloed positioning needs to be replaced by conjoined processes and 'digital coalition'.

This not only highlights the need to become more connected, dynamic, and customer-centric, but more importantly, mechanisms through which future AEC business will need to operate. This includes the need to think about new business strategies and models – from design to procurement and delivery, even the way goods are produced and delivered. Successful companies will be those that unleash their true potential, using business models that drive innovation and deliver evidence-based value. Those that do not do this will (more than likely) fall by the wayside. Therefore, AEC organisations will need to be highly competitive (perhaps more so than they are already), using factories and warehouses (physical and virtual) to leverage economies of scale (and expertise) to become much more streamlined, agile, and efficient. In doing so, they will be able to establish several new services and opportunities, especially through the deployment of cloud computing, big data, visualisation, artificial intelligence, machine learning, the internet of things, blockchain, etc. Data will undoubtedly be seen as the main asset –not only to inform decision-making, but also to drive innovation and facilitate continuous improvement. This will also enhance customer-experience analytics, providing new end-to-end services and servitisation opportunities; where, for example, significant growth-driven potential has already been evidenced in other sectors. In summary, the inertia underpinning Industry 4.0 provides AEC with many powerful opportunities to explore, nurture, and exploit. Some of these opportunities are presented throughout the following chapters.

1.2. Thematic overview of chapters

Chapter 2: AI-based architectural design generative BIM workspace for architectural design automation

This chapter concerns the integration and automation of design. It presents and builds upon a theoretical foundation that supports process integration (particularly at the conceptual design stage), including design representation, cognition, translation, and design integrity. Building Information Model (BIM) applications supporting design automation are explored, including their use in whole design integration. The concept of advanced Generative Design (GD) is presented as a significant opportunity to enhance the design experience. Core BIM and GD facets are identified and mapped into a generative BIM (G-BIM) framework for prototype development. This prototype was evaluated through multiple projects and scenarios, the results from which culminate in a valuable set of rubrics for further exploitation (cf. supporting the conceptual design stage using GD). Specific contribution also highlights the capabilities and opportunities provided through this prototype, from advanced collaborative features to the generation of optimised (and more purposeful) design solutions.

Chapter 3: Towards intelligent structural design of buildings: a BIM solution

This chapter outlines the challenges of design coordination and integration, especially between architectural and structural engineering practices. These different design approaches are examined, along with different BIM solutions and collaborative platforms. The main concept presented here is the need to provide automated synergy (given that these two disciplines are co-dependent). In this respect, these two approaches are examined in detail, noting the requirements needed for linking architectural models with structural engineering models. In doing so, an automated procedure and proof of concept is presented for discussion. This explores an automated approach that engages computational systems and toolsets into a solution that 'binds' architectural and structural models (for tall buildings). This prototype automatically generates, updates, and produces alternatives for structural models based on inputs from the architectural model. Findings highlight that solution generation can provide much richer optimum designs to meet set criteria. This work is acknowledged as being one of the first of its kind to automatically generate optimised structural design solutions based on architectural models.

Chapter 4: BIM and design for manufacturing and assembly

This chapter discusses the current state of integration of Design for Manufacturing and Assembly (DfMA) and BIM within AEC throughout the whole project lifecycle. The rationale of this chapter was to evaluate recurrent challenges such as low productivity and poor quality, high variances in predictability, along with greater building performance and energy use control. In doing so, a conceptual framework and BIM library for offsite manufactured components is presented for discussion. This work engaged a case study to demonstrate the implementation of BIM and DfMA. Findings present new and novel approaches for delivering synergy, highlighting the need to focus on the development of a new digital manufacturing–driven industry. This work also acknowledges the need for continuous improvement, highlighting a number of opportunities for exploitation, especially using a BIM-based DfMA approach to improve consistency and standardisation, reduce design discrepancies, reduce waste, improve safety, increase design flexibility, and maximise end-user value.

Chapter 5: Virtual reality–based cloud BIM platforms for integrated AEC projects

Chapter 5 reflects on the need to integrate collaborative design teams' project data to help coordinate the design, engineering, fabrication, construction, and maintenance of various trades (to facilitate project integration and interchange). In doing so, it evaluates a number of opportunities, including the implementation of BIM Level 3 (Cloud) as an innovative way of further enhancing the design, management, and delivery processes. This work proffers the need for change from 'traditional' approaches to those aligned with Integrated Project Delivery (IPD). This chapter also acknowledges that web-based platforms are particularly beneficial, as these are able to visualise, integrate, and share building components (in real time) and through geographically dispersed locations. A cloud-based virtual reality (VR) Construction Site Simulator was presented as a potential solution. This engaged a game environment supported by a web-based virtual reality cloud platform. Findings presented new insight and understanding into the development of training programs of this nature, particularly the use of the Unified Software Development Process and iterative phases of Elaboration, Construction, and Transition. Finally, this work offers new understanding and insight into the causal drivers and influences associated with successful decision-making design in non-collocated design teams, providing a stepping-stone for developing new relationship models in collaborative environments.

Chapter 6: XR-openBIM integration for supporting whole-life management of offsite manufactured houses

Chapter 6 advocates the need to understand the importance of integration from a granular level. In doing so, it critiques project integration across the AEC supply chain using offsite manufactured housing as an exemplar. This analysis included tools, technologies, and processes, and especially BIM and data models – noting that AEC needed to manage diverse project information. One of the major challenges in this area was interoperability, particularly the level of data compatibility, and how this affects collaboration and decision-making. This chapter argued that media-rich VR and augmented reality (AR) environments could help users better understand design solutions. It was proffered that a solution could be developed on the capabilities of openBIM and industry foundation class (IFC) schema, using the BIM server concept to provide concurrent multiuser engagement (with low-latency communication between applications). A prototype of an offsite manufactured integrated virtual showroom was presented for discussion. This exemplar used openBIM-Tango, BIM models, and data from IFCs. This enabled users to interact with these models though VR immersive and AR environments (including Google Tango-enabled devices). Findings highlight several new innovative approaches for interrogating data. It also presented wider AEC opportunities, noting that this concept could mitigate the need for advanced technical skills, as it allows decision-makers (with different skillsets or areas of expertise) to access and engage with data and information in a more accessible and meaningful way.

Chapter 7: A centralised cost management system: exploiting earned value management and activity-based costing within integrated project delivery

This chapter introduces the concepts of earned value management (EVM) and activity-based costing (ABC) within AEC. In doing so, it evaluates risk/reward sharing opportunities through IPD, advocating the need for a decentralised, automated, and secure financial platform for managing and controlling financial transactions. In pursuance of this goal, mathematical models for determining the three main IPD financial transactions were evaluated. This includes coding of reimbursed costs, profit and cost saving, and engagement of IPD smart contract functions. The development of this proof-of-concept prototype (Centralised Cost Management System) was validated through an IPD case project. Findings highlight the benefits of using this system to automate financial transactions, whilst also demonstrating several other advanced features. In summary, this work advocated the need for AEC to embrace IPD adoption as a means of resolving some of the existing financial challenges. Recommendations included the incorporation of technologies such as IBM's Blockchain Platform for IBM Cloud and Fabric v2.x for smart contract lifecycle management.

Chapter 8: Success factors driving cost management practices through integrated project delivery

This chapter evaluates the pivotal forces and concepts underpinning cost management practices in AEC. It promotes the need to support effective cost management practices, especially those that share cost data with all project stakeholders. Emphasis was placed on accuracy, timeliness, and transparency (to support governance and trust). From this focus, IPD was critically reviewed as a mode of project procurement, along with the application of cost management approaches and success factors underpinning these. This included the need for an integrated and resilient cost information system to support IPD. A questionnaire survey was conducted with leading IPD experts to identify challenges, priorities, and opportunities, especially factors driving the implementation of IPD. Findings revealed that ABC and EVM were particularly effective at identifying and appropriating costs. Moreover, that there was a greater need to develop these further in order to support accounting transparency. In pursuance of this task, the use of mathematical models could be generated to propagate equitable risk/reward distribution. Findings from this work endorse the role of BIM-enabled web-based management systems to not only enhance IPD-based cost management effectively but also encourage wider uptake.

Chapter 9: 4D BIM for structural design and construction integration

This chapter introduces the challenges of monitoring, inspecting, and evaluating work, particularly for concrete pouring. These issues were explored through literature, culminating in the need to support AEC with better monitoring/assessment measures, especially to support structural design, production processes, concrete pouring, integration, joint assessment, etc. This research observed that meticulous planning was required to achieve this goal, especially to ensure aesthetic and structural integrity (in order to mitigate structural defects and construction rework). Challenges included the need to manage several issues, not least design challenges, but also critical path dependencies and onsite operational constraints. In order to address these issues, a 4D BIM approach was presented as a solution. This used an automated concrete joint positioning (proof-of-concept) solution to help support design professionals and contractors. This engaged structural modelling information from Revit and spatial information on construction joints, linking these through Microsoft (MS) Excel and Matlab spreadsheets with Dynamo software. Findings presented significant benefits, including an automated system for optimising design solutions, a cost-effective and accurate methodology for addressing previous limitations, and a new way of designing construction joints and planning pours. These innovative solutions support the need for wider integration of structural design considerations with construction and site operational procedures. This novel application of BIM in structural engineering also highlighted how the different capabilities of various software applications could be integrated.

Chapter 10: BIM integrated project delivery: an automated earned value management–based approach

This chapter challenges current thinking on IPD, highlighting the need for robust and defendable systems to appropriate and manage project costs. In doing so, it presented a discussion on IPD and cost management practices used to determine the risk/reward ratio. This discussion included the concepts and application of EVM and ABC, especially in relation to supportive technologies such as BIM. It was proffered that AEC needed some form of solution that integrated IPD with BIM as an optimal approach for delivering construction projects. Acknowledging this knowledge gap, this chapter presented a bespoke model that could be used to exploit EVM – to structure the compensation approach in IPD (using ABC to optimise the cost structures). This innovative approach was expressly designed to exploit the capabilities of these techniques coupled with BIM to automate/optimise the process of IPD risk/reward sharing. Findings observed that the mathematical equations underpinning this risk/reward sharing approach could be used to strengthen IPD parties' relationships, especially through the EVM-Web grid, as this enabled project participants to track their costs more effectively. This research also suggested the need to incorporate future 4D/5D BIM platforms and developments with openBIM to further improve the accessibility, usability, and management of AEC digital data.

Chapter 11: Revolutionising cost structure within integrated project delivery: a BIM-based solution

Chapter 11 reflects on the cost structures underpinning IPD, highlighting the need to improve cost estimation, especially at the 'front end' of IPD projects (where project information is seldom fully available). This work explores several cost estimation approaches, methods, and tools currently used in IPD. In doing so, it presents a novel theoretical argument and new approach to enhance cost estimation. This approach incorporated target value design (TVD), ABC, and Monte Carlo simulation into an IPD cost structure within a BIM-enabled platform. A framework was developed to present the proposed methodology of cost estimation throughout all IPD stages. A case project was used to validate the practicality of this solution by comparing the profit-at-risk percentage for each party, using both traditional cost estimation and the proposed solution. Research findings highlighted the benefits of adopting such an approach as a workable solution for BIM–IPD integration. This produced reliable cost data from different sources and project delivery modes, noting that the use of BIM (as a means of developing a conceptual model to address client criteria) could enable costing professionals to build statistical models with higher levels of cost certainty and predictability.

Chapter 12: Dynamic sustainable success prediction model for infrastructure projects: a rough set–based fuzzy inference system

This chapter recognises the importance of being able to successfully implement sustainability in projects, particularly infrastructure projects. It acknowledges that whilst the definition of project success was subjective, it predominantly encompassed measures and criteria associated with time, quality, and cost. More importantly perhaps, it noted that several studies failed to address other success indicators associated with criteria, such as environmental compliance, building performance, client satisfaction, socio-political drivers, etc. To address this challenge, this study presents a decision support system (DSS) for evaluating and predicting project success against sustainability criteria. This used rough set theory (RST) for rules generation. The generated rulesets were filtered through a fuzzy inference system (FIS) to support the DSS. This tool was then tested and validated by applying data from a real infrastructure project. Research findings highlighted that the developed rough set fuzzy method was able to evaluate and predict project success through robust rulesets to support enhanced prediction. This tool also enabled decision-makers to dynamically evaluate and predict project success based on customisable sustainability criteria.

Chapter 13: Multi-objective optimisation to support building window design

This chapter investigates the concepts of configuring window systems design in office buildings, cognisant of such issues as energy performance, daylighting levels, visual comfort, etc. The need to produce better-quality evaluation tools to support the design process was highlighted. In this respect, a new multi-objective method of analysing and optimising the window system design process was presented for discussion. This system incorporated simultaneous consideration of multiple and conflicting design objectives, using rubrics based on the fundamental recognition that the process of optimising parameters on issues such as building energy loads via window system design can often reduce the quality of the view to outside (including the received daylight). This study developed a multi-objective method of assessment, using a reference room that was parametrically modelled against real climate data. A method of Pareto frontier and a weighting sum was applied for multi-objective optimisation to determine the best outcomes – ergo, one that balances design requirements and criteria. Findings present a new approach for quantifying the Quality of View in office buildings, one that balances energy performance and daylighting, thereby enabling and facilitating improved window design optimisation. This work provides decision-makers with a novel approach of window design evaluation based on performance criteria and desired outcomes.

Chapter 14: Artificial intelligence image processing for on-demand monitoring of construction projects

One of the continual challenges facing AEC is the need to monitor project performance. This chapter posits that whilst inspections and progress monitoring are a vital part of the process, in some instances, the actual process of comparing 'as-planned' with 'as-built' progress does not readily add any tangible intrinsic value to the process. It argues that for large-scale construction projects in particular, a better system is needed for monitoring and inspection. In this respect, a new framework and proof-of-concept was presented for discussion. This used AI-based Image Processing and Computer Vision for on-demand monitoring. This prototype also engaged ML, image processing, BIM, and VR, using the Unity game engine to integrate data from the original BIM models with as-built images. These were processed via various computer vision techniques, including object recognition and semantic segmentation (to identify different structural elements). Findings provide a unique insight into alternative approaches of monitoring and inspection through a 3D virtual environment. This prototype was proffered as being able to support project managers and the inspection team – to help them make better informed decisions, much quicker than through conventional approaches. Moreover, this work provides a technical exemplar for integrating ML with image processing approaches together with immersive and interactive BIM interfaces. The algorithms and programme codes presented could also help other specialists in different contexts/settings with issues of replicability.

Chapter 15: Digitalisation of Architecture, Engineering, and Construction: immersive technologies and unmanned aerial vehicles

This chapter investigates the concepts and applications of using Unmanned Aerial Vehicles (UAVs) and immersive technologies in AEC. In doing so, it presents a critical literature review of these areas and aligns key studies using meta-synthesis to focus on optimisation. This work examined immersive technologies and UAV technologies applications in order to evaluate and integrate these findings into a single context. The findings from this research were assessed and contextualised to AEC needs. Findings highlighted that whilst the uptake and use of UAVs and immersive technologies was steadily improving, there was still a greater need to accelerate these initiatives as part of the progress towards wider digital transformation. Several benefits and opportunities were discussed, showcasing potential applications of these technologies, noting the importance of integration. It was proffered that UAVs and immersive technologies could be used in conjunction with 4D BIM to assess project progress, undertake compliance checking of geometric design models, evaluate and control certain parts of construction projects remotely, undertake quality control, and help assess health and safety issues, etc. In summary, the opportunities presented were seemingly endless – with new avenues to explore, including thermal and acoustic sensors, links to developments in augmented reality (AR), mixed reality (MR), and BIM (including blockchain and distributed ledgers), all of which support AEC's transition to Industry 4.0.

Chapter 16: Optical code division multiple access–based sensor network for monitoring construction sites affected by vibrations

Chapter 16 presents a critical review on the need to engage effective and accurate sensors to monitor the structural health of large facilities in order to mitigate risk at the very early stages (before these risks develop further). In doing so, it evaluates a range of technologies and sensors currently deployed in AEC, specifically focussing on aspects such as vibration. From this study, literature highlighted the need to engage more accurate structural health-monitoring systems to support large-scale facilities such as modern high-speed railways and bridges. It was advocated that this required additional development in optical sensor networks (OSNs), as this could help mitigate challenges associated with conventional electric sensors (*cf.* their sensitivity to electromagnetic interferences); however, this chapter observed that existing fibre-optic infrastructures were not widely used by OSNs due to the lack of appropriate multiplexing techniques. Given this challenge, an Optical Code Division Multiple Access System (OCDMA)–based sensor network for monitoring construction sites affected by vibrations was developed and presented for discussion. This prototype supports vibration sensing of unequally distributed points, taking advantage of multiple wavelengths and spectral amplitude encoding optical code division multiple access (SAC-OCDMA) techniques. Findings highlighted that this prototype did not require traffic management or system synchronisation, and that this was resilient to performance degradation often caused by fibre non-linearities. These advantages were proffered as being particularly beneficial for AEC professionals wishing to improve their structural health monitoring systems' performance.

Chapter 17: Blockchain integrated project delivery: an automated financial system

This chapter evaluates the technologies and processes supporting IPD in AEC. In doing so, it advocated the need to incorporate more robust systems to support financial transactions, particularly concerning the appropriation of risk/reward sharing and deferral of parties' profit payments. Suggestions included the need to embrace some form of decentralised, automated, and secured financial platform, where project parties can monitor, control, and track financial transactions. In pursuance of this, blockchain technology was considered a viable solution, as it had the ability to discretely manage data transactions with a high level of fidelity and veracity. From this, a framework adopting blockchain technology for IPD projects was presented for discussion. This framework enabled core project team members to automatically execute financial transactions through IPD coding parameters of reimbursed costs, profit, and cost saving. This aligned to IPD smart contract functions. This proof-of-concept prototype was validated through an IPD case project and was presented for discussion. Findings highlighted significant improvements in control and monitoring, reinforced by BIM, the Hyperledger Fabric, and IBM Blockchain Platform free 2.0 beta. This prototype was acknowledged as being one of the first AEC frameworks to incorporate blockchain technology with IPD projects, with the express purpose of enabling core project team members to automatically execute financial transactions.

References

Ahuja, R., Sawhney, A., Jain, M, Arif, M., & Rakshit, S. (2020). Factors influencing BIM adoption in emerging markets – the case of India. *International Journal of Construction Management, 20*(1), 65–76. https://doi.org/10.1080/15623599.2018.1462445

Alaloul, M. S., Liew, M. S., Zawawi, N. A. W. A., & Kennedy, I. K. (2020). Industrial revolution 4.0 in the construction industry: Challenges and opportunities for stakeholders. *Ain Shams Engineering Journal, 1*(1), 225–230. https://doi.org/10.1016/j.asej.2019.08.010

CLC. (2018, July). *Procuring for value, construction leadership council*. Department for Business, Energy & Industrial Strategy. Retrieved May 29, 2021, from www.constructionleadershipcouncil.co.uk/wp-content/uploads/2018/07/RLB-Procuring-for-Value-18-July-.pdf

Day, M. (2019, October 9). The generation game. *AEC Magazine*, X3DMedia Limited. Retrieved May 28, 2021, from https://aecmag.com/features/the-generation-game/

Egan, J. (1998). Rethinking construction: The report of the construction task force. *HMSO*. ISBN:978-1851120949. Retrieved May 28, 2021, from https://constructingexcellence.org.uk/wp-content/uploads/2014/10/rethinking_construction_report.pdf

Elghaish, F., Abrishami, S., & Hosseini, M. R. (2020). Integrated project delivery with blockchain: An automated financial system. *Automation in Construction, 114*, 103182. https://doi.org/10.1016/j.autcon.2020.103182

Fruchter, R., Herzog, S., Hallermann, N., & Morgenthal, G. (2016). *Drone site data for better decisions in AEC global teamwork*. 16th International Conference on Computing in Civil and Building Engineering. www.see.eng.osaka-u.ac.jp/seeit/icccbe2016/Proceedings/Full_Papers/187-141.pdf

Goulding, J. S., & Pour Rahimian, F. (Eds.). (2019). *Offsite production and manufacturing for innovative construction: People, process and technology*. Taylor and Francis. ISBN:978-1-138-55068-1

Herr, C. M., & Fischer, T. (2019, April). BIM adoption across the Chinese AEC industries: An extended BIM adoption model. *Journal of Computational Design and Engineering, 6*(2), 173–178. https://doi.org/10.1016/j.jcde.2018.06.001

Leon, M. and Laing, R. (2021). A concept design stages protocol to support collaborative processes in architecture, engineering and construction projects. *Journal of Engineering, Design and Technology*. https://doi.org/10.1108/JEDT-10-2020-0399

Maskuriy, R., Selamat, A., Ali, K. N., Maresova, P., & Krejcar, O. (2019). Industry 4.0 for the construction industry – how ready is the industry? *Applied Sciences, 9*(14), 2819. https://doi.org/10.3390/app9142819

McKinsey. (2017, February). *Reinventing construction: A route to higher productivity*. McKinsey Global Institute. Retrieved May 29, 2021, from www.mckinsey.com/~/media/mckinsey/business%20functions/operations/our%20insights/reinventing%20construction%20through%20a%20productivity%20revolution/mgi-reinventing-construction-a-route-to-higher-productivity-full-report.pdf

Narain, A. (2020, May 25). Asia Pacific to drive the US$ 117.59 Bn global geospatial market in the AEC industry. *Geospatial World*. Retrieved May 28, 2021, from www.geospatialworld.net/blogs/geospatial-market-in-aec-industry/

Peansupap, V., & Walker, D. (2005). Factors enabling information and communication technology diffusion and actual implementation in construction organisations. *Electronic Journal of Information Technology in Construction, 10*, 193–218. Retrieved May 28, 2021, from www.itcon.org/2005/14

Pilechiha, P., Mahdavinejad, M., Pour Rahimian, F., Carnemolla, P., & Seyedzadeh, S. (2020). Multi-objective optimisation framework for designing office windows: Quality of view, daylight and energy efficiency. *Applied Energy, 261*, 114356. https://doi.org/10.1016/j.apenergy.2019.114356

Pour Rahimian, F., Chavdarova, V., Oliver, S., Chamo, F., & Potseluyko Amobi, L. (2019). OpenBIM-Tango integrated virtual showroom for offsite manufactured production of self-build housing. *Automation in Construction, 102*, 1–16. https://doi.org/10.1016/j.autcon.2019.02.009

Seyedzadeh, S., Pour Rahimian, F., Rastogi, P., & Glesk, I. (2019). Tuning machine learning models for prediction of building energy loads. *Sustainable Cities and Society, 47*, 101484. https://doi.org/10.1016/j.scs.2019.101484

2 AI-based architectural design generative BIM workspace for architectural design automation

2.1. Introduction

Architecture, Engineering and Construction (AEC) is a significant contributor to the growth, development and provision of worldwide infrastructure, assets and services. This contribution varies across countries, where for example in the UK, this represents approximately 6% of Gross Domestic Product (House of Commons Library, 2019), and arguably, almost double this (CIOB, 2020). This trend is continuing to grow, with GDP contributions ranging from 4%–78%. This significance is continuing to grow, as the sector as a whole has been trying to leverage change through transformation-driven initiatives across a number of areas. Given this significance, the sector as a whole has been trying to leverage change through transformation-driven initiatives across a number of areas. Whilst this change has been slow, things are now starting to improve; however, prior to discussing these changes (and the content of this chapter), it is useful to reflect on why these changes have been so slow. This is an important juncture for AEC, because knowing what has influenced (or stifled) progress has to be fully understood beforehand; otherwise, future changes are more than likely going to follow the same path. So, in order to do this, it is important to consider some of the underlying causes of this lack of progress. One major issue here is industry fragmentation. In this respect, the fragmentation of AEC is well recognised – the consequences of which have led to well-documented problems relating ostensibly to failures in communication and information processing (Egan, 1998; Latham, 1994). These failures have contributed to (in some instances) adversarial challenges, especially surrounding the nature of the different parties involved in a project (Forcade et al., 2007). This in turn has affected work-related communication issues, from the provision and veracity of design information (Cera et al., 2002; Fruchter, 1998) to process challenges across the project lifecycle. That being said, things are now changing, even given the increased nature and complexity of AEC projects. One of the key transformational levers here is the increased prevalence and use of advanced technology, particularly through web-based project collaboration technologies and project extranets. This embraces a raft of technologies, including shared immersive environments and cloud collaboration, Building Information Modelling (BIM), blockchain applications, digital twins, Artificial Intelligence (AI), Machine Learning (ML), Generative Design (GD), Extended Reality (XR) with digital assets [spatial computing], Virtual Design and Construction (VDC), and Building Energy Modelling (BEM) – to name but a few developments.

This chapter provides readers with a new approach for applying GD within BIM environments to support the conceptual design process; however, it is important to provide some background context to these areas in order to contextualise and ground this discussion. For example, whilst Information and Communication Technology (ICT) has led to dramatic

DOI: 10.1201/9781003106944-2

changes in terms of labour and skills (Fruchter, 1998), this has in turn also revolutionised production and design within AEC (Cera et al., 2002; Seyedzadeh et al., 2021). From a design perspective, new capabilities therefore became more prevalent, which required these to be harnessed (i.e. aligning these capabilities to real-life solutions), in order to predict the cost and performance of optimal design proposals (Petric et al., 2002). This was fundamentally important to enable design engineers to compare the quality and performance of solutions against previous iterations. This was reinforced by Goulding and Rahimian (2012), regarding the ability to experiment and experience decisions in a 'cyber-safe' environment, which allowed designers to pre-empt, mitigate, or reduce risks prior to construction. Consequently, the success of AEC projects was seen to be highly dependent upon the 'type', 'level', and 'quality' of the innovative communication exchange of various disciplines involved in the design and implementation phases.

One of the key debates in respect of advanced ICT adoption in AEC was the level of automation needed across a project's lifecycle (Frohm et al., 2008; Skibniewski, 1992; Rahimian et al., 2019). This debate included several exemplars, including offsite manufactured construction, as this utilised high product variety and significant variations in demand (Veenstra et al., 2006; Wikberg et al., 2010). This included flexible and reconfigurable manufacturing systems (Colombo & Harrison, 2008), effective/cohesive supply chains (Arif et al., 2005), and integrated and automatic modelling, simulation, and decision support systems (Fruchter, 1998); however, in order to facilitate, support, and leverage this automation required considerable integration. Gu and London (2010) asserted that this was unlikely to happen unless construction information was represented and managed throughout all stages of the project lifecycle, including the early conceptual design and planning processes. This was in part resolved by the introduction of BIM, but early BIM adoption was limited, and (arguably) did not really cover the full operational processes involved during the early stages of design and planning. In this respect, Rahimian et al. (2011) related this gap to the fact that conceptual design automation systems were still in their infancy. This stifled progression and also caused problems with transmission vis-à-vis data interoperability (Santos, 2009), which resonated with earlier issues raised by Fruchter (1998) on the disconnect and dissonance among various teams of designers concerning software and platform incompatibilities.

Acknowledging these challenges, this chapter explores methods in which BIM can now be more usefully employed: not as a representational tool for visualisation *per se*, but as a comprehensive support tool for the entire design process. In doing so, the specific focus of the research presented here is to showcase how the conceptual design process can be improved. A framework is presented for discussion. This was designed specifically to enhance designers' abilities in procuring highly novel and innovative solutions throughout the whole design process (including change management, model modifications, etc.). Whilst the methods introduced in this chapter are somewhat abstract in places regarding the level of detail they represent, the concepts are portrayed as viable solutions for exploring the future direction of computational design, where it is argued that designers often make vital decisions in the early phases of a project (Paulson, 1976). Given this, it was considered important to try and unpick this issue, rather than 'reinventing the wheel' so to speak. In doing so, GD was considered a potential solution, where the application of GD within existing tools was seen as a step-change solution for helping designers solve complex multi-criteria design problems. The rationale behind this thinking rested on the premise of designing a genotype of a building design (within a BIM application) at the early design stage. This would allow designers to generate new design alternatives by varying the pre-defined parameters based on the design constraints and associated requirements. The generated alternative population could then be amended and improved using BIM parametric features selected by designers (or the design team). Following this line of thought, it was envisaged that this framework would allow users to more fully exploit BIM's capabilities in the early design phases, especially in collaboration, parametric change management, simulation, and analysis.

~ This chapter presents the use and application of GD and BIM into a conceptual Generative BIM (G-BIM) framework for discussion. The proposed framework provides design solutions based on input data such as site data, constraints, and requirements. The first part of this chapter introduces the basics of conceptual design, followed by a critical review of design thinking within the design process. This is followed by a roadmap for conceptual design and computational support (at the early conceptual design stage). Existing tools and decision support mechanisms are discussed as part of this process. Development is then presented through two investigation steps: (i) studying the design process individually and (ii) evaluating the tools that can be used to support the early design stages. Technical issues are then presented, followed by a conclusion section that presents the main findings and future works.

2.2. Research methodological approach

In order to provide readers with a little background information and context supporting the development approach of this G-BIM framework, the following narrative highlights the main stages that led to the fruition of this framework (from a research methodological design perspective). In this respect, the first part of this process was framed using a literature review to identify current challenges, competing technologies, design requirements, and emerging new opportunities. This helped identify the specific knowledge gap that this work needed to address. In doing so, this led to the selection of an appropriate design strategy. This in turn was fine-tuned to tease out the philosophical underpinnings of the design theory continuum, which were then matched to the practical constructs of research practice (including the technology and tools used to deliver this). From this, core drivers were identified and mapped into the GD conceptual framework environment.

2.3. Literature review analysis

The literature review undertaken for this study encompassed both broad and deep analysis. In this respect, this process involved detailed analysis of the top ten journals in AEC, particularly those associated with design, ICT, and innovation. This was supplemented with various conference proceedings and core research databases in design and automation. The study used NVivo software to analyse the content of the selected publications using NVivo's Word Frequency Query. The minimum length for words in the frequency analysis was set to five, and the similarity scale was set to four out of five in order to increase focus and relevance. The results from this frequency analysis can be seen in Table 2.1.

During the literature review phase, it was considered important to capture the theoretical foundations of this study. In this respect, content analysis was adopted as a qualitative approach (Creswell, 2002), as this was able to uncover a deeper understanding of the subject matter, ergo the current state of computational support needed during the conceptual architectural design phase. The rationale behind this was to understand the main issues that needed to be captured and embedded into the theoretical framework, particularly as a method of automation for conceptual design. This was needed given the specific need to engage GD in this process. The identified core drivers and corresponding seminal authors can be seen in Table 2.2.

Table 2.1 Word frequency analysis

Word	Length	Count	Weighted (%)
construction	12	181772	0.43
design	6	142779	0.35
artefact	8	117323	0.32
architecture	12	109693	0.31
thinking	8	106158	0.25
CAD tools	5–13	104665	0.25
method	6	101403	0.24
BIM	5–11	101033	0.22
generative	10	100921	0.22
parametric	10	71977	0.20
create	6	65308	0.19
collaboration	13	63723	0.19
attributes	10	61388	0.17
system	6	57514	0.17
development	11	45604	0.16
environment	11	39701	0.15
figure	6	38654	0.14
building	8	36716	0.14
object	6	10586	0.13

Table 2.2 Research focus: analysis of core drivers

Subject	Description	Seminal Authors
Design research: conceptual design and design thinking	The process in which designers collaboratively author an assembly design	(Cross, 2007)
Modern design Opportunities	Computational support for design	(Narahara, 2007; Do & Gross, 2009; Johnson et al., 2009)
Generative design	Using a set of rules or an algorithm in order to generate designs (architectural forms)	(Biloria, 2018; Cera et al., 2002; Narahara, 2007; Leach, 2009; Roudavski, 2009)
Parametric design	Use of parameters to define a form and relations	(Fischer et al., 2005; Butz et al., 2005)
BIM	Intelligent model-based process	(Ibrahim 2004)
CAD tools	Computer-aided design tools	(Whyte et al., 2000; Cheon et al., 2012)
Knowledge sharing: collaboration	Collaborative design	(Cross & Clayburn, 1995; Cera et al., 2002)
Tool development	Techniques and models	(Johnson et al., 2006; Muehlbauer, 2018; Narahara, 2007; van Stijn & Gruis, 2020)

2.4. Framework development

Following the two main constructs of this study (i.e. information modelling and form generation), the main development focus for this framework centred on the integration of BIM and GD. In particular, to facilitate automation during the conceptual design stage, and specifically, to exploit the benefits of GD in this process. Given this approach, the framework structure was developed on the initial results of the literature review, supported by a detailed qualitative study by Abrishami et al. (2014). This followed standard process modelling concepts in order to develop the computational support framework. The framework was further developed into a working prototype based on a process-centred environment (Finkelsteiin, 1994). This was selected in order to be able to better capture, describe, and evaluate the evolving software development process. This framework culminated in the provision of three different levels: meta-process modelling, process model, and development iteration, where the meta-level (required information and key concepts) were classified in order to provide guidance for the development process (*cf.* Rolland, 1998). This framework is presented for discussion.

2.4.1. *Modern design opportunities*

From a design perspective, the focus of AEC design has increasingly moved away from contemporary architecture per se, with increased emphasis being placed on aesthetics, performance, structure, environment, construction, socio-economic drivers, cultural identity, etc. (Roudavski, 2009). This shift in design attitude has invited architecture (as a discipline) to radically re-think their approach, particularly conduits through which they can adopt new technologies to support this transition. In this respect, AEC designers have now started to adopt technology from fields such as industrial design, mechanical engineering, and product development (where performance tends to play a crucial role). This has been accompanied by new powerful computational design tools, including CATIA, Inventor, Digital Project, SolidWorks, Pro Engineer, etc. Moreover, practices have also started to adopt new enhanced computational design methods based on traditional approaches, but engaging new concepts such as genetic algorithms, parametric design, isomorphic surfaces, kinematics and dynamics, topological space, etc.

Acknowledging these developments, evolution, and transition of the industry, the success of AEC projects is still highly dependent on the decisions made during early conceptual design and planning processes, where 70–80% of production overheads are usually established (Paulson, 1976). This position has still not really changed. For example, tools that support advanced design planning, data-rich models (e.g. BIM) are now drawing design teams' attention (n.b. this was originally initiated by Eastman, 1999; Fischer, 2000) in order to meaningfully coordinate the fabrication of different building components. From an information definition perspective, Isikdag and Underwood (2010) defined BIM as an information management process throughout the lifecycle of a building that focuses on the collaborative use of semantically rich 3D information models. Whilst other definitions exist, the real issue here to appreciate is that BIM models contain rich geometric and semantic information on the building (design), and depending on the business need, different views/sub-models (e.g. Design, HVAC, FM) can be derived from these models. These models are now considered an integral part of the design process. Moreover, the use of BIM in the design of buildings has revolutionised AEC, most notably by enhancing team collaboration (Gu & London, 2010), improving project integration (Woo et al., 2004), leveraging better construction information flow (Ibrahim et al., 2004), helping documentation flow (Popov et al., 2006), and providing construction simulation for teamwork planning, clash prevention, and coordination interface (Fischer & Kunz, 2004). In line with these expectations, the UK Government announced its Government Construction Strategy, which included a mandate for the implementation of BIM Level 2 on all public projects by 2016 (BIM Task Group, 2013), where BIM Level 2 required digital building models to be shared/exchanged between parties in the design/

construction process for 2D/3D spatial coordination based on BS1192:2007. This initiative has been particularly successful, with outline criteria for BIM Level 3 now being adopted by many organisations.

Given these developments, unfortunately, some designers face challenges with technology-driven solutions. This may be rooted in 'siloed thinking' or through company-specific working approaches. A good example of this issue was highlighted by Bilda and Demirkan (2003), who identified that this gap was probably due to 'weakness' of the Computer-Aided Design (CAD) tools of that time in being able to support the intuitive design process that architects preferred in the early stage of the design lifecycle. This still seems to be a concern in some instances. Take Integrated Building Systems (IBS) as an example: architects engaging in IBS projects often have to accommodate numerous repetitive building components (with almost similar embedded information) during the modelling of prefabricated building projects. In this respect, they have to embed full-scale advanced manufacturing and rapid delivery processes within their designs (e.g. Design for Manufacturing and Assembly) to support BIM and data-rich models. In some cases, they simply do not have the tacit knowledge needed to fully leverage these benefits; however, things on the whole are now changing for the better.

Accepting these challenges, modern design approaches are now starting to make a significant impact on AEC. This research shares some of these developments using a working exemplar, which captures new conceptual design and computational support processes (within the early conceptual design stage of a project). This showcases how a BIM design environment can be integrated with new computational design methods in order to maximise opportunities. This proposed framework specifically exploits genetic algorithms in order to generate different alternatives, where the modification of the chosen alternative(s) engages parametric algorithms for solution generation (Abrishami et al. 2013). The following sections describe the features of this framework in more detail, the narrative of which identifies the different concepts and methods adopted.

2.4.2. Virtual reality applications

Early studies of using virtual reality (VR) in AEC tended to mainly use this as an advanced visualisation (or representational) medium; however, from around 1990 VR had become much more widely used, and by the year 2000 it was seen as an intuitive medium for designing 3D models; more specifically, it was starting to be used collaboratively throughout various phases of building projects (Whyte et al., 2000). Later iterations saw VR becoming embedded into mainstream design application to provide joint visualisation for improving the construction process (Bouchlaghem et al., 2005). This evolution also engaged construction planning and management, improving construction processes (Fischer, 2000) and linking 3D models to time parameters (Fischer & Kunz, 2004). This started to see the use of 3D immersive spaces in which 3D models were linked to databases [which hold inheritance characteristics], thereby allowing designs to be controlled through interactive databases, where Sampaio et al. (2010) noted the importance of incorporating VR 3D visualisation and decision support systems to perform real-time interactive visual exploration tasks.

In summary, VR applications became more mainstream and mature from 2000 onwards. Then 4D VR models started to become available in the market, and they were used to improve many aspects of construction (including communication) among project partners (Leinonen et al., 2003). Subsequently, design coordination started to evolve (Khanzade et al., 2007), which offered enhanced design creativity opportunities (Rahimian et al., 2011), along with wider integration aspects provided by BIM (Xie et al., 2011). This evolution is continuing, with 5D/BIM models [costs], 6D/BIM models [energy/sustainability], and 7D/BIM models [maintenance] now becoming mainstream delivery models.

2.4.3. *Building information modelling*

Construction projects are continuing to increase in scale, scope, and complexity (Cooke & Williams, 2009). In order to address some of these challenges, BIM has been proffered as a potential solution. For example, Suermann (2009) asserted that BIM was being used by many professionals, including designers, construction managers, and contractors, highlighting that they now had the ability to accomplish tasks more efficiently than ever before, paving the way for future construction professionals. Following this theme, a number of BIM-related services started to permeate the market. This increased proliferation has been supported by a number of governmental bodies around the world. In the UK, for example, the government (the largest procurement client of building and infrastructural development) mandated BIM Level 2 compliance by 2016 (Cabinet Office, 2011).

From a BIM definition perspective, there have been a number of attempts to codify BIM into an all-encompassing definition. Whilst it is argued that these definitions are still evolving, the two most common definitions from the UK and USA can be seen as follows: in the UK, the Construction Project Information Committee (CPIC) defined BIM as: "digital representation of physical and functional characteristics of a facility creating a shared knowledge resource for information about it forming a reliable basis for decisions during its lifecycle, from earliest conception to demolition" RIBA (2012). In the USA, the National BIM Standard (2007) defined BIM as "a digital representation of physical and functional characteristics of a facility. As such it serves as a shared knowledge resource for information about a facility forming a reliable basis for decisions during its lifecycle from inception onward" (NBIMS-US, 2007).

Notwithstanding these definitions (or derivations thereof), intrinsically a BIM model can be primarily seen as a 3D digital representation of a facility along with its core characteristics. In this respect, it consists of intelligent structural components, which include data attributes and parametric rules for each object. For instance, a window comprises certain materials; it has a shape and dimensions; it also has parametric links (e.g. to a wall) and contains other attributes (e.g. time, costs, etc.). These essential details are captured into objects, where the attributes are usually proportionate to that particular object (classification). Acknowledging this level of detail makes BIM an ideal candidate for managing project detail, as it provides a constant and coordinated view (and representation) of the digital model. It is therefore increasingly becoming a standard through which established communication and collaboration protocols are being operationalised.

2.4.4. *Generative evolutionary design*

From an AEC perspective, GD refers to any design practice where designers use a system (such as a computer program), which is set into motion with some degree of anatomy, which in turn contributes to (or results in) a completed work of art (Janssen, 2006; Janssen et al., 2006). The application of evolutionary algorithms in AEC have shown great promise, particularly as they are able to generate new and innovative design alternatives (Seyedzadeh & Rahimian 2021b). This has been particularly successful. Certain pioneers of this approach have advocated that this not only enhances the design process (through the generation of multiple complex forms), but also provides unique solutions with various details and layouts

Table 2.3 Developed tools

Non-Commercial Tools	Specifications	Tools
Constraint-based representation	The tool maintains the constraints and the integrity of the design.	SketchPad (Sutherland, 2003); The Sketcher (Medjdoub, 1999); CoDraw (Gross, 1992); BRIAR (Gleicher et al., 1991); ReDraw (Kolarevic, 1993)
Associative representations	Design relations constitute dependencies that are defined by the structure of the underlying model.	
Design grammar representations	Designs are represented by means of a vocabulary of shapes (defined by lines and labels) and a set of production rules; design relations as well as design transformations are encapsulated in those rules.	Discoverform (Carlson & Woodbury, 1990)
Hybrid representations	Combination of different representational models	SEED-Layout (Flemming & Chien, 1995); Floor Layout and Massing Study Programs (Harada, 1998); Performance Simulation Interface (Suter, 2000)
Commercial Tools	**Specifications**	**Tools**
Sector-specific representation	Industry-standard CAD tools	Revit (AutoDesk); GenerativeComponents (Bentley tools)

that would not have been possible without using GD. Several researchers have highlighted the benefits of using evolutionary design (Frazer, 2002; von Buelow, 2007; Janssen, 2006; Narahara et al., 2006). Moreover, architectural design has benefited from the application of generative algorithms by fundamentally adopting five different techniques, notably: genetic algorithm, cellular automata, L-systems, swarm intelligence, and shape grammars (Janssen, 2006). Indicative examples are presented in Table 2.3.

Given the importance and potential of GD, the emphasis of the research presented in this chapter is not to epitomise existing systems and approaches per se (vis-à-vis improvements), but rather to present the process of GD integration, highlighting how optimal solutions can be generated to support the design process. This support is fundamental, as the evolutionary design method uses evolutionary software systems (genetic algorithms) to enhance designers' ability during the design process. In this respect, evolutionary design is broadly recognised by its parametric evolutionary design and generative evolutionary design (Janssen, 2006).

2.4.5. Parametric evolutionary design

Parametric evolutionary design is often taken during the late design stages in order to find the 'best' solution to the design problem amongst different design alternatives. This requires a basic design concept to be established in advance. From this point, components are parameterised by designers for further improvement. The system 'evolves' these parameters at the last stage into generative alternative design solutions (Janssen 2006). Parametric design has been successfully adopted in a number of BIM applications; however, whilst parametric systems have matured into effective drawing tools, algorithm-based optimisation approaches and functional approximations of the fitness function are still evolving (Rasheed et al., 2005; Seyedzadeh & Rahimian 2021a).

2.4.6. Innovative opportunities

A number of exciting opportunities are now emerging in AEC. For example, the concepts and approaches of the evolutionary systems developed by Frazer (2002) [using AutoCAD and Sun's systems integrated with Micro Station] helped establish the foundations for integration. In particular, integrating evolutionary systems with advanced BIM modelling applications, where the generative process makes use of complex geometric functions through BIM. This has opened up discourse in this area to explore the future direction of computational design strategy – more specifically, debate on the general aspects of contemporary design in architecture practice. The following list provides a brief synopsis of this debate from literature:

- **Design Collaboration**: New technologies and systems (e.g. computer networking, video and computation integration, etc.) have created significant advanced opportunities for synchronous and asynchronous collaborative design (SCD and ASCD);
- **Sketch-pad Systems**: Computational support for sketching;
- **Computational Sketching Systems Integration into Augmented Reality Architectural Form:** Combining sketch-pad tools with real-time 3D information. Many examples are now emerging. One such example includes 3D environment information, enabling designers to better appreciate how their designs would perform on site. This includes the design support environment – ergo, proximity, temperature, brightness, humidity, wind direction, acoustics, etc.;
- **Digital Mock-ups (3D Sketching):** Application of 3D sculptures as a replacement for early design mock-ups, which also includes the use of Digital Twins;
- **Skills Development:** Skills are needed to support the design process. In particular, these skills should be aligned to intrinsic need at the early design stage, including the evolving technological solutions and tools underpinning this approach.

2.5. The conceptual framework for BIM integrated generative tools

As mentioned earlier in this chapter, the use of CAD in AEC design revolutionised the design process. This evolution is continuing, with new approaches presented almost every day. These tools have been openly acknowledged as being able to not only enhance designers' capabilities (especially in drafting and modelling), but also have helped designers procure highly innovative and effective design solutions. These tools have also made it possible for designers to work with ever more complex forms and complicated design tasks, including advanced drawing and editing (objects and properties), creating and manipulating free-form curves and surfaces, applying context-specific editing to objects, and optimising objects to meet specific criteria, e.g. lighting, rendering, etc. Algorithmic codes and scripts can now be integrated with CAD tools to support and enhance the design process. The next stage of these developments included the integration of BIM and GD. The following sections provide an overview of the framework environment and tool schema adopted in this research.

Having a single, flexible, and dynamic 3D environment that covers a wide range of architectural design requirements throughout the design process has often been seen as a vital prerequisite for designers. In this respect, generative evolutionary design can help assist designers in the early design stages, especially as BIM's parametric capabilities provide direct relationships with physical production processes. Acknowledging this ability, the proposed system aligns existing BIM process and concomitant flow of information to actively support AEC design. This framework and conceptual tool forms the schema for the final prototype, where this will actively engage GD methods into a single dynamic BIM environment. This study contributes to extant knowledge in this area by providing a 'stepping stone' for digital integration across all stages of AEC projects. In particular, it aligns to BIM Level 3 (Cloud), whilst also presenting a valuable set of rubrics for supporting the early design process, especially:

- **Creating models:** these link to all required information and details supporting the development process;
- **Establishing generative processes:** these are capable of controlling the variability of design outcomes and generation of designs (with required level of complexity). These are also capable of being able to generate alternatives (that differ significantly in terms of organisation and configuration);
- **Developing an innovative collaborative environment:** this enables designers to communicate more efficiently through the conceptual design phases (thereby enabling both short-term and long-term asynchrony);
- **Creating a computational design environment:** this supports sketches (either by scanning handmade sketches or by drawing-on-tablet technology) in both 2D and 3D environments;
- **Establishing a collaborative environment:** this facilitates true collaboration (from different geographical regions), enabling designers to edit, save, and improve sketches and designs through one shared space;
- **Facilitating incremental design:** this enables designers to take their sketches (2D and/or 3D) to the next level. This is particularly useful for testing and shaping ideas, thoughts, and designs in a 'safe' space, prior to incorporation into the final phases.

Given these rubrics, it is important to acknowledge that the integration of generative tools with information modelling (combined with advanced 3D knowledge-rich systems) has now started to open up new avenues of research for wider exploitation (Kocaturk & Medjdoub, 2011). This research presents one particular aspect of this: the incorporation of GD with BIM – specifically, the exploitation of parameters created within the early design stages. This work presents how generated solutions are produced in order to meet the design problem using the population of design alternatives. These are formulated through an algorithm (which contain design constraints, routines, and data files), where this changes the inputs [of the algorithm]. From this data, the final design can be altered accordingly – akin to creating a basic model based on 'Routines' and generating different design alternatives by adjusting the design parameters. For example, materials, fabrication constraints, and assembly logic can also be parameterised. The framework environment and tool schema for this approach can be seen in Figure 2.1.

Figure 2.1 presents the relationship between the environment and schema, where the generative system is capable of linking geometric behaviour patterns and performance properties of the system. The design environment (production/modification) constantly interacts with the generative system and BIM environment. This aligns to thinking on external behavioural tendencies and parameterisation, especially the exchange of an object with its specific environment (Hensel & Menges, 2008). This approach was adopted to capture, shape, and inform the process, whilst also being able to provide detailed granular data (to help identify the delimiters). The results from this approach defined the schema for this research model. From this, the conceptual framework for the generative BIM environment was developed (Figure 2.2).

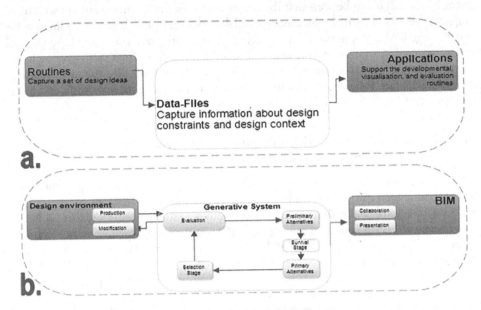

Figure 2.1 Framework environment (a) and tool schema (b)

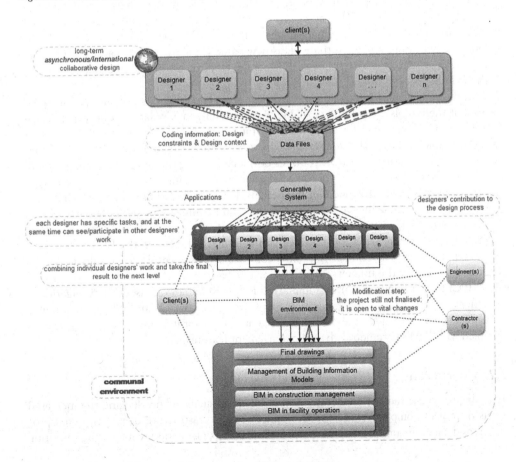

Figure 2.2 Conceptual framework for generative BIM environment

From Figure 2.2, it can be seen that the arrangements of this environment are structured to support the processes of a synchronous collaborative design – from the coding engine through to the BIM environment. This conceptual framework was used to develop a working prototype using a programming language embedded in Autodesk Revit. This approach also allowed the generative process to make direct use of the Revit modelling functions, particularly for visualisation (where feedback from the evolutionary process can be displayed in the Revit interface).

2.6. Technical implication issues: adoption and integration of generative BIM workspace within AEC

Existing evolutionary systems have ostensibly been formed based on source-code libraries or as programming toolkits. It is also widely acknowledged that heterogeneous parallel genetic algorithms are able to deal with a plethora of different operating systems (Alba et al., 2002). Given this, for the development of this conceptual environment, it was important to appreciate that the architecture and methods of data creation and retrieval have a significant direct impact on outcomes. In this respect, existing evolutionary tools have seldom implemented ready-made menu-driven systems. It was therefore important to make this G-BIM prototype as user friendly as possible, particularly to support the integration of evolutionary design. End users are not envisaged to be programmers or experts in genetic algorithms. Therefore, the proposed architecture adopted an approach similar to current tools such as Grasshopper® (a graphical algorithm editor). This approach follows recent developments in computational design, which have substantially changed the conventional design process (and by default, designers' way of working), such that: "This new paradigm aims to locate architectural discourse within a more objective framework when the efficient use of resources supersedes the aesthetic indulgence of works" (Leach, 2009).

Several widely available tools in the market are capable of handling detailed design processes, but these tools are not fully capable of purposefully manipulating conceptual design data. In order to overcome this challenge, the framework presented here was designed specifically to exploit this need by combining new concepts into a single BIM environment. From this, the G-BIM prototype encodes routines developed during the design brief (preparation stage), which forms the genotype for the generation step. This initially transforms a genotype into a phenotype (2D or 3D model of the design), followed by defining the representations, where the data-files specify the design constraints and context (i.e. site boundaries, minimum dimensions and distances, number of floors and spaces, etc.) (Figure 2.1a). The results generated are therefore based on routines and data-files. Thereafter, the BIM tool is used for representation and amendments. From a compatibility perspective, the results (ergo, the generated designs by the framework) will ultimately be translated into Industry Foundation Class (IFC) format (Figure 2.1b). In summary, this G-BIM framework uses genetic algorithms for conceptual design and form generation (population of alternatives). In doing so, it is able to harvest the advanced features of BIM tools, particularly for illustration and collaboration (coupled with BIM's parametric change management features).

2.7. Conclusion

This chapter presented the challenges facing AEC, especially acknowledging the increased levels of project complexity and design requirements, infrastructure demands, tiered processes, and data integration compliance issues. In this respect, clients are increasingly wanting to challenge the status quo through new unique and innovative designs. This opened up many new opportunities, particularly the exploitation of BIM, advanced modelling, simulation, visualisation, advanced design tools, and the use of machine learning, artificial intelligence, and GD applications. Some of these were highlighted from extant literature, including the need for AEC to develop these opportunities for wider exploitation. In doing so, this chapter presented the rubrics for a dynamic and flexible BIM application aimed specifically at the AEC market, which focussed specifically on the design requirements at the early stages of design. This work included the critical aspects required to support recent (computational) design paradigms, including algorithmic architecture, generative, and parametric design – all of which are capable of providing techniques for exploring and generating design solutions.

From these AEC challenges, several existing theoretical and technical gaps were explored, especially on the implementation of BIM tools and GD to support conceptual design. It was proffered that the implementation of new interfaces could lead to new approaches for using GD in order to help AEC designers explore different design solutions in a risk-free virtual environment. A conceptual framework was then presented for discussion. This highlighted the need to deliver seamless integration in order to fully exploit and leverage these benefits. The generative BIM framework highlighted the core structural arrangements along with its capability attributes. This framework allows designers to both analyse and optimise design solutions using GD (at the conceptual design stage) to: (i) provide techniques for exploring and generating new and innovative design solutions; (ii) create models with appropriate information and details needed for the development and support of design processes; and (iii) establish generative processes capable of controlling the variability of design outcomes. This generative BIM framework was able to accommodate designs with high levels of complexity (at each stage of the design process), and was also able to deliver viable design alternatives in line with the genetic algorithms and parametric design criteria. In this respect, these findings contribute to the wider understanding of AEC solution generation, especially the application of new concepts in computational design and architecture.

References

Abrishami, S., Goulding, J., Ganah, A., & Rahimian, F. (2013). *Exploiting modern opportunities in AEC industry: A paradigm of future opportunities* (pp. 321–333). AEI.

Abrishami, S., Goulding, J. S., Rahimian, F. P., & Ganah, A. (2014). Integration of BIM and generative design to exploit AEC conceptual design innovation. *Electronic Journal of Information Technology in Construction (ITcon)*, 19, 350–359 (Special Issue BIM Cloud-Based Technology in the AEC Sector: Present Status and Future Trends). www.itcon.org/2014/21

Alba, E., Nebro, A. J., & Troya, J. M. (2002). Heterogeneous computing and parallel genetic algorithms. *Journal of Parallel Distributed Computing*, 62(9), 1362–1385.

Arif, M., Kulonda, D., Jones, J., & Proctor, M. (2005). Enterprise information systems: Technology first or process first? *Business Process Management Journal*, 11, 5–21.

Bilda, Z., & Demirkan, H. (2003). An insight on designers' sketching activities in traditional versus digital media, *Design Studies*, 24, 27–50.

Biloria, N. (2018). Urban informatics: Decoding urban complexities through data sciences. *Smart and Sustainable Built Environment*, 7(1), 2–3.

BIM Task Group. (2013). *UK BIM task group website*. Retrieved October 2013, from www.bimtaskgroup.org

Bouchlaghem, D., Shang, H., Whyte, J., & Ganah, A. (2005). Visualisation in architecture, engineering and construction (AEC), *Automation in Construction*, 14(3), 287–295.

Butz, A., Fisher, B., Krüger, A., & Olivier, P. (2005). Multi-level interaction in parametric design. *Lecture Notes in Computer Science*, 3638, 151–162.

Cabinet Office. (2011, May). *Government construction strategy.* Retrieved October 2013, from www.gov. uk/government/uploads/system/uploads/attachment_data/file/61152/Government-Construction-Strategy_0.pdf

Carlson, C., & Woodbury, R. F. (1990). *Hands on exploration of recursive forms.* Carnegie Mellon University, Engineering Design Research Center.

Cera, C. D., Regli, W. C., Braude, I., Shapirstein, Y., & Foster, C. V. (2002). A collaborative 3D environment for authoring design semantics, *Computer Graphics and Applications,* IEEE, *22*(3), 43–55.

Cheon, S., Kim, B. C., Mun, D., & Han, S., (2012). A procedural method to exchange editable 3D data from a free-hand 2D sketch modeling system into 3D mechanical CAD systems. *Computer-Aided Design, 44*(2), 123–131.

CIOB. (2020). *The real face of construction 2020.* Socio-economic analysis of the true value of the built environment, The Chartered Institute of Building, UK. Retrieved from https://www.ciob.org/industry/research/Real-Face-Construction-2020

Colombo, A. W., & Harrison, R. (2008). Modular and collaborative automation: Achieving manufacturing flexibility and reconfigurability. *International Journal of Manufacturing Technology and Management, 14,* 249–265.

Cooke, B., & Williams, P. (2009). *Construction planning, programming and control* (3rd ed.). Wiley Blackwell.

Creswell, J. W. (2002). *Research design: Qualitative, quantitative, and mixed method approaches* (2nd ed.). Sage.

Cross, N. (2007). *Designerly ways of knowing* (chapterback ed.). Birkhäuser.

Cross, N., & Clayburn, C. A. (1995). Observations of teamwork and social processes in design. *Design Studies, 16*(2), 143–170.

Do, E. Y., & Gross, M. D. (2009). Back to the real world: Tangible interaction for design. *Artificial Intelligence for Engineering Design, Analysis and Manufacturing, 23*(3), 221–223.

Eastman, C. M. (1999). *Representation of design process.* Invited keynote Speech in Conference on Design Thinking, MIT.

Egan, J. (1998). *The Egan Report – rethinking construction, report of the construction industry taskforce to the deputy prime minister.* HSMO.

Finkelsteiin, F. (1994). *Software process modelling and technology* (J. Kramer & B. Nuseibeh, Eds.). John Wiley & Sons, Inc.

Fischer, M. (2000). *4D CAD-3D models incorporated with time schedule.* CIFE Centre for Integrated Facility Engineering in Finland, VTT-TEKES, CIFE Technical Report, Stanford University.

Fischer, M., & Kunz, J. (2004). The scope and role of information technology in construction, *CIFE Technical Report #156,* 1–8.

Fischer, T., Burry, M., & Frazer, J. (2005). Triangulation of generative form for parametric design and rapid prototyping. *Automation in Construction, 14,* 233–240.

Flemming, U., & Chien, S. (1995). Schematic layout design in SEED environment. *Journal of Architectural Engineering, 1*(4), 162–169.

Forcade, N., Casals, M., Roca, X., & Gangolells, M. (2007). Adoption of web databases for document management in SMEs of the construction sector in Spain. *Automation in Construction, 16,* 411–424.

Frazer, J. H. (2002). *Creative design and the generative evolutionary paradigm in creative evolutionary systems* (P. J. Bentley & D. W. Corne, Eds., pp. 253–274). Academic Press.

Frohm, J., Stahre, J., & Winroth, M. (2008). Levels of automation in manufacturing. *Ergonomia – an International Journal of Ergonomics and Human Factors, 30*(3).

Fruchter, R. (1998). *Internet-based web mediated collaborative design and learning environment.* Artificial Intelligence in Structural Engineering, Lecture Notes in Artificial Intelligence, Springer-Verlag.

Gleicher, M., & Witkin, A. (1991). *Creating and manipulating constrained models.* Technical Report CMU-CS-91-125, Computer Science Department, Carnegie Mellon University.

Goulding, J. S., & Rahimian, F. P. (2012). Industry preparedness: Advanced learning paradigms for exploitation. In A. Akintoye, J. S. Goulding, & G. Zawdie (Eds.), *Construction innovation and process improvement.* Wiley.

Gross, M. D. (1992). Graphical constraints in CoDraw. *Proceedings of the 1992 IEEE Workshop on Visual Languages*, 81–87. Retrieved from https://doi.org/10.1109/WVL.1992.275780

Gu, N., & London, K. (2010). Understanding and facilitating BIM adoption in the AEC industry. *Automation in Construction, 19*, 988–999.

Harada, M. (1998). *Discrete/continuous design exploration by direct manipulation*. Institute for Complex Engineered Systems, Carnegie Mellon University.

Hensel, M., & Menges, A. (2008). Versatility and vicissitude – performance in morpho-ecological design. *Architectural Design, 78*(2), 6–11.

House of Commons Library. (2019). Construction industry: Statistics and policy. *Briefing Paper Number 01432, December 2019, House of Commons, UK*. Retrieved from https://researchbriefings.files.parliament.uk/documents/SN01432/SN01432.pdf

Ibrahim, M., Krawczyk, R., & Schipporeit, G. (2004, September). *Two approaches to BIM: A comparative study*. Education and Research in Computer Aided Architectural Design in Europe (eCAADe) 2004 Conference.

Isikdag, U., & Underwood, J. (2010, May). *A synopsis of the handbook of research on building information modeling*. Proceedings of CIB 2010 World Building Congress at Salford.

Janssen, P. (2006). A generative evolutionary design method. *Digital Creativity, 17*(1), 49–63.

Janssen, P., Frazer, J., & Tang, M. X. (2006). A framework for generating and evolving building designs. *The International Journal of Architectural Computing, 3*(4), 449–470.

Johnson, G., Gross, M. D., & Do, E. Y. (2006). *Flow selection: A time-based selection and operation technique for sketching tools*. AVI'06: Proceedings of the Working Conference on Advanced Visual Interfaces, ACM.

Johnson, G., Gross, M. D., Hong, J., & Do, E. Y. (2009). Computational support for sketching in design: A review. *Foundations and Trends® in Human – Computer Interaction, 2*(1), 1–93.

Khanzade, A., Fisher, M., & Reed, D. (2007). *Challenges and benefits of implementing virtual design and construction technologies for coordination of mechanical, electrical, and plumbing systems on large healthcare project*. CIB 24th W78 Conference 2007, 205–212.

Kocaturk, T., & Medjdoub, B. (2011). *Distributed intelligence in design* (Hardcover ed.) Wiley-Blackwell.

Kolarevic, B. R. (1993). *Geometric relations as a framework for design conceptualization* (Ph.D. Thesis, Harvard University, Graduate School of Design).

Latham, M. (1994). *Constructing the team, joint review of the procurement and contractual arrangements in the UK construction industry, final report*. HSMO.

Leach, N. (2009). Digital Morphogenesis. *Architectural Design, 79*, 32–37.

Leinonen, J., Kähkönen, K., Retik, A. R., Flood, R. A., William, I., & O'Brien, J. (2003). New construction management practice based on the virtual reality technology. In A. A. Balkema (Ed.), *4D CAD and visualisation in construction: Developments and applications* (pp. 75–100). AEC.

Medjdoub, B. (1999, June 7–8). Interactive 2D constraint-based geometric construction system. *Proceedings of the Eighth International Conference on Computer Aided Architectural Design Futures*, 197–212.

Muehlbauer, M. (2018). Towards typogenetic tools for generative urban aesthetics. *Smart and Sustainable Built Environment, 7*(1), 20–32.

National BIM Standard-US (NBIMS-US). (2007). *United States, the national building information model standard*. Retrieved October 2013, from www.wbdg.org/pdfs/NBIMSv1_p1.pdf

Narahara, T. (2007, November 15–17). Enactment software: Spatial designs using agent-based models. Proceedings of the Agent 2007 Conference on Complex Interaction and Social Emergence, in association with North American Association for Computational Social and Organizational Sciences, Northwestern University.

Narahara, T. (2006). Multiple-constraint genetic algorithm in housing design, synthetic landscapes. *Proceedings of the 25th Annual Conference of the Association for Computer-Aided Design in Architecture*, 418–425.

Paulson, B. C. (1976). Designing to reduce construction costs. *Journal of the construction division, 102*, 587–592.

Petric, J., Maver, T., Conti, G., & Ucelli, G. (2002). *Virtual reality in the service of user participation in architecture.* CIB W78 Conference, Aarhus School of Architecture.

Popov, V., Mikalauskas, S., Migilinskas, D., & Vainiunas, P. (2006). Complex usage of 4D information modelling concept for building design, estimation, scheduling and determination of effective variant. *Technological and Economic Development of Economy, 12,* 91–98.

Rahimian, F. P., Ibrahim, R., Wirza, R., Abdullah, M. T. B., & Jaafar, M. S. B. H. (2011). Mediating cognitive transformation with VR 3D sketching during conceptual architectural design process. *International Journal of Architectural Research (Archnet-IJAR), 5*(1), 99–113.

Rahimian, F. P., Seyedzadeh, S., & Glesk, I. (2019). OCDMA-based sensor network for monitoring construction sites affected by vibrations. *Journal of Information Technology in Construction, 24,* 299–317.

Rasheed, K., Ni, X., & Vattam, S. (2005). Comparison of methods for developing dynamic reduced models for design optimization. *Soft Computing, 9*(1), 29–37.

RIBA. (2012). *BIM overlay to the RIBA outline plan of work.* Retrieved October 2013, from www.bdon-line.co.uk/Journals/2012/05/15/d/x/f/BIM_Overlay_RIBA_Plan_of_Work_Embargoed.pdf

Rolland, C. (1998). *A comprehensive view of process engineering Advanced Information Systems Engineering.* Proceedings of the 10th International Conference CAiSE'98, B. Lecture Notes in Computer Science 1413, C. Thanos Pernici, Springer.

Roudavski, S. (2009). Towards morphogenesis in architecture. *International Journal of Architectural Computing, 7*(3), 345–374.

Sampaio, A. Z., Ferreira, M. M., Rosário, D. P., & Martins, O. P. (2010). 3D and VR models in civil engineering education: Construction, rehabilitation and maintenance. *Automation in Construction, 19*(7), 819–828.

Santos, E. T. (2009). *Building information modeling and interoperability.* XIII Congress of the Iberoamerican Society of Digital Graphics – From Modern to Digital: The Challenges of a Transition.

Seyedzadeh, S., Agapiou, A., Moghaddasi, M., Dado, M., & Glesk, I. (2021). WON-OCDMA system based on MW-ZCC codes for applications in optical wireless sensor networks. *Sensors, 21*(2), 539.

Seyedzadeh, S., & Rahimian, F. P. (2021a). Multi-objective optimisation and building retrofit planning. In *Data-driven modelling of non-domestic buildings energy performance* (pp. 31–39). Springer.

Seyedzadeh, S., & Rahimian, F. P. (2021b). Building energy data-driven model improved by multi-objective optimisation. In *Data-driven modelling of non-domestic buildings energy performance* (pp. 99–109). Springer.

Skibniewski, M. J. (1992, June 3–5). *Current status of construction automation and robotics in the United States of America.* The 9th International Symposium on Automation and Robotics in Construction.

Suermann, P. (2009). Evaluating industry perceptions of building information modelling (BIM) impact on construction. *International Journal of IT in Construction, 14,* 574–594.

Suter, G. (2000). *A representation for design manipulation and performance simulation* (Ph.D. Thesis, School of Architecture, Carnegie Mellon University).

Sutherland, I. E. (2003). *Sketchpad: A man-machine graphical communication system.* Technical Report 574. University of Cambridge.

van Stijn, A., & Gruis, V. (2020). Towards a circular built environment. *Smart and Sustainable Built Environment, 9*(4), 635–653.

Veenstra, V., Halman, J., & Voordijk, J. (2006). A methodology for developing product platforms in the specific setting of the housebuilding industry. *Research in Engineering Design, 17,* 157–173.

Von Buelow, P. (2007). Advantages of evolutionary computation used for exploration in the creative design process. *Journal of Integrated Systems, Design, and Process Science, 11*(3), 3–16.

Whyte, J., Bouchlaghem, N., Thorpe, A., & McCaffer, R. (2000). From CAD to virtual reality: Modelling approaches, data exchange and interactive 3D building design tools, *Automation in Construction, 10*(1), 43–55.

Wikberg, F., Ekholm, A., & Jensen, P. (2010). *Configuration with architectural objects in industrialised house-building.* CIB W078 Proceedings.

Woo, J. H., Clayton, M. J., Johnson, R. E., Flores, B. E., & Ellis, C. (2004). Dynamic knowledge map: Reusing experts' tacit knowledge in the AEC industry. *Automation in Construction, 13,* 203–207.

Xie, H, Shi, W, and Issa, R. R. A. (2011).Using RFID and real-time virtual reality simulation for optimization in steel construction. *ITcon, 16,* 291–308.

3 Towards intelligent structural design of buildings
A BIM solution

3.1. Introduction

The effective design of buildings traditionally relies heavily on the successful collaboration of all project participants (Palomar et al., 2020). This is especially so with the growing complexity of modern architecture and the increased challenges of integration and collaboration among architects and engineers (Dawood et al., 2020; Ku & Pollalis, 2020; Oliver et al., 2020). In conventional design, architects and engineers tend to be treated as two separate disciplines (Leite, 2019), with limited integration taking place among them, especially during early stages (Liu et al., 2017). For example, the architect (or designer) begins with a conceptual design, which provides the building shape and layout needed to meet both the functional and aesthetic requirements (Abrishami et al., 2020). Thereafter, structural engineers uses this design to establish the structural systems and skeleton of buildings, particularly to meet safety and structural performance, including cost criteria, etc. (Hamidavi et al., 2020). Arguably, however, structural design requirements are not fully considered during the architectural design, where some have highlighted that the effects of architectural design on structural performance can be overlooked (Macdonald, 2018). A corollary of this is that architects then end up developing a design that is difficult to be made, that is structurally sound, and in some instances infeasible to be built (Hurol, 2013). This fragmented arrangement has several knock-on effects and repercussions, including information loss, duplication of data, inaccuracies, productivity loss, delays, cost overruns, increased litigation, and concomitant impact on quality (Leite, 2019).

In an attempt to prevent or mitigate some of these issues, it has been suggested that the design team and engineers should try and work more collaboratively in order to share experiences and domain expertise. Whilst it could be argued that terminologies such as 'design team' and 'engineers' is open to interpretation, for simplicity, this chapter uses the term architects (for the design team) and structural engineers (for engineers). Whilst this is not ideal, where derivations from this are needed, additional clarification will be provided to highlight niche differences. Notwithstanding this, the call for a collaborative system of working is needed, one that more meaningfully blends and optimises design alternatives and aesthetic requirements, with the structural considerations underpinning building performance. This requires a radical new way of thinking and working, where dynamic procedures can facilitate collaboration between the two teams (architects and engineers) during the whole design process, including the early stages (Ciribini et al., 2016).

Extant literature highlights these challenges, noting that research on developing an automatic system for generating and optimising different options for structural design is currently inadequate (Hamidavi et al., 2020). In this respect, structural engineers tend to

DOI: 10.1201/9781003106944-3

rely heavily on intuitive knowledge and experience. This often requires manually generating conceptual designs, then going through the process of trying to optimise these in line with performance requirements. Moreover, this is often seen as an iterative and cumbersome process, where engineers consider possible alternatives. That being said, things are starting to change, where for example methodologies like Building Information Modelling (BIM) have the potential to improve the structural design process through automation (Hamidavi et al., 2020; Liu et al., 2017; Sheikhkhoshkar et al., 2019). In addition, BIM can be used to enable designers to generate and optimise different structural models through automatic and parametric processes (Banihashemi et al., 2018); however, despite these capabilities, the use of BIM to facilitate structural design has remained somewhat unexploited (Hosseini et al., 2018; Leite, 2019; Vilutiene et al., 2019). Some opportunities have been explored, including the use of federated BIM models within a common data environment (CDE) to bridge the gap between various disciplines (Pärn et al., 2017), where the CDE acts as a central repository for construction project information. That being said, the use of these technologies has met certain challenges (Mignone et al., 2016; Oraee et al., 2019). The main issues seem to rest with challenges of automatically updating structural designs based on architectural design changes. This is also compounded by the issue that BIM federated models are not equipped with an intelligent structural optimisation function (Vilutiene et al., 2019).

Acknowledging these issues, this chapter presents a number of ideas and a solution for some of these challenges. This discussion aligns to the concept that structural design processes still resemble a "loosely coupled system" (Papadonikolaki, 2018). The discussion therefore will aim to bridge the gap between architects and structural engineers, by presenting a workable automated optimisation solution to integrate these two separate disciplines. Solutions of this type have already been advocated for in construction (Leite (2019); Vilutiene et al. (2019). The solution presented in this chapter can be considered a proof-of-concept (PoC), where this PoC (i) integrates architectural and structural models and (ii) facilitates automated intelligent structural model generation, exploration, and optimisation.

3.2. Contextual background

Maintaining the flow of information among various disciplines has been (and continues to be) challenging, especially given competing and frequently conflicting interests (Beghini et al., 2014). The traditional supply chain in Architecture, Engineering, and Construction (AEC) lacks integration, particularly between architects and structural engineers, where their working procedures are often fragmented and affected by a silo mentality (Durdyev et al., 2019). Arguably, these two groups predominantly work in isolation, and therefore recognise responsibility only for their portion of the work (Hurol, 2013; Pärn et al., 2018; Vilutiene et al., 2019). For example, traditionally, architects focus on the aesthetic aspects of buildings (building shape, etc.), while engineers give priority to the stability and performance of building structures (Hurol, 2013; Vilutiene et al., 2019). These distinctions have caused several problems, including non-productive activities (up to 15% of cost and time overruns), abundant change orders (accounting for between 60–90% of all variations), and inefficient communication (causing an additional 5–10% in cost and time overruns) (Kraatz et al., 2014).

The disaggregation and fragmentation between these two disciplines are examined further as follows.

3.3. Integration challenges

Several studies have recommended the use of technologies to facilitate collaboration among disparate team members, particularly through a single shared interface using federated models and a CDE (Emmitt & Ruikar, 2013; Merschbrock & Munkvold, 2014). The arrival of BIM emerged as an advanced methodology that provided the much-needed CDE, which also facilitated data sharing among various disciplines in design activities (Kuiper & Holzer, 2013). That being said, whilst BIM has (in many respects) been 'sold' as a panacea solution for integrating data, the same cannot be said for the integration of disciplines. For example, evidence indicates that BIM adoption in isolation has not bridged the gap between architects and engineers (Abrishami et al., 2015). This recurrent problem invited a new stream of research – to enhance collaboration through BIM-enabled projects (*cf.* Oraee et al. (2017). Of note here, whilst application of federated models can be used to create a common platform for various disciplines to share and integrate their models, many other problems still exist. The major problem here is the lack of intelligence and capability, as federated models are unable to translate architectural details changes into updated structural models, *cf.* Pärn et al. (2017).

Emphasising the role of contractual arrangements, Mignone et al. (2016) argued that the traditional legal structures underpinning construction projects needed to be modified. From this, Integrated Project Delivery (IPD) emerged. This redefined project delivery as an integrated process, where all disciplines work together as a single unit (Pishdad-Bozorgi, 2016). Whilst (arguably) IPD is relatively new and still evolving as a project delivery system, specific emphasis is placed on the involvement of key participants, particularly relationships and collaboration in the pursuit of mutual goals (Teng et al., 2017). Notwithstanding these benefits, few projects have adopted IPD due in part to cultural, financial, legal, and/or technological barriers (Hamzeh et al., 2019). Moreover, despite the availability of BIM, IPD, and CDE, it seems that the siloed positioning between architects and structural engineers is a problem that still needs to be addressed (Oraee et al., 2019). One solution, as Merschbrock and Munkvold (2014) argued, relies on developing data exchange frameworks and tools. In this respect, the development of new tools was also recommended as an effective solution to facilitate the integration of architectural-structural collaboration (Beghini et al., 2014).

In summary, therefore, several new tools and methodologies (including federated models) have undoubtedly improved processes; however, the structural design of buildings still remains a challenge. This is because, in many respects, extant engineering software tools tend to provide structural solutions for integration with architectural designs, but these do not actually compute or generate optimal designs (Chan, 2001). For example, structural design relies heavily on architectural design (Abrishami et al., 2015). Moreover, this includes the behaviour of lateral load response (particularly for tall buildings), as this is affected by the shape of buildings (Xie, 2014). For example, from a structural perspective, building design requires structural engineers to consider not only safety and functionality but also design optimisation (Foraboschi et al., 2014). In doing so, this would fulfil important requirements such as minimum structural cost, maximum structural efficiency, and maximum usable floor area (Chan, 2001). The ideal solution, therefore, would be an automated intelligent integrated procedure that was able to link architectural design with that of structural engineers, where this was capable of generating alternatives for each submitted architectural design (Abrishami et al. (2020). This opportunity is discussed in the following section.

3.4. Automated design integration

Automated design integration has been discussed for a number of years now. This includes the expressed needs of enabling architects to incorporate structural design requirements into their design, especially at the early stages of architectural design (Garber, 2009). This approach should also be capable of facilitating structural design processes, where architectural models are able to synchronise with structural ones, and where the system intelligently computes and translates any changes in the architectural model to alternations in the structural model (Hamidavi et al., 2020). Research into automated design integration provides a number of potential solutions (Table 3.1).

Table 3.1 Automated design integration studies

References	Focus	Objectives	Limitations
Baldock and Shea (2006)	Structural design optimisation	Reduce the structural mass of the bracing	Limited to bracing design for steel frameworks
Hofmeyer and Davila Delgado (2013)		Perform structural analysis and provide an indication of structural behaviour	• Limited collaboration with architectural models
Mueller and Ochsendorf (2015)		Address the need to consider both quantitative performance goals and qualitative requirements during conceptual design	• No function in BIM Only works with 2D frames
Vermeulen (2015)		Accelerate the process of topology and size optimisation	Limited to collaboration with architectural models
Richardson et al. (2013)		Provide a range of optimal bracing configurations based on evolving structural and architectural requirements	Limited to X-bracing and topology optimisation
Dutta et al. (2020)	Computer-Aided Lift Planning systems	Provide smart and optimal solutions for automatic crane lifting, supported by intelligent decision-making and planning algorithms along with computer graphics and simulation	• Focused on lift planning • Limited to structural design process
Zayed and Mohamed (2014)	Automatic Climbing System	A formwork to climb in various weather and height conditions	Focused on climbing systems and limited contribution to the structural design process
Leite (2019)	Wind-induced response analysis and design optimisation for rectangular steel tall buildings	Provide automated optimised design (minimum cost)	Limited to architectural models

References	Focus	Objectives	Limitations
Chunyu and Li (2018)	Generation of layout of high-rise building in urban landscape design	Automated generation of layout of high-rise buildings in urban landscape design	Limited to external elevation of high-rise buildings in urban landscape design
Wang and Mahin (2018)	Automated design and optimisation for seismic analysis	Performance-Based Earthquake Engineering: an automated tool that incorporates an optimisation engine and structural analysis software	• Limited in collaboration with architectural models • Focused on automatic design and optimisation of seismic analysis
Shahin (2019)	Adaptive building envelopes for energy performance	Control physical environmental factors (heat, light, sounds)	• Focused on energy performance • Limited contribution to the structural design process

In addition to the studies highlighted in Table 3.1, a number of studies have also focussed on the practical applications of architectural/structural synergy and evolutionary structural optimisation (Holzer et al., 2007). This includes the use of structural topology optimisation methods for generating optimal structural design (Stromberg et al., 2012; Tsavdaridis et al., 2015). Other streams of research include the automatic integration of computational methods for topology optimisation (Kazakis et al., 2017; Rahmani Asl et al., 2015).

In summary, literature reveals that these studies for the most part focus on: the optimisation of structural design (Hamidavi et al., 2020), energy analysis (Schlueter & Geyer, 2018; Seyedzadeh & Pour Rahimian, 2021a), the integration between architecture-structural design to facilitate collaboration (Mora et al., 2008), and interactive design processes (Elnimeiri & Almusharaf, 2010). Literature did not highlight cost-effective methodologies or toolsets that linked BIM to automated structural design systems, or systems for achieving acceptable optimal structural design. Given this, the following sections highlight the research design approach adopted for this study, along with the procedures followed.

3.5. Research methodology

The research methodological approach adopted for this study was grounded on the premise of supporting the proof-of-concept, where the PoC needed to follow a systematic procedure (Love et al., 2020). Acknowledging this, the work presented here follows a four-stage approach: (i) research need and requirement analysis, (ii) system design (and assumptions), (iii) solution development, and (iv) preliminary production assessment. Following this approach, the rationale was to move from assessing the needs of the industry to developing solutions, and then to testing the practicality of these proposed solutions through a case project. This resonates with similar approaches adopted elsewhere (Yan & Li, 2016).

To provide additional research context, modern tall structures are often complex, large-scale slender structures. Therefore, the design task of structural engineers in assessing tall buildings tends to be more complicated (compared with traditional building typologies), especially given the combined effects of various lateral loads (Chan, 2001). This chapter focuses on tall buildings in order to capture maximum complexity in developing the solution. In doing so, the assumption made here was that this approach could be replicable to other structures (complex/simple). Thus, the starting trajectory acknowledged that the nature of wind load was considered a critical factor, insofar as this predominantly controls the design of tall buildings (Bezabeh et al., 2018). Moreover, wind load is a major factor of structural optimisation in tall building design (Love et al., 2020; Sharma et al., 2018; Yan & Li, 2016; Zhao et al., 2020). In addition, the shape of tall buildings also plays a key role in the influence and effect of the wind load (Zheng et al., 2018).

Given the need to select an appropriate building for this work in order to assess such issues as performance, the Gherkin in London, UK was chosen as the case study project to demonstrate how architects and engineers could have used this PoC. Of note, the Gherkin is the informal name used for 30 St Mary Axe (previously the Swiss Re Building), which is a commercial skyscraper in London. This building was chosen because of its elongated, curved, rounded-end design. Furthermore, the unique shape of this building required structural engineers to work closely with the architects to achieve this curved form. The prototype presented in this chapter is therefore used to demonstrate the automation of synergy between architects and structural engineers using a BIM-based platform to reduce the design time. The Gherkin has two primary structures comprising the diagrid and the core. The diagrid is the main part of the structure, which resists lateral and gravity loads, and transfers these to the core section of the structure. Richardson et al. (2013) defined the diagrid as a series of triangles that combine gravity and lateral support into one, making the building stiff, efficient, and lighter, compared to a traditional high-rise building (Al-Kodmany, 2018).

3.6. Requirement analysis and solution generation

Following the four-staged methodological approach identified earlier, the research and requirements analysis phase used an online questionnaire to elicit a list of main issues and challenges the industry faced on structural design, analysis, and optimisation processes. Questions additionally focused on interoperability and collaboration between structural engineers and architects, where recipients were asked to suggest potential solutions to address these challenges. The rationale of this questionnaire was to validate the main issues emanating from literature, whilst also capturing practice-based issues that were prevalent.

This questionnaire was distributed to professionally accredited structural engineers of the Institution of Structural Engineers (IStructE), Institution of Civil Engineers (ICE), and the American Society of Civil Engineers (ASCE) in the UK and US. Before distribution, the questionnaire was pre-piloted with 30 academic staff and researchers from the University of Portsmouth (UK) for content, accuracy, and consistency. The questionnaire was then distributed to 378 professionally accredited structural engineers. From this, 105 completed questionnaires were returned, for a response rate of 27.7%, representing three groups: IStructE (60%), ICE (21%), and ASCE (19%). The results from this survey can be seen in Figure 3.1.

From Figure 3.1, participants highlighted the most challenging tasks associated with structural design, analysis, and corresponding optimisation processes. The main challenges were seen to be "interoperation with other disciplines" and "structural design automation", both of which returned a 52.8% importance level. These challenges correspond to findings from literature. Thus, in order to provide a deeper understanding of the causal factors underpinning these results, a further three questions were asked to identify potential solutions to these challenges (Figure 3.2).

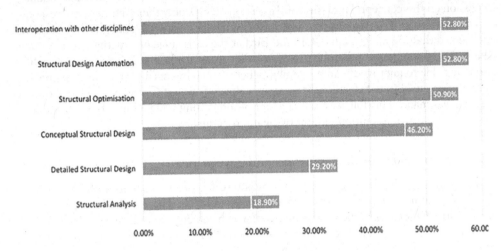

Figure 3.1 Areas in need of improvement based on responses to the questionnaire

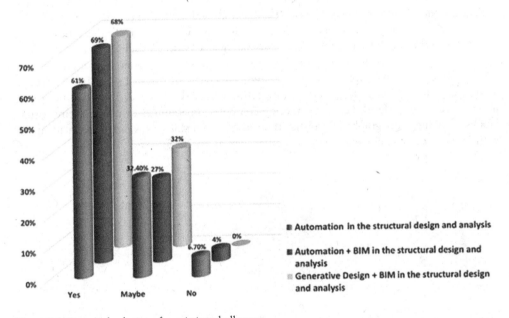

Figure 3.2 Potential solutions for existing challenges

From Figure 3.2, more than 61% of respondents stated that the use of automatic design would be a potential solution to the existing challenges of building design, analysis, and optimisation process. This rate increased to 69% in response to a question that asked about integration of automatic design through a BIM platform (as a potential solution). The third question invited respondents to share their experiences and familiarity with the concept of Generative Design (GD). For simplicity, GD is an iterative design process that uses advanced computer programming to generate design outputs that meet certain predefined constraints.

For further reading on GD, readers are referred to Singh and Gu (2012) for more detailed discussion on this concept. From this third question on GD, only the responses from respondents familiar with GD were considered for further analysis. Figure 3.2 highlights that respondents agreed that the integration of GD and BIM at the early stage of structural design would be a potential solution to solve the existing challenges in the building design process, whilst also being able to improve the interoperability between architects and structural engineers.

From this analysis, the decision to develop a state-of-the-art prototype to leverage these benefits was made. This followed the concept of integrating GD within a BIM platform, as this was also seen as a solution for improving interoperability.

3.7. System design: outline concepts and governance

The first stage in the development process was to establish the system design parameters, along with outline concepts and governance protocols. This followed the Robot Structural Analysis (RSA) approach, the arrangements of which can be seen in Figure 3.3.

From Figure 3.3, it can be seen that the main part of this arrangement concerns the prototype's ability to provide automated synergy. In this respect, the integration between structural engineers and architects was considered fundamental. It was also assumed that in order to achieve this goal, new tools and methods would be needed to not only address the technical challenges but also act as a segue to bring these two disciplines together.

To establish the automated synergy needed, this study drew upon the capabilities of visual programming in order to create the automated link between architectural and structural design procedures. In parametric BIM, the use of a visual programming interface can often replace the conventional elaborate coding associated with visual small blocks, which work independently to perform certain tasks (Rahmani Asl et al., 2015). Visual programming, therefore, enables designers to develop computer programmes by manipulating visual blocks – linking them graphically, rather than using coding, as this is much easier for users with limited programming experience (Banihashemi et al., 2018).

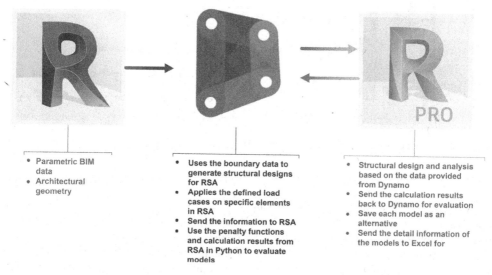

- Parametric BIM data
- Architectural geometry

- Uses the boundary data to generate structural designs for RSA
- Applies the defined load cases on specific elements in RSA
- Send the information to RSA
- Use the penalty functions and calculation results from RSA in Python to evaluate models

- Structural design and analysis based on the data provided from Dynamo
- Send the calculation results back to Dynamo for evaluation
- Save each model as an alternative
- Send the detail information of the models to Excel for

Figure 3.3 BIM RSA prototype development procedure

Recent developments in building design technology now offer designers several new opportunities to use visual programming tools. One of the main tools available is Grasshopper (Mirjalili et al., 2017). Dynamo is another product, which performs efficiently with Autodesk Revit (hereafter referred to as Revit) and extends the power of Revit by providing access to the Revit API (Application Programming Interface). In the design process, Dynamo synchronises the parametric data of the architectural model into the structural design platform in order to design different alternative structural models (Sheikhkhoshkar et al., 2019). The main advantage of this automatic synergy is that generated structural models are based on the boundary conditions, where changes in the architectural model are automatically updated in the structural models. This is particularly important, as structural engineers traditionally follow iterative processes (Seyedzadeh et al., 2020b).

From a system design perspective, this prototype was therefore developed on a BIM-based platform in order to facilitate data exchange and create synergy between structural and architectural models. Moreover, the use of BIM tools in a project such as this enables designers to use parametric data to model 3D geometric shapes (Banihashemi et al., 2018). Hence, this approach was used to generate alternative structural models using geometry parameters (from Revit) and rules (in Dynamo), the corollary of which automatically updates structural models according to changes made in architectural model parameters.

From a structural optimisation perspective, structural optimisation was considered the third most challenging task of the structural design process (Figure 3.1), where structural optimisation is a mathematical approach used to reduce the amount of material and, consequently, the impact on cost (Tsavdaridis et al., 2015).

Structural optimisation is generally demonstrated in the form of an equation by (Christensen & Klarbring, n.d.):

$$\{minimises\,/\,maximises\,f\,(x,y)\,with\,respect\,to\,x\,and\,y$$

$$subject\,to : \{behavioral\,constraints\,on\,y\,design\,constraints\,on\,x\,equilibrium\,constraints:$$

Where:

$f(x,y)$: **Objective function** : is used to classify the design and aims to either minimise or maximise the value. According to the aim of the designer, this value varies; it can measure weight, displacement, or stress. In this chapter, $f(x,y)$ is an objective function (effect of wind) to be reduced by using different design variables to generate different structural models.

x : **Design variable** : represents the geometric properties of the design, which can be altered during the optimisation process. The proposed prototype uses different x variables as design variables in predefined functions in order to generate different structural models.

y : **State variable** : is a function that defines (generates) the result of the structure for the specific x value. The proposed prototype uses y as a mathematical function to generate different structural models based on different x variables.

Structural optimisation can be classified into three optimisation methods, a brief description of which follows:

Shape optimisation: the shape of a structure can be optimised without changing the topology. In this chapter, the radius of each circle floor of the tower was considered as a design variable for shape optimisation.

Topology optimisation: the topology of a structure optimises the location of structural elements. In this chapter, the number of points at the circumference of floors determined the number and location of the diagrid of the tower.

Size optimisation: the typical size of a structure can be optimised (e.g. the thickness distribution of a beam or a plate). In this chapter, the parametric data of the Revit family was used to generate different cross sections for size optimisation purposes.

In summary, acknowledging the three optimisation methods, topology optimisation was seen as an effective method of facilitating the trade-off between architects' aesthetic criteria and engineers' engineering requirements. This method has been widely used in other fields – from mechanical to aerospace engineering (Sigmund, 2000).

3.8. Solution development

Solution development for this prototype started by mapping inputs, criteria, and outputs. A schematic process flowchart was then created to document this workflow (Figure 3.4).

Figure 3.4 presents the schematic process of this prototype. This focuses on solving the existing industry challenges identified earlier and captured through the questionnaire. From this, three optimisation methods (shape, topology, and size) were used to establish a BIM-based platform to use the parametric data in Revit and the mathematical predefined functions (in Dynamo) needed to generate, analyse, and optimise different options of structural models (following RSA).

The development procedure was initiated with the structural shape generation and optimisation by defining parametric shape variables and mathematical functions to create structural geometric entities (Figure 3.4). Thereafter, structural topology and size optimisation commenced, using a robust structural performance assessment process to generate different integrated options of structural designs during early stages. During this stage, based on shape

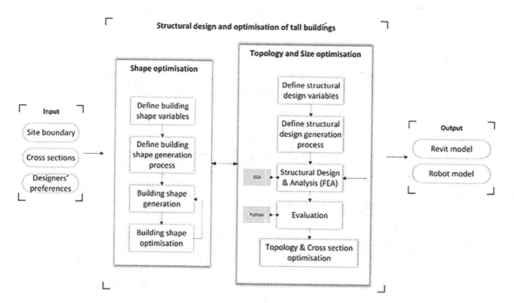

Figure 3.4 Structural design and optimisation: tall buildings schematic

variables, Dynamo mathematical functions define a structural model in Revit and RSA. From this, numerical variables and mathematical functions enable designers to vary the structural shape and structural topology of the building in order to optimise these quicker and easier.

From an input perspective (Figure 3.4), input data such as the site boundary and a list of various cross sections and designers' preferences are considered. Site boundaries are included to position the building in an accurate pre-defined location in order to prevent the building footprint from extending beyond the site boundary. The site boundary therefore is used in the shape design and optimisation section; hence, positions the structural design in the exact location of the architectural model. This site boundary therefore also dictates the location of the structural models. Following this, the input section of the prototype includes a list of various cross sections for use in the structural design and optimisation section (to optimise the structure in terms of elements' size optimisation). This enables designers to type the name of the cross sections into the relevant node to load the cross section in RSA, which generates the structural model by using the cross section. This also accommodates designers' preferences, which play a key role in defining mathematical functions and the whole architectural-structural design and optimisation process. These preferences also have an impact on the process of building shape generation, topology optimisation, cross sections size, and the code of structural analysis (e.g. Euro-Codes (EC), British Standards (BS) etc.), for structural evaluation purposes.

From a shape optimisation perspective, the shape of the building design and corresponding shape variables affect all parameters. Thus, designers' preferences (input data) affect the choice of mathematical functions needed to optimise the building parametrically (Figure 3.4). In the case presented here, the visual programming tool (Dynamo) was used to define the mathematical functions in order to design the building shape generation process. This process used the site boundary and building shape variables information to generate different alternative building shapes and structural models, where automatic optimisation was achieved by varying the variables. This approach was used to enable designers to generate different structural shapes within the required site boundary, whilst also being able to optimise the shape of the building design (by varying building shape variables).

From Figure 3.4, the structural topology and size optimisation process follows the shape optimisation process. This is because, any changes in the shape of the structural model is automatically affected by the topology arrangements. This creates a link between different stages of the prototype. It also helps improve collaboration between architects and engineers through an automated process using visual programming language tools (i.e. Dynamo). Topology and size optimisation of structural design are based on structural design variables. By changing structural design variables, different alternative structural models with different topology arrangements and various elements cross-section sizes are automatically generated. These structural design alternatives are then synchronised to architectural models based on designers' (engineers') preferences. In this respect, any changes or adjustments made to the architectural model (boundary conditions) or by the engineers' preferences (topology arrangement and/or elements' cross-section size) are automatically updated through a new generated model. Moreover, changing variables relating to the building shape, topology, and section size affects the mathematical functions. For example, the cross sections provide different element sizes to generate alternative structural models with different cross-section sizes (Figure 3.5).

From Figure 3.5, load sections can be analysed against design variables (inputs) to evaluate alternative structural models. Following this, any alterations of the building shape or topology arrangements, the system is able to provide designers with several alternative structural

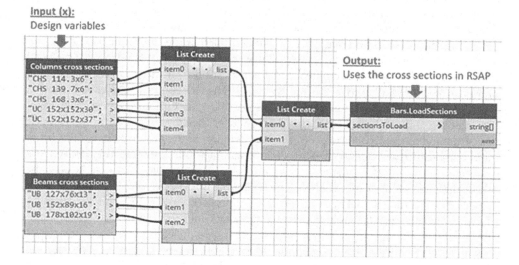

Figure 3.5 Load sections topology generator

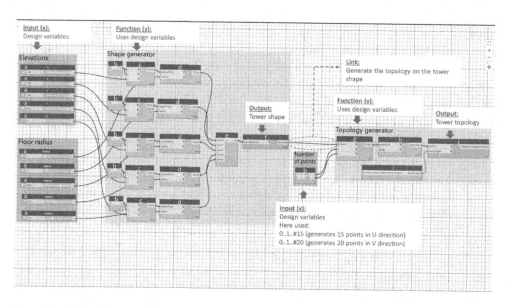

Figure 3.6 Tall building generator script (see accompanying notes)

models (to support the early stages of structural design), which in turn enables designers to select the optimum design (among all available options). This offers the design team conjoined synergy between architects and engineers, as the BIM-based platform provides them with alternative structural designs (with minimal human intervention). The combination of building shape optimisation and topology optimisation through this BIM-based platform is the starting point of this prototype.

This prototype includes four main parts: input, shape optimisation, topology and size optimisation, and output (Figure 3.6).

Notes for Figure 3.6:

a. **Slider node:** enables designers to adjust the design by changing the design variables. In this script, the design is generated by five elevations and floor radiuses.

b. **Code block node:** this node is used to provide numbers (where required).

c. **Point.ByCoordinates node:** this node uses three inputs as coordination in x, y, z direction and generates a point. These points were used as the centre point of five different circular floors in five different elevations.

d. **Circle.ByCentrePointRadius node:** this node uses a point and a number (radius) to generate circles (used as floors).

e. **List.Create node:** this node creates a list of inputs, in this case a list of circles.

f. **Surface.ByLoft node:** this node uses a list of curves (in this case a list of circles) and generates a surface (shape of the tower).

g. **Surface.PointAtParameter node:** this node uses a surface and two numbers (U and V) to generate points on a surface. In this case, this node generates points on the surface of the tower. This point is used as a placeholder to use different patterns on the shape of the tower, where changing the values of U and V changes the number of points, and consequently, changes the patterns on the shape of the tower.

h. **PointGrid.ToQuadSet node:** this node uses a set of four points to generate a pattern of quad on the shape of the tower.

i. **Family Type node:** this node enables designers to change the cross section of the elements used in the quad's patterns.

j. **AdaptiveComponent.ByPoint node:** this node uses a list of quad sets and a family type to generate the entire tower topology.

3.9. Preliminary production assessment

As previously discussed, the structural design of the Gherkin Tower relies on two primary structures to withstand loads, specifically: (i) the diagrid and (ii) the core. This arrangement can be seen in Figure 3.7.

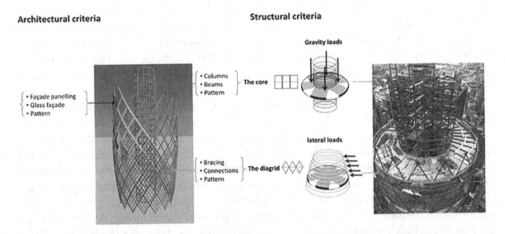

Figure 3.7 Integrated architectural-structural design

From Figure 3.7, the core is located at the centre of the building in order to resist major gravity loads. This also performs as a secondary structure for the diagrid, where the diagrid creates the structural shape of the Gherkin Tower. Therefore, this prototype had to consider these issues together. In doing so, the design and optimisation of the core and diagrid had to be dealt with separately. In this respect, two different mathematical functions were defined in Dynamo (which used parametric variables like diameter of the floor along the height of the tower), where any changes to parametric variables would be automatically reflected in the structural model. This required a series of input data (variables) to generate different structural design alternatives. Consequently, the development process needed to consider the objective function, shape optimisation process, topology optimisation process, and size optimisation.

The objective function is the main criterion used in the design of structures. Typically, these include optimisation, minimum weight, minimum cost, and maximum stability. In structural optimisation, the objective function should be formulated as a surrogate factor to explore other factors such as ductility, natural frequencies, P-D effects, buckling, and stability – put simply, to explore the most economic and feasible solution to the problem (Beghini et al., 2014). For this prototype, the use of precise formulation and functions for automatic structural design and optimisation was considered crucial to obtain a robust and defendable solution. This required an accurate and comprehensive problem formulation assessment, which not only evaluates the design problem but also captures the high levels of accuracy needed to support the evolutionary design process. In this case, maximum stability was considered as the objective function (to explore the most stable design of the Gherkin Tower against wind load). Acknowledging this, mathematical functions were defined in Dynamo using different sets of variables to design and analyse multiple options of the Gherkin Tower in RSA – to compare these results in terms of stability against wind load.

From a shape optimisation perspective, the shape of a building can often affect many things, from architectural aesthetics to structural integrity, structural safety, and stability. Therefore, decisions made on the shape of a building can have significant repercussions. This is why an agreement is needed between architects and structural engineers. This prototype helped facilitate this agreement by automating the structural design process based on the architectural model (boundary conditions). For instance, to improve safety and serviceability of tall buildings against strong winds, building shape optimisation in the early architectural design stage is considered as the most efficient method to create a wind-resistant design (Tang et al., 2014). This prototype followed this approach, enabling structural engineers and architects to work on a synchronised automatic platform. In doing so, this allows them to explore the optimum shape of the building, in particular, one that reduces the effect of wind loads on the building. An example of this can be seen in Figure 3.8, highlighting the cylindrical shape of the Gherkin Tower, where this shape allows the wind to move around the building to reduce the effect of wind load. Conversely, the cubic-shaped building has to resist this wind load.

In structural optimisation, some data are considered parameters (control variables) by means of which the structure varies until an optimum design with the desired properties can be achieved. These parameters can vary based on designers' requirements (Xie et al., 2005). In addition, parameterisation of the building shape design enables designers to define the building shape information as a set of variables and mathematical functions. This provides a holistic method of generating and assessing complex and irregular building shapes. For the prototype suggested here, building shape parameterisation was considered in two directions: horizontal and vertical. The horizontal direction represents the shape of the floor plan and the area of the building, and the vertical direction defines the height of the building. There are various types of base floor plans designed by parametric representation of geometry for

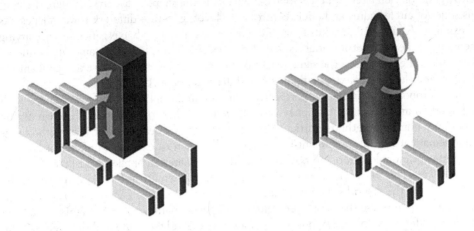

Figure 3.8 Effect of wind load on the shape of the building

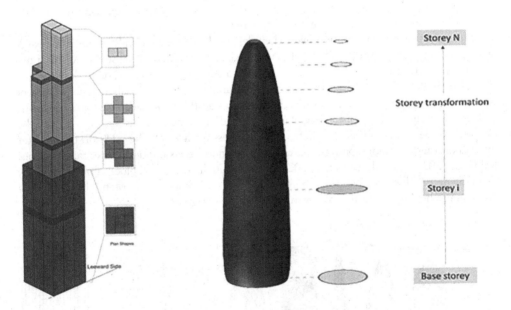

Figure 3.9 Horizontal and vertical transformation

both horizontal and vertical directions. For example, the geometry of the floor plan shape above the base floor can be transformed using different transformation functions (including no change translation, scaling, and rotation).

Figure 3.9 presents two types of horizontal and vertical transformations. The left side tower uses floor plan shapes with a series of squares, where the arrangements of the squares change with the height of the tower. Similarly, the right-side tower uses a circle floor base plan shape, where the radius of the circle changes with the height of the tower.

The solution proposed here uses a mathematical function in Dynamo to define the shape of the Gherkin Tower using two variables; that is, the diameter of circles and the level of the centre point of each circle. These variables help create the shape of the tower, where the various series of circle radius and circle centre point levels generate different tower shapes. As an example, Figure 3.10 presents four different options of tower shapes generated by varying the value of the parametric variables. All of the design options have the same site boundary constraints; they also have the same height of 179 m, the same base floor external diameter of 50 m, and the same maximum floor external diameter of 56.15 m.

From a topology optimisation perspective, the optimum option of the tower shape with architectural and structural efficiency was adopted. Topology optimisation is known as one of the most effective methods, particularly at the conceptual design stage. This method uses a mathematical and often gradient-based design approach to explore a number of options, including optimum location and boundary conditions, material, applied loads, etc. (Beghini et al., 2014). This prototype defined three different mathematical functions to design the core structure, the diagrid structure, and the façade panelling. Each approach had a particular formulation to solve the design problem, using specific parametric constraints (variables) to design different topology options. In this case, the diagrid structure and the architectural façade was designed by using the quad.diamond panel node from LunchBox Quad Grid by Face package (in Dynamo) to create diamond features. The core structure was designed using a surface panelling package to create rectangular features, where, by changing the variables of each function, the corresponding topology varies.

Different topology design options were then designed and analysed in RSA, and the results were used for optimisation (to choose the best model among different options). At this juncture it is important to note one consideration from this process. Even though design alternatives with more diagrid patterns might be preferable from a structural engineer's point of view, these designs may not be considered practical or aesthetically desirable for contractors, architects, or clients. In addition, the densely spaced diamond diagrids may also reduce the aesthetic value of the building, thereby making it less attractive. The addition of a greater number of diagrid diamond patterns may also affect energy consumption (Seyedzadeh & Pour Rahimian, 2021b; Seyedzadeh et al., 2020a), as potentially this could block sunlight, thereby reducing natural lighting within the building. These 'soft' issues could potentially cause conflict among the different stakeholders involved in the design process.

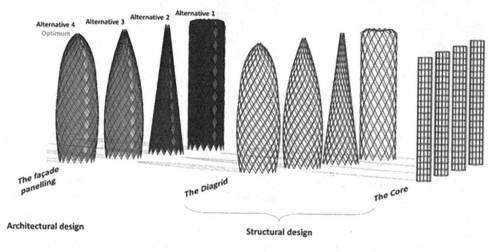

Figure 3.10 Gherkin tower shape optimisation

Notwithstanding these issues, Figure 3.11 presents four variations of structural topology designs. These are all linked to the architectural façade panelling design. From this, the architectural façade panelling design and the structural topology design (the diagrid and the core) are separated. This allows users to visualise the concept of automatic integrated design between architectural and structural models during the topology optimisation process.

After shape and topology optimisation, structural member size optimisation is the next stage needed for structural design performance evaluation. Size optimisation (or cross-section optimisation) is often used to explore the optimum cross-section area of column elements and beam elements in a frame, particularly to find the optimum thicknesses of plate elements, while also satisfying the design requirements (Beghini et al., 2014). For example, in this prototype, the diagrid and the core elements consist of parametric cross sections that can be adjusted to explore the optimum solution. Like shape and topology optimisation, this prototype engaged a mathematical function using a series of parametric variables to generate different structural models with various cross-section sizes (Figure 3.12).

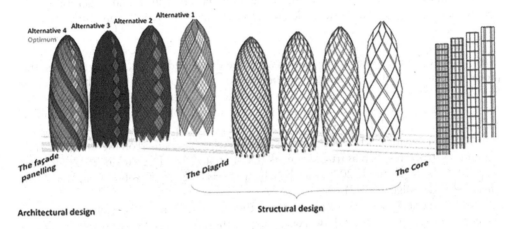

Figure 3.11 Gherkin tower topology optimisation

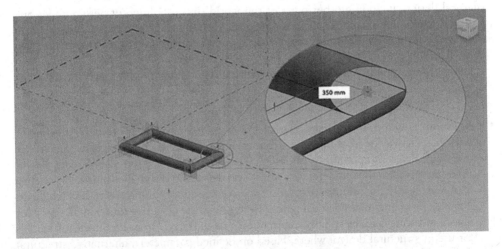

Figure 3.12 Parametric data (from Revit) size optimisation

3.10. Discussion

The findings from this work open up a number of issues for debate, not least the implications of this research in research and practice. From the outset, the challenges facing architects and structural engineers were shared, highlighting the need for synergy, conjoined thinking, and optimised solution generation; however, despite the advancements in BIM, federated models, and CDE, the design procedure of buildings was acknowledged as being predominantly a loosely coupled system. The corollary of this helped define the starting point for the development of this prototype – a much-needed toolset for integrating various actors across the design process. This prototype presents stakeholders with a new approach to design, one that minimises fragmentation and actively brings two disciplines together for one common purpose. This study stands out among similar ones, in that it provides the first tool of its kind that engages automated synergy and computational analysis procedures with architectural parametric data for multi-faceted optimisation of structural design. This extends the capabilities of federated BIM models. This also mitigates the need for multiple file exchange (between various disciplines), given that computational procedures and structural design analyses can both take place inside federated models and CDE. This is seen as one step further towards BIM level 3.

The findings from this study contribute to the field of structural design optimisation for, albeit in one dimension like shape, topology, and size (Tsavdaridis et al., 2015); however, this prototype considers all these dimensions simultaneously. In doing so, it provides a platform for 'merging' architects and structural engineers into one system. This aligns to the call to make architects' and structural engineers' practices inseparable, through a fully integrated architectural/structural design system (Hasançebi et al., 2009). The study also broadens the boundary of application, where for example previous work has focused on the structural optimisation of structures such as truss members (Dede et al., 2011; Degertekin, 2012; Gholizadeh, 2013; Hasançebi et al., 2009; Kaveh & Ilchi Ghazaan, 2015; Kaveh & Mahdavi, 2014; Miguel & Fadel Miguel, 2012).

From a practical application perspective, the findings from this study relate exclusively to tall building design; however, arguably, these findings could be rolled over to other types of structures, including infrastructure projects and complex structures. In this respect, this work would be particularly useful for practitioners active in various fields of structural design, including other parts of the world that use different design codes and standards, where, for example, solutions can easily be adjusted to accommodate different design codes for steel/aluminium, timber, and concrete structures from different regions (Figure 3.13).

The main issues and key findings from this work can be summarised through the following four points:

* **Solution**: Solution generation is seen as the main outcome. In this respect, it is important to acknowledge from the outset that architects and structural engineers invariably engage in different approaches to design. The challenge, therefore, is to create a solution that envelops these differences, whilst also delivering the challenge (the knowledge gap). Given this, the solution generation approach used for this prototype had to engage complex tasks and mathematical functions to satisfy the objective function, design variables, and state variables. This information was extracted from architectural models for use in structural design, where, based on defined parameters, alternative structural design solutions can be generated. This provides practitioners with extended levels of

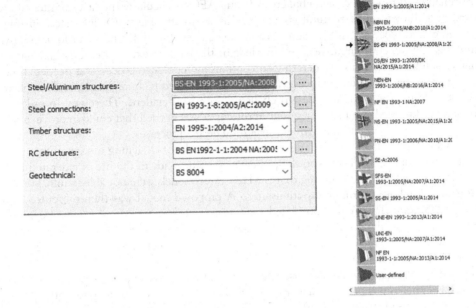

Figure 3.13 Different regional design codes

flexibility, enabling them to customise designs in line with project/client/performance demand.

- **Methodology**: The methodology adopted for the development of this prototype followed a four-stage approach. This helped establish system design parameters, along with outline concepts and governance protocols (following the RSA approach). Whilst this approach provided significant benefits and a wide range of structural design alternatives, this also attracted a number of constraints. For example, the need to consider the amount of development time needed for development and testing. This limitation also needs to embrace the notion of human error, especially to prevent designers from exploring all available solutions (as this would make the prototype somewhat counterproductive). That being said, this prototype overcame these challenges by carefully balancing designers' (architects and structural engineers) needs against core priorities and competing resources.

- **Constraints**: From a constraint perspective, this prototype focused on tall buildings only, limited to certain prescriptive parameters (identified earlier); however, this solution is very flexible and can be adapted to suit other scenarios. Moreover, as this model was developed using 'standard' tools and software, it supports most design practitioners' needs, and should therefore be more readily available for uptake.

- **Opportunities**: This prototype provides designers (architects and structural engineers) with new ways of working (collaboratively) to solve design challenges. The solution is cost effective (even for small businesses), as it seamlessly integrates with common platforms. A particular advantage here was the integration of GD with BIM, as this not only produced a number of efficiencies but also presents a unique solution for addressing interoperability issues.

3.11. Conclusion

This chapter introduced design challenges facing AEC, particularly the lack of integration between architects and structural engineers, which in part reflects (i) differences in fragmented design processes, (ii) disparate working practices, (iii) differences in silo mentality/ thinking, and (iv) cultural nuance. Acknowledging these issues, extant literature highlighted that an automatic system for generating and optimising the design process was needed; however, despite the advancements made in BIM and associated technologies, specific applications to address this structural design challenge were still evident. Therefore, in order to address this knowledge gap, a new system (prototype) was needed that could improve, automate, and optimise both the design and structural design processes.

This study provided a bespoke development approach for designing a new prototype to meet these challenges. This acknowledged developments made in the use of federated BIM models within a CDE, albeit noting that BIM federated models were not able to fully provide automated intelligent structural optimisation. A proposed model was then presented as a potential solution. This model considered a number of issues, ranging from design integration to requirements analysis and RAS prototype development procedures.

The solution provided here is considered a starting point (certainly in terms of technical complexity) using three optimisation methods: shape, topology, and size. These were used to establish a prototype using a BIM-based platform exploiting parametric data in Revit, which engaged mathematical predefined functions (in Dynamo) to generate, analyse, and optimise different options of structural models. This followed the RSA approach. Whilst this prototype was constrained to tall buildings (with predefined conditions and parameters), arguably these findings could be easily replicable to other contexts and settings. Hopefully, findings from this work will open up wider discourse for other potential development areas, particularly the journey to meet BIM Level 3, BIM Level 4, and quest to Industry 4.0. The real issue here is that of collaboration and raising awareness of the opportunities ahead. There are so many possibilities to explore – from enhanced design collaboration to the development of advanced innovative technologies, platforms, and bespoke programs. This simple example showcases how a prototype of this nature can transform the ways things are done: seamlessly automating procedures based on a set of variables, creating new and more efficient design alternatives, and opening up new, exciting, and novel GD-driven design alternatives. This solution does, however, come at a cost. The cost here is change, insofar as there needs to be a willingness to explore these opportunities. In some respects, the parallels being drawn here are something not too dissimilar from early adopters (*cf.* innovation). This change in thinking and approach will undoubtedly require new knowledge and skills, and possibly a new generation of designers with advanced programming and coding capabilities, to help AEC leverage these benefits.

References

Abrishami, S., Goulding, J., Pour Rahimian, F., & Ganah, A. (2015). Virtual generative BIM workspace for maximising AEC conceptual design innovation. *Construction Innovation*, 15(1), 24–41. https://doi.org/10.1108/ci-07-2014-0036

Abrishami, S., Goulding, J., & Rahimian, F. (2020). *Generative BIM workspace for AEC conceptual design automation: Prototype development*. Engineering, Construction and Architectural Management.

Al-Kodmany, K. (2018). The vertical farm: A review of developments and implications for the vertical city. *Buildings*, 8(2), 24. https://doi.org/10.3390/buildings8020024

Baldock, R., & Shea, K. (2006). Structural topology optimization of braced steel frameworks using genetic programming. In *Lecture notes in computer science* (pp. 54–61). Springer. https://doi.org/10.1007/11888598_6

Banihashemi, S., Tabadkani, A., & Hosseini, M. R. (2018). Integration of parametric design into modular coordination: A construction waste reduction workflow. *Automation in Construction, 88*, 1–12. https://doi.org/10.1016/j.autcon.2017.12.026

Beghini, L. L., Beghini, A., Katz, N., Baker, W. F., & Paulino, G. H. (2014). Connecting architecture and engineering through structural topology optimization. *Engineering Structures, 59*, 716–726. https://doi.org/10.1016/j.engstruct.2013.10.032

Bezabeh, M. A., Bitsuamlak, G. T., Popovski, M., & Tesfamariam, S. (2018). Probabilistic serviceability-performance assessment of tall mass-timber buildings subjected to stochastic wind loads: Part I – structural design and wind tunnel testing. *Journal of Wind Engineering and Industrial Aerodynamics, 181*, 85–103. https://doi.org/10.1016/j.jweia.2018.08.012

Chan, C.-M. (2001). Optimal lateral stiffness design of tall buildings of mixed steel and concrete construction. *The Structural Design of Tall Buildings, 10*(3), 155–177. https://doi.org/10.1002/tal.170

Christensen, P. W., & Klarbring, A. (n.d). Sizing stiffness optimization of a truss. In *Solid mechanics and its applications* (pp. 77–95): Springer. https://doi.org/10.1007/978-1-4020-8666-3_5

Chunyu, D., & Li, W. (2018). High building layout design based on emotional differential evolution algorithm. *Cluster Computing, 22*(S2), 5001–5007. https://doi.org/10.1007/s10586-018-2452-0

Ciribini, A. L. C., Mastrolembo Ventura, S., & Paneroni, M. (2016). Implementation of an interoperable process to optimise design and construction phases of a residential building: A BIM pilot project. *Automation in Construction, 71*, 62–73. https://doi.org/10.1016/j.autcon.2016.03.005

Dawood, N., Rahimian, F., Seyedzadeh, S., & Sheikhkhoshkar, M. (2020). *Enabling the development and implementation of digital twins: Proceedings of the 20th international conference on construction applications of virtual reality.* Teesside University. ISBN:9780992716127

Dede, T., Bekiroğlu, S., & Ayvaz, Y. (2011). Weight minimization of trusses with genetic algorithm. *Applied Soft Computing, 11*(2), 2565–2575. https://doi.org/10.1016/j.asoc.2010.10.006

Degertekin, S. O. (2012). Improved harmony search algorithms for sizing optimization of truss structures. *Computers & Structures, 92–93*, 229–241. https://doi.org/10.1016/j.compstruc.2011.10.022

Durdyev, S., Hosseini, M. R., Martek, I., Ismail, S., & Arashpour, M. (2019). Barriers to the use of integrated project delivery (IPD): A quantified model for Malaysia. *Engineering, Construction and Architectural Management, 27*(1), 186–204. https://doi.org/10.1108/ecam-12-2018-0535

Dutta, S., Cai, Y., Huang, L., & Zheng, J. (2020). Automatic re-planning of lifting paths for robotized tower cranes in dynamic BIM environments. *Automation in Construction, 110*, 102998. https://doi.org/10.1016/j.autcon.2019.102998

Elnimeiri, M., & Almusharaf, A. (2010). The interaction between sustainable structures and architectural form of tall buildings. *International Journal of Sustainable Building Technology and Urban Development, 1*(1), 35–41. https://doi.org/10.5390/susb.2010.1.1.035

Emmitt, S., & Ruikar, K. (2013). *Collaborative design management.* Routledge. https://doi.org/10.4324/9780203819128

Foraboschi, P., Mercanzin, M., & Trabucco, D. (2014). Sustainable structural design of tall buildings based on embodied energy. *Energy and Buildings, 68*, 254–269. https://doi.org/10.1016/j.enbuild.2013.09.003

Garber, R. (2009). Optimisation stories: The impact of building information modelling on contemporary design practice. *Architectural Design, 79*(2), 6–13. https://doi.org/10.1002/ad.842

Gholizadeh, S. (2013). Layout optimization of truss structures by hybridizing cellular automata and particle swarm optimization. *Computers & Structures, 125*, 86–99. https://doi.org/10.1016/j.compstruc.2013.04.024

Hamidavi, T., Abrishami, S., & Hosseini, M. R. (2020). Towards intelligent structural design of buildings: A BIM-based solution. *Journal of Building Engineering, 32*, 101685. https://doi.org/10.1016/j.jobe.2020.101685

Hamzeh, F., Rached, F., Hraoui, Y., Karam, A. J., Malaeb, Z., El Asmar, M., & Abbas, Y. (2019). Integrated project delivery as an enabler for collaboration: A Middle East perspective. *Built Environment Project and Asset Management, 9*(3), 334–347. https://doi.org/10.1108/bepam-05-2018-0084

Hasançebi, O., Çarbaş, S., Doğan, E., Erdal, F., & Saka, M. P. (2009). Performance evaluation of metaheuristic search techniques in the optimum design of real size pin jointed structures. *Computers & Structures, 87*(5–6), 284–302. https://doi.org/10.1016/j.compstruc.2009.01.002

Hofmeyer, H., & Davila Delgado, J. M. (2013). Automated design studies: Topology versus one-step evolutionary structural optimisation. *Advanced Engineering Informatics, 27*(4), 427–443. https://doi.org/10.1016/j.aei.2013.03.003

Holzer, D., Hough, R., & Burry, M. (2007). Parametric design and structural optimisation for early design exploration. *International Journal of Architectural Computing, 5*(4), 625–643. https://doi.org/10.1260/147807707783600780

Hosseini, M. R., Maghrebi, M., Akbarnezhad, A., Martek, I., & Arashpour, M. (2018). Analysis of citation networks in building information modeling research. *Journal of Construction Engineering and Management, 144*(8), 04018064. https://doi.org/10.1061/(asce)co.1943-7862.0001492

Hurol, Y. (2013). Ethical considerations for a better collaboration between architects and structural engineers: Design of buildings with reinforced concrete frame systems in earthquake zones. *Science and Engineering Ethics, 20*(2), 597–612. https://doi.org/10.1007/s11948-013-9453-4

Kaveh, A., & Ilchi Ghazaan, M. (2015). A comparative study of CBO and ECBO for optimal design of skeletal structures. *Computers & Structures, 153*, 137–147. https://doi.org/10.1016/j.compstruc.2015.02.028

Kaveh, A., & Mahdavi, V. R. (2014). Colliding bodies optimization method for optimum discrete design of truss structures. *Computers & Structures, 139*, 43–53. https://doi.org/10.1016/j.compstruc.2014.04.006

Kazakis, G., Kanellopoulos, I., Sotiropoulos, S., & Lagaros, N. D. (2017). Topology optimization aided structural design: Interpretation, computational aspects and 3D printing. *Heliyon, 3*(10), e00431. https://doi.org/10.1016/j.heliyon.2017.e00431

Kraatz, J., Sanchez, A., & Hampson, K. (2014). Digital modeling, integrated project delivery and industry transformation: An Australian case study. *Buildings, 4*(3), 453–466. https://doi.org/10.3390/buildings4030453

Ku, K., & Pollalis, S. N. (2020). 3D model-based collaboration and geometry control; research needs for contractual standards: A case study of the main street bridge, Columbus, Ohio. In *eWork and eBusiness in Architecture, Engineering and Construction* (pp. 641–649): CRC Press. https://doi.org/10.1201/9781003060819-97

Kuiper, I., & Holzer, D. (2013). Rethinking the contractual context for building information modelling (BIM) in the Australian built environment industry. *Construction Economics and Building, 13*(4), 1–17. https://doi.org/10.5130/ajceb.v13i4.3630

Leite, F. L. (2019). *BIM for design coordination.* Wiley. ISBN:9781119516019, 9781119515791

Liu, Y., van Nederveen, S., & Hertogh, M. (2017). Understanding effects of BIM on collaborative design and construction: An empirical study in China. *International Journal of Project Management, 35*(4), 686–698. https://doi.org/10.1016/j.ijproman.2016.06.007

Love, J. S., Taylor, Z. J., & Yakymyk, W. N. (2020). Determining the peak spatial and resultant accelerations of tall buildings tested in the wind tunnel. *Journal of Wind Engineering and Industrial Aerodynamics, 202*, 104225. https://doi.org/10.1016/j.jweia.2020.104225

Macdonald, A. J. (2018). Structure and Architecture. In: Routledge. https://doi.org/10.4324/9781315210513

Merschbrock, C., & Munkvold, B. E. (2014). How is building information modeling influenced by project complexity? *International Journal of e-Collaboration, 10*(2), 20–39. https://doi.org/10.4018/ijec.2014040102

Mignone, G., Hosseini, M. R., Chileshe, N., & Arashpour, M. (2016). Enhancing collaboration in BIM-based construction networks through organisational discontinuity theory: A case study of the new

Royal Adelaide hospital. *Architectural Engineering and Design Management*, *12*(5), 333–352. https://doi.org/10.1080/17452007.2016.1169987

Miguel, L. F. F., & Fadel Miguel, L. F. (2012). Shape and size optimization of truss structures considering dynamic constraints through modern metaheuristic algorithms. *Expert Systems with Applications*, *39*(10), 9458–9467. https://doi.org/10.1016/j.eswa.2012.02.113

Mirjalili, S. Z., Mirjalili, S., Saremi, S., Faris, H., & Aljarah, I. (2017). Grasshopper optimization algorithm for multi-objective optimization problems. *Applied Intelligence*, *48*(4), 805–820. https://doi.org/10.1007/s10489-017-1019-8

Mora, R., Bédard, C., & Rivard, H. (2008). A geometric modelling framework for conceptual structural design from early digital architectural models. *Advanced Engineering Informatics*, *22*(2), 254–270. https://doi.org/10.1016/j.aei.2007.03.003

Mueller, C. T., & Ochsendorf, J. A. (2015). Combining structural performance and designer preferences in evolutionary design space exploration. *Automation in Construction*, *52*, 70–82. https://doi.org/10.1016/j.autcon.2015.02.011

Oliver, S., Seyedzadeh, S., Rahimian, F., Dawood, N., & Rodriguez, S. (2020). Cost-effective as-built BIM modelling using 3D point-clouds and photogrammetry. *Current Trends in Civil & Structural Engineering-CTCSE*, *4*(5), 000599. https://doi.org/10.33552/CTCSE.2020.04.000599

Oraee, M., Hosseini, M. R., Edwards, D. J., Li, H., Papadonikolaki, E., & Cao, D. (2019). Collaboration barriers in BIM-based construction networks: A conceptual model. *International Journal of Project Management*, *37*(6), 839–854. https://doi.org/10.1016/j.ijproman.2019.05.004

Oraee, M., Hosseini, M. R., Papadonikolaki, E., Palliyaguru, R., & Arashpour, M. (2017). Collaboration in BIM-based construction networks: A bibliometric-qualitative literature review. *International Journal of Project Management*, *35*(7), 1288–1301. https://doi.org/10.1016/j.ijproman.2017.07.001

Palomar, I. J., García Valldecabres, J. L., Tzortzopoulos, P., & Pellicer, E. (2020). An online platform to unify and synchronise heritage architecture information. *Automation in Construction*, *110*, 103008. https://doi.org/10.1016/j.autcon.2019.103008

Papadonikolaki, E. (2018). Loosely coupled systems of innovation: Aligning BIM adoption with implementation in Dutch construction. *Journal of Management in Engineering*, *34*(6), 05018009. https://doi.org/10.1061/(asce)me.1943-5479.0000644

Pärn, E. A., Edwards, D. J., & Sing, M. C. P. (2017). The building information modelling trajectory in facilities management: A review. *Automation in Construction*, *75*, 45–55. https://doi.org/10.1016/j.autcon.2016.12.003

Pärn, E. A., Edwards, D. J., & Sing, M. C. P. (2018). Origins and probabilities of MEP and structural design clashes within a federated BIM model. *Automation in Construction*, *85*, 209–219. https://doi.org/10.1016/j.autcon.2017.09.010

Pishdad-Bozorgi, P. (2016). Case studies on the role of integrated project delivery (IPD) approach on the establishment and promotion of trust. *International Journal of Construction Education and Research*, *13*(2), 102–124. https://doi.org/10.1080/15578771.2016.1226213

Rahmani Asl, M., Zarrinmehr, S., Bergin, M., & Yan, W. (2015). BPOpt: A framework for BIM-based performance optimization. *Energy and Buildings*, *108*, 401–412. https://doi.org/10.1016/j.enbuild.2015.09.011

Richardson, J. N., Nordenson, G., Laberenne, R., Filomeno Coelho, R., & Adriaenssens, S. (2013). Flexible optimum design of a bracing system for façade design using multiobjective genetic algorithms. *Automation in Construction*, *32*, 80–87. https://doi.org/10.1016/j.autcon.2012.12.018

Schlueter, A., & Geyer, P. (2018). Linking BIM and design of experiments to balance architectural and technical design factors for energy performance. *Automation in Construction*, *86*, 33–43. https://doi.org/10.1016/j.autcon.2017.10.021

Seyedzadeh, S., & Pour Rahimian, F. (2021a). Building energy performance assessment methods. In *Data-driven modelling of non-domestic buildings energy performance* (pp. 13–30). Springer. https://doi.org/10.1007/978-3-030-64751-3_2

Seyedzadeh, S., & Pour Rahimian, F. (2021b). Machine learning for building energy forecasting. In *Data-driven modelling of non-domestic buildings energy performance* (pp. 41–76). Springer. https://doi.org/10.1007/978-3-030-64751-3_4

Seyedzadeh, S., Pour Rahimian, F., Oliver, S., Rodriguez, S., & Glesk, I. (2020a). Machine learning modelling for predicting non-domestic buildings energy performance: A model to support deep energy retrofit decision-making. *Applied Energy*, 279, 115908. https://doi.org/10.1016/j.apenergy.2020.115908

Seyedzadeh, S., Rahimian, F. P., Oliver, S., Glesk, I., & Kumar, B. (2020b). Data driven model improved by multi-objective optimisation for prediction of building energy loads. *Automation in Construction*, 116, 103188. https://doi.org/10.1016/j.autcon.2020.103188

Shahin, H. S. M. (2019). Adaptive building envelopes of multistory buildings as an example of high performance building skins. *Alexandria Engineering Journal*, 58(1), 345–352. https://doi.org/10.1016/j.aej.2018.11.013

Sharma, A., Mittal, H., & Gairola, A. (2018). Mitigation of wind load on tall buildings through aerodynamic modifications: Review. *Journal of Building Engineering*, 18, 180–194. https://doi.org/10.1016/j.jobe.2018.03.005

Sheikhkhoshkar, M., Pour Rahimian, F., Kaveh, M. H., Hosseini, M. R., & Edwards, D. J. (2019). Automated planning of concrete joint layouts with 4D-BIM. *Automation in Construction*, 107, 102943. https://doi.org/10.1016/j.autcon.2019.102943

Sigmund, O. (2000). Topology optimization: A tool for the tailoring of structures and materials. *Philosophical Transactions of the Royal Society of London. Series A: Mathematical, Physical and Engineering Sciences*, 358(1765), 211–227. https://doi.org/10.1098/rsta.2000.0528

Singh, V., & Gu, N. (2012). Towards an integrated generative design framework. *Design Studies*, 33(2), 185–207. https://doi.org/10.1016/j.destud.2011.06.001

Stromberg, L. L., Beghini, A., Baker, W. F., & Paulino, G. H. (2012). Topology optimization for braced frames: Combining continuum and beam/column elements. *Engineering Structures*, 37, 106–124. https://doi.org/10.1016/j.engstruct.2011.12.034

Tang, J., Xie, Y. M., & Felicetti, P. (2014). Conceptual design of buildings subjected to wind load by using topology optimization. *Wind and Structures*, 18(1), 21–35. https://doi.org/10.12989/was.2014.18.1.021

Teng, Y., Li, X., Wu, P., & Wang, X. (2017). Using cooperative game theory to determine profit distribution in IPD projects. *International Journal of Construction Management*, 19(1), 32–45. https://doi.org/10.1080/15623599.2017.1358075

Tsavdaridis, K. D., Kingman, J. J., & Toropov, V. V. (2015). Application of structural topology optimisation to perforated steel beams. *Computers & Structures*, 158, 108–123. https://doi.org/10.1016/j.compstruc.2015.05.004

Vermeulen, D. (2015). *Dynam (o) ite your design for engineers – part 5*. Revit BIM.

Vilutiene, T., Kalibatiene, D., Hosseini, M. R., Pellicer, E., & Zavadskas, E. K. (2019). Building information modeling (BIM) for structural engineering: A bibliometric analysis of the literature. *Advances in Civil Engineering*, 1–19. https://doi.org/10.1155/2019/5290690

Wang, S. S., & Mahin, S. (2018). High-performance computer-aided optimization of viscous dampers for improving the seismic performance of a tall steel building. *Key Engineering Materials*, 763, 502–509. https://doi.org/10.4028/www.scientific.net/kem.763.502

Xie, J. (2014). Aerodynamic optimization of super-tall buildings and its effectiveness assessment. *Journal of Wind Engineering and Industrial Aerodynamics*, 130, 88–98. https://doi.org/10.1016/j.jweia.2014.04.004

Xie, Y. M., Felicetti, P., Tang, J. W., & Burry, M. C. (2005). Form finding for complex structures using evolutionary structural optimization method. *Design Studies*, 26(1), 55–72. https://doi.org/10.1016/j.destud.2004.04.001

Yan, B., & Li, Q.-S. (2016). Wind tunnel study of interference effects between twin super-tall buildings with aerodynamic modifications. *Journal of Wind Engineering and Industrial Aerodynamics*, 156, 129–145. https://doi.org/10.1016/j.jweia.2016.08.001

Zayed, T., & Mohamed, E. (2014). A case productivity model for automatic climbing system. *Engineering, Construction and Architectural Management, 21*(1), 33–50. https://doi.org/10.1108/ecam-02-2012-0015

Zhao, L., Cui, W., Zhan, Y., Wang, Z., Liang, Y., & Ge, Y. (2020). Optimal structural design searching algorithm for cooling towers based on typical adverse wind load patterns. *Thin-Walled Structures, 151*, 106740. https://doi.org/10.1016/j.tws.2020.106740

Zheng, C., Xie, Y., Khan, M., Wu, Y., & Liu, J. (2018). Wind-induced responses of tall buildings under combined aerodynamic control. *Engineering Structures, 175*, 86–100. https://doi.org/10.1016/j.engstruct.2018.08.031

4 BIM and design for manufacturing and assembly

4.1. Introduction

Architecture, Engineering, and Construction (AEC) is widely recognised for its impact as a socio-political-economic driver (HM Government, 2017). For example, construction progress can be seen to be dependent upon the supply and availability of materials, resources, and skills – the culmination of which have ultimately influenced its evolution and subsequent success/failure (Spence & Kultermann, 2016). Moreover, as a sector, AEC is seen as a barometer of gross domestic product (GDP) and core influencer of prosperity and global competitiveness (World Economic Forum, 2016). Thus, decisions made (locally, nationally, and internationally) affect everything we do, from the type of projects procured to the materials and resources consumed and the wider impact of these on carbon use, sustainability, waste, etc. It is therefore important that AEC considers these implications and repercussions for the whole-life value of these services (Mills, 2019).

Despite the importance and contribution of AEC, historically a number of recurrent challenges have stifled progression, especially when compared to other sectors such as aerospace, pharmaceuticals, the automotive industry, etc. These challenges have been well documented in literature, especially concerning the high levels of fragmentation and poor levels of performance and productivity. More recently, in the United Kingdom (UK), issues such as low productivity, project delivery uncertainty, skills shortages, and a general lack of data transparency have been of concern (KPMG, 2016). Similar challenges have also been observed in most other countries around the world, including the need to deliver homes to meet the expanding population and housing crisis (World Economic Forum, 2016). In order to address these issues, AEC has pursued several change strategies, including novel approaches for delivering higher-quality homes in less time (Goodier & Pan, 2010).

Other sector challenges include issues surrounding 'process', where it has been acknowledged that many of these processes have not been revisited for some time now (Mills, 2019). The corollary of this has led to inefficient project planning and methodologies, low productivity, poor project predictability and uncertain delivery times, low-quality products, higher costs, and lower value. Skills shortages have also contributed to these challenges, where evidence suggests that this shortage is due (in part) to an ageing workforce and lower number of new entrants wishing to join the sector due to poor working conditions (Whysall et al., 2019). These issues have been captured in numerous reports. For example, Farmer (2016) observed that the fragmented sector and 'traditional' service delivery models were predominantly cost-focused rather than value-focused, but that these issues could be addressed through new approaches such as offsite. Anecdotally, offsite and Modern Methods of Construction (MMC) have been proffered as a viable solution for many years now (Arif et al.,

DOI: 10.1201/9781003106944-4

2012; Blismas et al., 2006; Egan, 1998; Ezcan et al., 2013; Farmer, 2016, Goodier & Pan, 2010; Latham, 1994; Nadim & Goulding, 2011).

In parallel with these issues, several new approaches have now emerged, including new tools and technologies to support design and construction. These include advancements in technology and data management, new manufacturing techniques, and advanced digitalisation and automation (Construction 4.0). From a housing perspective, a number of promising initiatives offer significant potential (The Housing Forum, 2019). Many other technological solutions have also emerged, including Building Information Modelling (BIM), Virtual Reality (VR), Digital Twins, and advanced discreet event simulation.

This chapter offers additional insight into some of these emerging areas. In doing so, it also presents a theoretical framework for discussion. This highlights an approach for creating sub-assemblies and component-based systems within a prefabrication construction process, specifically to integrate MMC with BIM and Supply Chain Management (SCM).

4.1.1. Digital tools for AEC

Digitalisation is continuing to reshape many industrial sectors, including AEC, where digital tools have been gradually implemented for designing, constructing, and operating buildings and infrastructure assets (Borrmann et al., 2018). These initiatives are also opening up many exciting opportunities for wider exploitation. One of these major developments has been with BIM. In this respect, several new approaches are now transforming the ways through which AEC leverages this digital platform, particularly through the integration of products and services (Alizadehsalehi & Yitmen, 2021; Elghaish et al., 2020; Mohammed Abdelkader, 2021; Newman et al., 2020; Rahimian et al., 2019; Seyedzadeh et al., 2021). Whilst several definitions of BIM are in extant literature, the following definition is adopted in this chapter, where the Construction Project Information Committee (CPIC) defined BIM as:

> digital representation of physical and functional characteristics of a facility creating a shared knowledge resource for information about it forming a reliable basis for decisions during its life cycle, from earliest conception to demolition.
>
> (CPIC, 2008)

As a digital tool, BIM can be broadly categorised as a computer-generated model for the planning, design, construction, and operational stages of a scheme/project (Eadie et al., 2013), where BIM is used to efficiently manage data (creation, maintenance, and utilisation) and information across the whole asset lifecycle by all stakeholders involved (Eynon, 2016). In this respect, this whole-life approach naturally involves people, processes, technology, and standardised processes, and is seen as a viable way of sharing information from one project phase to another (Dawood & Vukovic, 2015). Advocates of this have noted higher-quality coordination between stakeholders, greater productivity, and improved profit retention (Sun et al., 2017). These benefits have also been seen to include communication and coordination, sustainability, health and safety, and process efficiency savings (Hardin & McCool, 2015; Ibrahim et al., 2019; Mesároš & Mandičák, 2017; Saxon, 2016); however, the adoption and uptake of BIM in AEC seems to have been influenced by country-specific demand. This has changed over the last five years, with the majority of countries now accepting BIM as the preferred approach, in part promoted by governmental pressure. From a UK perspective, Borrmann et al. (2018) noted that the British government provided a noteworthy example of this type of approach, highlighting the importance of reducing costs, enhancing efficiency, and lowering the carbon footprint of construction projects, placing the UK "at the vanguard of a new digital construction era".

Reflecting on literature in this field, several studies have examined BIM in numerous project scenarios, including offsite, where, for instance, the synthesis of Off-Site Manufacturing (OSM) and BIM have been seen to serve as beneficial solutions in terms of improved AEC performance (Abanda et al., 2017). Examples include Ezcan et al. (2013), noting improvements in speed, modelling time, and quality of construction delivery using BIM; and Babic et al. (2010), who highlighted that the use of BIM with industrialised processes could support standardised BIM objects (in BIM object libraries) for greater design flexibility. Moreover, the concept of Design for Manufacturing and Assembly (DfMA) has also been useful in the delivery of OSM, especially with BIM, where this relationship has been seen to optimise the design and manufacturing processes, components, and assembly (Alfieri et al., 2020). Moreover, BIM can link DfMA activities (e.g. procurement, fabrication, transport, installation, etc.) to upstream activities such as briefing, appraisals, and conceptual design, thereby improving communication and collaboration with stakeholders (Building and Construction Authority & Wood, 2016).

Similar studies by Wang and Skibniewski (2019) evaluated BIM in the production of 3D printing models to support engineers and improve construction results. These types of evaluation are particularly useful, as BIM inherently captures rich geometric information. It has also been suggested that this could be blended with scheduling and assembly sequences to support 3D printing robots (Teizer et al., 2018), and several authors have highlighted this link between BIM and 3D printing (Ashraf et al., 2018; Sakin & Kiroglu, 2017; Wang & Skibniewski, 2019).

In summary, whilst a number of advanced digitalisation tools have now started to permeate the market, it is proffered that only a few of these have been purposefully aligned to BIM, OSM, and DfMA.

4.1.2. Offsite manufacturing within AEC

As mentioned earlier, the increased use and application of OSM and MMC in AEC is continuing to grow, evolve, and mature. Increasingly, BIM is now also starting to become part of organisational delivery platforms, where intrinsically, OSM provides prefabricated components (from a factory or manufacturing facility), which are then transported to site for assembly (Hu et al., 2019). In this respect, the type and level of assembly required on site is dependent upon the type of OSM used (as several options are available, from components to hybrid options, pods, and fully finished 'plug and play' solutions). Notwithstanding this, Abanda et al. (2017) explained the advantages of OSM compared to traditional methods. Benefits include improved quality, improved health and safety, better working conditions, higher tolerances, lower costs, improved productivity, lower labour re-works, reduced waste, consolidated processes, higher levels of sustainability, and greater reliability (Arif et al., 2012; Smith & Sweets, 2018; Durdyev & Ismail, 2019; Hu et al., 2019).

OSM projects tend to follow slightly different delivery approaches compared to traditional projects, particularly across the design, manufacturing, and construction phases. For example, they often use DfMA (Arif et al., 2012), which is especially suited to OSM, where it is noted that design techniques should be suitably selected and planned to make implementation much simpler (Arashpour et al., 2018). This should also be flexible in order to regulate design changes and accommodate levels of automation and standardisation. Intrinsically, whilst the level of OSM varies considerably depending upon the exact method used (Ginigaddara et al., 2019), each approach is based on the principle of assembled parametric components and modules. These require well-organised process control and management systems to be in place, especially to ensure the design and manufacturing plans coalesce (O'Connor et al., 2016). In this respect, the engagement of BIM with OSM has been seen to improve the design, communication, manufacturing, and assembly approaches (Arif et al., 2012).

OSM classification is still unfolding (Abosaod et al., 2010; Ayinla et al., 2019), including taxonomy and links with Industry Foundations Classes (IFC), for example. That being said, the Housing Forum (2019) guide indicated that several proposals aimed to encourage manufacturers to offer their systems through international and accessible standards such as Publicly Available Specifications (PAS). This standardisation is expected to guide specifiers, designers, and constructors to common and standardised components, thereby improving accessibility and uptake, whilst also reducing incompatibility risks.

In summary, the combination of OSM and BIM presents AEC with a number of valuable solutions to meet industry needs. This integration captures and blends the unique facets of each. For example, BIM supports high levels of accuracy, which directly supports the optimisation of design, manufacturing, assembly, and deconstruction (Abanda et al., 2017; Oliver et al., 2020). This resonates with the principles of DfMA used with OSM. It is therefore proffered that this alignment could also help solve many of the integration issues associated with technology, particularly with design changes and logistics (Ezcan et al., 2013). In this respect, BIM is particularly suited to this, as it is able to store specific information on attributes and components throughout the design, manufacturing, and assembly lifecycle processes.

4.1.3. *Design for manufacture and assembly*

DfMA is an accepted approach for OSM with AEC (Lu et al., 2021), where it can be used to engage with organisational processes to deliver designs in manufacturing and assembly (Gao et al., 2020), and thus reduce the level of onsite activity. This methodology emphasises the relevance of design for manufacturing and assembly of components, which ultimately form part of the final asset (Building and Construction Authority & Wood, 2016). Broadly speaking, there are two main types of DfMA, notably: design for manufacture (DfM) and design for assembly (DfA). DfM relates to the process of making individual parts, whereas DfA involves the ways of assembling them (Gao et al., 2020). The underlying concepts of DfMA are based on optimisation, where designers maximise the delivery process for clients. This naturally includes all activities, from concept to automation and logistics. Whilst AEC has only really started to embrace this approach more recently (Yin et al., 2019), the benefits are particularly encouraging with mass customisation or high repetition. This repeatability or mass customisation enables products to be delivered in volume, thereby embedding value into the production and delivery supply chains and delivery processes (Jensen et al., 2018). Given this, AEC has now started to look at meaningfully blending this approach with traditional delivery methodologies and digital design practices (of OSM) to radically improve productivity, costs, value, and time.

From a concept perspective, DfMA relies on the premise of standardisation, with repeatable processes and designs. Therefore, a key part of any decision (to adopt DfMA) is to establish if the level of standardisation is sufficient to add value to the process (and end product). The challenge here, therefore, is to assess whether this level of standardisation affects (or indeed compromises) the end product or indeed hinders functionality, the value proposition, etc. In this respect, Digital Built Britain (2017) advocated that solutions should be interrogated and refined through a process of rationalisation, standardisation, and optimisation. So, the decision to adopt DfMA requires some thought. For example, the needs and demands of designing, planning, adapting/optimising designs, automating, etc., particularly at the early stages, to enable the seamless production of components in their subsequent assembly onsite (Alfieri et al., 2020). This should also consider the methods by which these projects are being delivered, the offsite manufactured components used, and the planning/logistics processes involved (Pasquire & Connolly, 2003).

Whilst literature highlights that the use and application of DfMA can produce products faster, more safely, and more resource/cost effectively (RIBA, 2016; Buildoffsite, 2020), these benefits are contingent upon having effective systems and procedures in place to support them. For example, a series of 'teams' are required to be dedicated to this methodology. These may be engaged in producing one aspect of a building component (focussing on repetition) under the same conditions or across a platform of activities (focussing on productivity) to improve value, quality, etc. (Mills, 2019). This approach was endorsed by Milestone (2017), noting the additional impact on skills, particularly those needed to support technical advancement and innovation, and especially the "need to embrace more productive construction methods" (The Housing Forum, 2019).

In summary, whilst DfMA is partially founded on the assumption that lowest assembly costs can be streamlined, designed, and economically assembled (Gao et al., 2020), this is contingent upon not only the design of parts *per se* but also the ease of which these parts can be assembled (Stoll, 1986); however, the real challenge here concerns the effective use of BIM in this process, where DfMA can be more effectively managed through BIM. This includes a number of activities, from procurement to manufacturing, logistics, assembly, and construction processes (briefings, appraisals, conceptual design, etc.), but with the wider acceptance and understanding of contributing project stakeholders (Alfieri et al., 2020). Moreover, from a technology perspective, the engagement of BIM and DfMA requires a certain mindset, particularly to support the adoption and uptake of digital technologies into the manufacturing process (Staub-French et al., 2018). In doing so, data-rich DfMA models are seen as an essential part of this process, where "BIM has a role in making the project less risky by allowing the project team to simulate the construction virtually to identify potential pitfalls way before the actual construction begins" (Building and Construction Authority & Wood, 2016).

4.1.4. Design for deconstruction/disassembly

Further to the discussion on DfMA, a number of research initiatives have now started to investigate what would happen at the end of an asset's lifespan – in particular, solutions for dealing with deconstruction, disassembly, and disposal. This forms part of the wider AEC debate on sustainability. In this respect, DfMA can include a decommissioning process, as components or even whole buildings (in the majority of cases) can be reverse engineered to accommodate this – commonly known as Design for Deconstruction/Disassembly (DfD) (Buildoffsite, 2020).

DfD is increasingly being used to prompt designers to think about procedures supporting reuse and recycling, including preventive measures to avoid waste being unnecessarily produced (Barrit, 2016). This encouragement of thinking about DfD from the outset for end-of-life reuse is becoming very important within AEC (Adams et al., 2017). Traditionally, only two real options are available at the end of an asset's life cycle: demolition or deconstruction. Demolition is used as a fast approach to asset removal, whereas deconstruction is in many respects the polar opposite, requiring considerable thought on the recuperation of building materials for reuse, recycling, and remanufacturing (Akinade et al., 2020). Thus, DfD can be seen as a detailed process where assets are specifically designed to facilitate not only adaptation and renovation but also the reuse of building materials and components (Kanters, 2018). This approach requires an effective strategy to be engineered into the design from the outset. For example, this includes the building materials, connections, loads, etc. insofar as the intrinsic design supports deconstruction/disassembly, with the chosen materials being recyclable and harmless after the recycling process (Akinade et al., 2017). This requires developing a sustainable deconstruction plan that examines all of these factors, including cost, energy use, and carbon emissions (Akbarnezhad et al., 2014).

Other initiatives in this area include the Disassembly and Deconstruction Analytics System (D-DAS), a method of utilising information modelling and decision support tools to achieve effective end-of-life sustainability performance (Akanbi et al., 2019). This approach plays an important role at the design stage, where deconstruction strategies are developed and assessed in order to consider changes in the design and fabrication of the components (and the impact these may have on results). This type of thinking and approach supports the wider efforts of supporting the circular economy (CE), where DfD offers opportunities for developing components for reuse, remanufacture, or recycling (Guy et al., 2006).

In conclusion, DfMA and DfD have been particularly useful in addressing sustainability concerns. If appropriately designed from the outset, these approaches can minimise disposal and reduce end-of-life waste (Akanbi et al., 2019). Moreover, the impact of DfMA and DfA has the potential to deliver other benefits, including lower assembly and manufacturing costs, improved sustainability, and lower environmental impact.

4.2. Design for manufacture and assembly: framework development

This section presents a DfMA conceptual framework for discussion. This includes aspects of the different approaches analysed in this study, such as OSM, DfMA, and DfD, with BIM as the central connection point. This framework, highlighted in Figure 4.1, presents all links and dependencies, divided into eight stages following the Royal Institute of British Architects (RIBA) Plan of Work 2013. This plan was adopted in order to highlight the importance of the approaches that defined this new method of construction. A final stage 'end-of-life' section was added to accentuate that this process is cyclical, and component-based systems could be reutilised or reused, thereby encouraging DfMA.

To facilitate discussion, the framework development process is divided into the following four core parts:

- Preparation phase
- Design and pre-construction phases
- Construction/assembly to close-out phase
- Use and reuse/demolition phases

Figure 4.1(a) presents the conceptual framework (preparation and design); Figure 4.1(b) presents the conceptual framework (pre-construction and construction/assembly); and Figure 4.1(c) presents the conceptual framework (use and demolition/reuse).

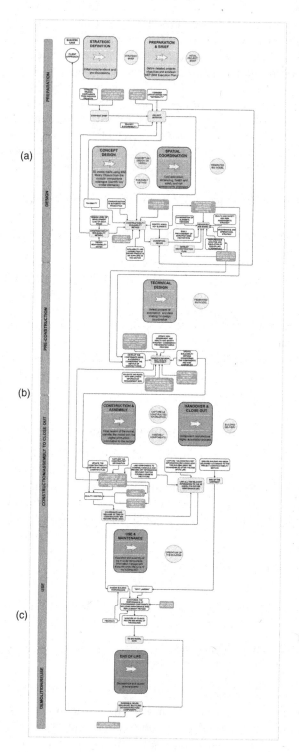

(a)

(b)

(c)

Figure 4.1 Conceptual framework

4.2.1. Preparation phase

During the preparation phase (strategic definition; preparation and brief), a strategic brief is developed using a BIM object library based on a set of components (that can be used across different multiple projects). This also considers how CE issues can be implemented. The strategic definition and preparation and brief stages help define the project objectives, including the requirements for DfMA, where the use of smart contracts are contemplated. A BIM Execution Plan (BEP) is then written to ensure that the asset is designed in accordance with the client's requirements. Along with the BEP, constructability issues are established.

4.2.2. Design and preconstruction phases

At the design phase (concept design; spatial coordination), the BIM object library (n.b. the BIM object library development is discussed later) is used to create a conceptual 3D model. Here, the design follows the DfMA and DfD approaches in order to obtain all of the benefits that these methods offer, such as mass construction, improved productivity, and end-of-life sustainability performance. Buildability is also included, along with the availability and capabilities of known products and suppliers, especially the use of standardisation to automate production. Subsequently, at the spatial coordination stage, a federated model is designed. This includes cost estimation, scheduling, health and safety, and risk assessment strategies. These strategies are based on OSM and additive manufacturing environments. This phase also establishes the manufacturing technique and defines the deconstruction plan.

During the technical design stage (pre-construction phase), the federated BIM model reaches the next level of development, where this includes Radio-frequency Identification (RfID) in selected components. Components are also defined with a higher Level of Detail (LoD) and Level of Information (LoI). This includes the process of automation and data sharing to facilitate design coordination. Together, the model is then validated following the Employers Information Requirements (EIR) prior to entering the construction/assembly phase.

4.2.3. Construction/assembly and close-out phases

During the construction/assembly stage, the final model and digital production strategy is forwarded to the chosen factory. Once the components have been produced and the quality control process completed, the components are then released to site in accordance with the logistics and buildability method. The assembly process then commences onsite. During this process, digital tools such as Internet of Things (IoT) enable stakeholders to track each step of the manufacturing, packing, logistics, and delivery processes. At the handover and close-out stage, all relevant information for the maintenance of the asset is linked to the 3D model for conformance. Of particular note, the information captured through these two stages can be linked to peripherals such as IoT-driven products, laser scanners, photogrammetry technology, or drones. Information from these services can then be analysed for predictive pattern matching (to help future projects) (Seyedzadeh & Rahimian, 2021).

4.2.4. Use and demolition/reuse phases

In the use and maintenance stage, the asset is continually monitored and aligned with the Facilities Management (FM) BIM model in order to keep this data up to date. Energy consumption and production is tracked, along with the performance of the components (using RFID where available). This continuous tracking also enables components to be analysed, thereby allowing repairs or replacements to occur much sooner than through conventional approaches. This is also particularly advantageous for components with fixed warranty periods. In this respect, this functionality can be embedded into smart contracts at the preparation and brief stage, along with the ownership of components. The use of OSM is also seen as being suited to this stage, as the use of standardised components with easy assembly/disassembly techniques more readily supports design changes. In the final stage of the conceptual framework, the end-of-life stage is presented. This covers how the asset and its components will be disassembled, reused, or recycled. This follows the principles of DfD and CE. In doing so, this helps minimise construction and demolition waste and also supports cradle-to-cradle initiatives (rather than cradle-to-grave).

4.3. Research methodology

The work presented in this chapter follows a mixed-method approach that captures both qualitative and quantitative data. Secondary data was gathered from extant literature in order to identify and explore state-of-the-art advancements in AEC and the manufacturing sector. An initial conceptual framework was developed from this exercise. In order to test and validate this framework, an exploratory case study was developed, where primary data was collected to evaluate performance, features, opportunities, and limitations.

The literature review process included an examination of a wide range of topics and information gathered from journal articles, books, reports, conference papers, and dissertations. Two stages were implemented in the literature review to raise the legitimacy and reliability of the data sources. The first stage used keywords search from databases, such as Springer, Scopus, Elsevier, etc. The second stage refined this process with pattern matching against core publications and reports in light of industry developments, legislation, and emerging technological solutions entering the market. This encompassed the use of proprietary industry databases such as BIM Task Group, Buildoffsite, Homes England, National Buildings Specifications, and Construction Leadership Council (CLC). These two stages used Nvivo software to organise and analyse non-numerical data. This was classified by topic, including relationships. Findings from this stage helped establish the initial conceptual framework. The exploratory case study phase was then undertaken to critique this framework. Primary data was captured from this case study and used to develop a prototype. This prototype was developed using Autodesk Revit software in order to obtain exhaustive information on the standardisation and automation details required for this new proposed method.

4.4. DfMA feasibility case study

4.4.1. Core components identification

The proposed framework introduces the utilisation of a BIM Objects Library formed by a set of core components. These components were specially selected for this case study and were identified by a process of rationalisation, standardisation, and optimisation (Figure 4.2).

During the rationalisation phase, analytical tools were applied to select similar elements and determine the degree of variation in order to satisfy a number of common solutions with a high degree of occurrence. The second stage involved redefining elements to achieve reliable layouts along with specified materials and other requirements. The third stage (optimisation)

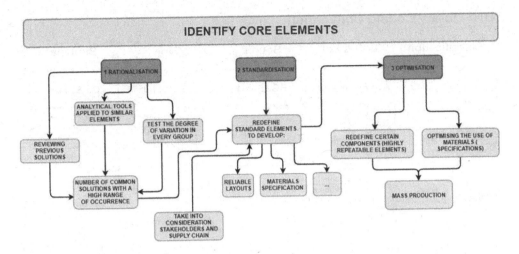

Figure 4.2 Identification of core elements

entailed analysing components in order to obtain repeatable elements, whilst also optimising the use of materials. The results from this three-stage process generated a set of components suitable for mass production.

4.4.2. BIM objects library development

BIM library components should be defined and classified in a format that enables and facilitates information transfer. This classification process should therefore be clear, readily understood, in a form readily understood by AEC professionals, and more importantly, each component should be uniquely named and described. Given this requirement, a library for international use and common data standards was adopted for developing the coding convention. Standards such as BS 8541–1, BS EN ISO 19650, BS EN ISO 13567, and UNICLASS 2015 were implemented, together with the American Institute of Architects (AIA) Framework, to indicate the Level of Detail (LoD) and Level of Information (LoI).

Figure 4.3 presents the BIM library coding convention created for this case study.

The general classification of components was primarily based on their functionality, ergo, structural elements, walls (non-structural), floors, and roofs. For the purpose of this study, components were designed to LoD300 (Detailed Design), although given the new method of construction, the BIM library should include LoD400 components (where information for manufacturing and assembly are specified). All components were treated as generic objects, insofar as their novelty or distinctiveness was not attributable to any specific library or manufacturer. Focusing on their type, components were classified through standards, Mechanical, Electrical, and Plumbing (MEP) and Aesthetics (where components included an aesthetic feature such as a door or window). Whilst developing this case study, a fourth type was added, where the same component presented MEP and Aesthetic characteristics. Of note, the subtype indicates the location of the component: external, internal, or assembled to foundations in the case of floors. In addition, as part of this case study, another coding system (alphabetically based) was added to similar components where needed in order to highlight different

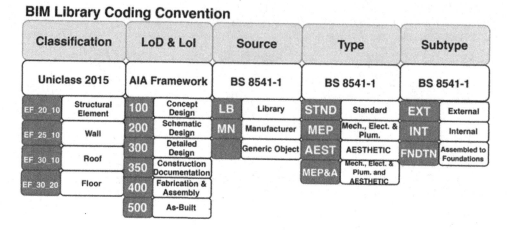

Figure 4.3 BIM library coding convention

attributes such as the length of walls, etc. From this, a set of standardised components was concluded. These were considered more optimal and favourable elements for use in the BIM library (given the proposed new method of construction). These components can be seen as follows:

- EF_20_10–300-AEST-EXT
- EF_20_10–300-AEST-INT
- EF_20_10–300-MEP&A-EXT
- EF_20_10–300-MEP-EXT
- EF_20_10–300-STND-EXT
- EF_20_10–300-STND-INT
- EF_20_10–300-MEP-INT
- EF_25_10–300-MEP&A -INT
- EF_25_10–300-STND -INT
- EF_25_10–300-MEP-INT
- EF_25_10–300-AEST-INT
- EF_30_10–300-STND
- EF_30_20–300-MEP
- EF_30_20–300-MEP-FDATION
- EF_30_20–300-STND
- EF_30_20–300-STND-FDATION

Table 4.1 presents the BIM library components description used in this case study.

4.4.3. *Technical specifications for standard type components*

Four main types of components were developed for this case study. Standard components were used as a basis for creating the other main types. Whilst structures were not expressly included in this case study, the walls, roofs, and floors were designed to comply with such issues as stability, logistics management, etc. In addition, materials and standardised

Table 4.1 BIM library components description

1	EF_20_10–300-AEST-EXT	**Element/Function**	Structural element
		LoD	Detailed design
		Source	Generic object
		Type	Aesthetic
		Subtype	Exterior
2	EF_20_10–300-AEST-INT	**Element/Function**	Structural element
		LoD	Detailed design
		Source	Generic object
		Type	Aesthetic
		Subtype	Interior
3	EF_20_10–300-MEP&A-EXT	**Element/Function**	Structural element
		LoD	Detailed design
		Source	Generic object
		Type	With Mech., Elect. and Plumbing and Aesthetic features
		Subtype	Exterior
4	EF_20_10–300-MEP-EXT	**Element/Function**	Structural element
		LoD	Detailed design
		Source	Generic object
		Type	With Mech., Elect. and Plumbing
		Subtype	Exterior
5	EF_20_10–300-MEP-INT	**Element/Function**	Structural element
		LoD	Detailed design
		Source	Generic object
		Type	With Mech., Elect. and Plumbing
		Subtype	Interior

(Continued)

Table 4.1 (Continued)

		Element/Function	Structural element	
6	EF_20_10–300-STND-EXT	**LoD**	Detailed design	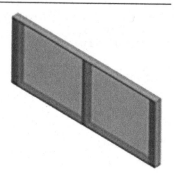
		Source	Generic object	
		Type	Standard	
		Subtype	Exterior	
7	EF_20_10–300-STND-INT	**Element/Function**	Structural element	
		LoD	Detailed design	
		Source	Generic object	
		Type	Standard	
		Subtype	Interior	
8	EF_25_10–300-MEP&A-INT	**Element/Function**	Wall (non-structural)	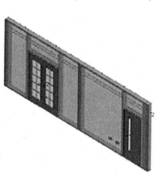
		LoD	Detailed design	
		Source	Generic object	
		Type	With Mech., Elect. and Plumbing and Aesthetic features	
		Subtype	Interior	
9	EF_25_10–300-STND-INT	**Element/Function**	Wall (non-structural)	
		LoD	Detailed design	
		Source	Generic object	
		Type	Standard	
		Subtype	Interior	
10	EF_25_10–300-MEP-INT	**Element/Function**	Wall (non-structural)	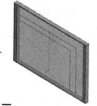
		LoD	Detailed design	
		Source	Generic object	
		Type	With Mech., Elect. and Plumbing	
		Subtype	Interior	
11	EF_25_10–300-AEST-INT	**Element/Function**	Wall (non-structural)	
		LoD	Detailed design	
		Source	Generic object	
		Type	Aesthetic	
		Subtype	Interior	

		Element/Function	Roof
12	EF_30_10–300-STND	LoD	Detailed design
		Source	Generic object
		Type	Standard
13	EF_30_20–300-MEP	Element/Function	Floor
		LoD	Detailed design
		Source	Generic object
		Type	With Mech., Elect. and Plumbing
14	EF_30_20–300-MEP-FNDTN	Element/Function	Floor
		LoD	Detailed design
		Source	Generic object
		Type	With Mech., Elect. and Plumbing
		Subtype	Assembled to foundations
15	EF_30_20–300-STND	Element/Function	Floor
		LoD	Detailed design
		Source	Generic object
		Type	Standard
16	EF_30_20–300-STND-FNDTN	Element/Function	Floor
		LoD	Detailed design
		Source	Generic object
		Type	Standard
		Subtype	Assembled to foundations

components were carefully chosen to satisfy standards and building regulations, ergo insulation and sound-proofing properties. Technical specifications of these are presented as follows:

- **EF_20_10–300-STND-EXT:** External standard structural wall components were designed to be 7.5 m long, within three structural columns, and a selection of layers (including insulation and waterproof membrane). For visualisation purposes, some layers were set up with a grade of transparency.
- **EF_25_10–300-STND-EXT:** Internal standard non-structural walls components were designed to be a maximum of 4 m long with layers that guaranteed the correct level of insulation. For visualisation purposes, some layers were set up with a grade of transparency.
- **EF_30_10–300-STND:** Standard roof was designed considering logistics and the most efficient way to transport and assemble this component. Components not exceeding 9–10 m were considered suitable for this purpose. For visualisation purposes, some layers were set up with a grade of transparency.
- **EF_30_20–300-STND:** Standard floors were designed following the same criteria for roofs regarding logistics. For visualisation purposes, some layers were set up with a grade of transparency.

4.4.4. *Prototype design based on DfMA*

Revit 2020 software was used to create a working prototype in order to corroborate the effectiveness of the components in the design stage. A set of components proposed in the

BIM library were used to design a two-bedroom house. This building was semi-detached and divided into two floors in order to utilise a variety of components.

Figure 4.4 presents the layout of the ground floor, and Figure 4.5 presents the layout of the first floor. This arrangement consisted of a living room, open plan kitchen to dining room, and a toilet. Prefabricated stairs were located in the living room, leading to the first floor. This consisted of two double bedrooms and a bathroom.

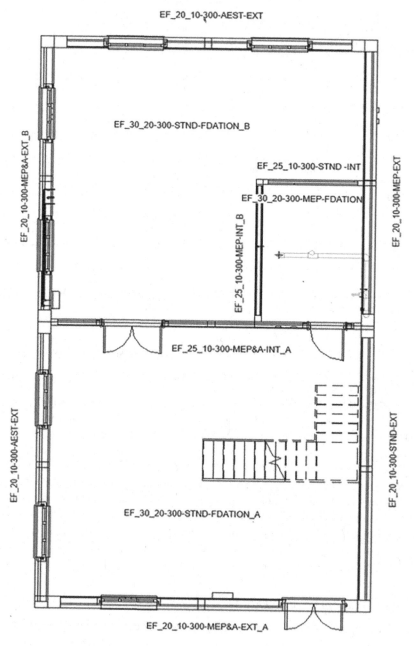

Figure 4.4 Ground floor, a two-bedroom house

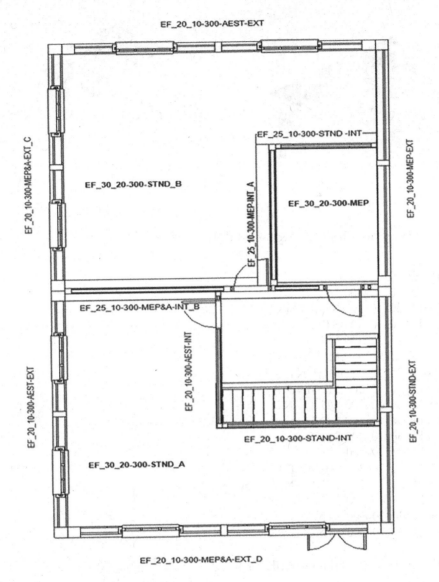

Figure 4.5 First floor, a two-bedroom house

Through Revit, components were renamed to add extra information; specifically, small changes were made to original components such as location or sizes. This extra information was added using letters A–D. The following set of components were used in this case study:

- 4 x EF_20_10–300-AEST-EXT
- 1 x EF_20_10–300-AEST-INT
- 1 x EF_20_10–300-MEP&A-EXT_A
- 1 x EF_20_10–300-MEP&A-EXT_B
- 1 x EF_20_10–300-MEP&A-EXT_C

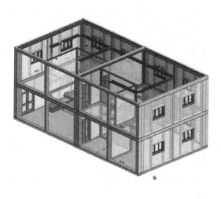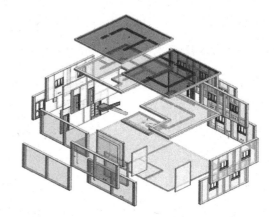

Figure 4.6 Two-bedroom house, perspective and displacements

- 1 x EF_20_10–300-MEP&A-EXT_D
- 2 x EF_20_10–300-MEP-EXT
- 1 x EF_20_10–300-STAND-INT
- 2 x EF_20_10–300-STND-EXT
- 1 x EF_25_10–300-MEP&A-INT_A
- 1 x EF_25_10–300-MEP&A-INT_B
- 1 x EF_25_10–300-MEP-INT_A
- 1 x EF_25_10–300-MEP-INT_B
- 2 x EF_25_10–300-STND -INT
- 1 x EF_30_10–300-STND_A
- 1 x EF_30_10–300-STND_B
- 1 x EF_30_20–300-MEP
- 1 x EF_30_20–300-MEP-FNDTN
- 1 x EF_30_20–300-STND-FNDTN_A
- 1 x EF_30_20–300-STND-FNDTN_B
- 1 x EF_30_20–300-STND_A
- 1 x EF_30_20–300-STND_B

The final prototype of this two-bedroom house can be seen in Figure 4.6.

4.5. Findings and discussion

4.5.1. General case study findings

During the development of this case study, a BIM objects component-based library was developed following a process of rationalisation, standardisation, and optimisation. This followed standards and protocols concerning coding convention. The use of this library mirrored the strategic definition stage and adopted the principles of DfMA. These findings aligned with standard agreements. From this, a conceptual design was developed utilising the BIM objects component-based library. Prior to the design of this library, key model components were identified and used to create this library. Findings from this case study helped achieve a deeper understanding of how the conceptual design phase aligns to DfMA principles. Specific findings are discussed further in the following sections.

4.5.2. BIM and DfMA strategy findings

The decision to adopt BIM methodology was made on the basis that this seemed to be the fundamental principle adopted in AEC. This was not only reinforced in literature but also acknowledged through several different studies and numerous worldwide governmental reports. Given this, the doctrine of BIM was uniquely embedded in all stages of the conceptual framework, particularly to ensure effective delivery. This also helped in managing information and decision-making through transparent coordination processes and a common data environment. From a findings perspective, this also supported the early involvement of manufacturer(s), enabling cross references to be made with the EIR as part of the design process. From this case study, a component-based BIM objects library was developed in accordance with standard coding convention, supported by a 3D model. For example, through the BIM library, relevant data can be edited and updated during the asset's lifecycle. In this respect, findings highlighted that BIM was particularly suitable for enabling this new method of construction.

That being said, it is equally important to acknowledge that (after analysing all data), successful implementation of a new method such as this (based on manufacturing) requires people with appropriate levels of skills and knowledge to make this happen. In particular, there is a specific need to engage stakeholders from the outset. Early engagement and collaboration is key in every part of the process. In this respect, BIM can help here, as this technology-driven solution is uniquely placed to support digital design and digital manufacturing methods.

From a technical perspective, the proposed component-based BIM library was perfectly suited to OSM, where the set of components proposed in this chapter were expressly designed to be standardised (to enable mass production). In doing so, this library supports automation, whilst also being flexible enough to incorporate some degree of customisation.

In this research has been proven that a component-based library offers the majority benefits of standardisation and also a degree of adaptability that the new method of construction needs to succeed. This approach provides clients with a series of choices to select from using a standardised set of components. This helps guarantee additional surety, cognisant of adaptability and replicability. The corollary of which can also help mitigate wider performance and productivity challenges.

In summary, findings from this research demonstrated that incorporating DfMA and DfD principles into the early stages of the project is possible. The component-based BIM library can be created to follow DfMA principles. In doing so, designs can be optimised to support assembly (and disassembly).

4.5.3. Discussion

This research reflected on the wider challenges facing AEC and the need to reflect on issues such as OSM, BIM, DfMA, etc. In doing so, it was evident from the outset that whilst a number of significant developments have been made in these areas, several areas still needed further work, particularly to harvest the benefits of OSM and DfMA with technology-driven tools such as BIM, IoT, blockchain, etc. The case study presented in this chapter highlighted a number challenges and opportunities. It is also important to note that not all of these issues could be resolved due to project scope and complexity, aesthetic requirements, logistics, component spans, design typologies, etc. Notwithstanding these issues, the use of parametric and generative design was considered a good starting point of departure for this study.

The development of this conceptual framework provided an opportunity to develop, test, and validate some of the theoretical underpinnings of this work. For example, the time spent designing the parametric BIM library was particularly beneficial, as it presented an opportunity to evaluate what could and could not be achieved. This was especially important, as AEC needs to have tools that are 'fit for purpose', especially when transitioning from traditional working practices to those that are more manufacturing-oriented. It was therefore important to not only capture and 'absorb' these into the finished product, but also to try and exploit these opportunities in line with AEC needs – cognisant of a number of high-level challenges, including process inefficiencies, waste, health and safety, communication, automation, predictability, quality, etc.

Reflecting upon these core challenges, the conceptual framework was designed to support collaboration and coordination, especially in the early design stage. In particular, the ability to simulate processes through 3D, 4D, 5D, and 6D BIM models was seen as particularly beneficial. That being said, it was acknowledged through this case study that in order to fully maximise these benefits, a certain degree of workforce upskilling would be required. This includes digital specialists and design teams conversant in DfMA. It is also recommended that these skillsets embrace blockchain, smart contracts (including integrated procurement methods), advanced digital platforms, and strategists capable of levering innovation from these new systems and technologies.

4.6. Conclusion

This chapter highlighted a number of recurrent challenges facing AEC. In doing so, OSM was suggested as one possible solution. This was expanded to include DfMA, including the need to embrace technologies such as BIM and GD. From this, a conceptual framework was presented for discussion, covering four main phases (preparation; design and pre-construction; construction/assembly to close-out; and use and reuse/demolition). These phases were discussed, along with the technical requirements needed. This included the creation of a BIM objects component-based library, along with a worked example prototype based on a two-bedroom house. Findings highlighted a number of significant advantages of using this approach. It also highlighted a few technical challenges, but (arguably) more important perhaps, the need for AEC upskilling. Whilst this work is still embryonic, it is therefore recommended that any generalisation, inference, or future replication is countered by the inclusion of additional test data and case study work in order to improve the veracity of these findings.

Finally, it is proffered that AEC is now entering a new technological era, where almost anything is possible. This statement is made from a somewhat halcyon perspective insofar as technological solutions are possible whenever or wherever a need exists. This requires considerable effort from all. The old adage of 'limited job security, harsh working conditions, and poor health and safety' may still exist in some parts of the industry; however, the obverse is equally true, evidenced through many innovative companies pioneering OSM, most of which are showcasing highly flexible and value-laden solutions. These companies have already resolved many of the challenges raised earlier in this chapter. Moreover, they are championing new products and divested services through OSM. This is very encouraging given the transition to Industry 4.0. The findings from this case study provide an important step in this direction, particularly through the use of bespoke BIM libraries, DfMA methodologies, and GD-driven solutions.

References

Abanda, F. H., Tah, J. H. M., & Cheung, F. K. T. (2017). BIM in off-site manufacturing for buildings. *Journal of Building Engineering, 14*, 89–102. https://doi.org/10.1016/j.jobe.2017.10.002

Abosaod, H & Underwood, Jason & Isikdag, Umit & Barony, S. (2010). *A classification system for representation of off-site manufacturing concepts through virtual prototyping.* 9th International Detail Design in Architecture Conference 2010: Innovative Detailing: Materials & Construction Methods for a Low Carbon Future, University of Central Lancashire. ISBN:13.978-1901922769

Adams, K. T., Osmani, M., Thorpe, T., & Thornback, J. (2017). Circular economy in construction: Current awareness, challenges and enablers. *Waste and* Resource Management, *170*, 15–24. https://doi.org/10.1680/jwarm.16.00011

Akanbi, L. A., Oyedele, L. O., Omoteso, K., Bilala, M., Akinade, O. O., Ajayi, A. O., Davila Delgado, J. M., & Owolabi, H. A. (2019). Disassembly and deconstruction analytics system (D-DAS) for construction in a circular economy. *Journal of Cleaner Production, 223*, 386–396. https://doi.org/10.1016/j.jclepro.2019.03.172

Akbarnezhad, A., Ong, K. C. G., & Chandra, L. R. (2014). Economic and environmental assessment of deconstruction strategies using building information modelling. *Automation in Construction, 37*, 131–144. https://doi.org/10.1016/j.autcon.2013.10.017

Akinade, O., Oyedele, L., Moteso, K., Aiayi, S. O., Bilal, M., Owolabi, H. A., Alaka, H. A., Ayris, L., & Looney, J. H. (2017). BIM-based deconstruction tool: Towards essential functionalities. *International Journal of Sustainable Built Environment, 6*, 260–271. https://doi.org/10.1016/j.ijsbe.2017.01.002

Akinade, O., Oyedele, L., Oyedele, A., Davila Delgado, J. M., Bilal, M., Akanbi, L., Ajayi, A., & Owolabi, H. (2020). Design for deconstruction using a circular economy approach: Barriers and strategies for improvement. *Production Planning & Control, 31*(100): 829–840. https://doi.org/10.1080/09537287.2019.1695006

Alfieri, E., Seghezzi, E., Sauchelli, M., Di Giuda, G. M., & Masera, G. (2020). A BIM-based approach for DfMA in building construction: Framework and first results on an Italian case study. *Architectural, Engineering and Design Management, 16* (4). https://doi.org/10.1080/17452007.2020.1726725

Alizadehsalehi, S., & Yitmen, I. (2021). Digital twin-based progress monitoring management model through reality capture to extended reality technologies (DRX). *Smart and Sustainable Built Environment, ahead-of-print(ahead-of-print).* https://doi.org/10.1108/SASBE-01-2021-0016

Arashpour, M., Wakefield, R., Abbasi, B., Arashpour, M., & Hosseni, R. (2018). Optimal process integration architectures in off-site construction: Theorizing the use of multi-skilled resources, *Architectural Engineering and Design Management, 14*, 46–59. https://doi.org/10.1080/17452007.2017.1302406

Arif, M., Goulding, J., & Rahimian, F. P. (2012). Promoting off-site construction: Future challenges and opportunities. *Journal of Architectural Engineering, 18*(2). https://doi.org/10.1061/(ASCE)AE.1943-5568.0000081

Ashraf, M., Gibson, I., & Rashed, M. G. (2018). *Challenges and prospects of 3d printing in structural engineering: Vol. 11.* 13th International Conference on Steel, Space and Composite Structures.

Ayinla, K. O., Cheung, F., & Tawil, A. (2019). Demystifying the concept of off site manufacturing method towards a robust definition and classification system. *Construction Innovation, 20*(2), 223–246. Emerald Publishing Limited. ISSN:1471-4175

Babic, N. C., Podbreznik, P., & Rebolj, D. (2010). Integrating resource production and construction using BIM. *Automation in Construction, 19*, 539–543. https://doi.org/10.1016/j.autcon.2009.11.005

Barrit, J. (2016). An overview on recycling and waste in construction. *Construction Materials, 169*, 49–53. https://doi.org/10.1680/coma.15.00006

Building and Construction Authority, & Wood, B. (2016). *BIM for DfMA (Design for manufacturing and assembly) essential guide.* Retrieved May 3, 2021, from www.corenet.gov.sg/media/2032999/bim_essential_guide_dfma.pdf

Blismas, N., Pasquire, C., & Gibb, A. (2006). Benefit evaluation for off-site production in construction, *Construction Management and Economics, 24*(2), 121–130. https://doi.org/10.1080/01446190500184444

Borrmann, A., Konig, M., Koch, C., & Beetz, J. (2018). *Building Information Modelling. Technology Foundations and Industry Practice*. Springer International Publishing. https://doi.org/10.1007/978-3-319-92862-3

Buildoffsite. (2020). *BIM and DfMA*. Retrieved May 3, 2021, from www.buildoffsite.com/publicationsguidance/bim-dfma/

Construction Project Information Committee (CPIC). (2008). *Drawing is dead, long live modelling*. Retrieved May 4, 2021, from www.cpic.org.uk/publications/drawing-is-dead/

Dawood, N., & Vukovic, V. (2015). *Whole lifecycle information flow underpinned by BIM: Technology, process, policy and people*. 2nd International Conference on Civil and Building Engineering Informatics.

Digital Built Britain. (2017). *Delivery platforms for government assets*. Retrieved May 4, 2021, from www.cdbb.cam.ac.uk/system/files/documents/delivery_platforms_screen.pdf

Durdyev, S., & Ismail, S. (2019). Offsite manufacturing in the construction industry for productivity improvement. *Engineering Management Journal, 31*(1), 35–46. https://doi.org/10.1080/10429247.2018.1522566

Eadie, R., Browne, M., Odeyink, H., McKeown, C., McNiff, S. (2013). BIM implementation throughout the UK construction project lifecycle: An analysis. *Automation in Construction, 36*, 145–151. https://doi.org/10.1016/j.autcon.2013.09.001

Egan, J. (1998). *Rethinking construction, UK department of the environment, transport and the regions*. HMSO. Retrieved May 1, 2021, from http://constructingexcellence.org.uk/wp-content/uploads/2014/10/rethinking_construction_report.pdf

Elghaish, F., Matarneh, S., Talebi, S., Kagioglou, M., Hosseini, M. R., & Abrishami, S. (2020). Toward digitalization in the construction industry with immersive and drones technologies: A critical literature review. Smart and *Sustainable Built Environment, ahead-of-print (ahead-of-print)*. https://doi.org/10.1108/SASBE-06-2020-0077

Eynon, J. (2016). *Construction manager's BIM handbook*. John Wiley & Sons. ISBN:978-1-118-89647-1

Ezcan, V., Isikdag, U., & Goulding, J. S. (2013, May 5–9). *BIM and off-site manufacturing: Recent research and opportunities*. CIB World Building Congress 2013, Queensland University of Technology.

Farmer, M. (2016). *The farmer review of the UK construction labour model: Modernise or die*. Construction Leadership Council (CLC). Retrieved May 3, 2021, from www.cast-consultancy.com/wp-content/uploads/2016/10/Farmer-Review-1.pdf

Gao, S., Jin, R., & Lu, W. (2020). Design for manufacture and assembly in construction: A review. *Building Research & Information, 48*(5), 538–550. https://doi.org/10.1080/09613218.2019.1660608

Ginigaddara, B., Perera, S., Feng, Y., & Rahnamayiezekavat, P. (2019, November 8–10). *Typologies of offsite construction*. Proceedings of the 8th World Construction Symposium: Towards a Smart, Sustainable and Resilient Built Environment, pp. 567–577.

Goodier, C., & Pan, W. (2010). *The future of UK housebuilding*. Loughborough University Institutional Repository: RICS. Retrieved May 5, 2021, from https://repository.lboro.ac.uk/articles/conference_contribution/The_future_of_offsite_in_housebuilding/9432068/1

Guy, B., Shell, S., Esherick, H. (2006). *Design for deconstruction and materials reuse*. 659 Proceedings of the CIB Task Group 39, pp. 189–209.

Hardin, B., & McCool, D. (2015). *BIM and construction management: Proven tools, methods and workflows*. John Wiley & Sons. ISBN:978-118-9427-5

HM Government. (2017). *Industrial strategy: Construction sector deal*. OGL. Retrieved May 5, 2021, from https://assets.publishing.service.gov.uk/government/uploads/system/uploads/attachment_data/file/731871/construction-sector-deal-print-single.pdf

Hu, X., Chong, H. Y., Wang, X., & London, K. (2019). Understanding stakeholders in off-site manufacturing: A literature review. *Journal of construction engineering and management, 145*(8), 03119003.

Ibrahim, H. S., Hashim, N., & Jamal, K. A. A. (2019). The potential benefits of building information modelling (BIM) in construction industry. In *IOP conference series: Earth and environmental science* (Vol. 385, No. 1, p. 012047). IOP Publishing.

Jensen, K. N., Nielsen, K., & Brunoe, T. D. (2018). Mass customization as a productivity enabler in the construction industry. In *IFIP international conference on advances in production management systems* (pp. 159–166). Springer.

Kanters, J. (2018). *Design for deconstruction in the design process: State of the art buildings.* Lund University. https://doi.org/10.3390/buildings8110150

KPMG (2016). *Smart construction report: How offsite manufacturing can transform our industry.* Retrieved May 15, 2021, from https://assets.kpmg/content/dam/kpmg/pdf/2016/04/smart-construction-report-2016.pdf

Latham, M. (1994). *Constructing the team.* Final Report of the Government/Industry Review of Procurement and Contractual Arrangements in the UK Construction Industry. HMSO. Retrieved May 15, 2021, from http://constructingexcellence.org.uk/wp-content/uploads/2014/10/Constructing-the-team-The-Latham-Report.pdf

Lu, W., Tan, T., Xu, J., Wang, J., Chen, K., Gao, S., & Xue, F. (2021). Design for manufacture and assembly (DfMA) in construction: The old and the new. *Architectural Engineering and Design Management, 17*(1–2), 77–91. https://doi.org/10.1080/17452007.2020.1768505

Mesároš, P., & Mandičák, T. (2017, October). Exploitation and benefits of BIM in construction project management. In *IOP Conference Series: Materials Science and Engineering* (Vol. 245, No. 6, p. 062056). IOP Publishing.

Milestone, N. (2017). *BIM and DfMA – the future of construction.* Retrieved May 3, 2021, from www.pbctoday.co.uk/news/bim-news/bim-dfma-future-construction/32775/

Mills, F. (2019, November). *Construction's digital Manufacturing Revolution.* Retrieved May 3, 2021, from www.theb1m.com/video/constructions-digital-manufacturing-revolution

Mohammed Abdelkader, E. (2021). On the hybridization of pre-trained deep learning and differential evolution algorithms for semantic crack detection and recognition in ensemble of infrastructures. *Smart and Sustainable Built Environment, ahead-of-print(ahead-of-print).* https://doi.org/10.1108/SASBE-01-2021-0010

Nadim, W., & Goulding, J. S. (2011). Offsite production: A model for building down barriers: A European construction industry perspective. *Engineering, Construction and Architectural Management, 18*(1), 82–101. https://doi.org/10.1108/09699981111098702

Newman, C., Edwards, D., Martek, I., Lai, J., Thwala, W. D., & Rillie, I. (2020). Industry 4.0 deployment in the construction industry: A bibliometric literature review and UK-based case study. *Smart and Sustainable Built Environment, ahead-of-print(ahead-of-print).* https://doi.org/10.1108/SASBE-02-2020-0016

O'Connor, J. T., O'Brien, W. J., & Choi, J. O. (2016). Industrial project execution planning: Modularization versus stick-built. *Practice periodical on structural design and construction, 21*(1), 04015014.

Oliver, S., Seyedzadeh, S., Rahimian, F., Dawood, N., & Rodriguez, S. (2020). Cost-effective as-built BIM modelling using 3D point-clouds and photogrammetry. *Current Trends in Civil & Structural Engineering-CTCSE, 4*(5), 000599.

Pasquire, C. L., & Connolly, G. E. (2003). *Design for manufacture and assembly.* 11th Annual Conference of the International Group for Lean Construction, pp. 184–194.

Rahimian, F. P., Seyedzadeh, S., & Glesk, I. (2019). OCDMA-based sensor network for monitoring construction sites affected by vibrations. *Journal of Information Technology in Construction, 24,* 299–317.

RIBA. (2016). *RIBA plan for work 2013: Designing for manufacture and assembly.* Plan of Work Royal Institute of British Architects, RIBA Enterprises Ltd. Retrieved May 5, 2021, from http://consig.org/wp-content/uploads/2018/10/RIBAPlanofWorkDfMAOverlaypdf.pdf

Sakin, M., & Kiroglu, Y. C. (2017). 3D printing of buildings: Construction of the sustainable houses of the future by BIM. *Energy Procedia, 134,* 702–711. https://doi.org/10.1016/j.egypro.2017.09.562

Saxon, R. (2016). *BIM for construction clients: Driving strategic value through digital information management.* National Building Specification Ltd. RIBA Enterprises. ISBN:10-1859466079

Seyedzadeh, S., Agapiou, A., Moghaddasi, M., Dado, M., & Glesk, I. (2021). WON-OCDMA system based on MW-ZCC codes for applications in optical wireless sensor networks. *Sensors, 21*(2), 539.

Seyedzadeh, S., & Rahimian, F. P. (2021). Building energy data-driven model improved by multi-objective optimisation. In *Data-driven modelling of non-domestic buildings energy performance* (pp. 99–109). Springer.

Smith, D., & Sweets, R. (2018). The standardisation dynamic: Could a design code for prefabricated housing help offsite take off?. *Construction Research and Innovation, 9*(2), 32–37. https://doi.org/10.1080/20450249.2018.1477471

Spence, W. P., & Kultermann, E. (2016). *Construction materials, methods and techniques*. Cengage Learning. ISBN:987-1-3050-8627-2

Staub-French, S., Poirier, E. A., Calderon, F., Chikhi, I., Zadeh, P., Chudasma, D., & Huang, S. (2018). *Building information modeling (BIM) and design for manufacturing and assembly (DfMA) for mass timber construction*. BIM TOPiCS Research Lab, University of British Columbia.

Stoll, H. W. (1986). Design for manufacture: An overview. *Applied Mechanics Review, 39*(9), 1356–1364. https://doi.org/10.1115/1.3149526

Sun, C., Jiang, S., Skibniewski, M. J., Man, Q., & Shen, L. (2017). A literature review of the factors limiting the application of BIM in the construction industry. *Technological and Economic Development of Economy, 23*(5), 764–779. https://doi.org/10.3846/20294913.2015.1087071

Teizer, J., Blickle, A., King, T., Leitzbach, D., Guenther, D., Mattern, H., & Konig, M. (2018). BIM for 3D printing in construction. In A. Borrmann, M. König, C. Koch, & J. Beetz (Eds.), *Building information modeling: Technologische Grundlagen und industrielle Praxis* (pp. 421–500). Springer International Publishing. https://doi.org/10.1007/978-3-319-92862-3

The Housing Forum. (2019). *MMC for affordable housing developers: A housing forum guide to overcoming challenges and barriers*. Retrieved May 3, 2021, from https://mmc.lhc.gov.uk/wp-content/uploads/2020/01/mmc-guide-2019.pdf

Wang, K., & Skibniewski, M. J. (2019, June 29–July 2). *Feasibility study of integrating BIM and 3D printing to support building construction*. Proceedings of the Creative Construction Conference 2019, Budapest University of Technology and Economics, pp. 845–850, ISBN 978-615-5270-56-7; https://doi.org/10.3311/CCC2019-116; https://repozitorium.omikk.bme.hu/handle/10890/13298

Whysall, Z., Owtram, M., & Brittain, S. (2019). The new talent management challenges of industry 4.0. *Journal of Management Development, 38*(2), 118–119. ISSN:0262–1711

World Economic Forum. (2016). *Shaping the future of construction: A breakthrough in mindset and technology*. Retrieved May 3, 2021, from http://www3.weforum.org/docs/WEF_Shaping_the_Future_of_Construction_full_report__.pdf

Yin, X., Liu, H., Chen, Y., & Al-Hussein, M. (2019). Building information modelling for off-site construction: Review and future directions. *Automation in Construction, 101*, 72–91. https://doi.org/10.1016/j.autcon.2019.01.010

5 Virtual reality–based cloud BIM platforms for integrated AEC projects

5.1 Introduction

Building Information Modelling (BIM) has demonstrated the need to integrate collaborative design teams' 'project data', not only to help coordinate the design, engineering, fabrication, construction, and maintenance of various trades, but also to facilitate project integration and interchange (Pour Rahimian et al., 2020). Numerous potential benefits have inspired several countries to consider the implications of implementing BIM Level 3 (Cloud) as an innovative way of further enhancing the design, management, and delivery process, ergo a paradigm shift is needed towards Integrated Project Delivery (IPD) (Elghaish et al., 2020; Newman et al., 2020; Oliver et al., 2020). Amongst the myriad of available innovative approaches, web-based platforms are particularly beneficial for integrating visualisation components. This also provides continuous sharing of relevant information for geographically dispersed end users.

However, at this juncture, it is important to revisit some of the major challenges pervading Architecture, Engineering, and Construction (AEC), in particular its level of fragmentation. This has been discussed in extant literature for a number of decades now, the corollary of which is well recognised. The consequences of this have led to several documented problems relating ostensibly to failures in communication and information processing and processes (Egan, 1998; Latham, 1994), which in turn have affected performance and efficiency (especially compared to other industries). These failures have also contributed to an increased proliferation of adversarial relationships among the different parties involved in projects (Forcade et al., 2007), which in turn has impacted the veracity of design information (Cera et al., 2002; Fruchter, 1998) within the project lifecycle. In essence, whilst some might argue that the nature and complexity of communication within AEC has changed significantly over the last 20 years, especially with advances in Information and Communication Technology (ICT) and the increased prevalence of web-based project collaboration technologies and project extranets, others disagree. What is clear though is that ICT has certainly helped to revolutionise production and design (Cera et al., 2002), which in turn has led to dramatic changes in terms of labour and skills (Fruchter, 1998); however, this ICT evolution has also brought with it added layers of complexity, particularly concerning the functionality and capabilities of such applications (and implementation thereof). This complexity embraces a number of areas, from design to planning, costing, simulation, etc. Just taking design as an example highlights how ICT and user participation can improve design proposals (Petric et al., 2002). This engagement is key, as it actively enables (and facilitates) designers to compare the quality of proposed design solutions against alternatives. This was reinforced by Goulding and Rahimian (2012), regarding the ability to experiment and experience decisions in a 'cyber-safe' environment, in order to mitigate or reduce risks prior to construction (Elghaish et al.,

DOI: 10.1201/9781003106944-5

2020; Parchami Jalal et al., 2020). Consequently, it is proffered here that the success of AEC projects is highly dependent upon the type, level, and quality of communication exchange among various disciplines involved in the design and implementation phases. This exchange also naturally embraces the support structures underpinning this communication.

It is contended here that the success of AEC projects is highly dependent on the decisions made during early conceptual design and planning processes, where 70–80% of the production overheads are incurred (Paulson, 1976). Among the tools available to support advanced design planning include data-rich models such as BIM (initiated by Eastman et al., 1974; Fischer, 2000), which allow design teams to coordinate the fabrication of different building components. This created innumerable advantages, particularly in advanced methods of construction such as offsite construction, including faster delivery, improved economic indicators, along with improved sustainability factors and enhanced safety measures (Nawari, 2012). Isikdag and Underwood (2010) defined BIM as the information management process throughout the lifecycle of a building, which focuses on the collaborative use of semantically rich 3D Building Information Models (BIMs). In this respect, BIMs contain rich geometric and semantic information about a building, where different views/sub-models such as Design; Heating, Ventilation, and Air-Conditioning (HVAC); and Facilities Management (FM) can be derived, depending on the business need. Seminal literature highlights that BIM can revolutionise AEC by enhancing team collaboration (Gu & London, 2010), improving project integration (Woo et al., 2004), leveraging better construction information flow (Ibrahim et al., 2004), helping documentation flow (Popov et al., 2006), supporting FM and reducing building maintenance costs (Seyedzadeh & Pour Rahimian, 2021; Wang et al., 2013), and providing construction simulation for teamwork planning, clash prevention, and coordination interface (Fischer & Kunz, 2004). These initiatives are in line with the UK's Government Construction Strategy, which included a mandate for the implementation of BIM Level 2 on all public projects by 2016 (BIM Task Group, 2013). [N.b. BIM Level 2 necessitates a digital building model for sharing/data exchange, including 2D/3D spatial coordination between parties in the design/construction process based on BS1192:2007.] Similar initiatives have been rolled out elsewhere, including Europe, North America, South America, Canada, Australasia, and China.

However, given these espoused benefits, there appear to be different levels of industry uptake worldwide (for a number of different reasons). For instance, a survey conducted by the Malaysian Construction Industry Development Board in 2005 highlighted a lack of interest in using new integrated and/or parametric design tools for building projects among architects. Later studies identified that this may have been due to the weakness of Computer-Aided Design (CAD) tools in supporting the intuitive design process that architects preferred, particularly in the early stage of the design lifecycle (Rahimian & Ibrahim, 2011). This was seen as a particular challenge, which created a gap at the beginning of the iterative cycles of the design process, where designers often handle numerous repetitive building components – with almost similar embedded information (Pour Rahimian et al., 2011).

Another emerging issue on this theme was the challenge of maintaining automation within (and through) the project lifecycle (Frohm et al., 2008; Skibniewski, 1992). This, for example, included advanced manufactured construction, which (by default) engaged high product variety and significant variations in demand (Veenstra et al., 2006; Wikberg et al., 2010). It also embraced flexible and reconfigurable manufacturing systems (Colombo & Harrison, 2008), along with effective/cohesive supply chains (Arif et al., 2005), which collectively included integrated, web-mediated, and automatic modelling, simulation, and decision support systems (Fruchter, 1998). On this theme, Gu and London (2010) asserted that changes were unlikely to happen unless construction information was represented and managed throughout all stages of the project lifecycle, including the early conceptual design and planning processes.

Until this juncture, previous efforts at incorporating BIM (as a solution) did not really cover the operation of such systems as an overarching integrated design and implementation platform. That being said, Ibrahim and Pour Rahimian (2010) acknowledged this gap, noting that conceptual design automation systems were still in their infancy, and that existing CAD interfaces required a lot of modelling and visualisation skills. In essence, designers' creativity and intuitive design reasoning seemed to be stifled when using these complicated interfaces. As such, designers still preferred to produce design ideas using traditional pen and paper media, leaving others to create digital versions of these design solutions through a completely separate process (Pour Rahimian et al., 2008). This in turn caused additional challenges, not least of which were problems of interoperability (Santos, 2009), especially data flows and platform disintegration among various teams of designers (Fruchter, 1998).

In summary, therefore, literature highlights the need to develop a viable solution – one which automates, integrates, and streamlines the whole design process. In this regard, Lee et al. (2013) advocated parametric design interfaces as a new paradigm of CAD, noting the potential of these for producing design alternatives, which were controlled by certain rules or limits, and thus, less dependent on the visualisation skills of designers. In this respect, supporting design creativity is not new – see the use of synectics as an idea seeding technique (Blosiu, 1999), or the incorporation of personal characteristics in the generation of design alternatives (Kim & Kang, 2003).

Cognisant of these challenges and potential opportunities, the work presented in this chapter presents a game-like (an interactive rule-based system capable of being controlled by multiple users in real time) parametric site simulator tool for discussion. This uses a web-based virtual reality interface and cloud BIM platform. The work focusses on the need to support multi-dimensional, data-rich modelling processes used by AEC in order to deliver effective integrated projects. This web-based Virtual Reality (VR) solution is discussed, highlighting the adapted Unified Software Development Process of specifying this platform (which employed iterative phases of elaboration, construction and transition). This study presents new understanding and insight into the causal drivers and influences associated with successful decision-making design in non-collocated design teams. Findings from this work form a stepping-stone in the development of new relationship models in AEC collaborative environments, particularly through the use of gaming interfaces.

5.2 Cloud-based BIM in AEC

From a development perspective, BIM has been described as a model-based design process that adds value across the entire lifecycle of a building project (Autodesk®, 2011). It is considered an intelligent integrating modelling tool for building design and construction, which allows data sharing with vested parties and stakeholders. In this respect, it has been advocated as being a key implementation vehicle and principal design delivery method for sharing building information data during the design and construction processes. For example, the information contained in a BIM model comes in various formats, and in this respect, these different formats need to be exchanged in an efficient way (Santos, 2009); however, exchanging data can often be quite complex and challenging, due in part to software incompatibility, different specifications, categorisation, format requirements, etc. That being said, these issues are not insurmountable. Addressing these issues can create interoperable systems that can help data modelling migrate among different teams with minimal data loss and with improved optimal accuracy (Fruchter, 1998).

Conventionally, prior to the adoption of BIM, architects (and the wider design team) tended to create 3D models merely for visualisation purposes, and certainly not using these as data-rich intelligent models. This was perceived as a wasted opportunity, especially as BIM models contained rich information needed to support information exchange amongst various team members. Arguably, in some instances, BIM was seen as a pseudo-communication and social networking tool for designers; however, things started to change as BIM became more mainstream. Succar (2011) explained various stages of BIM adaptation and introduced three major levels, namely: modelling, collaboration, and integration. Prior to this, the Australian Institute of Architects (2009) allocated the traditional production of two-dimensional documentation as stage zero in modelling implementation stages, the result of which presented four capability stages. The Australian Institute of Architects (2009) proposed a model that divided these stages into two sub-divisions, making each stage more specific, along with capability definitions. According to this model, BIM level 1 was defined as 3D modelling (stage 1A) with intelligent modelling (stage 1B). Intelligent modelling also included attached data, whilst 3D modelling was merely for visualisation purposes only. BIM level 2 referred to the ability of two or more computer systems or software applications to exchange formats following an agreed-upon standard (to make better use of the information delivered). This is now frequently defined as an interoperability system that allows users to respond to the delivered model and to customise it based on requirements, specification, and needs, utilising the nD's modelling concept: 4D, time; 5D, cost; and 6D, facility management.

Given these developments, BIM is now considered more than a representation tool or means for developing a model or prototype to generate intelligent input. Additional benefits embrace several other issues, including facilitating project teams to engage in innovative contractual relationships and new project delivery strategies. For example, BIM level 3 offers an innovative way to excel in construction management through such approaches as Integrated Project Delivery (IPD), where the uses of such an approach helps create a greater team ethos, through transparent communication procedures, wider integration measures, and consensus building platforms. This has been called the "future of BIM". In this respect though, one of the main barriers to achieving this goal is that of BIM interoperability (Santos, 2009).

Interoperability refers specifically to incompatibility issues. This predominantly relates to non-cognate products and software applications that have been built and designed without specific mechanisms for data transfer and sharing. Thus, in order to address this challenge of 'incompatibility', a number of vendors have created a solution to this by converting their BIM model into a neutral object-based file format; i.e. a format that is not controlled by any particular vendor, thereby creating a new platform for exchanging data. The whole ethos of interoperability refers to the ability to exchange/share information between separate computer programs without any loss of content or meaning (Aranda-Mena & Wakefield, 2006). According to Succar (2009), interoperability can be seen as a linear workflow that allows simultaneous interdisciplinary changes to be shared through a single file-based sharing approach.

In the single operational file-based sharing model (Succar, 2009), once the building information model (1) is complete, it can then be exported to the interoperable model, BIModel (v1), to allow another process of modelling to take place. This interoperable model (v1) captures both geometry and properties of the BIModel (1), and thus facilitates the sharing of information. This interoperable model (v1) can then be imported to the BIModel (2) to allow modelling process to take place. This procedure can be repeated for other modelling processes until the project is completed. The capability of this interoperability system allows BIM to take one step further forward in improving interdisciplinary collaboration among project team members. Moreover, this approach is seen as a stepping-stone for web-based platforms, and is particularly beneficial for integrating and sharing visualisation components across geographically dispersed areas with different parties (end users).

One of the most referred-to industry standards (IEEE-1516) for large-scale modelling and simulation is the High-Level Architecture (HLA), which was originally introduced by the U.S. Department of Defense (Kuhl et al., 2000). Zhang et al. (2012) advocated that this system could integrate various simulation applications and provide a standard architecture for interconnectivity, interoperability, and reusability. In addition, Uygun et al. (2009) posited that by integrating various approaches and applications in computer simulations (using a unique framework, functional rules, and common interfaces), this could support flexible distributed simulations. Moreover, this could contribute to the reduction of software costs by supporting the reuse of simulation models, including providing an infrastructure for managing the run-time of these simulations.

On a similar theme, Wang et al. (2014b) proposed a structured methodology for integrating Augmented Reality (AR) technology with BIM, particularly in order to overcome issues relating to the limited sense of immersion and real-time communication of BIM within virtual environments. In this respect, Wang and Dunston (2013) developed a tangible Mixed Reality (MR) interface that facilitated non-collocated collaboration for problem-solving and design error detection; and Abrishami et al. (2013) proposed adopting Generative Algorithm (GA) with BIM to leverage integration into the conceptual design phases. Other notable areas of work in this area include Hou et al. (2013), who developed a platform for controlling building components assembly procedures in order to improve accuracy and reduce errors, and Wang et al. (2014a), who adopted a more overarching approach, advocating the need for the development of a computer-mediated remote collaborative design support system – specifically, to leverage distributed cognition and help capture the non-collocated team's knowledge (which is typically distributed in memories, facts, objects, individuals, and tools).

Given these challenges, this research extends the findings of previous studies in this area by placing specific emphasis on the measures needed to support decision-making processes in the construction stages. This uses a novel approach of an interactive and interoperable simulation platform to support non-collocated design teams using game-like VR environments blended to social sciences theory (social rules) and behavioural science theory (decision science/communication science). In essence, the aim of this work was to provide a flexible, interactive, safe learning environment for practicing new working conditions with respect to Off Site Production (OSP) in general and Open Building Manufacturing (OBM) in particular, without the do-or-die consequences often faced on real construction projects. Hence, a VR interactive learning environment was sought that builds upon the multi-disciplinary, practice-based training concept (Alshawi et al., 2007). In this context, the prototype presented in this chapter discusses this concept – in particular, the need to enable (and facilitate) disparate stakeholders (with different professional specialisations) to engage with the various aspects of OSP concepts. This approach was adopted in order to help overcome the problem of the 'compartmentation' of knowledge (Mole, 2003). Another important feature that needed to be incorporated was that of flexibility. The prototype had to be flexible enough to allow 'anytime, anyplace' learning, so as not to be constrained to a particular place or time for learning to take place. This chapter presents the system development process, including the features and the capabilities of the developed web-based, game-like VR construction site simulator.

5.3 Development approach

5.3.1 Background

Wellings and Levine (2010) noted that the high-level use of game technologies in emerging educational schemes was becoming an indispensable part of the next generation's lives in modern societies. In this respect, literature highlights the various benefits of using advanced digital media in education. For example, Wellings and Levine (2010) suggested transforming existing text-based lessons into problem-based learning platforms (where collaboration training helps the development of solutions for real-world problems). This approach proffered that this was only possible using immersive visualisation and simulation environments embedded in games and interactive interfaces (to increase the engagement of trainees and trainers). On this theme, Thai et al. (2009) argued that educational digital games offered an opportunity to empower the engagement of trainees, particularly to help transform teaching and learning into a new stage. The ACS (2009) summarised the benefits of interactive game-like immersive environments as (i) exploring knowledge by clicking on objects with linked information; (ii) strengthening education by providing a repository of aids, tools, etc., associated with nD objects; (iii) offering collaborative workspaces, e.g. nD informal discussion forums; (iv) providing traditional instructor-based education via a distance delivery method; and (v) enhancing simulated learning by modelling a process or interaction that closely imitates the real world (in terms of outcomes). With respect to AEC, Su et al. (2013) advocated the use of VR and game environment technologies to support training. This included training on construction equipment operation, which not only facilitated hands-on practice (as a vital component of this type of training), but also prevented the risks of personal injury, equipment damage, and health and safety challenges.

With these issues in mind, the aim of the development of this simulator was to embrace real-life issues facing OSP construction projects. This was of particular importance, as it naturally embraced the need to acquire new skills. Moreover, this development also needed to appeal to professionals by making it engaging and challenging, placing (immersing) learners in a real work context with real OSP problems, in order for them to generate solutions to these problems (through the simulator). Given this task, as part of the development process, it was decided to model OSP concepts against a real construction project. This decision was based on the need to govern the authenticity of the learning environment, but also to ensure relevance. In this context, one of the underpinning cornerstones of this prototype was to 'challenge' learners. In this respect, it was agreed that the learning simulator would do this by [deliberately] allowing 'things to go wrong'. Hence, this helped develop deeper informed learning, following 'learning through experimentation' or 'learning by doing'.

Following the concepts of 'learning by doing', it was important to consider the type and number of 'scenes' that needed to be captured within the simulator, as these 'scenes' would need to accurately reflect the construction site. Moreover, these scenes captured the target audience (learners), which in this case was construction professionals, e.g. project managers, construction managers, architects, designers, commercial managers, suppliers, manufacturers, etc. The environment (construction site) also needed to be realistic, as this was the main domain through which all unforeseen issues and problems would occur, e.g. inadequate early decision-making, logistics, weather, faulty work, etc. These issues would then trigger consequences and implications, particularly in respect of time, cost, resources, etc. In this respect, a decision was taken to govern learning through the following three streams:

- **Learner autonomy** – allowing learners to make all decisions;
- **Interactivity** – creating an environment that provides full feedback on all decisions taken, including the implications on the overall project (cost, time, resources, health and safety, etc.);
- **Reflection** – facilitating user reflection, where learners are able to defend their decisions based on feedback provided. This powerful option enables deeper learning, particularly being able to identify means and measures of avoiding/mitigating potential problems in the future, such as (i) OSP strategies e.g. Design for Manufacture Logistics and Assembly (DfMLA); (ii) business processes, procurement/contractual arrangements, project management, quality assurance, etc.; (iii) health and safety procedures; (iv) supply chain integration; and (v) new manufacturing technologies, open systems, etc.

5.3.2 VR simulator development concept

The main concept of the simulator was based on its ability to run scenarios through a VR environment in order to address predefined training objectives (and three learner streams identified). In this respect, learning was driven by problems encountered in this environment, supported by a report critique on learners' choices, rationale, and defence thereof. In this respect, the development encompassed two phases. Phase I embodied the development of the various scenarios, including the generation of reports, etc., and Phase II included the 'intelligence' components and assessment engine underpinning these scenarios, and the need to subsequently 'interrogate' learners on their choices and level of understanding.

5.3.3 VR simulator development framework

The simulator development framework encompasses four main activities: identify training objectives, develop scenario(s), develop the VR environment, and evaluation/validation of the prototype (see Figure 5.1). This framework required extensive input from the construction industry in order to not only secure relevance but also help govern the authenticity of each of these stages.

From Figure 5.1, the underpinning rationale of this framework was predicated on two primary needs: training objectives and scenario development, the details of which are highlighted as follows:

Training Objectives: The training objectives were seen as the governing principles underpinning the simulator. In this respect, the formation of these objectives established the direction of travel needed to deliver the outcomes needed. The first stage in the identification of these objectives was gathered from a synthesis of seminal literature covering the potential risks and threats facing OSP and OBM. The findings from this process were then refined and prioritised with representatives from AEC, covering in particular construction, manufacturing, and design. This knowledge and insight was fundamental, as it not only provided a conduit for focus but also helped identify what learners needed to understand and fully appreciate. This also supported the criteria and protocols needed to deal with such problems (through the simulator), and consequently, the learning outcomes needed to mitigate these challenges in practice. In this context, seven primary risks were identified: (i) late design changes, (ii) loss of factory production (or production capacity), (iii) unpredictable planning decisions and designs

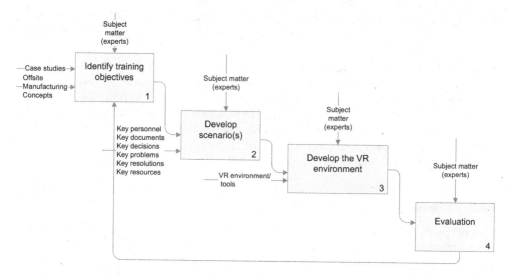

Figure 5.1 VR environment development framework

(alignment with OSP), (iv) tolerances (all), (v) supplier failure (delivery issues), (vi) manufacturer bankruptcy or insolvency, and (vii) stakeholder changes (incorporation of alternative manufacturers).

Scenario Development: The scenarios were specifically developed to challenge learners. In particular, to expose learners to new working conditions and issues that they were likely to face on real construction projects employing new OSP/OBM concepts. Given this, it was considered important to challenge learners to think about the routes of these problems (i.e. what caused them in the first place), rather than just reacting to these problems. This concept was adopted to provoke learners to think 'proactively' about future OSP projects. In essence, to pre-empt issues before they became a problem. Following this mandate, the scenario development process was predicated on this philosophy. This relied on identifying all (or at least the main) possible problems/issues that were traditionally associated with OSP practice. These were colloquially referred to as Problem 1, Problem 2, etc. Figure 5.2 presents the Scenario Implementation Concept, which details these activities and problems. For example, Problem 1 (Activity 2) is invoked through the simulator, the consequences of which offer the learner with four decisions, from which each decision then has a concatenating action, trigger, and outcome. Depending on the action chosen, the programme schedule is affected. This includes times, costs, and resources, including bottlenecks, delivery schedules, links to manufacturers, etc. The system then generates a final report on learner performance, summarising all decisions made, the impact on the project, and an opportunity to reflect on these decisions (in line with the learning outcomes).

These scenarios were specifically used to simulate how OSP operates in real life, in order to provoke learners to think 'how' and 'why' things go wrong, but more importantly, what needs to be considered beforehand in order to prevent these problems from occurring in the first place. This reflection is enacted through a debriefing post-mortem session, where

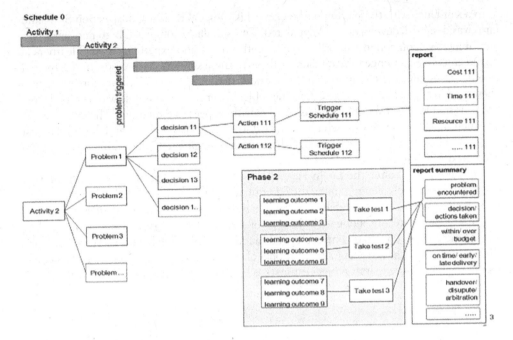

Figure 5.2 Scenario implementation concept

decisions made in the simulator are discussed in person. This debriefing session also enables wider debate to take place on the finer nuances underpinning the decisions made at the time concerning the problems and choices selected. This option was embedded into the simulator in order to distinguish between 'being immersed' within the environment and the process of critical reflection that takes place outside of the VR environment (De Freitas & Oliver, 2006).

Prior to running a VR session, various information and data has to be input into the system in order to populate the scenario. This data includes: (i) construction site type – location, site, scale, constraints, and layout; (ii) project type – primary use of building (e.g. commercial or residential), budget allocated, type of structure, special layout and planning, etc.; (iii) manufacturer type – scope, production capacity, variances, location, costs, maintenance, etc.; (iv) equipment hire – type, number, size, capacity, assembly rate, labour, hire rates, etc.; and (v) work plan – schedule, anticipated interruptions/problems, logistics, handling, etc. This information was sourced from several live (and recently completed) OSP projects, the data of which was mapped into a relational database.

In essence, the simulator was designed to provide learners with (i) simulation of site operations with real life, fast-tracked time scales; (ii) generation of reports based on all decisions made, including their direct influence on project costs and risks; (iii) an opportunity to save their learning session and the ability to reload this at a later date; (iv) randomised challenges – running unpredictable 'scenario directions/alterations' randomly based on pre-defined interruptions and problems (meaning that every session was unique); (v) feedback stages, where transitional waypoints could be cross-examined between different phases; and (vi) an opportunity to impart their feedback (on their performance) with experts through a critical reflection session.

In accordance with these criteria, the Game-Like Virtual Reality Construction Site Simulator was designed, developed, and populated. This simulator comprised both non-immersive and immersive components in order to give both novice and experienced AEC learners an opportunity to experience OSP challenges through these simulated scenarios. In this respect, in order to minimise interruption on the learners' reasoning process and sensory perception (Seyedzadeh et al., 2021), a scaled-down Graphical User Interface (GUI) was designed; ergo, it was as simple and straightforward as possible with respect to data input. The GUI was also designed to be accessible through any standard web browser. The logic and flow of data can be seen in Figure 5.3. The first stage of this interface [login] includes an option to capture account details (new/existing).

From Figure 5.3, after the Login Page, the Welcome Page presents users with information on the game (challenge), including additional criteria such as the selection of available construction sites (i.e. rural, suburban, and urban projects). This includes different available projects in each category, along with the type of scenario for the game (e.g. late design decisions), and a list of qualified contractors and manufacturers available for each project. After their initial selection, learners have additional choices to make on categories such as equipment selection before a draft project estimate is generated. Upon approval, the game is then

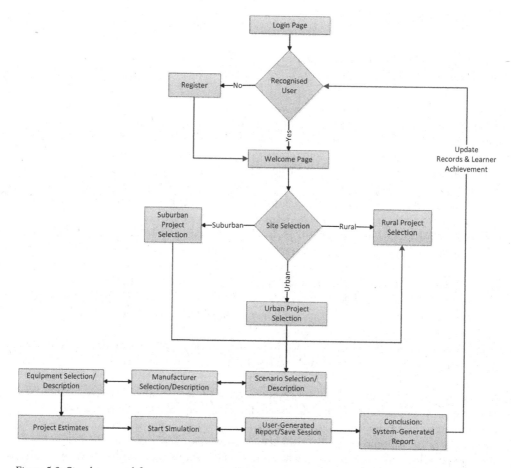

Figure 5.3 Simulator workflow

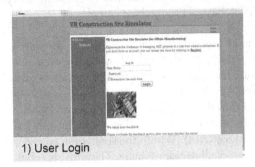

1) User Login

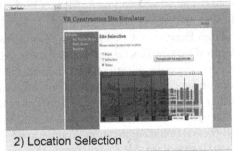

2) Location Selection

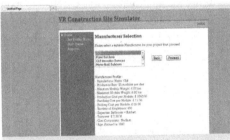

3) OSP System Selection

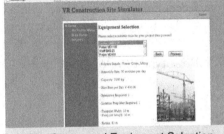

4) Scenario Selection

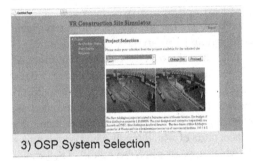

5) Manufacturer Selection

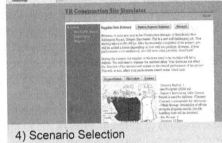

6) Site Set-up and Equipment Selection

7) Summary of Decisions and Project Estimates

8) Launch the VR Simulation Session

Figure 5.4 VR environment selection screens

commenced. Incidentally, all choices and decisions made are time-stamped (along with their registration details) and recorded in a MySQL database. This database can also be accessed through the project platform.

From Figure 5.3, following the project estimates session (and after completing the initial decision-making process through the interactive ASP.Net Web Forms), learners are then able to commence the 4D stimulated training session. This starts with a 'walkthrough' to experience and appreciate the complexity of the project. At this stage, the application provides users with a summary of the project and contract, and runs the simulation of the project within an immersive and interactive environment. This environment was developed in Quest3DTM VR programming Application Programming Interface (API). Within the simulation environment, learners are able to experience the outcomes of all decisions made. They are also challenged by unexpected events designed according to the selected scenario. Decisions therefore have to be made at these intervention points.

The simulation runs in a fully immersive 4D environment, where learners are able to navigate the interior and exterior spaces of the project site. At various points in the scenario, they are also able to interact with different elements of the simulator in order to retrieve further information (e.g. technical specifications, videos on selected OSP construction systems/details, project data, etc.); however, in order to keep learners on track, the simulator provides them with monitoring tools. These reveal additional information such as project time, latest assembled module, cumulative costs of the project, team communication, etc. Monitoring and communication tools are embedded in different parts of the main interface, parts of which are also embedded into a virtual Personal Digital Assistant (PDA) interface (which appears when prompted). The simulator constantly records and tracks all learner actions, storing this information in a database. Upon completion, learners are presented with a conclusion page, which reveals their performance. Figure 5.4 presents a number of different screenshots from this simulator.

5.4 Prototype development

5.4.1 Development concept

The first stage of the prototype development lifecycle was used to establish the project requirements and priorities (from Phase 1), represented in an ontological structure. The generic structure and content needed for the knowledge objects was then formulated with wrappers using metadata. The next stage established object classes and their hierarchy to satisfy multiple abstractions (in compliance with extranet metadata). The final stage developed the human/computer interaction (HCI) in compliance with Norman's 7 Principles, concerning protocols, accessibility, standardisation, etc. (to make the system as robust and reliable as possible). Intrinsically, the developed system included simulated scheduling of the project, association of the 3D models and building blocks for the project lifecycle, supply chain analysis monitoring for each building block or activity, management of delays in material delivery by sending emails to managers (internal and external), and generation or report schedules to appreciate project outcomes.

5.4.2 System architecture

Existing VR interfaces have ostensibly been formed based on one single idea: creating 3D models and incorporating them with some pieces of information so that both 3D models and information are editable through an interactive real-time interface (Pour Rahimian & Goulding,

2010). Contrarily, they tend to differ from each other based on their architecture and the utilised methods for data creation and retrieval; however, data creation and retrieval methods in VR interfaces can be investigated from two different perspectives, namely: (i) creating 3D bodies of constructional elements *per se* and (ii) defining characteristics of these elements.

Whilst creating 3D objects directly in VR environments is not impossible, these are usually created in CAD applications, as doing so in a VR environment is often more cumbersome and time consuming. Consequently, VR interfaces convert CAD models into VR elements. In terms of transforming design elements from CAD into VR, practitioners use three de facto approaches. Whyte et al. (2000) noted three approaches for this translation as being: (i) straightforward translation approach – importing the whole environment from CAD to VR; (ii) library-based approach – putting the elements of construction in the library of the VR environment, then 'calling' them as necessary; and (iii) database-oriented approach – using a central database to control the module characteristics. Here, the database utilises both CAD and VR environments as graphical interfaces, and can be characterised as a combination of computer graphics and web programming. Given that the simulator presented here for discussion used a database-oriented approach, then option 3 was selected, as this was the only way to facilitate learner engagement from multiple remote locations.

Anecdotally, VR interfaces in AEC vary considerably, predominantly based on the method of manipulating the objects within the environment and the adapted programming method used in the VR interface. Currently, three major programming applications are used by VR programmers: (i) 3D Application Programming Interfaces (APIs), (ii) Virtual Reality Modelling Language (VRML) and 3D web technologies, and (iii) recent commercialised object-oriented VR programming packages. In this respect, 3D APIs (e.g. Open GL and Direct 3D) are principal environments for VR programming in C++ and Visual Basic. Falling into the category of computer graphics, they are capable of either creating models directly inside the space and/or importing them from CAD applications. These are considered perfect programming environments for creating Win32 console applications (used for developing computer games). That being said, the integration of such interfaces with web programming is often quite difficult in complicated works.

The first iterations of VRML and 3D web technologies were made as a division of Open Inventor; thereafter, they have become the international standard for 3D web modelling. These applications provide a variety of facilities for manipulating immersive library-based web interfaces; however, they ostensibly lack the capability of integration, particularly with inter-related databases (as they are essentially not database-oriented applications). More recently, however, commercialised object-oriented VR programming packages now contain built-in modelling environments for creating VR spaces directly or importing them from CAD applications. Such VR programming applications also contain logical libraries that can be used to define behavioural links among the objects along with other phenomena. Although the architecture of such applications is predominantly based on APIs of C++, in some aspects, they can offer a higher-level abstraction for programmers. Several commercial VR programming applications are available on the market. For the purposes of this exercise, Quest3D™, EON Reality™, and Virtools™ were investigated. The outcomes of these applications could all be directly deployed into Visual C++ and Visual Basics web programming platforms (EON Reality Inc., 2008). This made them all extremely flexible in terms of integration with VR programming (which is a part of computer graphics), along with web programming and data mining. They are also supported by full Software Development Kits (SDKs) in order to help programmers create advanced building blocks and prototypes inherent to need (to create additional functionality and/or behaviour not originally provided by the host application).

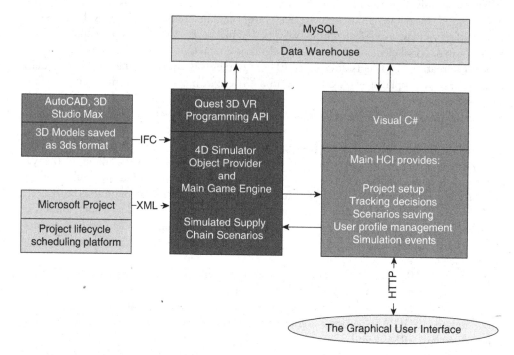

Figure 5.5 Simulator system architecture

In addition, SDKs allow programmers to integrate interfaces with particular VR input/output devices, e.g. Head-Mounted Displays (HMDs), data gloves, etc. Given this increased flexibility and functionality, the training simulator design team employed a database-driven approach as the main development platform. This used structured modelling phases and API-based programming that linked 3D objects to datasets through a web environment. This helped associate objects to schedules and activities (4D visualisation) in order to optimise and deliver real-time learning outcomes.

From a design modelling perspective, this study modelled all elements and components of the construction site simulator into either AutoCAD™ or 3D Studio Max™. The scenarios then were scheduled in the MS Project™ environment. All VR programming tasks were performed in the Quest3D™ environment, engaging Active Server Pages (ASP.Net™) web development tool using C# programming language to develop the user interface. The final part of this work used a MySQL™ database (which was compatible with both programming environments) that was installed on a server in order to track, manage, and transmit learner data. The adapted Unified Software Development Process enabled the simulator to be iteratively tested to diagnose and troubleshoot functionality, compliance, grouping, integration, maintenance, version control, and validation. This included modifying or amending the interface to include any additional fields/delimiters identified in Phase 1. The architecture of this simulator can be seen in Figure 5.5.

5.4.3 3D modelling of site elements in AutoCAD and 3D Studio Max environments

Figure 5.5 presents the system architecture for the training simulator prototype. From this, AutoCAD and 3D Studio Max were employed as the primary geometrical modelling platforms. Different construction site elements, including both permanent (e.g. structural/ architectural element and building blocks) and temporary (e.g. scaffolding and site barriers) construction objects, were created in detail; however, models for construction machinery and operational elements (e.g. tower cranes and trucks) were downloaded from open CAD forums and websites to fast-track development.

In order to optimise system performance, models were developed at a very primitive level (e.g. beams, columns, walls, building blocks). This helped facilitate loading and final assembly within Quest 3D as the 4D simulation tool. This approach enabled quicker regeneration and repetition of single models in geometrical arrays and placed lower demand on graphical memory. The consequence of this is that several bitmaps were used (to give the illusion of secondary details), rather than actually creating them as 3D models. The use of bitmaps also helped the simulation look more realistic, particularly the visualisation of textures of different materials (e.g. concrete, wood, etc.). All models were then subsequently converted into the 3DS file format (in order to be readable by Quest 3D) and saved in the same folders as their respective bitmaps.

5.4.4 *Project scheduling within the MS Project™ environment*

Microsoft Project was used to schedule project timelines and activities against the developed scenarios. For each scenario, temporal relationships amongst different assembly tasks were planned from the commencement of the project to the day of completion. This also factored in random occurrences of 'unexpected' problems and interruptions during the project lifecycle in order to provide additional learner challenge. These schedules formed the basis of data sequencing amongst different assembly activities, spanning different constructional tasks and their time constraints (start time, finish time, float, etc.). These schedules were converted to MySQL databases in order to be accessible for both C# and Quest3D programming tools.

Figure 5.6 identifies an array of the various building blocks used in this simulator. This includes the 3D position of the building blocks, sequence of assembly, delivery times, and machinery involved (including transportation and assembly). This data was stored in a relational database, along with costs, labour, and association with each task. This approach allowed learners to modify or update scheduling data through ASP.Net interfaces.

5.4.5 *Data warehouse in MySQL™ environment*

MySQL was selected as the main platform for hosting all databases used in this project. This supported Quest3D with an SDK, which was also accessible in the MS Visual Studio environment through Devart DotConnect™ and ADO.Net™. The relational database comprised 44 different tables (created in MySQL) to manage information regarding manufacturers, equipment, labour, and costs associated with different tasks, schedules for different scenarios, etc.

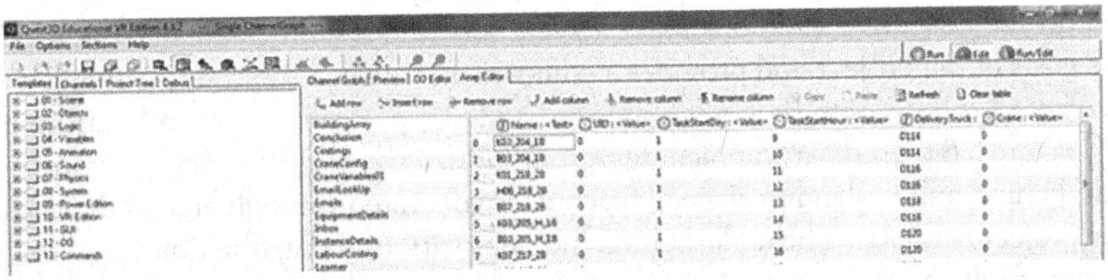

Figure 5.6 Sample project array within Quest3D imported from MySQL database

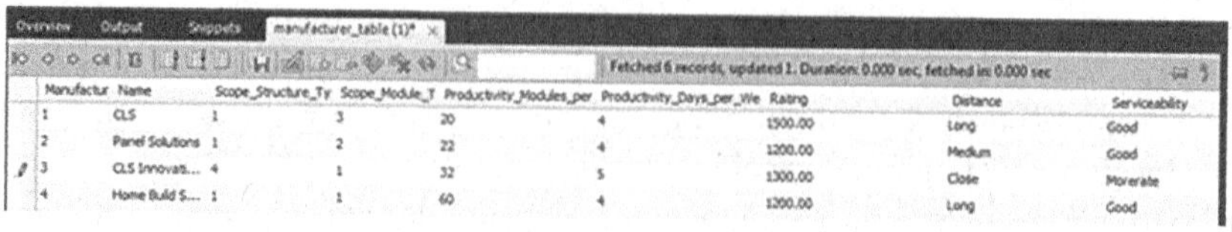

Manufactur	Name	Scope_Structure_Ty	Scope_Module_T	Productivity_Modules_per	Productivity_Days_per_We	Rating	Distance	Serviceability
1	CLS	1	3	20	4	1500.00	Long	Good
2	Panel Solutions	1	2	22	4	1200.00	Medium	Good
3	CLS Innovati...	4	1	32	5	1300.00	Close	Morerate
4	Home Build S...	1	1	60	4	1200.00	Long	Good

Figure 5.7 Manufacturers' information table within MySQL environment

This approach also provided learners with full control of project data, through web forms and directly in the simulated project environment. Figure 5.7 presents a sample of data and criteria associated with different manufacturers, including name of manufacturer, scope of their work, type of structure they produce, number of units they are able to deliver per day, fees charged per unit, distance from the selected project, and level of customer satisfaction of the particular manufacturer; however, additional fields could also be added if needed.

5.4.6 *Main human/computer interaction interface with visual C#*

One of the main challenges of developing an HCI interface is that of visual appeal and level of engagement. In this respect, the designed ASP.Net interface used C# to help learners select, justify, and defend project parameters within the simulated project site. For example, it was important to support learners in this process so that they could evaluate details of any potential planning or assembly processes directly through the user interface. In this respect, ASP.Net web forms were used as the initial GUIs to transfer messages from learners to the system in order to gather data regarding specifications of the desired construction process. These web forms enabled learners to control the type and sequence of construction tasks in order to make decisions on the type and level of machinery required. After these forms have been populated, the system provides learners with an estimate of project costs and time. The simulator is then fully populated and ready for activation. Simulation was generated through the ASP.Net environment as an embedded object, by calling an external object exported by Quest3D. For example, an embedded object could be called by executing the following Hypertext Markup Language (HTML) code:

```
<body>
  <form id="form1" runat="server">
    <embed height="720" width="100%" checkupdate="1"
    type="application/quest" src="Quest3D_v59a.q3d" documenturl="~/
    index.htm" id="Quest3DObject">
  </form>
</body>
```

Based on the collective decisions made by learners, the system calculates the total duration and costs of the project and provides a comparison between the results achieved against those expected. The system then generates a detailed report on the overall performance of learners. Figure 5.8 presents a graphical representation of the project estimate, along with the C# code-behind and design interface. The chart is generated through analysed data derived from the project database. The code-behind includes various string functions used to retrieve information from the project, including type, category, productivity, etc. This information is stored in the project database and is adjusted according to decisions made. These strings

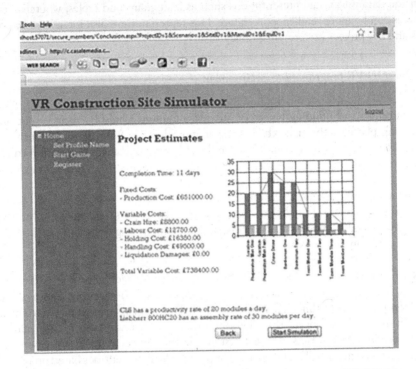

Figure 5.8 Design interface (top) and C# code-behind (bottom)

convert the data into visual presentations such as histograms and tables, whereby key facts and figures can be more readily appreciated. This representation also includes a breakdown of costs and key project performance indicators.

5.4.7 Quest3D VR programming environment

Geometrical 3D models of constructional elements were imported into Quest3D to provide the basic entities and building blocks of VR programming. Quest3D is an object-oriented programming platform through which programming logic is formed using interconnected logical building blocks. In the context of the development of this simulator, the structure of the programme architecture comprised four main components: (i) static 3D models of the construction site, including tower cranes, trucks, land, surroundings, and supporting elements (which do not change from one learner to another – e.g. scaffolding); (ii) building blocks as the dynamic 3D models of the project; (iii) project schedules for controlling all events in the assembly and delivery process; and (iv) monitoring tools for keeping control of project timelines and resources.

Figure 5.9 presents a representation of the static elements connected to the interface. These were directly generated using 3DS and bitmap files (which appear at particular locations); however, tower cranes and trucks were programmed to perform the desired animations at certain points of time. The modules of all static objects were directly connected to the project interface, except for the additional tower cranes and trucks (which may or may not have been hired by learners using the system). In this respect, modules were connected to the interface through an IF Toggle Channel in order to 'call' the entities based on preferences stored in the database. An additional IF Toggle Channel was also connected to the main interface in order to facilitate 'switching views' between various locations (e.g. the interior of kitchens, bathrooms, and hallways).

In terms of the assembly of dynamic objects, the system relies on the project schedule imported from the MySQL database (Figure 5.6) for the assembly sequence and project timeline. Based on the given sequence, system checks for assembly permission for each module are undertaken seamlessly in the background. For example, if the project time coincided with the time allocated to a particular module, then the system would run all related animations

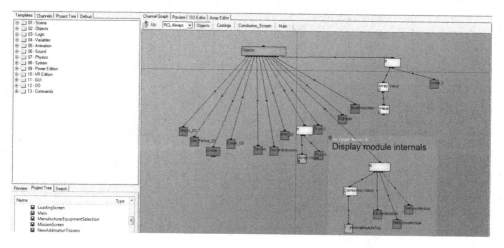

Figure 5.9 Static objects connected to the interface

connected to the delivery and assembly of that module. That being said, as previously mentioned, from time to time, random events are initiated (to provide greater depth of engagement on learner demand). These random events temporarily interrupt learner progression, requiring direct intervention before progress can recommence. These interruptions (triggers) could range from minor learner engagement to a more detailed intervention required though email to a nominated party. In the case of the latter, a trigger channel reinitiates the performance of system subject to delivery of emails to the database. In this respect, delays or inappropriate decisions would (more often than not) result in an increased project cost, along with an extended project completion time.

5.5 Conclusion

This chapter highlighted the need to revisit some of the major challenges pervading AEC, particularly in respect of supporting OSP initiatives. Given this challenge, a cloud-based VR Construction Site Simulator was proffered as a potential solution, providing learners with direct experience of OSP within a simulated environment. The developed simulator presents learners with an opportunity to experience OSP challenges in a risk-free environment. Moreover, to learn from the decisions they make and how these affect project success (or failure). These decisions include many aspects of project delivery, from design to process, logistics, plant/equipment, supply chain issues, etc.

This chapter explained the development process of this simulator, using an adapted Unified Software Development Process, which employed the iterative phases of elaboration, construction, and transition. The VR environment and coding approach highlighted the capabilities of this simulator, including information storage, use of string functions, decision criteria, and reporting functionality.

Findings from this work contribute to the AEC body of knowledge in a number of areas, perhaps most importantly in the identification and application of decision-making criteria in VR immersive environments and how learners respond to specific challenges (planned or otherwise). These findings also provide new insight and understanding into collaborative environments and the application of game interfaces. Whilst it is acknowledged that 'measuring' or assessing learner performance is not new *per se*, this simulator offers wider reflectivity by capturing direct learner involvement. This approach is bespoke insofar as this is personalised to each learner: the decisions they take, the actions they make, the time they take to do this, the emails they send, the rationale they present to defend decisions, etc. This insight and understanding has a direct impact on organisational behaviour and the social constructs that affect decision-making.

However, it is also important to acknowledge that simulators of this type are after all just simulators. Given this, further work is needed to strengthen the veracity of findings presented so far – in particular, how roles/levels affect critical decision-making, how different professions respond to challenges (and why), the impact of pedagogy and learning traits (Goulding & Syed-Khuzzan, 2014), or how the direct application of the Charrette Test Methodology (Clayton et al., 1998) can be used to reinforce rigour in the testing and validating process. All of these issues provide fertile points of discussion and avenues to explore in future works.

References

Abrishami, S., Goulding, J. S., Ganah, A., & Rahimian, F. P. (2013). *Exploiting modern opportunities in AEC industry: A paradigm of future opportunities.* Paper presented at the AEI 2013: Building Solutions for Architectural Engineering – Proceedings of the 2013 Architectural Engineering National Conference. https://doi.org/10.1061/9780784412909.031

ACS. (2009). *3D learning and virtual worlds*. ACS.

Alshawi, M., Goulding, J. S., & Nadim, W. (2007). Training and education for open building manufacturing: Closing the skills gap. In A. S. Kazi, M. Hannus, S. Boudjabeur, & A. Malon (Eds.), *Open building manufacturing: Core concepts and industrial requirements*. ManuBuild in collaboration with VTT – Technical Research Centre of Finland.

Aranda-Mena, G., & Wakefield, R. (2006). *Interoperability of building information – myth or reality?* Paper presented at the the 6th European Conference on Product and Process Modeling.

Arif, M., Kulonda, D., Jones, J., & Proctor, M. (2005). Enterprise information systems: Technology first or process first? *Business Process Management Journal, 11*(1), 5–21. https://doi.org/10.1108/14637150510578692

Australian Institute of Architects. (2009). *National building information modelling (BIM) guidelines and case studies*. Australian Institute of Architects.

Autodesk®. (2011). *Realizing the benefits of BIM, Autodesk® building information modeling*. http://images.autodesk.com/adsk/files/2011_realizing_bim_final.pdf

BIM Task Group. (2013). *UK BIM task group website*. www.bimtaskgroup.org

Blosiu, J. O. 1999. Use of synectics as an idea seeding technique to enhance design creativity. Systems, Man, and Cybernetics. IEEE SMC'99 Conference Proceedings. 1999 IEEE International Conference. https://doi.org/10.1109/ICSMC.1999.823365

Cera, C. D., Reagali, W. C., Braude, I., Shapirstein, Y., & C.Foster. (2002). A collaborative 3D environment for authoring design semantics. *Graphics in Advanced Computer-Aided Design, 22*(3), 43–55.

Clayton, M., Kunz, J., & Fischer, M. (1998). *The Charrette test method* (p. 120). Stanford University Press.

Colombo, A. W., & Harrison, R. (2008). Modular and collaborative automation: Achieving manufacturing flexibility and reconfigurability. *International Journal of Manufacturing Technology and Management, 14*(3/4), 249–265.

De Freitas, S., & Oliver, M. (2006). How can exploratory learning with games and simulations within the curriculum be most effectively evaluated? *Computers & Education, 46*(3), 249–264.

Eastman, C., Fisher, D., Lafue, G., Lividini, J., Stoker, D., & Yessios, C. (1974). *An outline of the building description system*. Institute of Physical Planning Research Report.

Egan, J. (1998). *The Egan report – rethinking construction, report of the construction industry taskforce to the deputy prime minister*. HSMO.

Elghaish, F., Matarneh, S., Talebi, S., Kagioglou, M., Hosseini, M. R., & Abrishami, S. (2020). Toward digitalization in the construction industry with immersive and drones technologies: A critical literature review. *Smart and Sustainable Built Environment, ahead-of-print*(ahead-of-print). https://doi.org/10.1108/SASBE-06-2020-0077

EON Reality Inc. (2008). *Introduction to working in EON studio*. EON Reality Inc.

Fischer, M. (2000). *4D CAD-3D models incorporated with time schedule*. CIFE Centre for Integrated Facility Engineering in Finland, VTT-TEKES, CIFE Technical Report.

Fischer, M., & Kunz, J. (2004, February). *The scope and role of information technology in construction*. CIFE Technical Report #156. Center Integra.

Forcade, N., Casals, M., Roca, X., & Gangolells, M. (2007). Adoption of web databases for document management in SMEs of the construction sector in Spain. *Automation in Construction, 16*, 411–424.

Frohm, J., Stahre, J., & Winroth, M. (2008). Levels of automation in manufacturing. *Ergonomia – an International journal of ergonomics and human factors, 30*(3).

Fruchter, R. (1998). Internet-based web mediated collaborative design and learning environment, in artificial intelligence in structural engineering. In *Lecture notes in artificial intelligence* (pp. 133–145). Springer-Verlag.

Goulding, J. S., & Rahimian, F. P. (2012). Industry preparedness: Advanced learning paradigms for exploitation. In A. Akintoye, J. S. Goulding, & G. Zawdie (Eds.), *Construction innovation and process improvement* (pp. 409–433). Wiley-Blackwell. https://doi.org/10.1002/9781118280294.ch18

Goulding, J. S., & Syed-Khuzzan, S. (2014). A study on the validity of a four-variant diagnostic learning styles questionnaire framework. *Journal of Education + Training, 56*(2–3), 141–164. https://doi.org/10.1108/ET-11-2012-0109

Gu, N., & London, K. (2010). Understanding and facilitating BIM adoption in the AEC industry. *Automation in Construction, 19*(8), 988–999. www.sciencedirect.com/science/article/B6V20-5186RB4-1/2/bddc75d5549b2fa69008ff545875682a

Hou, L., Wang, X., Bernold, L., & Love, P. E. (2013). Using animated augmented reality to cognitively guide assembly. *Journal of Computing in Civil Engineering, 27*(5), 439–451.

Ibrahim, M., Krawczyk, R., & Schipporiet, G. (2004). *Two approaches to BIM: A comparative study.* Paper presented at the Proceedings of eCAADe.

Ibrahim, R., & Pour Rahimian, F. (2010). Comparison of CAD and manual sketching tools for teaching architectural design. *Automation in Construction, 19*(8), 978–987. www.scopus.com/inward/record.url?eid=2-s2.0-78149282987&partnerID=40&md5=5679ee25ec3b7eb4ea5764e569505b05

Isikdag, U., & Underwood, J. (2010). *A synopsis of the handbook of research on building information modeling.* Paper presented at the Proceedings of CIB 2010 World Building Congress. www.cib2010.com/

Kim, Y. S., & Kang, B. G. (2003). *Personal characteristics and design-related performances in a creative engineering design course.* Paper presented at the 6th Asian Design Conference.

Kuhl, F., Weatherly, R., & Dahmann, J. (2000). *Creating computer simulation systems: An introduction to the high level architecture.* Prentice-Hall.

Latham, M. (1994). *Constructing the team, joint review of the procurement and contractual arrangements in the UK construction industry, final report.* HSMO.

Lee, J., Gu, N., Ostwald, M., & Jupp, J. (2013). Understanding cognitive activities in parametric design. In J. Zhang & C. Sun (Eds.), *Global design and local materialization* (Vol. 369, pp. 38–49). Springer. https://doi.org/10.1007/978-3-642-38974-0_4

Mole, T. (2003). *Mind your manners.* Paper presented at the Proceedings of CIB W89 International Conference on Building and Research Bear.

Nawari, N. (2012). BIM standard in off-site construction. *Journal of Architectural Engineering, 18*(2), 107–113. https://doi.org/10.1061/(ASCE)AE.1943-5568.0000056

Newman, C., Edwards, D., Martek, I., Lai, J., Thwala, W. D., & Rillie, I. (2020). Industry 4.0 deployment in the construction industry: A bibliometric literature review and UK-based case study. *Smart and Sustainable Built Environment, ahead-of-print*(ahead-of-print). https://doi.org/10.1108/SASBE-02-2020-0016

Oliver, S., Seyedzadeh, S., Rahimian, F., Dawood, N., & Rodriguez, S. (2020). Cost-effective as-built BIM modelling using 3D point-clouds and photogrammetry. *Current Trends in Civil & Structural Engineering-CTCSE, 4*(5), 000599. https://doi.org/10.33552/CTCSE.2020.04.000599

Parchami Jalal, M., Yavari Roushan, T., Noorzai, E., & Alizadeh, M. (2020). A BIM-based construction claim management model for early identification and visualization of claims. *Smart and Sustainable Built Environment, ahead-of-print*(ahead-of-print). https://doi.org/10.1108/SASBE-10-2019-0141

Paulson, B. C. (1976). Designing to reduce construction costs *Journal of the construction division, 102*(4), 587–592.

Petric, J., Maver, T., Conti, G., & Ucelli, G. (2002). *Virtual reality in the service of user participation in architecture.* Paper presented at the CIB W78 Conference, Aarhus School of Architecture.

Popov, V., Mikalauskas, S., Migilinskas, D., & Vainiunas, P. (2006). Complex usage of 4D information modelling concept for building design, estimation, scheduling and determination of effective variant. *Technological and Economic Development of Economy, 12*(2), 91–98.

Pour Rahimian, F., & Goulding, J. S. (2010, November 4–5). *Game-like virtual reality interfaces in construction management simulation.* A New Paradigm of Opportunities 9th International Detail Design in Architecture Conference, University of Central Lancashire, pp. 247–257.

Pour Rahimian, F., Ibrahim, R., & Baharudin, M. N. (2008, August 26–28). *Using IT/ICT as a new medium toward implementation of interactive architectural communication cultures.* International Symposium on Information Technology. https://doi.org/10.1109/ITSIM.2008.4631984

Pour Rahimian, F., Ibrahim, R., Wirza, R., Abdullah, M. T. B., & Jaafar, M. S. B. H. (2011). Mediating cognitive transformation with VR 3D sketching during conceptual architectural design process. *Archnet-IJAR, International Journal of Architectural Research, 5*(1), 99–113.

Pour Rahimian, F., Seyedzadeh, S., Oliver, S., Rodriguez, S., & Dawood, N. (2020). On-demand monitoring of construction projects through a game-like hybrid application of BIM and machine learning. *Automation in Construction, 110*, 103012. https://doi.org/10.1016/j.autcon.2019.103012

Rahimian, F. P., & Ibrahim, R. (2011). Impacts of VR 3D sketching on novice designers' spatial cognition in collaborative conceptual architectural design. *Design Studies, 32*(3), 255–291. www.scopus.com/inward/record.url?eid=2-s2.0-79953058047&partnerID=40&md5=268a4ca902c749a349e-9a705270cedf8

Santos, E. T. (2009). *Building information modeling and interoperability.* Paper presented at the XIII Congress of the Iberoamerican Society of Digital Graphics – From Modern to Digital: The Challenges of a Transition.

Seyedzadeh, S., Agapiou, A., Moghaddasi, M., Dado, M., & Glesk, I. (2021). WON-OCDMA system based on MW-ZCC codes for applications in optical wireless sensor networks. *Sensors, 21.* https://doi.org/10.3390/s21020539

Seyedzadeh, S., & Pour Rahimian, F. (2021). Building energy performance assessment methods. In *Data-driven modelling of non-domestic buildings energy performance* (pp. 13–30). Springer. https://doi.org/10.1007/978-3-030-64751-3_2

Skibniewski, M. J. (1992, June 3–5). *Current Status of construction automation and robotics in the United States of America.* The 9th International Symposium on Automation and Robotics in Construction, pp. 17–26.

Su, X., Dunston, P. S., Proctor, R. W., & Wang, X. (2013). Influence of training schedule on development of perceptual – motor control skills for construction equipment operators in a virtual training system. *Automation in Construction, 35*(0), 439–447. https://doi.org/10.1016/j.autcon.2013.05.029

Succar, B. (2009). Building information modelling framework: A research and delivery foundation for industry stakeholders. *Automation in Construction, 18*(3), 357–375.

Succar, B. (2011). *Organizational BIM; how to assess and improve your organization's BIM performance.* Paper presented at the Revit Technology Conference Australasia, V1.1, session 3 part B.

Thai, A. M., Lowenstein, D., Ching, D., & Rejeski, D. (2009). *Game changer: Investing in digital play to advance children's learning and health.* John Ganz Cooney Center.

Uygun, Ö., Öztemel, E., & Kubat, C. (2009). Scenario based distributed manufacturing simulation using HLA technologies. *Information Sciences, 179*(10), 1533–1541. https://doi.org/10.1016/j.ins.2008.10.019

Veenstra, V., Halman, J., & Voordijk, J. (2006). A methodology for developing product platforms in the specific setting of the housebuilding industry. *Research in Engineering Design, 17*(3), 157–173.

Wang, X., & Dunston, P. S. (2013). Tangible mixed reality for remote design review: A study understanding user perception and acceptance. *Visualization in Engineering, 1*(1), 1–15.

Wang, X., Love, P. E., Kim, M. J., & Wang, W. (2014a). Mutual awareness in collaborative design: An augmented reality integrated telepresence system. *Computers in Industry, 65*(2), 314–324.

Wang, X., Truijens, M., Hou, L., Wang, Y., & Zhou, Y. (2014b). Integrating augmented reality with building information modeling: Onsite construction process controlling for liquefied natural gas industry. *Automation in Construction, 40,* 96–105.

Wang, Y., Wang, X., Wang, J., Yung, P., & Jun, G. (2013). *Engagement of facilities management in design stage through BIM: Framework and a case study.* Advances in Civil Engineering.

Wellings, J., & Levine, M. H. (2010). *The digital promise: Transforming Learning with innovative uses of technology.* John Ganz Cooney Center.

Whyte, J., Bouchlaghem, N., Thorpe, A., & McCaffer, R. (2000). From CAD to virtual reality: Modelling approaches, data exchange and interactive 3D building design tools. *Automation in Construction, 10*(1), 43.55.

Wikberg, F., Ekholm, A., & Jensen, P. (2010). *Configuration with architectural objects in industrialised house-building.* Lund University.

Woo, J.-H., Clayton, M. J., Johnson, R. E., Flores, B. E., & Ellis, C. (2004). Dynamic knowledge map: Reusing experts' tacit knowledge in the AEC industry. *Automation in Construction, 13*(2), 203–207. www.sciencedirect.com/science/article/B6V20-4B4S5GT-2/2/d6b47a38aff72ec00846e6d9616f9f2b

Zhang, X., Wang, H., Ma, H., & Wang, H. (2012). The research of digital proving ground simulation system based on HLA. *Procedia Engineering, 29,* 3624–3630. https://doi.org/10.1016/j.proeng.2012.01.542

6 XR-openBIM integration for supporting whole-life management of offsite manufactured houses

6.1 Introduction

AEC has increasingly encountered rising complexity in terms of the design and construction of projects, the corollary of which is exacerbated by the multitude of parties involved in project ventures. Clients, owners, investors, end users, and facility managers have to work efficiently and productively along with architects, engineers, and contractors to successfully complete a project. The smooth operation of these processes necessitates iterative information exchange, as well as effective means of communication, particularly when trying to convey concepts and principles across different disciplines – from financial management and spatial planning to Operations and Maintenance (O&M). The aim here is to ensure that the common body of diverse information is comprehensible to everyone at every level of the organisational structure of a project.

The implementation BIM has quickly gathered momentum through its promise of structure, organisation, and efficiency of work processes and associated tools (Newman et al., 2020; Pour Rahimian et al., 2020). The benefits of BIM for helping the integration of architectural, construction, or manufacturing projects have been widely highlighted in seminal literature (Abanda et al., 2017; Arayici et al., 2018; Cha & Lee, 2015; Damen et al., 2015; Sebastian, 2010); however, given these advantages, there are also a number of inherent flaws in the structure of this semantically rich modelling technology. For example, the accumulation of diverse data from different disciplines through consecutive phases can often increase file sizes and structure complexity, thus impeding the smooth flow of information exchange and archiving interoperability (Arayici et al., 2018; Seyedzadeh & Pour Rahimian, 2021a). Moreover, increased sophistication of recent BIM models can sometimes exceed the comprehension of 'typical' stakeholders, clients, end users, owners, and other non-engineering professionals – the corollary of which prevents them from being effectively involved in various design development and project implementation review processes (Walasek & Barszcz, 2017), due to the so-called "black-box effect", which refers to a system without transparency and legibility for everyone.

Advanced visualisation could potentially address these problems, by further enhancing comprehension of non-technical people from naturally technical BIM models. Yet, despite improvements in orthogonal drawings, current BIM technologies are primarily designed as management tools or repositories of interrelated descriptive and 3D information, which limits their visualisation capacities. VR and AR, however, are widely proven as enabling technologies to address these issues, by helping clients experience design in a 3D space, to assess such factors as lighting, aesthetics, etc. to plan for future maintenance and make decisions for future needs (Chi et al., 2013; Dong et al., 2013; Koch et al., 2014; Mansuri & Patel, 2021;

DOI: 10.1201/9781003106944-6

Moshtaghian et al., 2020; Petrova et al., 2017; Rasmussen et al., 2017). Conversely, currently available immersive experience solutions on the market offer only partial opportunities for building design integration. They also require advanced technological skills from the user. As a result, users with no prior advanced training and the necessary hardware are limited in what they can view and achieve with these packages. Additionally, existing software solutions offer limited integration of BIM, which provides minimal data on construction materials, services, or costs – an important part of interaction.

As a response to this functionality gap between BIM and immersive technologies, this study presents an interface that integrates the two, thereby streamlining the design processes to provide a comprehensive pared-down BIM system. The aspiration of this interface was to be fully agnostic towards the diverse BIM editing tools, which can become a source of input in order to offer synchronised concurrent user accessibility with low latency to promote active collaboration. To illustrate the operational principles of this system, this project was based around offsite manufactured self-build housing, where the manufacturing company presents house designs in a virtual environment directly from their BIM models, allowing their clients to walk through and customise a range of home features remotely.

6.2 Related studies

This study is concerned with four groups of literature to form a theoretical and methodological basis for its development. These four area are: (i) OpenBIM standards in order to provide the project with a reliable open-source platform to facilitate unlimited coding for the integration of BIM models with other environments, i.e. VR and AR; (ii) parametric and generative design interfaces as opposed to the traditional geometric Computer-Aided Design (CAD) tools and their impact on how designers think and collaborate with each other; (iii) the emerging VR and AR technologies and their role in facilitating intuitive and immersive interfaces for users to interact with design solutions without the need for a higher level of skills or discipline knowledge for understanding design solutions, collaborating with professional team members, and contributing to the decision-making processes throughout the project life cycle; and (iv) the impact of these high-tech technologies on the people involved in the teams and the ways through which these technologies change behavioural interactions with each other. The following subsections elaborate on these four areas in more detail, highlighting how seminal literature has approached these issues.

6.2.1. OpenBIM standards and the IFC format

The exponentially rising interest in BIM technologies in AEC has led to an acute necessity for a common vehicle of exchange between disciplines to promote and deliver effective management of vast quantities of data (Dawood et al., 2020). Research has already noted a lack of interoperability as one of the main obstacles to the wider adoption of BIM in AEC (Walasek & Barszcz, 2017), where the role of BIM can be seen as an enabler of efficient collaboration and communication, but is frequently hindered by the diversity and heterogeneity of the output information. The implementation of a universal platform for referencing processes and information throughout the lifecycle of an asset could potentially eradicate the issues related to software incompatibility. As an attempt to tackle this issue, the buildingSMART alliance developed an openBIM approach based around open standards and formats (Pauwels et al., 2017). While the emphasis in buildingSMART is on data and technology, less so on people and processes (Arayici et al., 2018), the openBIM concept initiates a collaborative approach among a wide range of disciplines, where different users can use specialised software and store diverse data in the same unified model. The aim of this movement was to establish a common platform for sharing information in a common language, enabling participation and engagement in the project regardless of the tools employed by different practitioners (Smart BIM Solutions, 2012).

The Industry Foundation Classes (IFC) format is the primary vendor-neutral data model developed by buildingSMART (formerly the International Alliance for Interoperability), where the majority of the BIM authoring software supports the IFC import/export feature. The IFC schema is an extensible object-oriented data model divided into base entities and sub-entities. This format classifies and gives structure to the data, thereby allowing different sets of information to be extracted easily. The inherent data consistency of the format aims to eliminate the need to input the same information multiple times, thus minimising errors and allowing the project team to focus on the compatibility of their workflows.

This common IFC structure has the potential to serve as a vehicle to achieve agnosticism towards BIM editing tools, where team members and companies do not need to be selected based on the software they use but on their actual knowledge and capabilities. This universal approach to collaboration in the process of building information management can promote further improvement of the online product supply information within the market, so that accurate product data (in terms of both characteristics and 3D objects) can be directly imported into BIM projects. The accuracy of product catalogues modelled in 3D can further release the potential for parametrisation and subsequent reuse of models and information, to not only improve the efficient use of materials, but also reduce the need for advanced technical skills from end users. The different needs of professionals involved in a project throughout its lifecycle tends to necessitate multiple types of geometry, properties, and relations; so, the IFC construct preserves the data structures inherited from its native format – geometry, logical relations, and attributes of the 3D elements; however, in order to accommodate all of these types, the IFC format is highly redundant. Consequently, coordinating and communicating building data becomes increasingly complicated, despite the compatibility of formats. The resulting high complexity therefore leads to lower usage among professionals, where the limited understanding of import and export features typically causes translational errors, thereby impeding the intent for interoperability (Afsari et al., 2017; Pauwels et al., 2017). As a response to this inherent dysfunction of the IFC exchange, many studies (e.g. Choi & Kim, 2016, 2017; Lee et al., 2018) proposed various code-checking logics to ensure open BIM interoperability and data consistency in order to minimise BIM data loss when transferring to IFC data structure from one library to another. They emphasised the necessity of a user-friendly interface to tackle the problem of complexity through logical rule compositions that automatically check the accuracy and integrity of the exchanged BIM data during translation. The semantics of the heterogeneous building information is verified against a set of rules that need to be satisfied in order to determine if a model is correct. In this respect, Arayici et al. (2018) highlighted the necessity to improve open standards for an integrated BIM practice and adopted a common format to describe information, as well as its syntactic, semantic, and topological relationships, in a database readable for all tools in the industry. In addition, Shi et al. (2018) devised an automatic method for comparing IFC files based on their content and analysis of the hierarchical structure, rather than visual checks and manual counts, in order to keep track of changes during the lifecycle of construction projects. As a way of ensuring the quality of BIM data, Choi and Kim (2016, 2017) developed an open BIM-based building code checking process to assess BIM data in IFCs. The purpose of these systems was to establish open BIM interoperability in order to minimise BIM data loss when transferring to IFC data structure through predefined specifications.

6.2.2. Parametric and generative design

The increased complexity of projects in the AEC industry now requires advanced skills to interact with design tools as a result of low engagement levels with architects, engineers, end users, clients, owners, and other professionals. This lack of engagement has had an impact on the number of design flaws and construction errors missed during the process, thereby increasing the end costs of these projects. BIM was introduced to improve the efficiency of the AEC industry by providing a means to access and evaluate data over a comprehensive body of information, thereby laying the groundwork for better organisation, development, and procurement of building projects. The main advantage of BIM is that the information directly relates to 3D elements, making it easier to comprehend and visualise. As a result, the object-based, data-rich environment of BIM opens the possibility for quick and interactive methods of design generation and development (Cao et al., 2015; Park, 2011; Yuan et al., 2018).

From a philosophical design perspective, Oxman (2017) discussed a new paradigm in design thinking, where technological advances (and their dissemination within the industry) could be brought together through cognitive and computational efforts in the design process. The algorithmic reasoning based around rules and instructions was a way of thinking that created this form. Writing code was a means of building design expression; however, this required not only spatial comprehension but also an understanding of the formal expression of design actions and visual outcome in code. The design process is no longer made exclusively through single decisions on individual objects or processes, but rather through a matrix that simultaneously covers a larger aspect through a network of systematic relationships.

The potential of repositories of digital information as drivers for design actions was also key in the research of Ercan and Elias-Ozkan (2015), which explored performance-based design through databases, evolutionary algorithms, and parametric tools. Their proposed workflow integrated simulation tools and embedded performance measures to support the design process, where intuition, experience, or theoretical knowledge in isolation could not effectively support decision-making.

Parametric modelling techniques, based on constraints and variables, define and diversify the options for design through different parameters (e.g. materials and energy performance) (Seyedzadeh & Pour Rahimian, 2021b; Seyedzadeh et al., 2020). Predefining the logical relations between geometric entities allows them to move or change as per the values of the established parameters or iteration. This ensures further flexibility in the design process to enable practitioners to choose the most suitable solution from the generated options. This method is exemplified through the development of a solar shading device whose design iterations were automatically generated through customisation/parametrisation, checked against a number of different design criteria to determine the optimal solution in terms of performance.

In addition to optimising the performance of building elements, parametric design presents a significant potential for improving the efficiency of materials, thereby reducing waste (Craveiro et al., 2017; Monizza et al., 2017). Craveiro et al. (2017) aimed to enhance efficiency and effectiveness using a computational method for the design and fabrication of functionally, grading building components to improve both material use and structural efficiency. Monizza et al. (2017) also strived to improve mass customisation through parametric and generative design techniques and developed an algorithm to reduce waste of raw materials and facilitate the manufacturing process for glue-laminated timber (GLT). Issues related to resource constraints (Liu et al., 2015), supporting prefabrication as a means of optimising production, transportation, and installation (Khalili & Chua, 2013; Liu et al., 2018), and modular coordination (Banihashemi et al., 2018) was also addressed through parametric rules and 3D models in the research (Khalili & Chua, 2013). Their proposed workflow used the IFC format as a data structure that retained the relationships of geometry and topology, which then used analysis and checks against constructability rules to facilitate the design of precast elements. The proposed method strived to achieve a higher level of prefabrication as it was not based on the generation of individual elements design, but rather on solutions for groups of elements.

On a similar theme, Banihashemi et al. (2018) presented a workflow that integrated parametric design into modular coordination using a generative algorithm that generated modules compliant with predetermined design constraints. The parameters included in the algorithm automatically calculated certain aspects of the design based on user input, in this case, construction waste. They highlighted the necessity to create structured well-developed building models, which effectively supported planning and manufacturing through parametric design, as a way of defining and managing the relations between BIM objects. Similarly, Yuan et al. (2018) noted that prefabricated building designs were not fully integrated with BIM products and libraries, and thus hindered the potential of parametric design. In response, their study focused on establishing an information model better suited for manufacturing. They acknowledged the limitations of BIM as a tool, which was primarily developed for non-prefabricated buildings, and introduced the concept of parametric design oriented to support Design for Manufacture and Assembly (DfMA), indicating the capacity of the IFC format as an exchange vehicle. This method leveraged the key capabilities of BIM and provided a meaningful structure for establishing products into separate components, supported with discrete feedback mechanisms using dynamic 3D BIM.

6.2.3. Mobile interfaces and immersive technologies

The issues of increased levels of technical complexity and a need to engage customers and stakeholders to visualise/grasp spatial problems solutions with limited or no prior technical training is a known challenge (Liu et al., 2018). Despite BIM's significant potential for data accumulation and organisation, it only offers limited immersive capabilities within building designs. Therefore, additive technologies such as AR or VR are required, which as a by-product are also able to minimise the potential of data loss with BIM. The increased tendency towards the integration of these technologies has also aimed to increase labour productivity, given the more efficient use of information (Li et al., 2017). Real-time interaction through AR or VR also allows users to transcend from their physical location into the environment of the entire project, which then opens a possibility for more effective interaction and simulation analysis of projects (Wang et al., 2013).

Effective client engagement is particularly important for housing projects, particularly with non-professional end users (who may have limited spatial comprehension), who may not be able to fully understand conventional architectural drawings (which is crucial for end user satisfaction). In this respect, Damen et al. (2014) recognised the need for participatory design to support decision-making, including the potential of BIM's parametric capabilities as a vehicle for integrated design development of collective self-organised housing. Their approach to client inclusion was based on a seamless translation of parametric design in the physical world through AR/VR interfaces. Wang et al. (2014) also investigated the possibility for a conjunction between BIM and AR, to provide an onsite tool that integrated BIM data in a physical context, thus helping users construct mental models of the building design. Additionally, Ren et al. (2016) facilitated architects and engineers during construction through a developed mobile AR system that integrated lightweight BIM data and 3D registration.

In summary, while AR is an effective way of overlaying digital information over a real physical context whilst also being able to identify clashes with existing structures onsite, VR also holds great promise for end users' engagement because it is a powerful tool for visual communication of building designs. This can also be seen as a powerful and dynamic feedback initiator. Feedback is seen as a key component, where Rasmussen et al. (2017) developed a case study for furnishing design using BIM as a visualisation tool that integrated with VR to assist communication and decision-making. This was seen as a powerful method of collaboration between designers and users; however, despite various attempts of bringing together BIM and immersive technologies, the compatibility of formats and workflows remain an obstacle to progress. One example of this was exemplified by Du et al. (2018), who recognised issues of interoperability between BIM data and VR tools, including data transfer difficulties. This work focused on the need to synchronise information to ensure optimal decision-making using a cloud-based synchronisation system mapped to BIM metadata from Autodesk Revit (Oliver et al., 2020).

6.2.4. *Media richness and social presence*

Effective communication management is crucial for the success of construction projects, a critical skill for design professionals (Norouzi et al., 2015), and a cornerstone function of project management (Sarhadi, 2016; Wu et al., 2017; Zulch, 2014). In this respect, ineffective communication has often been identified as a source of conflict in construction projects (Jaffar et al., 2011; Mitkus & Mitkus, 2014), including communicative ineffectiveness in the form of poor documentation and lack of trackability (Lester, 2017). Given that cross-disciplinary qualitative assessments are typically seen as the main causes behind construction project delays Odeh and Battaineh (2002), where Assaf and Al-Hejji (2006) found that although perceptions of primary sources of delays vary by practitioners' roles in the construction industry, communication failures play a critical role in project overruns.

Discussing social presence in the context of virtual environments, Biocca (1997) suggested that the minimum level of social presence occurs when the users feel that a form, behaviour, or sensory experience indicates the presence of another intelligence, which broadly translates to the amount of social presence or degree to which the user feels access to the intelligence, intentions, and sensory impression of another (Rahimian et al., 2019; Seyedzadeh et al., 2021). Biocca (1997) highlighted the usefulness utility of this description, citing Husserl (1973) in relation to a human's capacity for empathic perception of another through nonverbal cues, indicating that the measure should be the level in which communication between one person and another feels like face-to-face. For high-level discussion on presence, Biocca (1997) noted two additional forms of presence: (i) telepresence – the experience of being present there in a location in the now, and (ii) co-presence – the feeling of being in the company of others at a fundamental level, not necessarily encompassing all aspects of social presence, citing Schroeder (2002).

BIM has been seen as a major step forward in how communication is practiced in AEC, thereby improving efficiency, openness, accountability, and the richness of the information transferred between parties; however, it has not made significant changes to the immediacy and latency aspects of communication. Given this, its richness cannot be fully utilised at level 2 of BIM (N.b. Levels relate to the UK Government's Construction Strategy 2016–2020 published in 2016). Moreover, BIM level 3 requires the greater application of Computer-Mediated Communications (CMC) technologies; however, it is posited here that for innovation to flourish, it should be based on an open standard rather than merely accommodating post-process generation of open standard files. Given this position, the project presented in this chapter does not actively solve all of these problems *per se*, but rather serves as a foundation for developers to extend their preferred interfaces (Revit or MS Word, etc.) for wider integration with other components.

6.3 Project framework

The first phase of the project development presented in this chapter required a thorough understanding of the complexities involved, with not only the technology *per se* but also the processes and actors involved. This required the engagement of an extensive literature review to diagnose and prioritise AEC grand challenges and underlying issues. This served to determine potential routes on how different technologies can be utilised to achieve smoother collaboration and higher levels of interoperability. Interoperability and portability ensures innovation and optimal market growth in software and platform development. It is no longer a nicety to be included at the discretion of the vendor (Haile & Altmann, 2018). Stifling innovation not only creates vendor-lock, but it also prevents new products and services from reaching the market. There is therefore a greater need to incentivise products and services to maximise innovation. On this theme, Blind (2013) noted that standardisation could promote commercial innovation and provide a platform for researchers and other actors to diffuse and expand ideas, products, and services. Vendor-lock, however, can be counterproductive for providers and users (Haile & Altmann, 2018). Similarly, Gasser and Palfrey (2007) argued that interoperability and standardised development interfaces were beneficial to innovation, as this minimised vendor constraints.

From the literature, there is a paucity of knowledge on concurrent multiuser, low-latency BIM synchronisation, albeit recognition on the need for such research. In this respect, Du et al. (2018) developed a Revit API-based solution, similarly linking vector CAD with the Unity discrete environment. Liu et al. (2018) also used Revit API for leveraging rich information from the BIM model into an algorithm in order to optimise panel use and panel cutting in offsite construction; however, their project differs from the one presented in this chapter in both implementation and targeted features. The framework developed in this study supports interoperability features through IFC-Centric and Component Object Model (COM) services, whilst their project uses an on-the-fly asset importing mechanism to accommodate asset distribution, including avatars for social presence in terms of co-presence and telepresence to improve the sense of "being there" (Schroeder, 2002). Generally, their solution provides better integration with existing systems at the expense of interoperability and bidirectional interactions, and better asset distribution at the expense of transaction loads. This project also focuses on openBIM principles and IFC, rather than Revit or any vendor-specific product, due to its interoperability capability (Blind, 2013; Gasser & Palfrey, 2007).

Continuous movement was identified but not yet implemented in either project. As discussed previously, this may induce some form of discomfort, or cause conflicts between interacting users (particularly with many concurrent connections); most importantly, it may also result in fluctuating perceptions of immersion. Given this, the work presented here posits that social presence considerations beyond visualisation of interactions are future areas for consideration. The rationale for this is that the perception of presence is fundamental to the purpose of virtual environments, and arguably more significant than visual realism improvements (Lee & Nass, 2003) given base levels of visual realism in all virtual environments. This chapter acknowledges Du et al. (2018) practical applications, but wider recognition of the heterogeneity of potential interfacing systems is also important. Finally, one could also argue that column manipulation could be considered risky given the level of design flexibility needed in practical situations. In summary, extant literature provides significant advancements in BIM integration research. In this respect, the aim of the work presented hereafter is an attempt to showcase one method for integrating openBIM data into immersive environments, so as to streamline the design process and provide a pared-down BIM system that presents the right information to the right people at the right time.

Using this philosophical standpoint as the development basis, the second phase of this work focuses on developing an integration strategy for the openBIM, VR, and AR technologies. The most suitable formats for the purposes of this project were determined through preliminary technical tests. The third phase is therefore devoted to prototype design, highlighting the framework development process using an agnostic approach towards openBIM editing tools (using the vendor-neutral IFC file format). An iterative process of testing, reviewing, and troubleshooting ensured design fixity (as part of the validation process to ensure interoperability). Finally, feedback data collection was conducted via usability testing. The potential impact of an integrated virtual showroom prototype was identified by revisiting literature to evaluate results in the context with offsite manufactured housing.

6.4 Prototype development

6.4.1. *Project overview, BIM library, and IFC database*

From the outset, a virtual showroom prototype was developed in consecutive stages beginning with the 3D modelling of a kit home in BIM environment. The project used Norscot Joinery's kit home design called Assynt 4-Bed, which was the main industry partner and one of the funders of this project. (In the following sections of this chapter, Norscot Joinery Ltd will be referred to as Norscot.) Full design information on the Norscot Assynt 4-Bed was obtained from floor plans, elevations, and rendered images. Given Norscot's wide portfolio of products, the BIM library of the project was expanded to include Norscot doors, windows, and wall types, modelled in 3D as Revit families.

From this, the modelled BIM assets were integrated into a game engine environment, where they were coupled with interactive functionalities that allow users to explore and manipulate building elements within a 3D space. The game engine setting contained all features and mechanics, which was then split into two different flows – a virtual reality application and an augmented reality interface that benefitted from recently developed Google Tango technology. This allowed for two different immersive experiences of the home design: (i) in virtual space through a desktop device and a VR headset, and (ii) in augmented reality through a Tango-enabled mobile device.

The BIM model and its library served as the basis of an IFC database, which was created using PostgreSQL. The IFC objects were linked to Unity as meshes through an ifc.cs module, where a method for converting triangle meshes into faces was developed, including an experimental mode for identifying joined faces. This method was considered invaluable for future energy simulation integration.

BIM integration facilitates building model modification indirectly through Unity's environment using the purpose-built C# BIM library for IFC models. The feature was designed to bind BIM and Unity building models such that information from the BIM model could be retrieved for the Unity environment, using equivalent objects and interactions, where furnishings or materials in Unity would be reflected in the BIM model. Development was carried out in C# for both Unity and IFC library components. The IFC file format was chosen for interoperability between environments, referencing the library and all BIM packages implementing ISO 16739:2016. This is in contrast to existing software collectively sharing similar functionality with limited interoperability through development with vendor-specific APIs. Interactions between Unity and IFC geometry were delegated to an intermediary environment for translation between Unity's discrete and IFC's continuous/functional entity definitions, which provided data structure and coordinate system resolution within differing environments.

6.4.2. System architecture

Previous research investigated the use of BIM and VR technologies, albeit with noticeable gaps in conceptual frameworks. Once the appropriate software and equipment were identified, the next critical choice was to select the most suitable format that would allow a transition from BIM to VR with a minimum loss of data and geometry. The flexibility of the chosen file formats that transfer 3D information (as well as the appropriate handling of the data) were considered key for the successful integration of BIM models into a game engine environment. Autodesk Revit offered a reasonable range of output formats, yet certain technical challenges existed regarding the transition of the 3D model to a VR environment and virtual showroom. Figure 6.1 illustrates the process flows for BIM data conversion and integration within a virtual environment that were adopted in this project.

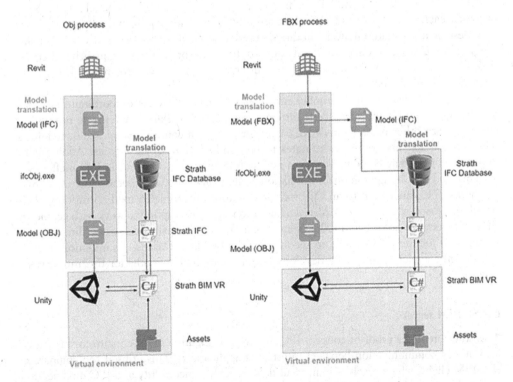

Figure 6.1 Process flows for BIM data conversion and integration within a virtual environment

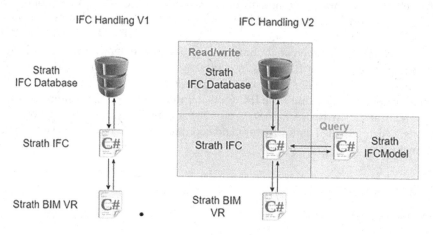

Figure 6.2 File handling flow options

The initial strategy for BIM integration into VR was based on Autodesk's FBX format. The 3D model was exported before being delivered to Unity for VR functionality development. Materials mapping was the main issue in that conversion process, since the FBX import appeared not to retain any textures when loaded into the game engine environment; so brick, wood, glass, or any other material would be visualised as a plain white solid surface. This meant that the underlying causes and potential solutions had to be further investigated in detail. Given that BIM and VR technologies are not necessarily developed for the same purposes, each treats files and materials differently. Revit, a modelling and editing tool, uses Autodesk or user-created material libraries stored in a specific directory on the user's computer, while Unity references materials directly from the project. Consequently, when an FBX file is imported into Unity, it does not automatically carry over the materials from the application that created it.

Another issue was Unity's primary use of interdependent relative coordinate systems, which required foreknowledge of the imported models since relative systems do not always correlate with other nests. Unity's XZY coordinate system caused several issues, requiring mesh coordinates and world coordinates to be handled with separate tools. Additionally, imported objects may be bound to pivot points, thus making world translations ineffective. As a method of floating-point error mitigation, scaling differs between vector and unity environments. IFC conversion to an OBJ format was also susceptible to mild corruption when translating faces, which was considered a limitation when attempting to work with geometry. BIM formats were not created to be bidirectional; hence, the precision of coordinates must be retained for Revit to be able to reconstruct the building. Database transaction latency and design complexity led to the development of the inline C# model for IFC interactions (Figure 6.2).

6.4.3. BIM *server*

This study created a proof-of-concept IFC BIM server (Figure 6.3) for concurrent multiuser, low-latency communication between interfacing applications. This utilised the companion IFC BIM client, which resolves numerical data structure uncertainty and IFC modification

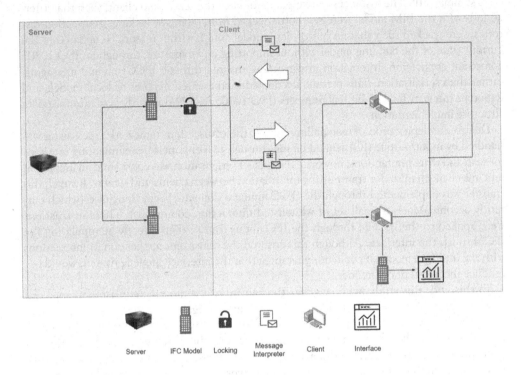

Figure 6.3 BIMServer/BIMClient relationship model

concurrency through a simple DSL. The server manages multiple IFCFile instances with command history tracking for incremental partial or complete updating, without replacing the source IFC file. Where a single IFCConnection is active, it can connect and effectively handle basic requests from more additional client connections than would be practical for a team. The impetus for development beyond agnostic BIM interfaces creates a framework for high media richness in virtual environments – particularly beneficial for professional use. Social software innovation can also result from making architecture more accessible, especially with immersive tools.

The server provides a framework for connecting multiuser concurrency and data concurrency between interfacing applications by implementing the BIMClient with multi-IFC instance management. It was designed to accommodate the development of BIM interfaces in most applications with programming interfaces. Connected applications were allocated read/write named pipes in the form of a BIMServerConnection. These were assigned to the IFCFile instance's associated IFCConnection whenever an IFCCommand. Load request was made for an existing IFC model (either opening a new IFCConnection or linking to the existing instance, where the model has been opened elsewhere).

Concurrency between server and client instances was managed through transaction histories and the pipe-locking system, such that interactions from an interface were approved and applied centrally prior to distribution between other clients. Of note, messaging between clients and interfaces does not implement any form of serialisation; instead, using IFCCommand transactions, which replicate actions applied to local IFCFile instances – with the central server instance serving as a mechanism for maintaining consistency between the server and client.

In summary, IFCFile instances were created for both the server and client, such that interactions with the BIM model were not bottlenecked from piping or locking. Transaction histories were tracked such that an out-of-date model could, within reason, be updated to the current state of the building model without importing and exporting through an ISO 10303 compliant application. Interaction integrity was ensured through IFCCommand messaging, rather than serialisation. This permits decoupled incremented updates to local models and data type integrity for 128-bit real numbers (ISO 10303:41 1994) regardless of interface data structure implementation.

Due to the deprecation of marshalling in C#, the transaction framework's messaging was handled by input/output (IO) named piping through two channels per connecting interface for read and write interactions; however, IFCObject serialisation was considered an inefficient and unsafe mechanism for transmitting information between clients and servers. Instead, this transfer was implemented through the IFCCommand domain-specific language (which currently accommodates a small set of whitelisted interaction commands). These instructions were applied to the IFCFile through the IFC library, rather than directly manipulating the IFC through the interface. Although this ensured the consistent application of interactions with the interface model, it could not guarantee that the interface implementation would not produce inconsistent commands.

To this end, this study acknowledges that an interface message consistency approval mechanism was required. Initial connections were delegated through a single connection pipe. Clients request a pipe alias for a dedicated BIMServerConnection, which is then opened by the server upon request. The client then creates the BIMClientConnection, which has two NamedPipeClientStreams for reading and writing messages with threaded, asynchronous messaging for client-server transactions. This method contrasts with server message distribution, which is threaded with synchronous messaging and facilitated through the BIM[Server|Client]Connection locking mechanism (which prevents clients from sending messages, where the server is processing an IFC interaction message).

Linking clients to models and distributing commands from the server was achieved through unique keys assigned to the IFCConnection and BIMClientConnection. When the server receives a message from a client, it uses the Message.Parameters.ClientKey to identify the linked IFCConnection. Once identified, the server distributes a message with Message.Parameters.Lock active message to each linked client before validating and applying the IFCCommand. Once clients receive a lock request, the connection lock state is changed to ConnectionLockState.Read, disabling the client write named pipe and associated asynchronous outgoing message queue handler. In contrast to the client, server message interpretation and application, whilst threaded for each IFCConnection, this is synchronously processed by disabling message retrieval during processing; however, server message distribution post-process is not processed in the same way, and therefore, is dependent on sequential message transmission. As illustrated in Figure 6.4, messages are currently limited to the following properties, which are not necessarily omnipresent:

- **Client key:** A BIMClient unique identifier linking messages and commands to the client;
- **Sender confirm key:** A unique message identifier linking server responses to initial Message objects on the initiating client;
- **Receiver confirm key:** A unique identifier for action inverse to those associated with the sender confirm key, where messages either originate from the server or communication is multi-staged;

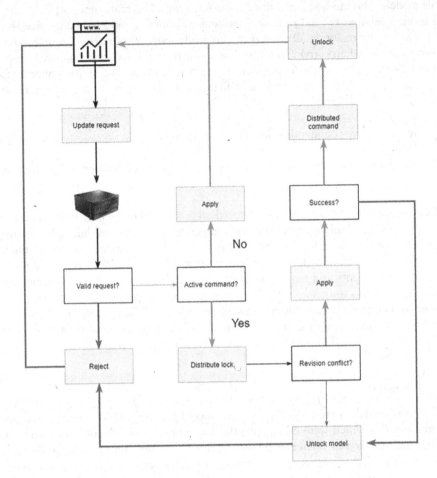

Figure 6.4 BIM server command transaction model

- **Response:** The returned value from either client or server once a message has been processed;
- **Expects response:** Whether the message expects a response and resides in the awaiting response list – effectively asking the server for a promise;
- **Success:** A flag to identify whether a message was processed successfully where validation is required;
- **Lock:** Sent from the server, this is used to inform clients whether it should lock or unlock its write pipe;
- **IFCParameters:** A delimited list of parameters for IFCCommands;
- **Error:** A supporting message for situation where an error (defined by Success = false) occurs.

Commands from interfaces are interpreted by the BIMServer instance and applied through IFC namespace models and support classes. A simple DSL was created for messages in place of serialisation to reduce network load and standardise interactions between interfaces and

IFCFile models. This method maintained consistency with IFC real numbers (which are sensitive to manipulation beyond precision loss and will break a model if adherence to their rules is not strictly maintained); however, this study found that utilising serialisation was counterproductive when carrying out insertions, which would not accommodate non-prefab insertions. Nevertheless, these features are currently complemented by pre-imported assets in the Unity environment. Commands are split into the following two categories and transaction types:

- Passive and active commands – depending on whether the server requires locking during processing, and
- Client or group for messages – for individual clients or all connected to a given IFCConnection.

IFCConnection IFCFile models are mirrored on connected clients, such that marshalling was not required and building model interaction performance was not diminished by relying on a remote instance. Initially, this appeared to be an inconvenience; however, as the locking mechanism was planned, it became apparent that IFCFile-level locking would not meet all locking requirements and may result in concurrency issues. Although this issue was not fully investigated, this study was concerned about the impact this may have on many concurrent user connections through the marshalling mechanism. The primary application of the mirrored model was read-only subtree construction, delegating interactions with data to the IFCVector library, UnityIFC class, and the game's core Unity library.

6.4.4. IFC model

IFC data was stored in an EXPRESS schema model represented in single record entries with arbitrary IDs, which were non-sequentially listed in the file. Record associations were predominantly tracked as reference ID lists in the parameters, representing a loosely coupled graph structure of coordinates, entities, functions, layers, and properties; however, some property associations such as IFCRELASSOCIATESMATERIAL descendants to IFCBUILDINGELEMENTPROXY records were linked via the IFCMATERIAL name in the element proxy parameters. The record defines three generic values: record ID, category, and parameters – where the category identifies the type of object and dictates parameter types. Data was encoded under rules of the American Standard Code for Information Interchange (ASCII), resulting in a very large file compared to binary CAD representations (which does not support all characters supported by BIM interfaces, represented as byte literals).

There are two primary physical entity definition structures: coordinate-based planes and profiles. The former tends to share points and materials with similar objects, whereas the latter does not share points but may share materials. Materials were stored by name only and not exported with the model. Any importing interface must have an associated material to render the model, as created in the interface. Colours related to materials are stored as separate records, which may be shared. Entity-specific coordinate systems may also share localisation. World coordinates and axis direction vectors whose axes are not necessarily parallel to those of the world coordinate system or other entities, profile entities have secondary direction vectors to orientate extrusions. The model structure supports basic revision histories tracking the organisation and person that last edited these entities or properties. Figure 6.5 presents a sample graph subset of an IFCBEAM, I-Beam definition, showing the record ID and category taken from the IFC – exclusively representing structural entities.

```
383336 IFCBEAM
   41 IFCOWNERHISTORY
     38 IFCPERSONANDORGANIZATION
       35 IFCPERSON
       37 IFCORGANIZATION
     5 IFCAPPLICATION
     1 IFCORGANIZATION
 383306 IFCLOCALPLACEMENT
   122 IFCLOCALPLACEMENT
     32 IFCLOCALPLACEMENT
       383366 IFCLOCALPLACEMENT
         383365 IFCAXIS2PLACEMENT3D
           6 IFCCARTESIANPOINT
       31 IFCAXIS2PLACEMENT3D
         6 IFCCARTESIANPOINT
     121 IFCAXIS2PLACEMENT3D
       6 IFCCARTESIANPOINT
   383305 IFCAXIS2PLACEMENT3D
     383303 IFCCARTESIANPOINT
     19 IFCDIRECTION
     13 IFCDIRECTION
 383334 IFCPRODUCTDEFINITIONSHAPE
   383331 IFCSHAPEREPRESENTATION
     100 IFCGEOMETRICREPRESENTATIONSUBCONTEXT
       97 IFCGEOMETRICREPRESENTATIONCONTEXT
         94 IFCAXIS2PLACEMENT3D
           6 IFCCARTESIANPOINT
         95 IFCDIRECTION
                               383329 IFCPOLYLINE
       9 IFCCARTESIANPOINT
       383327 IFCCARTESIANPOINT
   383325 IFCSHAPEREPRESENTATION
     102 IFCGEOMETRICREPRESENTATIONSUBCONTEXT
       97 IFCGEOMETRICREPRESENTATIONCONTEXT
         94 IFCAXIS2PLACEMENT3D
           6 IFCCARTESIANPOINT
         95 IFCDIRECTION
     383315 IFCEXTRUDEDAREASOLID
       383311 IFCISHAPEPROFILEDEF
         383310 IFCAXIS2PLACEMENT2D
           383308 IFCCARTESIANPOINT
         27 IFCDIRECTION
       383314 IFCAXIS2PLACEMENT3D
         383312 IFCCARTESIANPOINT
         11 IFCDIRECTION
         21 IFCDIRECTION
       19 IFCDIRECTION
```

Figure 6.5 Sample graph subset of an IFCBEAM, I-Beam definition showing record ID and category taken from IFC exclusively representing structural entities

6.4.5. Unity planar geometry

Unity, in addition to using a discrete world, is not specifically designed for planar geometry, and subsequently was not considered ideal for several functionalities of the prototype (where visualisation was considered an enhancement rather than function-critical, in contrast to environmental modification functionality). Game engines (including Unity) almost ubiquitously use some form of discrete polygon mesh, typically triangle meshes, for both rendering and associated data structures. Beyond hardware specialisation in working with triangles, they have convenient properties that accommodate efficient rendering and raycasting algorithms. Triangle vertices are coplanar (which is effectively a guarantee for reducing the complexity of operations) and accommodate barycentric coordinate systems. In terms of computer graphic engines, these systems are most notably useful for interpolation, creating more efficient point-in-boundary calculations.

Triangles, however, were not well-suited to the operations required for this project. In particular, challenges with surface relationships, as well as decoupling between vertices and triangle definitions (which prevent links to IFCFaces). They can also hinder profile-dependent evaluation, including simple surface area calculations and voids definitions, creating an absence of triangles rather than closed boundaries. These features are fundamental partition rendering modifications, since Unity materials are shared with the mesh (not implicit surfaces), with diegetic interface attachment and envelope parametric modelling. This in turn requires knowledge of coplanar triangles, surface relations, and void awareness.

This research resolved this challenge through the creation of an IFC library Facer, which constructs faces from vertex and triangle arrays, identifies boundary polylines, creates shared vertices, and surface relationships retains the vertices, triangle indices, nested triangle relationships, and directional properties (used by Unity for rendering). By creating the relationships between Unity and IFC.Facer.FaceSets, this study was able to develop the desired functionality, whilst taking advantage of Unity's triangle system. This was achieved through a simple edge-detection algorithm, which translated triangles into Face objects. This process involved identifying coplanar and connected subsets using shared vertices to find linked line segments. Segments are present on two triangles, which were filtered, and segment vertex associations were used to create the anti-clockwise closed polyline. Surface voids were identified by repeating the process with remaining line segments. Finally, the closed polygons were sorted by area to separate the outer boundary along with void representations.

Through linking Unity triangles with the Facer.Triangle objects, the library only requires triangle calculation if meshes are procedurally generated independent from any existing mesh. FaceSet-to-Unity entity conversion was achieved by iterating over the set and generating MeshFilters from the replicated triangles. Facing was primarily created for envelope operations, which did not require offsetting out with parametric modelling. This meant that there was no value in breaking the relationships between the existing Unity mesh and procedurally generated surface, thereby reducing the need for recalculation or adding MeshColliders. Although often frustrating, Unity Vector3 coordinates are read-only. This meant that any operation carried out on the Face could not be simplified; where it was possible to carry it out directly on the existing mesh, the Unity mesh vertex list was recreated whilst the Face.PolyBoundary vertices were modified directly.

Void inner surfaces were used to identify nested windows and doors by finding their centre points and evaluating point-inside-boundary for XY and YZ compounded profiles. Linking between PolyBoundary and MeshFilter objects coupled with knowledge of envelopes and their embodied fixtures allowed basic parametric modelling offsets to be achieved with relatively low time complexity; however, the library does not yet support void insertion to existing MeshFilter objects, creating constrains such that this type would currently require destruction of the existing MeshFilter and subsequent redefinition of the current integration strategy.

IFC model construction and Unity linking

Linking Unity and IFC entities was decided to be a run-time (rather than prep-time) or one-off event, such that bidirectional transactions between models could be accommodated. This meant that functionality intended for post-completion extensions to the project would not require refactoring. Extending the IFC file and revision histories were considered as alternatives but were deemed unnecessary and overcomplicated for the planned functionality.

Unity models differ from representations in IFC files in several ways:

- The naming convention changes depending on the way in which the model was generated and whether the Unity model was imported from. obj or. fbx.;
- Unity entities are discrete representations of IFC entities, which lose precision and functional constraints (predominantly with arcs and other abstract geometric shapes);
- Unity's coordinate system is XZY where IFC is standard XYZ;
- Data structures are changed from 128-bit real as implemented, with many IFC record properties as per ISO 16739:41 to 32-bit float.

These representations prevent standard one-to-one mapping through name, geometry, or spatial similarities and can require multiple validations to link entity definitions in both environments. Although indexing the IFC model was constructed to reduce latency between Unity entities and their associated IFC model representations, it was not possible to maintain suitable framerates within the Unity environment synchronously (which led to semi-lazy, asynchronous linking). This study was, however, able to link entities in a relatively straight-forward fashion for all models. Revit families (not exported from the BIM model using fuzzy lookups) presented several additional issues, including ambiguous naming, which required further consideration of the information available in the IFC model.

BIM models without families

Naming was the primary linking mechanism for entities, where Unity import models were produced without families in the BIM interface export. Through testing imports with differing sources, it was discovered that naming changes during Unity import were not consistent across all sources, notably differing where the import model was imported as. obj or. fbx. There were five recurring changes that were not a symptom of character encoding:

- **White space to colon**: White space characters, whether space or tab were replaced with colons, including multicharacter instances;
- **White space to underscore**: White space characters, whether space or tab were replaced with underscores, including multicharacter instances;
- **Case sensitivity**: Casing was not guaranteed to be retained between IFC and Unity models, although this may be a symptom of using intermediary software such as Twinmotion or 3DMax;

- **Multicharacter replacement of colons and white space characters**: Several instances of substrings were replaced with either individual characters, as per previously mentioned amendments, in some cases with characters in others, substrings;
- **Arbitrary inclusion or exclusion of square brackets**: Square brackets present in either the IFC or Unity model where not guaranteed to be present in their complementary model entity names.

Due to these differences, a simple or partial string-matching process did not appear to be practical. The end solution for name-based linking procedurally generated regular expressions, which guaranteed either the name that was as expected (or was at least consistent with one or more of the previously mentioned amendments). The secondary objective of making the C# IFC model ignorant of the IFC schema, to the possible extent without losing functionality, led to the initial inclusion of the fastIndexedRecords array. This model stored all records from the IFC file in their original order. Since IFC record IDs are not contiguous, this did not represent a one-to-one indexing of indices; however, it was identified to be the most efficient means of carrying out $O(n)$ operations without knowledge of the IFC record IDs.

The operations carried out by the C# IFC model were predominantly read/modify (with rare write operations), which were present in batch operations. This suggested that the only concern for using an IFCObject[] over List<IFCObject> was the required time during parsing or writing to regenerate the array. The house model used for development had 1.24 million records, where using the regular expression-based parser required 17 seconds to parse the model. The time was increased by 1 second to generate the fast search array, which was deemed insignificant in comparison to the expected benefits from its intensive use; the current parser that does not implement regular expression reduced parse time to 6.5 seconds in C# and approximately 3.6 seconds in its C++ equivalent.

During stress testing, it was discovered that a failed lookup took approximately 7 seconds using the regular expression-based name linking, which was not user-friendly, where a user would either have to wait for the duration to be notified of a failure or return to the entity having carried out other operations in the interim. The fastIndexedRecords array lookup was replaced with a top-level record lookup, where top-level were identified predominantly as those having no parents, excluding IFC representation layers.

This lookup method reduced the number of operations required for lookups by around 95–97% across tested models. This was achieved through extension to the parser with a temporary Dictionary<int, List<IFCObject>>, which tracked child-parent relationships. It was initially expected that this would require hard-coded exemptions for IFCCARTE-SIANPOINT omission to reduce load; however, parse duration was increased by less than 2 seconds, and its inclusion did not appear as a merit due to the negligible benefits this alternative provided. With the top-level fuzzy lookups in place, worst-case lookups were reduced to around 2.5 seconds.

Further improvements were made to this feature by introducing a semi-eager component to linking entities, where the linking process was threaded during environment initialisation, such that all entities could be linked in the background regardless of the user's activity or the entities they viewed whilst in the environment. Unity does not permit interactions with game objects directly through secondary threads, thereby preventing run-time referencing of game object and components from the linking thread. This constraint was remedied through extension of the CADObject to decouple useful component properties from their Unity entities, such that they could be referenced with the main thread. Concurrency was retained through further extension to CADObject to incorporate entity modification functionality through its instances, as an intermediary between Unity and IFC in a manner that did not deviate from the original plan of implementation. This, however, required extension to the synchronous component of initialisation to generate CADObject instances in order to extract meaningful properties prior to threading, although this did not incur significant overheads.

BIM models with families

Building models exported with families proved to be more challenging due to game object nesting in the Unity environment, as families are nested n-depth depending on the entity and context. For example, furniture may be nested import->model->family->type->object->entity compared to windows import->model->windows->entity. This meant that the project needed either to remove ignorance from the linking process entirely or to create a mechanism for identifying entities, where nesting depth and end entity naming was not guaranteed to be meaningful to the IFC model.

Removing ignorance would have resulted in failing to meet the secondary objective of this project, so an alternative was created to the existing name-linking method. This constructed the family name by traversing the Unity model hierarchy upwards until the model entity was identified, constructing a regular expression for family-equivalent lookups. Although this did not prove useful, it is worth noting that families naming was near-consistent (in comparison to the convention changes that arose with family-less model hierarchies).

Construction of family lookup patterns was further complicated by redundant top-level name components, which were relevant only within the context of the family set (not the IFC entity name itself). The final lookup method for models with this type of structure extended the family-less method to add lazy matching of top and bottom-level Unity game object names, which increased the worst-case lookup duration, but was mitigated with top-level searching in place.

Familied models can often result in several IFC candidates for individual Unity entities due to ambiguous naming in both models. This required additional filtering to retrieve the correct representation from the IFC model. After working through the IFC model, this research was unable to identify a match criterion that would accommodate linking without referencing geometric definitions, leading to the first utilisation of the IFCCARTESIANPOINT entity type. Given that IFC and Unity worlds are not identical, nor are their respective data structures, this method would inevitably become intensive as the model complexity increased or instances of specific entities increased. The process would deviate from localising insertion points dealing with further discrepancies in the Unity environment. As a response to this, the study constructed a localisation method utilising bounding boxes, such that centre points could be used without translating full geometries from IFC to Unity – resolving data structure precision or mapping discrete features to functional IFC definitions; however, this method is not necessarily the best approach in the system's current state given that bounding boxes for IFC entities are lazily evaluated. That being said, with interacting entities where the base IFC record is known, all child retrieval requests approached through the record dictionary posed no concerns for performance beyond generation of the bounding box, which is cached upon creation, reducing future linking operations.

Although the application design was constrained such that pre-processing was not required, the constraint was primarily for Unity environment considerations and bidirectional transactions. In future iterations of the IFC library, semi-lazy pre-processing of boundary boxes are likely to become an extension to IFC files generated for the application, which is expected to be important for larger models, including the 7.8 million-record model used for stress testing.

Strategy for linking Unity and IFC

The process of linking IFC files to Unity with agnostic BIM integration faced several challenges from the Unity model as import modifications dependent on import source file structure and type prevented simple name-based linking. This was resolved using procedurally generated regular expressions, and later a bounding box solution was used for ambiguously named entity instances. Maintaining usable performance required numerous changes to the library, as functionality for linking and other operations were introduced. These were universally resolved through refinement of IFC record indexing and dictionaries, which increased the load on the parsing stage of the C# IFC model construction. These changes were later mitigated by replacing the regular expressions-based parser with a character-by-character explicit comparison to mitigate the increased load from indexing, thereby reducing the parsing time significantly. In doing so, this study confirmed that whilst the interpreted languages considered, JS, Ruby, and Python were favoured (for developer friendliness and flexibility), they were not adequately capable of performing as necessary for IFC-based modelling. The linking process is heavily dependent on a semi-eager approach to lazy evaluation, which proved a common requirement throughout the IFC library development and working with Unity; however, the prototype presented here was able to achieve links between both models without vendor-locking the application with Revit or other existing BIM packages.

A flexible representation of the IFC format was created in C# to enable querying with the IFC model, based on the extent in which the IFC Category objects were defined in the namespace of a referencing application. The model was initially designed with query performance at the centre of its focus, which led to several caching strategies being implemented. A dictionary was created to record ID queries, where the IFC record was known in advance, which became the subscriptible reference for the IFCFile class. A standard list was created for queries with either unknown related IFC records or those where multiple results were expected, and a dictionary of lists for each IFC category present in the model (Figure 6.6).

All IFC records have a standard for property construction, their ID, category and parameters, child IDs, numeric values, strings, enums, and comments. The model was designed to be easily extended without modifying the core model construction for class-dependent

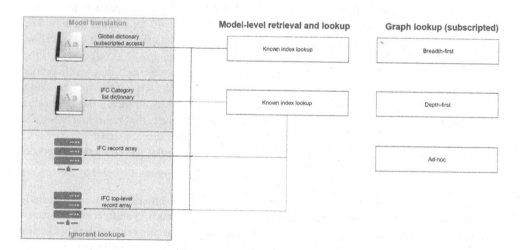

Figure 6.6 IFCFile lookup indexing model

constructors. Using C# meant that run-time object type delegation without hardcoding cases was not straightforward. The recommendation from the C# documentation is the use of the Activator class, which is designed specifically for this type of action; however, Activator could not be practically implemented for IFC files given that record counts were in millions. This study built an alternative using Linq and the ConstructorInfo class to cache the standard constructor for descendant of the IFCObject and IFCGeometricObject classes into an IFC category keyed dictionary for the CreateIFCObject factory method (which retrieves the most appropriate constructor of a given category). This approach did not suffer notable performance difference from the switch case alternative and remained a moderately low contributor to file parsing time.

Entity manipulation and IFC exporting

The IFCFile library supports manipulation of properties and classification information found in the base IFC model. This is extensible to accommodate manipulation or otherwise to interact with information present in the model. The IFCFile model representation, being graph-based, scopes queries and property discoveries such that the developer can request a property with a known IFC* identifier and graph traversal ensures the correct instance is identified. The process had several variations that determine the direction of the initial traversal, the desired target, and element property index or indices required. An example of how to move, retrieve, and update the U-Value of a wall representation can be seen as follows.

Moving the element (relative position):

- With the IFCWall:IFCObject either use the IFCFile[<#>] subscripted access method with the fifth property of the IFCWall object or retrieve the first IFCLOCALPLACEMENT via a depth search;
- With the IFCLOCALPLACEMENT either use the IFCFile[<#>] subscripted access method with the second property of the IFCLOCALPLACEMENT to retrieve the first result from either an ad-hoc or breadth search for IFCAXIS2PLACEMENT3D;
- Apply arithmetic operation to the three sub-properties of the only property in the IFCARTESIANPOINT identified either via IFCFile[<#>] subscripted access with the IFCAXIS2LOCALPLACEMENT3D's first property or retrieval of the only IFCCARTESIANPOINT via breadth, depth, or ad-hoc search.

Retrieving and updating wall element U-Value:

- From the IFCWall:IFCObject call, traverse up the tree to find the element's parents;
- Search the parent objects for the associated IFCRELDEFINESBYPROPERTIES object;
- With the IFCRELDEFINESBYPROPERTIES either use the IFCFile[<#>] subscripted access with the sixth property or retrieve the only IFCPROPERTYSET via a breadth, depth, or ad-hoc search;
- With the IFCPROPERTYSET perform a breadth or ad-hoc search for the only IFCPROPERTYSINGLEVALUE objects and iterate over them, comparing their first property to the string "ThermalTransmittance";
- Where an IFCPROPERTYSINGLEVALUE is found, the U-Value is available and can be retrieved as the first sub property of the fourth property. To update this property simply replace the fourth property with the new U-Value.

Model exporting

The BIMServer section best describes the bidirectional aspect of the IFCFile library, where actions are applied directly to the base IFC model as record updates or insertions. This means that where a change in the Unity environment (which is translated to IFC) is made, the corresponding IFC records are updated. For example, the U-Value of an entity may be updated, e.g. changing record 687 from its starting value to 0.55, the change on the IFCObject and subsequent string representation in the IFC file would be as follows:

Record update:
Ifc[687].properties[2].value = "IFCTHERMALTRANSMITTANCEMEASURE(0.55)"
Original (ifc[687].toString()):
687 IFCPROPERTYSINGLEVALUE('ThermalTransmittance',$,IFCTHERMALTRANS
    MITTANCEMEASURE(1.),$);
Amended (ifc[687].toString()):
687 IFCPROPERTYSINGLEVALUE('ThermalTransmittance',$,IFCTHERMALTRANS
    MITTANCEMEASURE(0.55),$);

6.5 Virtual Showroom Interface

6.5.1. *Limitations and future development considerations*

The BIMServer accommodates desired multiuser, low-latency interactions between IFC BIM models and interfacing applications, including concomitant data concurrency features; however, this does not have a formal interface integrity unit-testing mechanism and is therefore open to incompatible implementation of BIMClient, which can result in conflicts between users' expectations. Moreover, this is also not ISO 16739:2016 compliant, making it unclear to what extent functionality can be extended without requiring manual mirroring of changes in a compliant system; however, given the original intent was to provide interactivity and visualisation, this was not seen as a major detractor.

The server has moderate model concurrency assurance but does not support entity freezing where a user is interacting with an entity. In addition, the semi-eager linking methodology in the Unity environment is not well suited for this type of functionality. Whilst this creates a more immersive environment by maintaining user functionality in isolation, it also minimises fluctuations in media richness whilst also preserving the sense of telepresence; however, Lu et al. (2014) identified perceived interactivity that may be a reasonable proxy for real interactivity; and similarly, Anandarajan et al. (2010) noted the subjective nature of perceived richness and the social influence impact of effectiveness. Given this, the asynchronous command process is procedural (not declarative), meaning that without an extension to command/message history management, users' interactions could result in ghost operations. For instance, consider an example where a furnishing element located at (0, 0, 0) is moved by user A 1000 mm along the Y axis; at the same time user B moves the furnishing element 1000 mm along the X axis, finishing their command after user A, resulting in a new location of (1000, 1000, 0). This generates a solution that neither user desired.

Although the server was designed with multiuser interaction in mind, the development did not focus on social presence. Moreover, it is currently unclear how the BIMClient could be used to support co-presence beyond audio communication given it is an abstract system with no influence over the selection of interfacing applications. The concern here is that different levels of engagement and immersion may have a profound impact on interpersonal interaction and discussion. Whilst there is nothing preventing users from using third-party audio communication applications, the separation between BIM and communication servers prevents balancing delays to minimise sensory conflict – entities moving seconds before or after intent has been expressed by a user. Similarly, whether audio communication is integrated or indirect, the server does not support agent representation in connected environments, preventing capable interfaces from utilising binaural audio, which Ballestero et al. (2017) suggested was an important mechanism for instilling a sense of immersion.

Functionality supported by the IFC library is exclusive of building services design and the constraints imposed by the profession. This raises concerns regarding which actions should be available to users, and perhaps more importantly, how their impact can be evaluated. Inference is made here regarding modifications that may invite consequences. which may not be practical or cause conflicts that are not readily accessible to users. Similarly, atmospheric effects from modifications cannot be enforced, and depending on the interface, may not be representable. This may result in vastly different environment representations renditions being presented to users, resulting in unnecessary conflict between users. For example, UK Building Regulations (Approved Document F, section 5, Table 5.1a) on extract ventilation rates sets conditional design requirements on intermittent and continual extract rates, dependent on the type and location of food preparation fixtures and changing envelope configurations and service system loads. For example, replacing glazing may change light transmittance, and internal rendering changes may affect perceived lighting comfort. Whilst not particularly a concern for domestic buildings *per se*, this can subsequently diminish the effectiveness using multiple non-uniform interfaces.

6.5.2. *Integration strategy for* VR

For the purposes of this project, Unity served as the main medium of application development as it allowed applications to be quickly complied. It also provided access to a large online user knowledge base, whilst also being able to accommodate changes from virtual to augmented reality through simple adjustments.

Controls in virtual reality

The primary VR headset used for this project was the HTC Vive. This was fitted with scanners to detect a very wide space for play while accurately tracking the player and the controls. HTC Vive comes equipped with two controllers and two wall-mountable scanners that allow users to move around extensively within their personal real-world space. Vive's controllers consist of five buttons, four of which ('Menu', 'Touchpad', 'Grip', and 'Trigger') also support the assignment of custom functions. There were, however, more functions that required buttons than there were buttons available, meaning some functions had to be restricted to one specific controller – left or right – for situational purposes.

The 'Menu' button on either controller was used to access the on-screen teleport menu, which was displayed in the form of a floor plan of the building. On both controllers, the 'Touchpad' was used to activate a pointer to interact with the virtual environment. On the left-hand controller, the touchpad is used to 'teleport' or move around the virtual world, whereas the right-hand controller's touchpad is used to interact with user interface elements. This intentional splitting of interface and world interaction was implemented to stop users from teleporting into walls or other areas whilst interacting with interfaces. The 'Grip' button was used to 'hold' and carry objects such as furniture (once they were highlighted with the pointer).

Movement in virtual reality

While movement in virtual space can sometimes be problematic, this study explored several solutions to these problems. The method applied varied depending on how the virtual showroom worked and how the designer intended the user to move within and interact with the world around them. Moving around in a virtual environment is very intuitive, albeit users are tethered to a computer with the VR headset and wires. Freedom of movement in the real-world space can also vary due to furniture or walls. Within the Norscot Virtual Showroom, users can traverse an entire house, as well as a small section of the exterior shown as a garden-like area. Although the space was quite large, the garden was implemented with a fence around the house to act as a visual barrier for users. Instead of being allowed to roam all over the space, users were confined to this restricted space around the house. This was implemented to minimise 'mis-clicks', which would result in the user being teleported too far away from the house, along with the potential of experiencing momentary confusion. Figure 6.7 illustrates the overall view of the prototype, as well as developers' view in Unity.

A 'teleport' function was added to Vive's controller to allow users to simply point at a surface, typically the floor/ground, and instantly 'move' to that location. This meant that although users could still move in the real world, if they needed to travel farther than they could move in reality or even walking through the showroom from a seated position, then they could do so without trouble. This solution also meant that users did not experience motion sickness. If, for example, a button was added that allowed the user to 'walk' in VR without actually moving in the real world, this would cause motion sickness as users would see themselves moving in VR, but this would not physically feel like moving in the real world. This, however, caused problems in some scenarios, where users randomly or accidentally

Figure 6.7 Prototype development in Unity

clicked the teleport button, moving them to a location they didn't intend to move to, leaving them momentarily confused until they got their bearings again. Another issue that came with teleporting was that when users were interacting with interfaces or moving objects, they would accidentally teleport into the object or a wall. In order to minimise accidental and unintentional teleporting by users, e.g. teleporting while intending to click and move objects, this function was restricted to only the left-hand controller, which could not interact with any user interfaces or objects. Further to this issue, the teleport function allowed users to teleport into walls and objects. Because of this, users were often left disoriented and were forced to move in the real world or quickly move in-app, particularly as all they would see was a blank object blocking their view. There are functions available to address these challenges.

Interfaces in virtual reality

Interfaces within VR applications can often be complicated to get right the first time, and in not doing so, can often cause many issues. In the Norscot Virtual Showroom project, interfaces were made to be diegetic as opposed to non-diegetic. Diegetic interfaces are interfaces that exist within the world or environment that are not entirely obvious, such as a light switch button on a wall. Non-diegetic interfaces such as Heads Up Display (HUD), which covers the entire screen, can be found in many games and apps. The most important reason for keeping interfaces diegetic in VR is so that users' vision is not 'blocked' by an interface, as there is only a limited field of view available when using a VR headset. Alongside this issue, having diegetic interfaces means that users are more immersed in the surrounding virtual world without being disrupted by a wall of text, icons, or images.

Many VR apps also keep interfaces confined to the controllers, where users can simply lift their hand, look at the controller, and see one primary interface (to allow them to carry out all the actions they need). This is traditionally more effective than keeping interfaces on walls or on the ground, as users have the choice of seeing an interface or not, much like a person using a watch on their wrist.

Differences exist between objects that are interactive and those that are not. Designers of VR apps can employ many engagement strategies, ranging from a user interface hovering over whatever object is interactive to highlighting an object only when users 'touch' it with their controller. To maintain immersion within this prototype, no specific interfaces were used to indicate if an object was moveable, meaning objects such as furniture that were moveable had their outline highlighted with a certain colour when users pointed at the object with a controller pointer. This allowed users to easily differentiate between what was interactive and what was not (without the need to read the entire interface). Likewise, interfaces such as the material picker, colour painter, and light switches had buttons that were clearly visible to users.

Another problem facing many designers of VR applications is that text can often be quite hard to read in VR. Users therefore have to move their head around to read text, rather than directly with their eyes; moreover, the distance from the piece of text can also affect the visual quality. In this project, the text was only contained within interfaces such as the material picker and IFC interface, which mostly consisted of trigger buttons or drop-down menus (Figure 6.8).

Teleport Menu: Users can press the "Menu" button, which activates a floor plan teleport menu (Figure 6.9). This interface displays a floor plan of the house. From here, users select a level and click a specific room to teleport into. This interface was attached

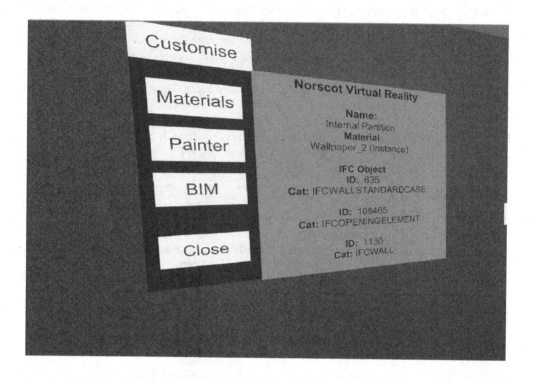

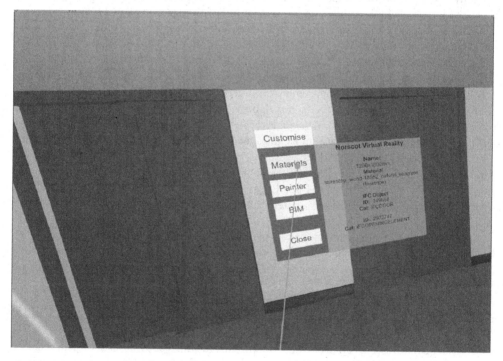

Figure 6.8 Interaction with BIM data

Figure 6.9 Teleportation interface in virtual reality

to the user's head within the application. By placing the interface on the user's head, they are able to easily identify and use the interface without losing it. One issue that arose with this interface was that because the interface was placed a set distance away from the user's head, if they were very close to a wall, the interface would often get lost behind it. As the building used in this prototype was quite confined, this issue occurred often during usability testing; however, a solution was found to stop the interface from rendering behind any objects.

Light Switch: This was a simple interface consisting of an easily recognisable white box with text above labelled "Light Switch" (Figure 6.10). Users simply point and click to turn lights on and off. This interface was placed on walls in areas where one might typically find a light switch, in order to allow for immersion and not create an intrusive

Figure 6.10 Light switch interface

object. In other words, the interface only 'existed' when users were looking for it or just happened to spot it.

Customisation Menu: This was the most extensive interface, accessible by clicking the trigger button on the left controller. This interface was placed on every wall and could be customised so that users can choose specifically how they wanted to decorate each wall. The interface (by default) was hidden to allow users to immerse themselves in the showroom first, where they would be able to activate and deactivate the interface as they wished. This meant that users had no interface permanently placed onto a wall that would restrict their view of it. The freedom of transition between interface and no interface also helped users to exist in two different game states – immersion and customisation. Figure 6.11 shows typical examples of interaction with wall finishes in an immersive environment.

Interactions in the VR application

The difference between objects that are interactive and those that are not was difficult to present in a VR setting. Designers of VR apps often employ many strategies to address this, ranging from obvious user interfaces that hover over whatever object was interactive to high-lighting an object only whilst the user 'touches' the object with their controller. To maintain

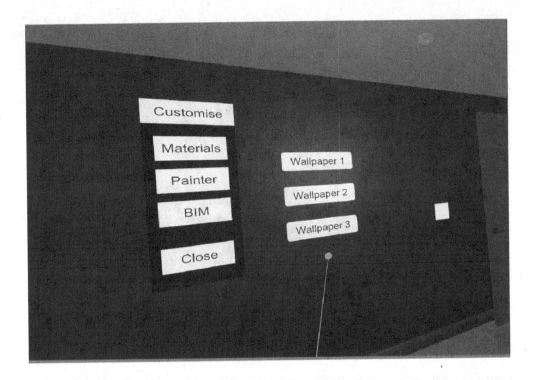

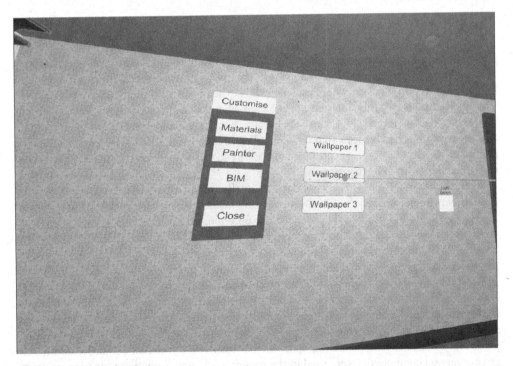

Figure 6.11 Interaction with wall finishes in the immersive environment

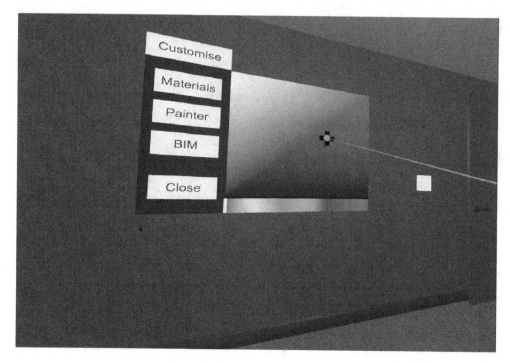

Figure 6.11 (Continued)

immersion within this prototype, no specific interfaces were used to indicate if an object was moveable, meaning objects such as furniture that were moveable had their outline highlighted with a certain colour when users pointed at the object with their controller's pointer. This allowed users to easily differentiate between what was and was not interactive (without the need to read an interface). Likewise, interfaces such as the material picker, colour painter, and light switches had buttons that were clearly visible to users. Of course, a natural behaviour when seeing a button is to 'press' it, and users can easily do so via the 'trigger' button on the controller (with prior instruction).

As the house in this prototype was relatively small, a much wider environment was created to maintain immersion, rather than just creating a small boxed area in which users could see below the ground itself (if they stepped out far enough); however, this caused additional problems, as users could travel several metres away from the house (intentionally or unintentionally). To address this, a small fence was placed around the house. The primary purpose of this fence was to act as a visual aid to users, allowing them to understand their distance limit, whilst still keeping them immersed (as a fence is a normal item to see in relation to a house). As a result, the prototype was more aesthetically pleasing, as foliage such as bushes and trees are visually more appealing than leaving empty or open spaces. All of these additional features allowed for a more immersive experience, allowing users to freely roam within the confined limits and being able to navigate this space as they might do in the real world. No additional environmental components were used in the background, such as roads, mountains, or other terrain. The strategy behind this was to limit distractions. Conversely, it could be argued that adding low-detail backdrops might increase the level of immersion. This is something to be explored at a later date as part of future AR integration development strategies.

The Google Tango platform was used to run the augmented reality experience of the off-site manufactured kit home showroom prototype. Currently, the application is supported by certain mobile devices, so the *Lenovo Phab 2 Pro* was used for this project. Tango is a mobile application developed by Google that uses the two cameras on the back of mobile devices in order to take accurate measurements of 'motion tracking' and 'depth perception' to provide a smooth and engaging augmented reality experience. Tango also offers an 'area learning' feature that remembers where objects were placed in relation to users' position, preventing them from being lost or 'drifting' away when they turn and walk in a different direction or hold the device at an awkward angle. Whilst Google Tango is still a relatively new technology, it presents significant potential through its technical and usability capabilities.

As part of the transition from BIM to immersive experiences, the 3D model of the house (already imported in Unity for virtual reality) could also be simply re-imported into a separate augmented reality–enabled version of Unity; however, Unity does not support virtual and augmented reality simultaneously. Fortunately, all information is retained in the transfer of assets across VR and AR Unity projects, so no further work was required to support this transition.

Controls and interfaces in augmented reality

Due to the nature of mobile applications, touchscreens, and lack of controllers (as opposed to VR, controls and interfaces within AR) there was a need to keep the interface mostly as non-diegetic – as a HUD type of interface. This meant that multiple buttons on the screen would carry out similar actions to those of the VR version. Because of this, a certain degree of immersion was lost; however, this was not considered detrimental, as the interface was designed to maximise both form and function. For example, the mobile phone's touchscreen was the only appropriate form of input needed to perform actions within the virtual showroom environment, where the controls or buttons on the interface consisted of a teleport menu, zoom in/out buttons, and move forward/backward buttons. The AR version of the virtual showroom also featured the same customiser and light switch interface as the VR version. As with the VR version of the showroom, these interfaces allowed for slightly more immersion and realism by allowing users simply to use their fingers to, for example, change the paint colours on the walls, by using a gradient and tap, or to turn a light on and off.

Movements in augmented reality

Users had more freedom of movement in the AR version than in the VR version, as no position scanners were needed to track them; however, this did not mean that walls and furniture in the physical world were not a problem, as real-world obstacles could still block users from reaching their desired location within the showroom. For this reason, a teleport menu was added, consisting of one "Teleport Menu" button that could expand into several buttons, each signifying a teleport to the centre point of every different room in the showroom (Figure 6.12). This feature was also useful in situations where users got lost or confused, helping them 'reset' their location. In addition to this, the "Move Forward" and "Move Backward" buttons moved users to a set distance forward or backward. This was useful whenever users were blocked from reaching their desired location due to real-world obstacles. The "Zoom In" and

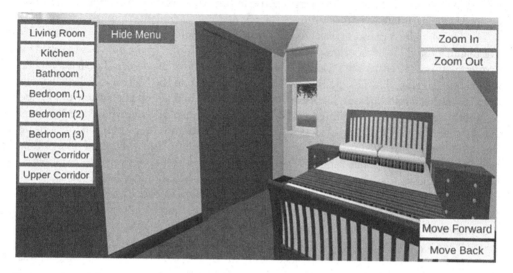

Figure 6.12 Mobile version of the showroom and teleportation menu in AR

"Zoom Out" button was also implemented to allow users to adjust their camera's field of view. The feature was added based on the users' feedback during usability test trials.

In addition to this, the "Move Forward" and "Move Backward" buttons moved users to a set distance forward or backward respectively, with each button's press. This was useful where users were blocked from reaching their desired location due to the real-world environment, but the teleport feature could not help. The "Zoom In" and "Zoom Out" button was implemented to allow users to adjust their camera's field of view. With each button press, the camera would either zoom in or out a set distance but would not move in any direction. This feature was added after users requested it, so they then had freedom to adjust their field of view to be able to see objects closer or view full rooms in one screen.

Interactions in the AR application

As there was no 'pointer' in the AR interface, the portability of objects was not as readable for users as it was in VR. In other words, users had to 'guess' which objects were moveable and tap and hold them. This was not so much of an issue as all 'stand-alone' objects (not walls or staircases) were made moveable in this version of the showroom. Once users tapped on a moveable object, it would then hover in the air, allowing them to position it to a desired location by physically walking, while still holding the object. This study found that users trying to find out which objects were moveable allowed them to experience a feeling of experimentation, thereby generating a higher sense of immersion.

6.6 Usability test

At the final stage, the prototype was tested, validated, and refined by conducting a series of usability test trials. These trials engaged 20 project participants from different age groups and varied VR/gaming experience levels. From these trials, nine participants were selected from designers, managers, and customers of Norscot, who volunteered themselves to contribute to this study. An additional 11 more participants were invited to attend the test during the Scottish Innovation of The Year 2018 Award Ceremony. All individual and group briefing tests started with a general discussion of the project, followed by a detailed briefing. This enabled every participant to familiarise themselves with the technology in order for them to use the hardware correctly. Each participant was then given 10 minutes to experience the virtual showroom, navigating through different spaces to experience different features of the software. Feedback was captured throughout this process. Once the tests were complete, each participant was invited to complete a usability test questionnaire comprising 10 questions regarding their experience from using the system. For each question, users were asked to rank their experience of using a specific feature from 1 (Difficult—Confusing) to 5 (Very Easy—Intuitive). The range of questions varied from the ease of teleporting through the showroom or climbing a staircase to customising the interior such as changing the type of wallpaper or manipulating features imported from the IFC file, such as U-Values or types of windows. Finally, each user was requested to answer an open-ended question regarding their overall experience. This included any specific challenges they faced and future development wishes. Figure 6.13 shows the specific setting and environment for these usability tests.

Regarding the VR environment, all users appreciated the developed VRE and considered it an engaging way to show how a house would look and feel. Movements and interaction inside the showroom were found to be quite smooth, and no individual user raised issues of 'motion sickness' or 'jittering'. More importantly perhaps, users did not experience challenges

Figure 6.13 Specific setting and environment of the conducted usability tests

Figure 6.13 (Continued)

of adapting back to the normal environment after testing had been completed. Regarding interactions implemented in the VR environment, users found that turning the lights on/off was very easy and intuitive. Wallpaper changes were also straightforward, while paint colour changes were somewhat less obvious.

The study also found that some parts of tutorial sections for the VR environment (Figure 6.14) were a little too heavy with text, which proved difficult to read by some testers. The testing process also revealed that many users wanted to 'rush into the action' rather than wait for tutorials to end. In order to mitigate this challenge, a number of supportive graphics were subsequently included.

Regarding the AR environment, feedback from the users and clients indicated that they would like the app to be usable on both iOS (mobile operating system created and developed by Apple Inc. exclusively for its hardware, e.g. iPhone devices) and mainstream Android platforms. In other words, they were not too particular on the extra capabilities that Tango provided, such as the possibility of physically walking in order to navigate in the virtual showroom, although they recognised these were very useful functions. All other functionality and operational issues seemed to meet their needs. The only real challenge concerned the two-button Zoom function to change the camera's field of view, which appeared rather difficult to use. In order to mitigate this issue, a potential solution was provided by allowing users to just use finger-pinch movement to zoom in and out or display the value of the field of view number to users as an additional piece of information.

Additional sample participant feedback and solutions can be seen in Table 6.1.

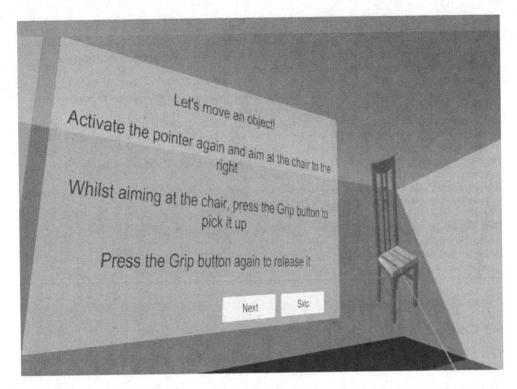

Figure 6.14 Tutorial section of the VR application

Table 6.1 Virtual reality mode usability test feedback

Feedback	Solutions
Users were confused between the left and right controllers.	Change the models of the controllers in VR so that users can visually differentiate between them.
Client wanted the colour picker/ gradient to be changed to a 'swatch' interface.	Add an interface of buttons to display the colours (and possibly names) of set paint colours. These colours/ materials could be loaded from a library.
Interface elements such as buttons and text were too 'basic'.	Include custom interface and supporting elements (buttons/ fonts). Possible inclusion of third-party software.
Tutorial text was difficult to read and understand.	Adjust accordingly. Support from assistant in interim.
Grip and Touchpad buttons were hard to find due to controllers' design.	Additional support needed to help users understand the controls beforehand.
Use of touchpad alone was easy, but added functionality created confusion in movement/interfaces.	Additional support needed to help users understand the controls beforehand.
Menu and Teleport button were often mixed up.	Additional support needed to help users understand the controls beforehand. Possible inclusion of some form of animation or highlighting (on a button in VR) to visually indicate the buttons.

(Continued)

Table 6.1 (Continued)

Feedback	Solutions
Teleport Menu (Floor Plan Menu) was hard to see in tight spaces – got lost inside the walls.	Add this interface to the controller rather than the headset.
Colour picker/painter interface was hard to use with head movement.	Gradient interface changed to swatches to mitigate head movement.
Holding down the Touchpad whilst using the Trigger was a common issue.	The controller's pointer can be made to automatically activate when hovering over an interface, rather than via manual input.
Headset was quite heavy and cables were a nuisance.	Consider future headsets that feature wireless VR capabilities.
Teleporting with the controller led to some participants teleporting into walls.	Allow the pointer to only recognise the floor as a teleport location.
Difficulty in appreciating room types' identification on the floor plan.	Add names on the floor plan image.
Inability to adjust the distance of a moveable object.	Add a button/function on the controller to allow users to change the distance of an object.

6.7 Conclusion

This chapter highlighted the feasibility of creating an abstract standardised interface for IFC modelling. The research presented addressed the functionality gap between BIM and immersive technologies using an interface that integrates 3D building information within an immersive environment. From extant literature, it was noted that despite the wide adoption of BIM in AEC, data management and software interoperability issues still existed. These challenges were significant and real, impeding the smooth collaboration, integration, and interoperability of AEC professionals. For example, interoperability has been known to be a significant factor in spurring innovation and competition amongst vendors. The work presented in this chapter highlighted some of these issues, noting that BIM integration need not be constrained to a particular proprietary interface, nor does it require the interfacing application be a strict vector CAD package or even a discrete geometry virtual environment.

This study is a response to the need for a universal platform that streamlines design processes, placing emphasis on addressing clients' needs. The challenge was to maximise engagement and immersion, especially with clients and those untrained in spatial comprehension (as is often the case in AEC). This project extended the synergy between technologies, focussing on the integration of immersive technologies to tackle the challenges of perception imposed by conventional drawings. Findings revealed that VR representation of BIM models also improved the immersion into building designs and facilitated the comprehension of complex spatial problems relating to physical form and non-physical constraints.

Despite several attempts by other studies at developing VR add-ons for the major BIM applications (especially Autodesk Revit), little research in the market actively integrated openBIM with VR environments. Given this lack, the work presented here specifically aimed to address this integration, since leveraging interoperability was seen as a major opportunity to maximise engagement. Another innovative element stemming from this project was the potential for users to transition away from bulky and expensive headset-based VR systems towards a less expensive and (arguably) more user-friendly Google Tango interface. This interface also had the capability of remembering proximity space, whilst also being able to track users' movements in 3D, as well as integrating the physical environment with VR. The project demonstrated the process of linking VR representation to openBIM information in real-time, offering data-rich content through a virtual showroom to enable users to engage in early decision-making throughout the development stages. This showroom was available for use on both VR and AR platforms to enable users to engage with the system in a way that better suited their needs.

One of the main findings emanating from this project was the need to investigate how powerful interactive design and visualisation applications such as this can help AEC stakeholders address their needs. In this specific instance, the client body was deemed to include a house builder, small contractor, or prospective homeowner. This prototype facilitated an interactive and immersive design environment, which enabled them to see inside the home – to truly visualise the environment and interact with this program to make design decisions; through this engagement it was perceived that might enable potential customers to be more likely to proceed to purchase. In this respect, virtual showrooms of this nature could be used for both marketing and streamlining of the purchasing process, ensuring homes are delivered to clients' expectations on a right first-time basis. This system could also be used to streamline sales processes by minimising costs for clients and the businesses (minimal design changes/ iterations prior to sign-off); however, whilst such showroom prototypes do not necessarily replicate the full functionality of existing BIM platforms *per se*, they instead offer an ability to combine advanced interactive visualisation with the power of BIM. Moreover, the real intelligence captured here is access to all relevant BIM data, so that the impact of design changes can be seen on such issues as building performance, aesthetics, costs, etc.

This fully working proof-of-concept prototype includes all aspects of integration between openBIM, VR, and the Tango environments. The next stage of this process will be to develop this application commercially, providing an automatic platform to import various kinds of BIM files through a common user environment. As an agnostic pared-down openBIM system, this does not discriminate between BIM authoring tools and is therefore not tethered or constrained by proprietary systems. Here, building information is taken straight from an authoring tool and is transformed through IFC conversion into a database, which is then visualised on the Unity platform as a game engine. In this respect, VR representation of BIM data improves immersion into building designs and also facilitates the comprehension of complex spatial problems related to physical form and non-physical constraints. AR opens a further possibility of eliminating bulky equipment, particularly as mobile AR applications are now able to increase the awareness of real-world space and sizes, where, for example, a site walkthrough (using an AR app) allows users to see conflicts between the physical site and the desired design.

This agnostic approach allows media richness to be fully exploited, whilst also allowing users to appreciate the full functionality of low-latency interaction. Future development opportunities include fully incorporating the concepts of social presence theory – to not only balance concurrent entity interaction with telepresence, but also explore the psychological processes and social orientation underpinning motivation and collective cohesion.

Acknowledgements

The authors would like to gratefully acknowledge the generous funding received from Innovate UK, Scottish Funding Council, Construction Scotland Innovation Centre (CSIC), and Norscot Joinery Ltd for this award-winning project, which eventually received the buildingSMART International Awards 2018: Special Distinction and Scottish Knowledge Exchange Awards 2018: The Innovation of the Year Award. We also would like to acknowledge the contributions of Veselina Chavdarova, Stephen Oliver, Farhad Chamo, and Lilia Potseluyko Amobi to the underpinning research presented in this chapter.

References

Abanda, F. H., Tah, J. H. M., & Cheung, F. K. T. (2017). BIM in off-site manufacturing for buildings. *Journal of Building Engineering*, *14*(Supplement C), 89–102. https://doi.org/10.1016/j.jobe.2017.10.002

Afsari, K., Eastman, C. M., & Castro-Lacouture, D. (2017). Javascript object notation (JSON) data serialization for IFC schema in web-based BIM data exchange. *Automation in Construction*, *77*(Supplement C), 24–51. https://doi.org/10.1016/j.autcon.2017.01.011

Anandarajan, M., Zaman, M., Dai, Q., & Arnzie, B. (2010). Generation Y adoption of instant messaging: An examination of the impact of social usefulness and media richness on use richness. *IEEE Transactions on Professional Communication*, *53*(2), 132–143. https://doi.org/10.1109/TPC.2010.2046082

Arayici, Y., Fernando, T., Munoz, V., & Bassanino, M. (2018). Interoperability specification development for integrated BIM use in performance based design. *Automation in Construction*, *85*(Supplement C), 167–181. https://doi.org/10.1016/j.autcon.2017.10.018

Assaf, S. A., & Al-Hejji, S. (2006). Causes of delay in large construction projects. *International Journal of Project Management*, *24*(4), 349–357. https://doi.org/10.1016/j.ijproman.2005.11.010

Ballestero, E., Robinson, P., & Dance, S. (2017). Head-tracked auralisations for a dynamic audio experience in virtual reality sceneries. *International Congress on Sound Vibration*. www.iiav.org/archives_icsv_last/2017_icsv24/content/papers/papers/full_paper_386_20170329170651697.pdf

Banihashemi, S., Tabadkani, A., & Hosseini, M. R. (2018). Integration of parametric design into modular coordination: A construction waste reduction workflow. *Automation in Construction*, *88*, 1–12. https://doi.org/10.1016/j.autcon.2017.12.026

Biocca, F. (1997). *The cyborg's dilemma: Embodiment in virtual environments*. Paper presented at the Proceedings Second International Conference on Cognitive Technology Humanizing the Information Age. https://doi.org/10.1109/CT.1997.617676

Blind, K. (2013). *The impact of standardization and standards on innovation: Compendium of evidence on the effectiveness of innovation policy intervention*. Manchester University Press.

Cao, D., Wang, G., Li, H., Skitmore, M., Huang, T., & Zhang, W. (2015). Practices and effectiveness of building information modelling in construction projects in China. *Automation in Construction*, *49*(Part A), 113–122. https://doi.org/10.1016/j.autcon.2014.10.014

Cha, H. S., & Lee, D. G. (2015). A case study of time/cost analysis for aged-housing renovation using a pre-made BIM database structure. *KSCE Journal of Civil Engineering*, *19*(4), 841–852. https://doi.org/10.1007/s12205-013-0617-1

Chi, H.-L., Kang, S.-C., & Wang, X. (2013). Research trends and opportunities of augmented reality applications in architecture, engineering, and construction. *Automation in Construction*, *33*(Supplement C), 116–122. https://doi.org/10.1016/j.autcon.2012.12.017

Choi, J., & Kim, I. (2016). *Development of openBIM-based interoperability system for code checking system*. Paper presented at the Green and Smart Technology.

Choi, J., & Kim, I. (2017). *A methodology of building code checking system for building permission based on openBIM*. Paper presented at the ISARC 2017 – Proceedings of the 34th International Symposium on Automation and Robotics in Construction.

Craveiro, F., Bartolo, H. M., Gale, A., Duarte, J. P., & Bartolo, P. J. (2017). A design tool for resource-efficient fabrication of 3d-graded structural building components using additive manufacturing. *Automation in Construction*, *82*(Supplement C), 75–83. https://doi.org/10.1016/j.autcon.2017.05.006

Damen, T., MacDonald, M., Hartmann, T., Di Giulio, R., Bonsma, P., Luig, K., Sebastian, R., & Soetanto, D. (2014). *BIM based collaborative design technology for collective self-organised housing*. Paper presented at the 40th IAHS World Congress on Housing. Sustainable Housing Construction.

Damen, T., Sebastian, R., MacDonald, M., Soetanto, D., Hartmann, T., Di Giulio, R., Bonsma, P., & Luig, K. (2015). The application of BIM as collaborative design technology for collective self-organised housing. *International Journal of 3-D Information Modeling (IJ3DIM)*, 4(1), 1–18. https://doi.org/10.4018/IJ3DIM.2015010101

Dawood, N., Rahimian, F., Seyedzadeh, S., & Sheikhkhoshkar, M. (2020). *Enabling the development and implementation of digital twins*. Proceedings of the 20th International Conference on Construction Applications of Virtual Reality, Teesside University. ISBN:9780992716127

Dong, S., Feng, C., & Kamat, V. R. (2013). Sensitivity analysis of augmented reality-assisted building damage reconnaissance using virtual prototyping. *Automation in Construction*, 33(Supplement C), 24–36. https://doi.org/10.1016/j.autcon.2012.09.005

Du, J., Zou, Z., Shi, Y., & Zhao, D. (2018). Zero latency: Real-time synchronization of BIM data in virtual reality for collaborative decision-making. *Automation in Construction*, 85, 51–64. https://doi.org/10.1016/j.autcon.2017.10.009

Ercan, B., & Elias-Ozkan, S. T. (2015). Performance-based parametric design explorations: A method for generating appropriate building components. *Design Studies*, 38(Supplement C), 33–53. https://doi.org/10.1016/j.destud.2015.01.001

Gasser, U., & Palfrey, J. G. (2007). *Breaking down digital barriers: When and how ICT interoperability drives innovation*. Berkman Center Research Publication, Social Science Research Network (SSRN). https://doi.org/10.2139/ssrn.1033226

Haile, N., & Altmann, J. (2018). Evaluating investments in portability and interoperability between software service platforms. *Future Generation Computer Systems*, 78, 224–241. https://doi.org/10.1016/j.future.2017.04.040

Husserl, E. (1973). *Cartesian mediations: An introduction to phenomenology*. Martinus Nijhoff. www.24grammata.com/wp-content/uploads/2011/11/Husserl-Cartesian-Meditations-24grammata.com_.pdf

Jaffar, N., Tharim, A. H. A., & Shuib, M. N. (2011). Factors of conflict in construction industry: A literature review. *Procedia Engineering*, 20, 193–202. https://doi.org/10.1016/j.proeng.2011.11.156

Khalili, A., & Chua, D. K. H. (2013). IFC-based framework to move beyond individual building elements toward configuring a higher level of prefabrication. *Journal of Computing in Civil Engineering*, 27(3), 243–253. https://doi.org/10.1061/(ASCE)CP.1943-5487.0000203

Koch, C., Neges, M., König, M., & Abramovici, M. (2014). Natural markers for augmented reality-based indoor navigation and facility maintenance. *Automation in Construction*, 48(Supplement C), 18–30. https://doi.org/10.1016/j.autcon.2014.08.009

Lee, K., & Nass, C. (2003). *Designing social presence of social actors in human computer interaction, computer human interaction 2003* (Vol. 5) ACM.

Lee, Y.-C., Eastman, C. M., & Solihin, W. (2018). Logic for ensuring the data exchange integrity of building information models. *Automation in Construction*, 85(Supplement C), 249–262. https://doi.org/10.1016/j.autcon.2017.08.010

Lester, E. I. A. (2017). Chapter 42 – conflict management and dispute resolution. In *Project management, planning and control* (7th ed., pp. 393–399). Butterworth-Heinemann. https://doi.org/10.1016/B978-0-08-102020-3.00042-5

Li, X., Wu, P., Shen, G. Q., Wang, X., & Teng, Y. (2017). Mapping the knowledge domains of building information modeling (BIM): A bibliometric approach. *Automation in Construction*, 84, 195–206. https://doi.org/10.1016/j.autcon.2017.09.011

Liu, H., Al-Hussein, M., & Lu, M. (2015). BIM-based integrated approach for detailed construction scheduling under resource constraints. *Automation in Construction*, 53, 29–43. https://doi.org/10.1016/j.autcon.2015.03.008

Liu, H., Singh, G., Lu, M., Bouferguene, A., & Al-Hussein, M. (2018). BIM-based automated design and planning for boarding of light-frame residential buildings. *Automation in Construction*, 89, 235–249. https://doi.org/10.1016/j.autcon.2018.02.001

Lu, Y., Kim, Y., Dou, X., & Kumar, S. (2014). Promote physical activity among college students: Using media richness and interactivity in web design. *Computers in Human Behavior*, 41, 40–50. https://doi.org/10.1016/j.chb.2014.08.012

Mansuri, L. E., & Patel, D. A. (2021). Artificial intelligence-based automatic visual inspection system for built heritage. *Smart and Sustainable Built Environment, ahead-of-print* (ahead-of-print). https://doi.org/10.1108/SASBE-09-2020-0139

Mitkus, S., & Mitkus, T. (2014). Causes of conflicts in a construction industry: A communicational approach. *Procedia – Social and Behavioral Sciences, 110*, 777–786. https://doi.org/10.1016/j.sbspro.2013.12.922

Monizza, G. P., Rauch, E., & Matt, D. T. (2017). Parametric and generative design techniques for mass-customization in building industry: A case study for glued-laminated timber. *Procedia CIRP, 60* (Supplement C), 392–397. https://doi.org/10.1016/j.procir.2017.01.051

Moshtaghian, F., Golabchi, M., & Noorzai, E. (2020). A framework to dynamic identification of project risks. *Smart and Sustainable Built Environment*. https://doi.org/10.1108/SASBE-09-2019-0123

Newman, C., Edwards, D., Martek, I., Lai, J., Thwala, W. D., & Rillie, I. (2020). Industry 4.0 deployment in the construction industry: A bibliometric literature review and UK-based case study. *Smart and Sustainable Built Environment, ahead-of-print* (ahead-of-print). https://doi.org/10.1108/SASBE-02-2020-0016

Norouzi, N., Shabak, M., Embi, M. R. B., & Khan, T. H. (2015). The architect, the client and effective communication in architectural design practice. *Procedia – Social and Behavioral Sciences, 172*, 635–642. https://doi.org/10.1016/j.sbspro.2015.01.413

Odeh, A. M., & Battaineh, H. T. (2002). Causes of construction delay: Traditional contracts. *International Journal of Project Management, 20* (1), 67–73. https://doi.org/10.1016/S0263-7863(00)00037-5

Oliver, S., Seyedzadeh, S., Rahimian, F., Dawood, N., & Rodriguez, S. (2020). Cost-effective as-built BIM modelling using 3D point-clouds and photogrammetry. *Current Trends in Civil & Structural Engineering-CTCSE, 4* (5), 000599. https://doi.org/10.33552/CTCSE.2020.04.000599

Oxman, R. (2017). Thinking difference: Theories and models of parametric design thinking. *Design Studies, 52* (Supplement C), 4–39. https://doi.org/10.1016/j.destud.2017.06.001

Park, J. (2011). BIM-based parametric design methodology for modernized Korean traditional buildings. *Journal of Asian Architecture and Building Engineering, 10* (2), 327–334. https://doi.org/10.3130/jaabe.10.327

Pauwels, P., Zhang, S., & Lee, Y.-C. (2017). Semantic web technologies in AEC industry: A literature overview. *Automation in Construction, 73* (Supplement C), 145–165. https://doi.org/10.1016/j.autcon.2016.10.003

Petrova, E. A., Rasmussen, M., Jensen, R. L., & Svidt, K. (2017). *Integrating Virtual Reality and BIM for End-user Involvement in Building Design: A case study.* Paper presented at the The Joint Conference on Computing in Construction (JC3) 2017. https://doi.org/10.24928/JC3-2017/0266

Pour Rahimian, F., Seyedzadeh, S., Oliver, S., Rodriguez, S., & Dawood, N. (2020). On-demand monitoring of construction projects through a game-like hybrid application of BIM and machine learning. *Automation in Construction, 110*, 103012. https://doi.org/10.1016/j.autcon.2019.103012

Rahimian, F. P., Seyedzadeh, S., & Glesk, I. (2019). OCDMA-based sensor network for monitoring construction sites affected by vibrations. *Journal of Information Technology in Construction, 24*, 299–317.

Rasmussen, M., Gade, A. N., & Jensen, R. L. (2017). *Bridging the gap between actors and digital tools in a knotworking design process.* Paper presented at the When Social Science meets BIM and LEAN (Vol. 1, Chapter 5, p. 3).

Ren, J., Liu, Y., & Ruan, Z. (2016). Architecture in an age of augmented reality: Applications and practices for mobile intelligence BIM-based AR in the entire lifecycle. *DEStech Transactions on Computer Science and Engineering (ICEITI)*. https://doi.org/10.12783/dtcse/iceiti2016/6203

Sarhadi, M. (2016). Comparing communication style within project teams of three project-oriented organizations in Iran. *Procedia – Social and Behavioral Sciences, 226*, 226–235. https://doi.org/10.1016/j.sbspro.2016.06.183

Schroeder, R. (2002). Social interaction in virtual environments: Key issues, common themes, and a framework for research. In R. Schroeder (Ed.), *The social life of avatars: Computer supported cooperative war.* SDK. https://doi.org/10.1007/978-1-4471-0277-9_1

Sebastian, R. (2010). Integrated design and engineering using building information modelling: A pilot project of small-scale housing development in The Netherlands. *Architectural Engineering and Design Management*, 6(2), 103–110. https://doi.org/10.3763/aedm.2010.0116

Seyedzadeh, S., Agapiou, A., Moghaddasi, M., Dado, M., & Glesk, I. (2021). WON-OCDMa system based on MW-ZCC codes for applications in optical wireless sensor networks. *Sensors*, 21. https://doi.org/10.3390/s21020539

Seyedzadeh, S., & Pour Rahimian, F. (2021a). Building energy data-driven model improved by multi-objective optimisation. In *Data-driven modelling of non-domestic buildings energy performance* (pp. 99–109). Springer. https://doi.org/10.1007/978-3-030-64751-3_6

Seyedzadeh, S., & Pour Rahimian, F. (2021b). Building energy performance assessment methods. In *Data-driven modelling of non-domestic buildings energy performance* (pp. 13–30). Springer. https://doi.org/10.1007/978-3-030-64751-3_2

Seyedzadeh, S., Rahimian, F. P., Oliver, S., Glesk, I., & Kumar, B. (2020). Data driven model improved by multi-objective optimisation for prediction of building energy loads. *Automation in Construction*, 116, 103188. https://doi.org/10.1016/j.autcon.2020.103188

Shi, X., Liu, Y.-S., Gao, G., Gu, M., & Li, H. (2018). IFCdiff: A content-based automatic comparison approach for IFC files. *Automation in Construction*, 86(Supplement C), 53–68. https://doi.org/10.1016/j.autcon.2017.10.013

Smart BIM Solutions. (2012). *SmartBIM solutions open BIM & closed BIM*. Retrieved February 28, 2017, from http://ckegroup.org/thinkbimblog/wp-content/uploads/2012/04/BIM-vs-OPENBIM.pdf

Walasek, D., & Barszcz, A. (2017). Analysis of the adoption rate of building information modeling [BIM] and its return on investment [ROI]. *Procedia Engineering*, 172(Supplement C), 1227–1234. https://doi.org/10.1016/j.proeng.2017.02.144

Wang, X., Kim, M. J., Love, P. E. D., & Kang, S.-C. (2013). Augmented reality in built environment: Classification and implications for future research. *Automation in Construction*, 32, 1–13. https://doi.org/10.1016/j.autcon.2012.11.021

Wang, X., Truijens, M., Hou, L., Wang, Y., & Zhou, Y. (2014). Integrating augmented reality with building information modeling: Onsite construction process controlling for liquefied natural gas industry. *Automation in Construction*, 40(Supplement C), 96–105. https://doi.org/10.1016/j.autcon.2013.12.003

Wu, G., Liu, C., Zhao, X., & Zuo, J. (2017). Investigating the relationship between communication-conflict interaction and project success among construction project teams. *International Journal of Project Management*, 35(8), 1466–1482. https://doi.org/10.1016/j.ijproman.2017.08.006

Yuan, Z., Sun, C., & Wang, Y. (2018). Design for manufacture and assembly-oriented parametric design of prefabricated buildings. *Automation in Construction*, 88, 13–22. https://doi.org/10.1016/j.autcon.2017.12.021

Zulch, B. G. (2014). Communication: The foundation of project management. *Procedia Technology*, 16, 1000–1009. https://doi.org/10.1016/j.protcy.2014.10.054

7 A centralised cost management system

Exploiting earned value management and activity-based costing within integrated project delivery

7.1. Introduction

Integrated project delivery (IPD) has been characterised as the early, collaborative, and collective engagement of key stakeholders throughout all phases of the project delivery process (Ahmad et al., 2019). For example, traditional forms of IPD (such as alliancing) can be implemented without Building Information Modelling (BIM); however, newer forms of IPD are being defined in relation to their integration with BIM (Rowlinson, 2017), which it is proffered can help facilitate much smoother data exchange between project packages and parties, and consequently, is much more in line with IPD's aims and objectives (AIA, 2007).

The integration of BIM and IPD can help improve the anticipated outcomes emanating from projects, particularly within the design and construction process, which includes (but is not limited to) such issues relating to cost/profit, scheduling, return on investment, safety, productivity, and associated relationships (Elghaish et al., 2019a). In this respect, IPD relies on the principles of open pricing techniques and fiscal transparency amongst participants (Ahmad et al., 2019). This is especially important, as project stakeholders (such as designers and contractors) typically assess and determine their profit and shared risks according to the mandated deviations agreed between actual and target costs (AIA, 2007). Given this level of transparency, successful delivery of a project through IPD can sometimes be challenging, as IPD requires parties to fulfil a wide range of requirements (Fischer et al., 2017). Of these requirements, the IPD compensation model, also called risk/reward compensation approach, is of cardinal importance (Ma et al., 2018). This has been described as a key principle of IPD (Zhang et al., 2018) and has been acknowledged as playing a pivotal role in stimulating creativity, motivating collaboration, and sustaining performance (Zhang & Li, 2014b). In this respect, the proportion of risk and reward must therefore be shared/allocated to all project participants within the core project team in order to facilitate joint project control (Fischer et al., 2017). Thus, designing the risk and reward model (hereafter referred to as compensation approach) is an important part of this delivery model, as its foundations are predominantly cost-driven (AIA, 2007).

It is important to acknowledge at this juncture that IPD should only be considered as an approach if it meets the requirements of project parties. In this respect, the formulation of the cost structure of IPD may require significant improvements in order to avoid obfuscated profits in any of the estimated costs (Allison et al., 2018), which is somewhat counterproductive, as one of the main reasons for using IPD is to maximise trust amongst project parties (Ma et al., 2018). Given that risks/rewards are not specifically shared individually for IPD core team members (AIA, 2007; Pishdad-Bozorgi & Srivastava, 2018), then any errors in calculating individual costs for each work package could be misrepresented, thereby

DOI: 10.1201/9781003106944-7

influencing the profit-at-risk value of each IPD team member. On this theme, one of the main characteristics of IPD is the ability to defer the allocation of parties' profits until all project works are completed. This in itself invites a discussion on the challenges associated with the implementation speed of IPD, since this requires all members to attend all meetings even if their works have been completed at an early stage in the project (Roy et al., 2018). As such, using Information and Communication Technology (ICT) has been seen as a vital support mechanism, to not only facilitate communication *per se* but also to share information among project parties regardless of their geographical location.

A review of literature in this field shows several trends of research emanating from these topics. Of these, a major part of research is now starting to explore the potential of available tools and techniques, particularly with Earned Value Management (EVM) and Activity-Based Costing (ABC) within IPD (Hosseini et al., 2018). These studies, for the most part, stop at providing an outline of how these methods and techniques really add value to IPD's risk/reward sharing mechanism (Pishdad-Bozorgi & Srivastava, 2018). Literature also highlights the importance of BIM integration with IPD practices (Allison et al., 2018; Fischer et al., 2017; Rowlinson, 2017). For example, the challenges of integration are explored through a stream of studies, including financial challenges, the difference in cost accounting between participants, and the lack of a risk/reward sharing mechanism that can be accepted by all participants (Kahvandi et al., 2018). Whilst this work is particularly valuable, it stops short of providing clear workable methodologies that demonstrate BIM tools/dimensions and an interrelationship with the IPD stages in practical terms (Roy et al., 2018).

In order to start to uncover some of these issues, this chapter presents a discussion on the design and development of an automated cost control model for IPD projects. This integrates ABC into EVM by engaging mathematical equations that support EVM to determine risk/reward for the owner and all non-owner parties. The EVM approach is extended by a grid that allocates the output of its Cost Performance Ratio (CPR) and Schedule Performance Ratio (SPR). This enables all parties to subsequently track their performance through the system. The EVM-web system includes two kinds of reports: (i) a graphical report that shows previous performance along with current project performance; from this, each milestone is presented as a star inside the EVM-grid, which is divided into four zones ('optimal zone', 'neutral zone', 'risk zone', and 'crisis zone'); and (ii) a metrics report that shows three main values (reimbursed costs, profit, and cost saving) for both owner and non-owner parties.

7.2. Information and communication technology in construction management

From an ICT perspective, Jacobsson and Linderoth (2010) highlighted the need to share a wide range of information within Architecture, Engineering, and Construction (AEC). Coincidentally, the need for integrative ICT solutions have been espoused for many years now, particularly to address (i) integration between design and production (construction stage), and (ii) to facilitate communication among different disciplines (teams) whether internal (the same organisation) or across different organisations (Söderholm, 2006).

Whilst some progress has been made, BIM has been considered as one potential solution to help this integration happen (Latiffi et al., 2013). For example, BIM has now become mandatory in many countries, thus both increasing the take-up of this approach and also supporting the rate of ICT adoption more readily (Eadie et al., 2013). In this respect, ICT web-based management systems are considered 'proven tools', enabling parties to work more efficiently and effectively, including cost control activities (Ozorhon et al., 2014). For example, Li et al.

(2006) developed and tested web systems for managing and displaying project performance using the EVM method. This web system was used for data management in construction, particularly for the application of Map-Based Knowledge Management (MBKM) for contractors (Lin et al., 2006). This use of ICT in data management helped facilitate a greater understanding of digitalising knowledge as a 'map', presented graphically as symbols with embedded data. This made it much easier for users to communicate through specific symbols, minimising the need for redundant text (Wexler, 2001). Web systems have also been used for monitoring cost/schedule projects (Chou et al., 2010). Moreover, the use of EVM enables the display of scheduling and costs simultaneously, thereby enabling stakeholders to better appreciate, understand, and track their tasks more readily (Li et al., 2006).

7.3. Cost management implications within BIM and IPD

In moving towards more efficient project delivery, the ultimate goal is having a database of information that is available to all project participants, with confidence in its accuracy, universal utility, and clarity (Oraee et al., 2017). From this aspiration, one of the main drivers for adopting BIM is its ability to integrate and manage project documents and stages (i.e. design, planning, and costing) within a single/dynamic context to secure the proper exploitation of available information (Abrishami et al., 2015); however, in order to be able to do this, BIM design elements should contain the appropriate required information, including design or management (Banihashemi et al., 2018), to acquire smartly designed elements, rather than traditional 3D components (Pärn & Edwards, 2017). The rationale behind this is that BIM users should be capable of acquiring all the required information from a single BIM element in order to make informed decisions (Elghaish et al., 2019a).

Four-dimensional modelling (4D BIM) can embed progress data into 3D model objects by adjusting the task-object relationship (Hamledari et al., 2017). In doing so, the application of 4D BIM has now led to more readily accessible workflows, greater on-site management efficiency, and enhanced constructability (Hartmann et al., 2008). In relation to cost management, the use of BIM has been acknowledged as one of the most efficient AEC tools for increasing productivity (Wang et al., 2016a). The next iteration of BIM, colloquially termed 5D BIM (Aibinu & Venkatesh, 2013), had the ability to incorporate quantities from 3D models, which enabled cost consultants to include productivity allowances and pricing values (Lee et al., 2014). For example, the cost estimating process starts by exporting data from 3D models into the BIM-based cost estimating software (e.g. CostX®) in order to prepare quantity take-off. From this, the Bills of Quantities are generated and exported to an external database (Aibinu & Venkatesh, 2013). If required, prices and productivity allowances can also be added to project schedule preparation (Lee et al., 2014). Such incorporation and automation has been seen to shorten the quantity take-off processing time (especially if design changes are made), thereby helping to fast-track projects (Wang et al., 2016a).

In summary, applying cost estimation in IPD provides a number of significant benefits (AIA, 2007; Elghaish et al., 2019a). In this respect, it is particularly useful for tracking and scrutinising events, thereby enabling core team members to appreciate their profit and shared benefits/risks, according to deviations between actual and target costs (Zhang & Li, 2014b). That being said, it is important to acknowledge that the compensation approach structure must be capable of drawing upon effective methods in order to determine cost overrun proportions, cost underruns, and any saving in the total budget under the agreed cost (Elghaish et al., 2019a). The rationale behind this is because the risk/reward proportion relies on the degree of achievement during the entire project (Pishdad-Bozorgi & Srivastava, 2018). Given this, the compensation approach has two limits: (i) the direct, indirect, and overhead costs (which can be nominated as agreed cost), and (2) the profit-at-risk percentage after estimating the agreed cost (AIA, 2007; Zhang & Li, 2014b). On this basis, the precise determination of risk perception is critical for ensuring the agreed compensation structure is implemented correctly so that the risk/reward ratio is fairly allocated among project participants. For example, participants carrying higher proportions of uncertain works can be compensated with higher profit-at-risk percentage (Das & Teng, 2001). Table 7.1 presents a summary of the main challenges of IPD cost management.

Table 7.1 IPD cost management challenges

Stage	Challenges	References
Cost Estimation Challenges	Accounting system can be unclear or unreadable to all IPD core team members due to different backgrounds.	(Roy et al., 2018)
	Given that the Target Value Design (TVD) is a part of the IPD approach, therefore continuous estimation feedback is vital to accomplish the pre-construction IPD stages to making proper decisions.	(Allison et al., 2018; Zimina et al., 2012)
	Given that LIMB-2 represents the overhead costs in addition to the profit-at-risk percentage, a detailed estimation technique is needed to ensure that the contractor does not hide any profit into overhead costs.	(Ashcraft, 2011)
	There is a misunderstanding in risk contingency accounting where sophisticated pricing is presented as critical barriers to IPD implementation.	(Ashcraft, 2012; Liu et al., 2013)
Cost Budget and Control (Risk/Reward sharing) Challenges	Utilising BIM can improve the traditional cost/ scheduling processes; however, most of the research does not consider the variances amongst project delivery approaches.	(Lu et al., 2016)
	Given that the IPD approach stages do not include a tender stage to select the optimal bid, a methodology framework for developing a cash flow system using BIM tools within documentation and buyout stage is needed.	(Wang et al., 2016b)
	Some aspects of implementing IPD (particularly, that of financial management) can act as a major barrier, e.g. sharing risk/reward requires an automated/immutable system to record achieved profit, cost saving, and reimbursed monetary values for each member due to the fact that IPD core team members cannot receive their profits/rewards until all project works have been delivered.	(Ashcraft, 2012; Zhang & Li, 2014a)
	"Cost and schedule are relatively easy to measure. If there are early profit distributions, however, there must be a method for comparing progress achieved to the progress required at that milestone. This will invariably involve some level of estimating using a modified earned value calculation with claw-back and true-up provisions."	(Ashcraft, 2011)

(Continued)

Table 7.1 (Continued)

Stage	Challenges	References
	Due to all participants sharing their profit/risks (regardless of their timeline of executing works), an automated system is therefore required to ensure that all profits/risks are moved to the profit/risk pools accurately.	(Allison et al., 2018; Roy et al., 2018)
BIM and IPD Integration Challenges	IPD, TVD, and BIM are regarded as a winning combination for improving project delivery success; however, very limited research is available to validate the positive aspects of these relationships by providing workable solutions that appeal to practitioners.	(Do et al., 2015; Pishdad-Bozorgi et al., 2013)
	Limited workable methodologies are available for demonstrating the interrelationships among BIM tools/dimensions and IPD stages in practical terms.	(Allison et al., 2018; Holland et al., 2010)
	There are significant issues regarding how BIM is specified, the precise processes for developing BIM communication standards, and how BIM should be managed and administered.	(Glick & Guggemos, 2009)

7.4. Earned value management

EVM is a quantitative project management technique that can be used for measuring project progress. It can also be used to provide project participants with early warning signs, particularly where a project is running 'over budget' or 'behind schedule' (PMI, 2013). This is particularly useful, as Khamooshi and Abdi (2016) provided evidence of EVM being successfully applied to several real-life projects and being able to deliver accurate cost/schedule metrics. Similarly, Naeni et al. (2011, p. 764) noted that "earned value technique is a crucial technique in analysing and controlling the performance of a project". EVM can therefore be seen as an effective tool for managing and measuring costs through CPR and SPR values, when for example, in traditional projects, the granularity of activities embedded in the project schedule are represented using the Work Breakdown Structure (WBS), and project costs are represented through the Cost Breakdown Structure (CBS). This arrangement has been acknowledged as a challenge concerning the accurate implementation of EVM (Pajares & López-Paredes, 2011), where it was acknowledged that any EVM system needed to be smarter, incorporate more advanced capabilities, be able to correlate data from multiple sources, and be able to automatically generate cost control reports (Lipke et al., 2009). In this respect, it has also been advocated that interoperability (cf. various data sources) needed to be resolved in order to build federated project cost control sheets, which could be addressed using advanced technologies and visualisation techniques (Chou et al., 2010).

7.5. Activity-based costing

Costing for construction projects can sometimes be overly complicated. This complexity is due in part to the interface of multiple activities, fragmentation of the process, disparate project structures, and project participants, a corollary of which can lead to an increase in overhead activities/costs (Mignone et al., 2016). In this respect, the main processes of costing uses various approaches, including Resource-Based Costing (RBC), which relies on the resources' cost, and

Volume-Based Allocation (VBA), which is based on allocating costs directly to objects/activities, regardless of the cost structure – ergo, direct, indirect, and overhead costs (Holland & Jr, 1999); however, the cost allocation processes can often attract cost distortions, usually due to conflation of all indirect costs into one, which distorts pricing (Miller, 1996). So, acknowledging these challenges, ABC is considered one such solution for dealing with these distortions, by allocating costs of multi-pools and determining the overhead activities (and associated costs needed to transform resources into activities) to deliver the final product (Kim & Ballard, 2001). In this respect, the ABC approach is able to measure costs based on activities, linking cost drivers to impact measures of the product or service (Tsai & Hung, 2009). In doing so, ABC is able to improve the efficiency and accuracy of cost-related information and further monitor and control project costs (Tsai et al., 2014). This becomes particularly relevant in a collaborative working environment (like IPD) that engages multiple stakeholders (Kim et al., 2016).

7.6. Research methodology

The research methodological approach adopted in this chapter commences with a literature review to highlight the research gap in terms of the capabilities of proposed methods and processes – specifically, how ABC, EVM, and BIM can be integrated in order to automate financial transactions within IPD. As such, the methodological approach commences by investigating measures for integrating ABC into EVM; in particular, the development of mathematical formulas and associated metrics for IPD, where proposed formulas show the risk/reward values for each party, including the entire project performance. The methodology then presents a data visualisation technique based on EVM, where a 'proof-of-concept' framework is developed to test the applicability of the framework. The following tools/approaches were used in this process:

- Microsoft Access (for developing the database) – the process was strengthened using macros and the Visual Basic (VB) programming language to automate processes;
- Caspio (a cloud platform for building online database applications) was used to develop a set of web pages for sharing data;
- A website was developed and linked with the data server to automate/synchronise the data sharing process;
- Integration of 4D and 5D BIM into the web-based management system supported by EVM-grid web pages.

7.7. Framework development

The development of this framework was divided into three stages. The first stage involved building a robust cost structure of IPD based on ABC, using a proposed Centralised Cost Management System (CCMS). The second stage involved developing EVM-based ABC mathematical formulas to determine the risk/reward values (to determine three main financial transactions: reimbursed costs, profit, and cost saving). The third stage involved demonstrating how BIM and web-based information systems could be effectively utilised.

7.7.1. Centralised cost management system alignment to IPD

One of the main challenges in developing a system of this nature was to ensure it effectively captured the system requirements. In this respect, CCMS can be seen as a cost management

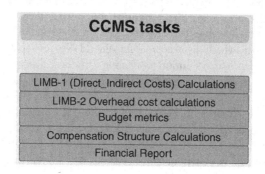

Figure 7.1 CCMS IPD user interface

system for bridging the gap in IPD cost management practices. The main interface of CCMS is divided into five main areas (Figure 7.1).

7.7.2. *Development of EVM based on ABC extensions*

A set of mathematical formulas based on EVM and ABC were considered in order to determine the risk/reward values for owner and non-owner parties. These were developed to provide the due reimbursed costs, cost savings, and profit for both the owner and non-owner parties (see equations 10, 11, and 12). These equations help enable payment automation in the CCMS model. This approach was adopted to enhance transparency and trust among IPD core team members.

An EVM-grid was developed to display the outcome of EVM's CPR and SPR, which divides the project into four zones, where each zone represents a different case (Figure 7.2). Potential project cases are then actuated on the grid, whilst considering the X-axis (as the schedule) and the Y-axis (as the cost); each zone can then be divided into small squares around the planned point. The main four zones are (1) the cost and schedule outcomes are positive – this case is therefore deemed 'optimal'. In this research, the cost was assumed as a critical parameter. In zone (2) when the cost was positive but the schedule was negative, the case is called 'neutral'; however, if the outcome of the schedule performance was significantly negative, then it might be that the cumulative parameter was very close to the risk zone (which would need investigating). Similarly, through zone (3), if the cost performance was negative and schedule performance was positive, this zone according to the mentioned assumption would be the 'risk' zone. Finally, zone (4) represents the 'crisis' zone, where the outcome of both cost and schedule are negative.

From an operational perspective, users commence the process by determining the values of the CPR and SPR before entering them into the grid (as a positive or negative percentage). This codifies the project situation at each milestone (or for each package). In this respect, the financial expert (typically the Quantity Surveyor) marks the square in accordance with CPR and SPR percentages in order to determine the cumulative progress throughout the project execution stages. Thereafter, the 'profit-at-risk' percentage is shared in accordance with the output of the developed EVM-based IPD grid.

From a development perspective, Elghaish et al. (2019b) demonstrated a risk/reward sharing model for the IPD approach. This model was extended in this research by enhancing the

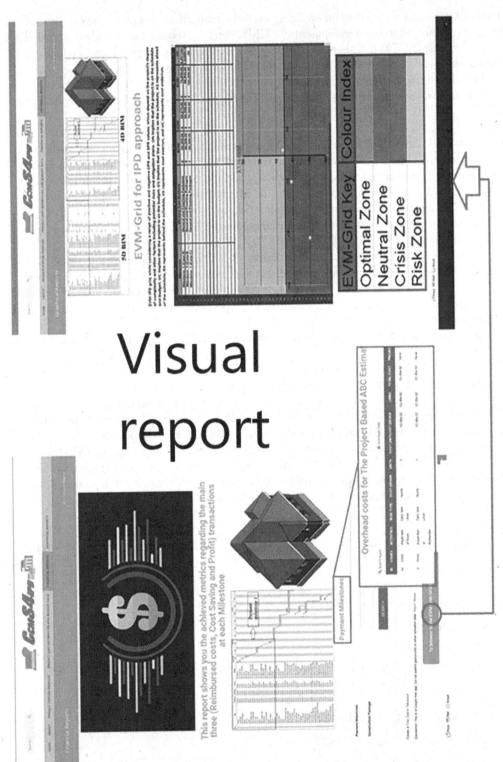

Figure 7.2 CCMS financial reporting pages

automation process (as it relies on applying a set of equations according to the outcome of the EVM); however, in the development of CCMS, users are required to provide the Earned Value Outcome (EVO) and other cost values, where subsequently the profit, cost saving, and reimbursed costs are determined automatically through CCMS.

The risk/reward sharing approaches based on EVM and ABC are supported by a number of equations to support the scenarios. An overview of these equations can be seen as follows:

- Equation 1 shows the EVO representing the schedule and cost performances. Equation 2 presents the adjusted EVO, considering the P@R% (since this shows whether the performance is greater or less than the P@R%), and is used to determine the project case. Equation 3 is another adjustment to decide whether there are costsavings (Reward) or not. This equation is structured as a conditional equation, so that if the adjusted EVO ≥ 0, then the results would be the value of the adjusted EVO; otherwise, the value would be considered zero.
- After determining the project case, equations 4, 5, and 6 were developed to determine the value of achieved rewards (in the direct and indirect costs). Equation 4 determines the total value of the reward if there is a cost saving in the direct and indirect costs. Equations 5 and 6 are used to calculate the proportions for both the owner and non-owner parties.
- Equations 7 and 8 determine the cost savings for overheard costs (based on the ABC sheet).
- Equations 9 and 10 are used to calculate the summation of the reward for both owner and non-owner parties (for direct, indirect, and overhead costs).
- Equation 11 calculates the reimbursed costs according to the project case. This is a conditional equation based according to the EVO4Profit and two sub-equations for determining the reimbursed costs, i.e. if EVO4Profit > 0 and if EVO4Profit < 0.
- Equation 12 determines the profit as a conditional equation according to EVO4Profit value against the P@R%. This relates to two sub-equations: one equation relates to the case where the entire LIMB-1 (Profit) is to be paid, and another equation relates to the case where part of this profit was consumed as a cost.

$$EVO$$
$$= ([CPI]$$
$$* [SPI])$$
(1)

Where EVO represents Earned Value Outcome

$$Adjusted\, EVO$$
$$= [P@R\, per]$$
$$- (1 - [EVO])$$
(2)

$$EVO4Profit = IIf ([Adjusted\, EVO] >$$
$$= 0, [Adjusted\, EVO], 0)$$
(3)

Where EVO4Profit is Earned Value Outcome for Profit

$$MV \text{ } for \text{ } R \text{ } for \text{ } each \text{ } party(LIMB-1)$$
$$= IIf\left(\left[EVO4Profit\right]\right.$$
$$> [P@R per],([PLIMB-1]$$
$$-[Actual \text{ } LIMB-1]),0\Big) \tag{4}$$

Where MV for R for each party (LIMB-1) represents Monetary Value for Reward for each owner and non-owner parties and LIMB-1 is the direct and indirect cost

$$Reward \text{ } For \text{ } Owner \text{ } (LIMB-1)$$
$$= \left[MV \text{ } for \text{ } R \text{ } for \text{ } each \text{ } party(LIMB-1)\right] \tag{5}$$
$$*[PoO]$$

$$Reward \text{ } For \text{ } non-Owner(LIMB-1)$$
$$= \left[MV \text{ } for \text{ } R \text{ } for \text{ } each \text{ } party(LIMB-1)\right] \tag{6}$$
$$*[PoNO]$$

Where PoNO or PoO is the proportion of sharing cost saving for non-owner parties/owner

$$CSoOC \text{ } for \text{ } NO$$
$$= \left([CSoOOA]\right. \tag{7}$$
$$+\left([CSoOPA]*NoARP\right)\Big)$$

$$CSoOC \text{ } for \text{ } O = \left([CSoOPA]\right. \tag{8}$$
$$*[OARP]$$

Where CSoOC for NO represents cost saving of overhead cost for non-owner parties, CSoOOA represents cost saving of overhead organisation activities, CSoOPA represents cost saving of overhead project activities, and NoARP/OARP is the non-owner/owner agreed reward percentage

$$TR4O = \left(\left[Reward \text{ } For \text{ } Owner(LIMB-1)\right]\right. \tag{9}$$
$$+[CSoOC \text{ } for \text{ } O]$$

$$TR4NO = \left[Reward \text{ } For \text{ } non-Owner \text{ } parties(LIMB-1)\right] \tag{10}$$
$$+[CSoOC \text{ } for \text{ } NO]$$

Where TR4O/TR4NO is total reward for owner/non-owner parties

$$Reimbursed\ Cost = IIf\Big([EVO4Profit] > 0, \big([TCS] - \big([Profit]$$

$$+\big[MV\ for\ R\ or\ RD\ for\ each\ party\ (LIMB-1)\big] + \big[CSoOC\ for\ NO\big]$$

$$+\big[CSoOC\ for\ O\big]\big)\big), \big(\big([TCS] - [Profit]\big) + \big[DC\ above\ TCS\big]\big)\Big) \qquad (11)$$

$$Profit = IIf\Big([EVO4Profit] >= [P@R\ per], [LIMB-3],$$

$$\big([EVO4Profit]*10*[LIMB-3]\big)\Big) \qquad (12)$$

Where TCS represents total compensation structure

7.7.3. *Development of the web interface for displaying project data*

The web-based management system was developed through six primary web pages. Figure 7.2 and Figure 7.3 present three pages, representing three functional pages (cost estimation and budgeting page – image 1 in Figure 7.3; financial report and graphical report presented in Figure 7.2). The remaining three pages are not functional (Figure 7.4), representing issues such as the 'Home' page (including information about the purpose and rubrics of this platform), 'About' page (includes information from the framework, in order to demonstrate how the cost estimation, budgeting, and control tasks were developed), and the 'Project parties profiles' page (identifying the profiles of IPD core team members). This arrangement provides a comprehensive source of information on all cost management tasks, including the proposed risk/reward models. This arrangement was structured to increase trust and transparency amongst all IPD core team members.

The three pages used for displaying and managing the cost management data (including cost estimation, budget, and the risk/reward for each party) are based on the EVM outcomes (Figure 7.2), where the data is stored on a server (MS Access database).

Figure 7.2 depicts the financial metrics web page for each party, showing the 4D/5D BIM data. From this, each party can use the search function using the name of a 'Package' (i.e. general package) to show the financial metrics used for different payment milestones. The financial metrics present three main transactions (reimbursed costs, cost saving, and profit). Given that the profit/risk is shared regardless of individual performance, the achieved values of the three financial transactions are presented individually to maximise transparency, trust, and collaboration amongst IPD core team members (without the need to attend all regular meetings). The generic values of the three transactions also show project progress. In this respect, the equations (1 to 12) highlighted earlier are used to integrate ABC into EVM to calculate the risk/reward sharing models, including the outcome of payment milestones (see Figure 7.2). In addition, a formal report can be retrieved by each party as required. This also includes collective details of all parties' performance against the financial metrics agreed upon. If required, parties are also able to share this report with others in order to facilitate a wider understanding of total project performance.

Figure 7.2 presents a snapshot of the web-data page of the EVM-grid. This includes the calculation parameters used; however, in order to protect access to sensitive data, each party is required to provide their username and password to access this information. In this respect, all cost-driven data (e.g. risk/reward values) is updated directly and is seamlessly aligned to serve various IPD stages of the project. Moreover, during the buyout and documentation

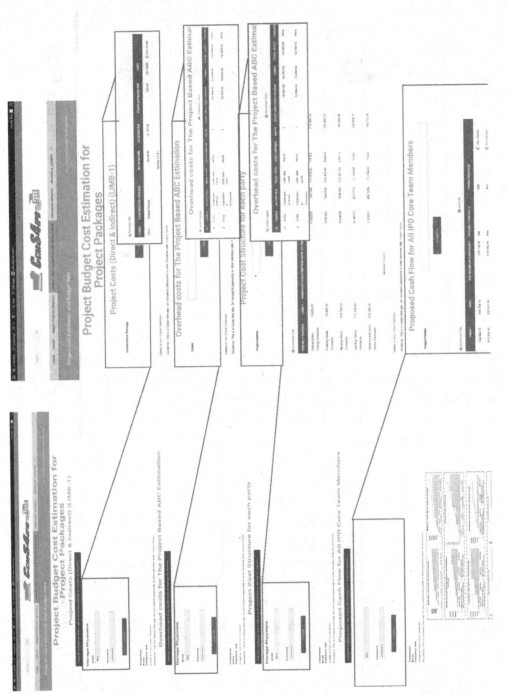

Figure 7.3 CCMS cost management content

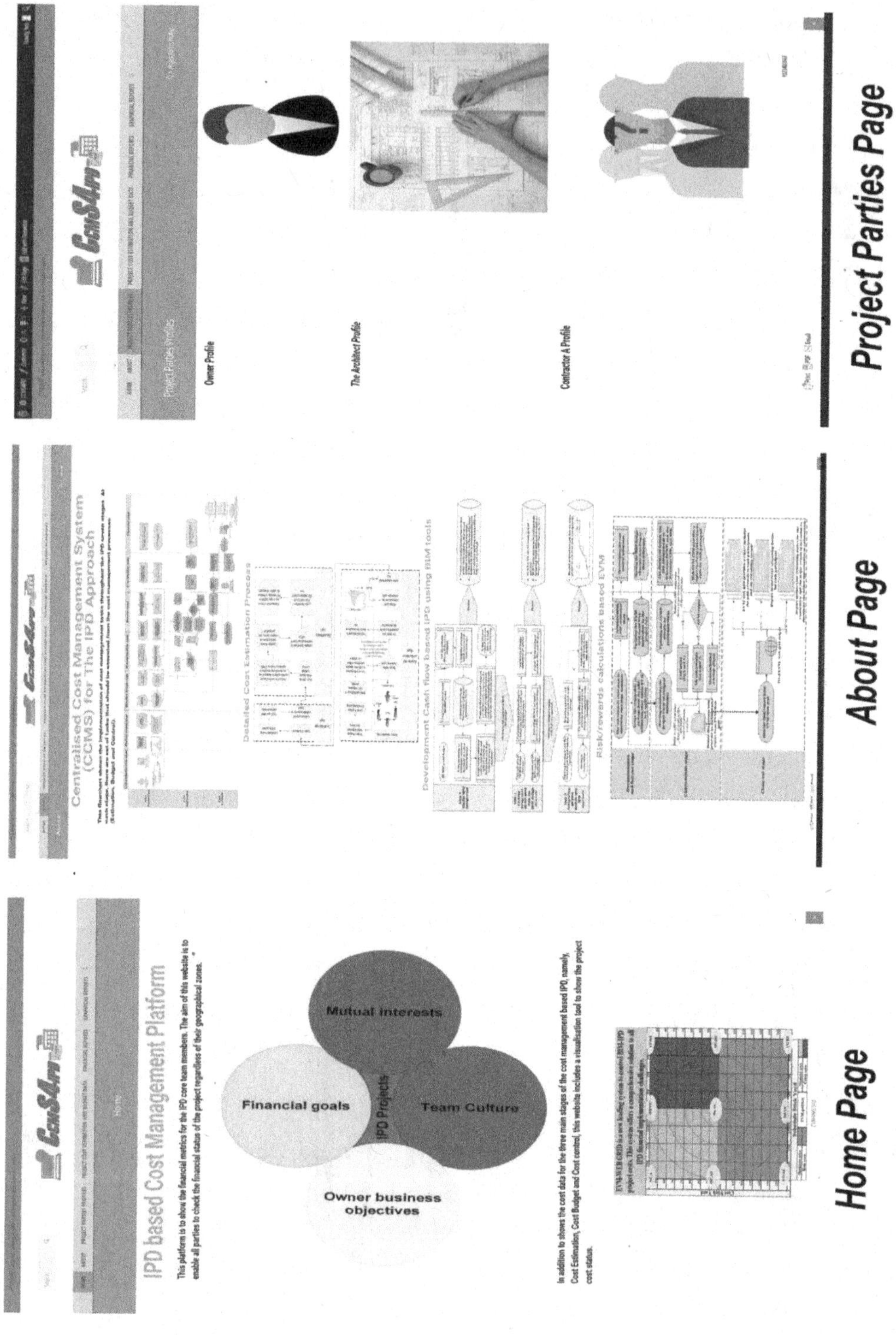

Figure 7.4 CCMS non-data page

stages, the web page Project Cost Estimation and Buyout Data presents all data needed to make appropriate decisions.

Figure 7.3 presents a snapshot of the cost estimation and budgeting tasks used for IPD in three forms, namely: Limb-1 (direct and indirect costs), Limb-2 (ABC sheet for estimating the overhead costs), the cost structure of each trade package in the project (the cost-plus P@R%), and the budgeting values including the estimated minimum and maximum cash flows. The web page structure was designed to enable database searches using different parameters, such as the construction package for Limb-1, the code and project parties for Limb-2, and the project parties for Limb-3. This functionality enables parties to retrieve specific data they require, regardless of their attendance at IPD core team members' meetings. Moreover, the legibility of data was considered specifically to allow both cognate and non-cognate parties (from various backgrounds) to understand the structure of this data. This helped ensure consistency in matters relating to privacy, sensitive data, costing data, and authentication (username/password) before displaying data (Figure 7.3, image 1), where usernames and passwords are allocated by specified users from the server admin team.

Figure 7.4 presents the CCMS non-data page. This covers the following pages Home page, About page, and Project Parties page. Given the nature of IPD, it was important to acknowledge these three areas in order to present the logical arrangement of the system. Anecdotally, from a design perspective, one of the design parameters considered important to members was the capability of being able to add or remove IPD core team members (even in 'live' projects). This functionality was embedded to accommodate members who might have completed their work packages early. The concept of providing details of project parties (Figure 7.4, image 3) resonates with IPD's rationale of being able to build and support sustainable relationships from the outset.

7.8. Conclusion

This chapter presented a new structured approach for managing financial tasks within IPD. The literature review evaluated IPD's cost management process to identify high-impact areas for exploitation, along with weak points that needed to be addressed. The context used aligned specifically to previously accepted financial approaches such as ABC and EVM, with the intention of aligning and integrating these through BIM into a single dynamic process.

This study is novel in several ways, insofar as it presents (i) an innovative grid for locating both the CPR and SPR ratios (to provide a picture of the project's position in terms of cost and schedule); (ii) a system for integrating the EVM-grid with the ABC estimating method to optimise the cost structure (to accurately reflect the compensation structure); (iii) specific models for dealing with risk/reward sharing, offering maximum transparency of data from the ABC sheets (to distinguish between direct and overhead cost savings); (iv) a framework for distinguishing between the sustaining/organisation level and the project level; and (v) a web-based EVM-grid that allows project parties to track their work packages, whilst also being able to monitor total project performance.

In summary, the web-based management system (CCMS) provides an interactive interface for tracking project cost data across all IPD stages. The structure and arrangements of this enables project parties to check their work package costs (direct, indirect, and overhead costs) using the 'name of the construction package' or their 'party name'. The entire compensation structure can also be viewed, along with accompanying data. CCMS is also able to generate financial reports (profit, cost savings, and reimbursed costs at each payment milestone). The layout and format of this information is aligned to accommodate parties with disparate backgrounds in order to facilitate more meaningful engagement and comprehension.

References

Abrishami, S., Goulding, J., Pour Rahimian, F., & Ganah, A. (2015). Virtual generative BIM workspace for maximising AEC conceptual design innovation: A paradigm of future opportunities. *Construction Innovation, 15*(1), 24–41.

Ahmad, I., Azhar, N., & Chowdhury, A. (2019). Enhancement of IPD characteristics as impelled by information and communication technology. *Journal of Management in Engineering, 35*(1), 04018055. https://doi.org/10.1061/(ASCE)ME.1943-5479.0000670

AIA. (2007). *Integrated project delivery: A guide.* Retrieved May 20, 2021, from www.aia.org/resources/64146-integrated-project-delivery-a-guide

Aibinu, A., & Venkatesh, S. (2013). Status of BIM adoption and the BIM experience of cost consultants in Australia. *Journal of Professional Issues in Engineering Education and Practice, 140*(3), 04013021. https://doi.org/10.1061/(ASCE)EI.1943-5541.0000193

Allison, M., Ashcraft, H., Cheng, R., Klawens, S., & Pease, J. (2018). *Integrated project delivery: An action guide for leaders.* Integrated Project Delivery Alliance (IPDA), Center for Innovation in the Design and Construction Industry (CIDCI), Charles Pankow Foundation. Retrieved May 20, 2021, from https://hdl.handle.net/11299/201404

Ashcraft, H. W. (2012). *The IPD framework* (pp. 1–25). Hanson Bridgett LLP. Retrieved May 20, 2021, from www.hansonbridgett.com/~/media/Files/Publications/NegotiatingIntegratedProjectDeliveryAgreement.pdf

Ashcraft, Jr., H. W. (2011). Negotiating an integrated project delivery agreement. *Construction Law, 31*, 17.

Banihashemi, S., Tabadkani, A., & Hosseini, M. R. (2018). Integration of parametric design into modular coordination: A construction waste reduction workflow. *Automation in Construction, 88*, 1–12. https://doi.org/10.1016/j.autcon.2017.12.026

Chou, J.-S., Chen, H.-M., Hou, C.-C., & Lin, C.-W. (2010). Visualized EVM system for assessing project performance. *Automation in Construction, 19*(5), 596–607. https://doi.org/10.1016/j.autcon.2010.02.006

Das, T., & Teng, B.-S. (2001). A risk perception model of alliance structuring. *Journal of International Management, 7*(1), 1–29. https://doi.org/10.1016/S1075-4253(00)00037-4

Do, D., Ballard, G., & Tommelein, I. D. (2015). *An analysis of potential misalignments of commercial incentives in integrated project delivery and target value design.* Proceedings of the 23rd Conference of the International Group for Lean Construction, pp. 277–286.

Eadie, R., Browne, M., Odeyinka, H., McKeown, C., & McNiff, S. (2013). BIM implementation throughout the UK construction project lifecycle: An analysis. *Automation in Construction, 36*, 145–151.

Elghaish, F., Abrishami, S., Abu Samra, S., Gaterell, M., Hosseini, M. R., & Wise, R. (2019a). Cash flow system development framework within integrated project delivery (IPD) using BIM tools. *International Journal of Construction Management*, 1–16.

Elghaish, F., Abrishami, S., Hosseini, M. R., Abu-Samra, S., & Gaterell, M. (2019b). Integrated project delivery with BIM: An automated EVM-based approach. *Automation in Construction, 106*, 102907.

Fischer, M., Ashcraft, H. W., Reed, D., & Khanzode, A. (2017). *Integrating project delivery.* John Wiley & Sons Inc. ISBN:978-0-470-58735-5

Glick, S., & Guggemos, A. (2009, April). *IPD and BIM: Benefits and opportunities for regulatory agencies.* Proceedings of the 45th ASC National Conference.

Hamledari, H., McCabe, B., Davari, S., & Shahi, A. (2017). Automated schedule and progress updating of IFC-based 4D BIMs. *Journal of Computing in Civil Engineering, 31*(4), 04017012. https://doi.org/10.1061/(ASCE)CP.1943-5487.0000660

Hartmann, T., Gao, J., & Fischer, M. (2008). Areas of application for 3D and 4D models on construction projects. *Journal of Construction Engineering and Management, 134*(10), 776–785. https://doi.org/10.1061/(ASCE)0733-9364(2008)134:10(776)

Holland, N. L., & Jr, D. H. (1999). Indirect cost categorization and allocation by construction contractors. *Journal of Architectural Engineering, 5*(2), 49–56. https://doi.org/10.1061/(ASCE)1076-0431(1999)5:2(49)

Holland, R., Messner, J., Parfitt, K., Poerschke, U., Pihlak, M., & Solnosky, R. (2010). *Integrated design courses using BIM as the technology platform.* Proceedings of the National Institute of Building Sciences: Annual Meeting of EcoBuild America Conference.

Hosseini, M. R., Maghrebi, M., Akbarnezhad, A., Martek, I., & Arashpour, M. (2018). Analysis of citation networks in building information modeling research. *Journal of Construction Engineering and Management, 144*(8), 04018064. https://doi.org/10.1061/(ASCE)CO.1943-7862.0001492

Jacobsson, M., & Linderoth, H. C. (2010). The influence of contextual elements, actors' frames of reference, and technology on the adoption and use of ICT in construction projects: A Swedish case study. *Construction management and Economics, 28*(1), 13–23.

Kahvandi, Z., Saghatforoush, E., Ravasan, A. Z., & Mansouri, T. (2018). An FCM-based dynamic modelling of integrated project delivery implementation challenges in construction projects. *Lean Construction Journal, 2018*, 63–87.

Khamooshi, H., & Abdi, A. (2016). Project duration forecasting using earned duration management with exponential smoothing techniques. *Journal of Management in Engineering, 33*(1), 04016032. https://doi.org/10.1061/(ASCE)ME.1943-5479.0000475

Kim, Y. W., & Ballard, G. (2001, August). *Activity-based costing and its application to lean construction.* Proceedings of the 9th Annual Conference of the International Group for Lean Construction, National University of Singapore.

Kim, Y.-W., Han, S.-H., Yi, J.-S., & Chang, S. (2016). Supply chain cost model for prefabricated building material based on time-driven activity-based costing. *Canadian Journal of Civil Engineering, 43*(4), 287–293. https://doi.org/10.1139/cjce-2015-0010

Latiffi, A. A., Mohd, S., Kasim, N., & Fathi, M. S. (2013). Building information modeling (BIM) application in Malaysian construction industry. *International Journal of Construction Engineering and Management, 2*(4A), 1–6.

Lee, S.-K., Kim, K.-R., & Yu, J.-H. (2014). BIM and ontology-based approach for building cost estimation. *Automation in Construction, 41*, 96–105. https://doi.org/10.1016/j.autcon.2013.10.020

Li, J., Moselhi, O., & Alkass, S. (2006). Internet-based database management system for project control. *Engineering, Construction and Architectural Management, 13*(3), 242–253.

Lin, Y.-C., Wang, L.-C., & Tserng, H. P. (2006). Enhancing knowledge exchange through web map-based knowledge management system in construction: Lessons learned in Taiwan. *Automation in Construction, 15*(6), 693–705.

Lipke, W., Zwikael, O., Henderson, K., & Anbari, F. (2009). Prediction of project outcome: The application of statistical methods to earned value management and earned schedule performance indexes. *International Journal of Project Management, 27*(4), 400–407. https://doi.org/10.1016/j.ijproman.2008.02.009

Liu, M., Griffis, F., & Bates, A. (2013). *Compensation structure and contingency allocation in integrated project delivery.* ASEE Annual Conference and Exposition, Conference Proceedings.

Lu, Q., Won, J., & Cheng, J. C. (2016). A financial decision making framework for construction projects based on 5D building information modeling (BIM). *International Journal of Project Management, 34*(1), 3–21.

Ma, Z., Zhang, D., & Li, J. (2018). A dedicated collaboration platform for integrated project delivery. *Automation in Construction, 86*, 199–209. https://doi.org/10.1016/j.autcon.2017.10.024

Mignone, G., Hosseini, M. R., Chileshe, N., & Arashpour, M. (2016). Enhancing collaboration in BIM-based construction networks through organisational discontinuity theory: A case study of the new Royal Adelaide Hospital. *Architectural Engineering and Design Management, 12*(5), 333–352. https://doi.org/10.1080/17452007.2016.1169987

Miller, J. A. (1996). *Implementing activity-based management in daily operations.* The University of Michigan, John Wiley & Sons. ISBN:9780471040033

Naeni, L. M., Shadrokh, S., & Salehipour, A. (2011). A fuzzy approach for the earned value management. *International Journal of Project Management, 29*(6), 764–772. https://doi.org/10.1016/j.ijproman.2010.07.012

Oraee, M., Hosseini, M. R., Papadonikolaki, E., Palliyaguru, R., & Arashpour, M. (2017). Collaboration in BIM-based construction networks: A bibliometric-qualitative literature review. *International Journal of Project Management, 35*(7), 1288–1301. https://doi.org/10.1016/j.ijproman.2017.07.001

Ozorhon, B., Karatas, C. G., & Demirkesen, S. (2014). A web-based database system for managing construction project knowledge. *Procedia-Social and Behavioral Sciences, 119*, 377–286.

Pajares, J., & López-Paredes, A. (2011). An extension of the EVM analysis for project monitoring: The cost control index and the schedule control index. *International Journal of Project Management, 29*(5), 615–621. https://doi.org/10.1016/j.ijproman.2010.04.005

Pärn, E. A., & Edwards, D. J. (2017). Conceptualising the FinDD API plug-in: A study of BIM-FM integration. *Automation in Construction, 80*, 11–21. https://doi.org/10.1016/j.autcon.2017.03.015

Pishdad-Bozorgi, P., Moghaddam, E. H., & Karasulu, Y. (2013). *Advancing target price and target value design process in IPD using BIM and risk-sharing approaches.* 49th Associated Schools of Construction Annual Internation Conference.

Pishdad-Bozorgi, P., & Srivastava, D. (2018). *Assessment of Integrated Project Delivery (IPD) Risk and Reward Sharing Strategies from the Standpoint of Collaboration: A Game Theory Approach.* Paper presented at the Construction Research Congress 2018. New Orleans, Louisiana. https://doi.org/10.1061/9780784481271.020

PMI. (2013). *A guide to the project management body of knowledge (PMBOK® guide)* (5th ed.). Project Management Institute. ISBN:978-1-935589-67-9

Rowlinson, S. (2017). Building information modelling, integrated project delivery and all that. *Construction Innovation, 17*(1), 45–49. https://doi.org/10.1108/CI-05-2016-0025

Roy, D., Malsane, S., & Samanta, P. K. (2018). Identification of critical challenges for adoption of integrated project delivery. *Lean Construction Journal*, 1–15.

Söderholm, A. (2006). *Kampen om kommunikationen-Om projektledningens Informationsteknologi.* Research Report. Royal Institute of Technology.

Tsai, W.-H., Yang, C.-H., Chang, J.-C., & Lee, H.-L. (2014). An activity-based costing decision model for life cycle assessment in green building projects. *European Journal of Operational Research, 238*(2), 607–519. https://doi.org/10.1016/j.ejor.2014.03.024

Tsai, W. H., & Hung, S.-J. (2009). A fuzzy goal programming approach for green supply chain optimisation under activity-based costing and performance evaluation with a value-chain structure. *International Journal of Production Research, 47*(18), 4991–5017. https://doi.org/10.1080/00207540801932498

Wang, K.-C., Wang, W.-C., Wang, H.-H., Hsu, P.-Y., Wu, W.-H., & Kung, C.-J. (2016a). Applying building information modeling to integrate schedule and cost for establishing construction progress curves. *Automation in Construction, 72*, 397–310. https://doi.org/10.1016/j.autcon.2016.10.005

Wang, Q., Mei, T., Kong, L., & Xiao, Y. (2016b). Incentive compensation structure for cost control of construction project based on IPD-Ish in China. *Journal of Construction Engineering and Management*, 101–108.

Wexler, M. N. (2001). The who, what and why of knowledge mapping. *Journal of Knowledge Management, 5*(3), 249–264.

Zhang, L., Cao, T., & Wang, Y. (2018). The mediation role of leadership styles in integrated project collaboration: An emotional intelligence perspective. *International Journal of Project Management, 36*(2), 317–230. https://doi.org/10.1016/j.ijproman.2017.08.014

Zhang, L., & Li, F. (2014a). Risk/reward compensation model for integrated project delivery. *Inzinerine Ekonomika-Engineering Economics, 25*(5), 558–567.

Zhang, L., & Li, F. (2014b). Risk/reward compensation model for integrated project delivery. *Engineering Economics, 25*(5), 558–567.

Zimina, D., Ballard, G., & Pasquire, C. (2012). Target value design: Using collaboration and a lean approach to reduce construction cost. *Construction Management and Economics, 30*(5), 383–398.

8 Success factors driving cost management practices through integrated project delivery

8.1. Introduction

Integrated Project Delivery (IPD) is characterised by early, collaborative engagement of key stakeholders throughout all phases of a project (Ahmad et al., 2019; Ashcraft, 2014; Bensalah et al., 2019; Newman et al., 2020). Compared to traditional methods of project delivery, such as design-bid-build, construction management at-risk, and design-build, IPD is regarded as a superior delivery mode (Asmar et al., 2016; Manata et al., 2018). For example, evidence shows that IPD has the potential to improve 14 key metrics of project performance, including quality, scheduling, communication management, and cost performance (Ahmad et al., 2019; Asmar et al., 2016). Moreover, it is advocated that IPD can also facilitate trust among project participants, in that it fosters greater open pricing and transparency (Ahmad et al., 2019).

Whilst acknowledging these benefits, the IPD approach is not commonly adopted (Hamzeh et al., 2019; Pishdad-Bozorgi, 2017), particularly as a number of barriers have been identified that can hinder widespread adoption (Ghassemi & Becerik-Gerber, 2011; Sun et al., 2015), with IPD requiring extensive support systems (Fischer et al., 2017). Failure to establish these support systems from the outset can erode successful delivery of IPD projects (Durdyev et al., 2019). The required support systems include the need to establish fair IPD compensation models, full and effective information sharing, responsive decision-making regimes, and suitable liability waivers between stakeholders (Kent & Becerik-Gerber, 2010; Smith et al., 2011). Of the support systems, the IPD compensation model (also known as risk/reward compensation) is of primary importance (Ma et al., 2018), as this plays a pivotal role in stimulating creativity, motivating collaboration, and sustaining performance (Liu & Bates, 2013b; Zhang & Li, 2014). The compensation model identifies costs – direct, indirect, and overhead – and perhaps more significantly, allocates profit-at-risk percentage compensation across project participants. An agreed-upon, fair IPD compensation model is thus a vital precondition to successful project delivery (Allison et al., 2018b; Durdyev et al., 2019; Elghaish et al., 2020a; Zhang & Chen, 2010). Moreover, comprehensive cost management practices are the mainstay to IPD compensation models (Elghaish et al., 2019a; Elghaish et al., 2019b). Consequently, researchers have attempted to identify those factors that directly or indirectly affect the success of cost management practices in IPD projects. A brief description of these factors follows.

It is important to appreciate from the outset that IPD cost management systems must be integrated and resilient to the loss of cost information throughout all stages of the project (Ma et al., 2018; Zhang & Li, 2014). The cost structure must also flag potential hidden profits within the estimated costs Allison et al. (2018b). This transparency is essential to foster

DOI: 10.1201/9781003106944-8

trust between stakeholders (Ma et al., 2018; Pishdad-Bozorgi & Srivastava, 2018). Moreover, according to Roy et al. (2018), all participants must be continuously involved and engaged in any decision-making.

On reflection, existing cost management literature tends to be rather narrow in scope, with each study focusing on select aspects of cost management systems. Absent from literature is the detailed examination of the precise factors that drive the success of cost management practices in IPD. This represents a significant knowledge gap (Durdyev et al., 2019; Elghaish et al., 2020a). This chapter aims to address this gap by identifying the antecedents to the successful design of cost management practices in IPD projects. This work also includes Building Information Modelling (BIM)–enabled IPD projects.

8.2. Contextual background

8.2.1. Integrated project delivery

The term IPD refers to a project delivery approach that integrates a range of project dimensions, including people, organisations, and business structure, right from the conceptualisation stage (Mesa et al., 2016; Singleton, 2010). Kent and Becerik-Gerber (2010) argued that IPD's main objective was to eliminate fragmentation that often results when a project is led by a single entity, such as 'master builder,' over the entire project stages. In essence, IPD attempts to mobilise participants' resources to maximise value and minimise waste (Kent & Becerik-Gerber, 2010). As an example, studies show that projects employing IPD have been successful in minimising defects associated with dimensional and geometric variations, and as a result, improve the energy performance of buildings (Fischer et al., 2017; Seyedzadeh & Pour Rahimian, 2021; Talebi et al., 2020a, 2020b).

The equitable sharing of risk and reward sits at the financial heart of the IPD approach (AIA, 2007; Allison et al., 2018a). Achieving this balance requires a continuous cost estimation feedback loop over a pre-detailed design stage (Allison et al., 2018a). In this respect, several techniques have been recommended in order to optimise cost management practices of IPD projects (Allison et al., 2018b; Durdyev et al., 2019; Elghaish et al., 2019a, 2019b). A summary of these follows.

8.2.2. Earned value management

Earned Value Management (EVM) is a quantitative project management technique that can be used to measure project progress, whilst also providing early warning signs of potential budget overruns and schedule delays (Pajares & López-Paredes, 2011; PMI, 2013). On this theme, Khamooshi and Abdi (2016) highlighted that EVM was successful at delivering accurate cost and schedule metrics. According to Naeni et al. (2011, p. 764) "earned value technique is crucial technique in analysing and controlling the performance of a project". Conventional project scheduling approaches such as Work Breakdown Structure (WBS) are often mapped against actual outcomes and expenditure; however, the EVM approach requires a more focussed approach, particularly on implementation (Pajares & López-Paredes, 2011). Some have argued that EVM should be smarter and equipped with sufficient capabilities to be able to synthesise data from multiple sources – to automatically generate cost control reports (Lipke et al., 2009). In this respect, greater interoperability is needed, with federated project cost control sheets supported by dedicated technologies and visualisation tools (Chou et al., 2010).

8.2.3. *Activity-based costing*

Resource-Based Costing (RBC) is a major traditional cost accounting method. This relies on Volume-Based Allocation (VBA), in which the cost of resources is directly allocated to objects, regardless of the accounting cost structure distribution of direct, indirect, and overhead costs (Elghaish et al., 2019b). In many respect, traditional methods often fail to find the key decision variables that affect the total cost, particularly overhead costs (Kim et al., 2016). Activity-Based Costing (ABC) minimises this distortion by allocating costs through multi-pools. Thus, this method determines the overhead costs needed to transform resources into activities that deliver the final product (Kim & Ballard, 2001; Wang et al., 2010). The ABC approach measure costs based on activities and links cost drivers to impact measures to products or services (Tsai & Hung, 2009). It can therefore improve the efficiency and accuracy of cost-related information to monitor and control project costs (Tsai et al., 2014). This is particularly applicable in collaborative working environments (such as IPD), where multiple stakeholders have an impact on cost drivers (Kim et al., 2016).

8.2.4. *4D/5D BIM automation*

Integrating BIM into daily construction activities can facilitate automatic updating of site information and, as such, can enhance productivity, strengthen relationships amongst stakeholders, whilst also engendering improved levels of trust (Dawood et al., 2020; Oliver et al., 2020; Omar & Dulaimi, 2015). For example, BIM 4D automation can improve the quality of collected data and reduces human interference in the data collection process (Hamledari et al., 2017; Hartmann et al., 2008). Similarly, 5D BIM provides an effective methodology for cost data collection and analysis of construction projects (Aibinu & Venkatesh, 2013; Lee et al., 2014; Wang et al., 2016).

Acknowledging these developments, automated data collection methods have improved significantly over the years and have benefitted from the introduction of various kinds of technologies, such as barcoding, 3D laser scanning, and photogrammetry (El-Omari & Moselhi, 2011; Turkan et al., 2012, 2013). Whilst these developments are recognised and supported, Eastman et al. (2011) observed that BIM-based cost management platforms should be able to perform all cost-related processes. To date, this is still a challenge; however, research studies have considered various means for improving cost management practices of IPD projects (Hosseini et al., 2018; Wang et al., 2016). A summary of these follows.

8.3. IPD literature and the research gap

Numerous studies on the theme of IPD are available for review from extant literature. An overview of these can be seen (presented chronologically) in Table 8.1. From this, studies relating to cost management practices of IPD have, for the most part, attempted to develop tools and techniques that improve these costing practices; however, these studies do not explicitly focus on the critical success factors of cost management practices. Moreover, studies that have focussed on identifying the success of IPD projects also have a number of shortfalls. For example, Kahvandi et al. (2018) identified factors that promoted the success of IPD projects, but only from a general managerial perspective, which did not take into account cost management practices. Similarly, Tillmann et al. (2017) described the success of cost estimation practices of IPD projects, but did not provide solutions, nor consider how overhead resources were allocated.

Table 8.1 Previous studies on the topic of IPD

Authors	Contribution and limitations
Elghaish et al. (2020b)	Provided a new approach to develop a fair compensation structure of IPD-based BIM and Activity-Based Costing.
Elghaish et al. (2019a)	Developed a methodology to develop the project budget through estimating the minimum and maximum potential cash inflow (to enable project parties to make the right decision before the construction stage commence).
Elghaish et al. (2019b)	Presented a fair model for estimating three main transactions (reimbursed cost, profit, and cost saving) in IPD projects.
Kahvandi et al. (2018)	Explored various key critical success factors, largely from a managerial perspective, with limited attention to cost estimation issues.
Pishdad-Bozorgi and Srivastava (2018)	Developed a model to share risks and rewards using game theory approach, particularly for cases in which project cost exceeded the profit-at-risk percentage. This study provided an overview of the model with future empirical research needed to assess its practicality and quantify its impacts.
Alves et al. (2017)	Presented various techniques commonly used for Target Value Delivery (TVD) and applicable to the IPD context.
Tillmann et al. (2017)	Discussed the underlying mechanisms of cost estimation within IPD-oriented projects and explored the factors that influence success. This did not embrace the tactics of allocating overhead resources.
Ballard et al. (2015)	Recommended a set of procedures to enhance the chance of success in IPD cost estimation processes. Although the authors acknowledged that following TVD principles was a critical success factor, no explicit technique or procedure was recommended to make practical recommendations.
Zhang and Li (2014)	Developed a risk/reward compensation mechanism by combining risk perception and Nash Bargaining Solution (NBS) techniques. This model did not consider the method of sharing actual risk/reward amongst participants or the impact of IPD compensation structure in successful profit/cost saving sharing.
Zhang and Li (2014)	Combined risk perception and Nash Bargaining Solution (NBS) techniques to formulate a risk/reward compensation model. This model did not cover all possible types of engineering data.
Liu and Bates (2013a)	Articulated a probabilistic contingency calculation model to predict proper contingency for minimising cost overrun. The mechanisms for sharing pain/gain percentages remained unexplored.
Pishdad-Bozorgi et al. (2013)	Discussed the potential of integration among TVD, BIM, and IPD cost estimation.
Ross (2003)	Presented a risk/reward sharing model. The risk/reward ratio was measured by the Overall Performance Score (OPS), which is a scale between 0 and 100, where 0 to 50 represents the pain scope and 50 to 100 represent the gain range.

Acknowledging the findings presented in Table 8.1, and other literature available on IPD, it seems that the mechanisms by which effective cost management practices are developed still need to be described, including the antecedents to success in cost management practices. The following work presented in this chapter aims to address this need.

8.4. Research method

The research method adopted in this study was undertaken using a questionnaire survey approach engaging purposive sampling. Such sampling entailed "identification and selection of information-rich cases related to the phenomenon of interest" (Palinkas et al., 2015, p. 533). Individuals knowledgeable and experienced in regard to the topic were chosen (Etikan et al., 2016), where participants fulfilled a set of qualifying criteria (Palinkas et al., 2015). Specifically, they (i) possessed theoretical and practical knowledge of BIM, (ii) had a strong level of understanding of IPD, and (iii) were able to assess cost management tools and methods – both traditional and innovative.

An online questionnaire was designed to identify the antecedents of success for cost management practices of IPD projects. The questions sought to check and assess the status quo of cost management methods and validate the effectiveness of some solutions in dealing with IPD cost management. Regarding sampling, purposive sampling was used – defined as non-random sampling where members of the target population with predefined qualifications meet certain practical criteria, such as accessibility, proximity, and availability (Etikan et al., 2016). The sampling criteria for this study required participants to have the three basic criteria highlighted earlier as a precursor. A pilot study was conducted with six BIM and IPD experts, located in the UK. The engagement and analysis of their responses confirmed that the designed questionnaire was fit for purpose.

Questionnaires were sent out and the data collected. Reliability of the collected data was assessed, returning a Cronbach Alpha coefficient (CA) of 0.854. This indicated that all the items in the questionnaire were relevant to the research (Field, 2013). The details of this questionnaire follow.

8.5. Participants' profiles

In total, 50 participants were selected for this study, where 40% were academics (lecturers or researchers), 20% were quantity surveyors, with the remaining 40% selected from differing backgrounds (Figure 8.1).

Figure 8.2 presents the range of experience of participants, where the majority of participants recorded experience of between 1 and 5 years (46%), and 10% of participants had experience exceeding 11 years.

Figure 8.3 illustrates participants' familiarity with IPD concepts and processes, where 46% had a high level of understanding and 28% had an intermediate level of understanding, meaning that the majority of participants were well versed in IPD issues and processes.

From an antecedents of success perspective, factors facilitating IPD success are explored in the following two sections. The first section represents the characteristics of the IPD approach, while the second explores how existing IPD characteristics can be further enhanced.

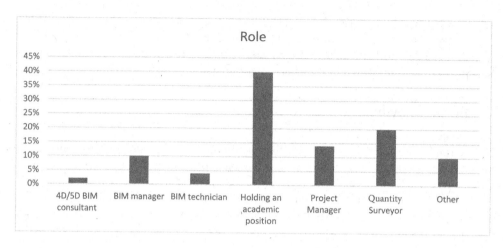

Figure 8.1 Role of participants

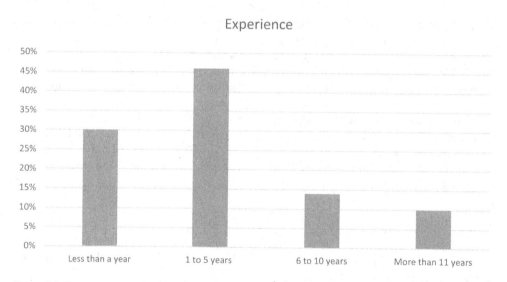

Figure 8.2 Experience ranges of participants

From literature, four main success factors associated with IPD-based cost management processes were identified (Table 8.2). Participants were asked to rank these factors in terms of the advantage they brought to IPD. The first factor, 'early involvement of all participants from the design stage', attracted 30%, followed by 'open pricing technique' (as there is no tender stage in IPD) at 26%. The next category was 'fair compensation approach' at 20%, and the final factor was the 'allocation of responsibilities and risks', ranked at 18%. 'Other factors' came in at 6%. Thus, the initial conclusion drawn was that the four identified factors captured represented the majority of possible influential factors.

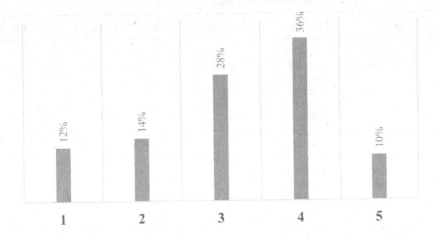

Figure 8.3 Participants' experience of IPD

Table 8.2 The IPD success factors

Factors	Frequency	Percentage	Valid Percentage	Cumulative Percentage
Tendering stage – using an open pricing technique	13	26.0	26.0	32.0
Allocation of responsibilities and risks should be clear and understandable	9	18.0	18.0	50.0
Compensation approach (Risk/Rewards sharing) should be flexible	10	20.0	20.0	70.0
Early involvement of all participants	15	30.0	30.0	100.0
Other	3	6.0	6.0	6.0
Total	50	100.0	100.0	

8.6. Improving IPD implementation

Literature revealed enabling procedures for improving the effectiveness of cost management systems. In this respect, participants were asked to evaluate these procedures. These findings are presented in Table 8.3, which includes the descriptive analysis of ten factors. These factors were further categorised into four categories, namely: ABC and EVM integration, cost estimation and budgeting, risk and reward sharing, and general. These factors are ranked from low to high, according to respondents' assessments. A discussion on each of these categories follows.

ABC and EVM integration category: Participants were asked to measure the applicability of a set of proposed features of EVM and ABC, enhanced with certain extensions. The first factors (F1 and F2) related to integrating ABC into EVM in order to develop mathematical models able to calculate risk/reward monetary values for the owner and non-owner parties. The next factor was to develop an automated platform where the developed mathematical models could be implemented automatically. The mean values for both factors were 3.36 and 3.38, respectively. The third factor (F8 in ranking) related to the integration between ABC

Table 8.3 Proposed recommendations for improving IPD cost management practices

Factors	Category	Questions	Range	Minimum	Maximum	Mean	Std. Deviation
F1	ABC and EVM Integration	Integrating EVM into IPD can easily facilitate its implementation regarding sharing risk/reward between owner/non-owner parties	4	1	5	3.36	.964
F2	ABC and EVM Integration	Using an automated model to show due payments for all parties based on their achievement against planned values	4	1	5	3.58	.906
F3	Cost Estimation/ Budgeting	Providing a separate cash flow for each participant including the proposed proportional cash based on agreed profit at risk percentage	3	2	5	3.60	.700
F4	Cost Estimation/ Budgeting	Adopting ABC to develop a list of activities to enable a reliable cash curve (S-curve) by considering (direct, indirect, and overhead)	3	2	5	3.64	.749
F5	Risk/reward sharing and ICT	Developing an EVM-based web report to enable project tracking by all participants, including easy access from different devices	4	1	5	3.64	.942
F6	Cost Estimation/ Budgeting	Utilising ABC in order to clearly identify the different sources of overhead cost	4	1	5	3.68	.935
F7	Risk/reward sharing and ICT	A fair allocation system with clear implementation models for enhancing and implementing IPD	4	1	5	3.72	.927
F8	ABC and EVM Integration	Adapting EVM with ABC to identify risk/ reward sharing fairly through developed mathematical models for all potential cases	4	1	5	3.82	.896

Factors	Category	Questions	Range	Minimum	Maximum	Mean	Std. Deviation
F9	Risk/reward sharing and ICT	Providing an EVM-grid to locate the Cost Performance Ratio (CPR) and Schedule Performance Ratio (SPR) to determine the holistic view of project progress	3	2	5	3.86	.808
F10	General	Using a comprehensive process for cost management within all IPD stages to increase implementation and minimise waste (time and resources)	4	1	5	3.98	.820

as a cost estimation tool to optimise overhead costs and EVM as a cost control tool to enable calculation of the realised cost saving for each party. This factor was rated highly by respondents, attracting a mean value of 3.82.

Cost estimation and budgeting category: Three factors were associated with the use of ABC to optimise cost structure and enhance trust among IPD team members (F3, F4, and F6), where F3 and F4 concerned the feasibility of developing a new budgeting system to present different cash curves (direct, indirect, overhead, and accumulative), based on AB. This was because conventional mechanisms for developing project budgets did not comply with IPD heuristics. F3 proposed the use of ABC to provide further details in terms of minimum and maximum profit boundaries for each party. This was to enable parties to make optimal decisions, particularly during the IPD buyout stage. Moreover, F4 proposed the development of project activities-based ABC to enable tracking of overhead activities. This attracted mean answers of 3.6 and 3.64 respectively, with experts submitting 'agree' and 'strongly agree' replies to both questions. The third factor (ranked as F6) discussed the role of ABC for optimising the overhead cost during the IPD buyout stage through a determination of the trade package that consumed significant parts of the overhead resources. In this scenario, IPD team members could move activities to create an overlap in overhead activities across different packages to maximise the utilisation of overhead resources. The mean score of responses was 3.68.

Risk and reward sharing, and ICT category: Factors in this category were highly ranked, reflecting its importance to IPD. These were ranked as F5, F7, and F9, with mean responses of 3.64, 3.72, and 3.86. These factors facilitated the sharing of information among IPD core team members with minimum human interference to maximise trust and transparency, which was also facilitated by a tool for visualising EVM metrics.

General category: From this, F10 represented the development of a comprehensive cost management framework for the IPD approach by combining all other nine factors. The mean score for this factor was very high (3.98).

8.7. Discussion

All of the IPD success factors identified in the questionnaire were ascribed a high degree of importance by respondents. Specifically cited were that "there was no tender stage, but rather, an open book pricing technique", "the allocation of responsibilities and risks were clear", "the compensation of risk adjusted reward was flexible", and "there was early involvement of all participants". Therefore, all of these features were required and considered benefits of the IPD approach. Conversely, where parties did not accept or want to adopt all of these features, the desired objectives for employing IPD could be accordingly reduced or minimised.

Analysis of the ten factors presented as potential enhancements of the cost management process for the IPD approach revealed that there was a need to develop a comprehensive cost management framework. The second category concerned the sharing of risks and rewards across ICT utilisation. Respondents recommended using a visualisation tool to show the outcome of EVM and thereby facilitate better understanding of the cost performance outcomes from all IPD team members. Moreover, adopting a web-based management system that shared data among IPD team members was expected to enhance trust and thereby facilitate timely information exchange, which in turn elevates project management outcomes. In this regard, BIM was recommended by both industrial and academic experts as integral to the IPD process (Rowlinson, 2017). Furthermore, 5D BIM was deemed particularly suited to handling cost elements (direct, indirect, and overhead costs). In this respect, BIM-based cost management within the IPD approach is still uncommon (compared with traditional applications). Therefore, it is important to appreciate the enabling modifications that will be required. The recommended improvements of BIM-based cost management were to enhance the integration of 4D and 5D BIM to develop detailed cost budgets that display the compensation structure (estimated cost and profit-at-risk percentage) for each party, both individually and cumulatively, across the entire project. This would provide the necessary transparency and enable all parties to make informed decisions prior to the buyout stage.

The second most important category concerned cost estimation and budgeting improvements. Cost structure was one of the critical identified issues of IPD cost management (Pishdad-Bozorgi & Srivastava, 2018; Roy et al., 2018; Seyedzadeh et al., 2021), where the allocation and distribution of cost overheads presents a major concern to project stakeholders (Kreuze & Newell, 1994; Kumar & Mahto, 2013). Indeed, IPD requires relatively greater overheads in order to accommodate the management involvement of several parties across all project stages (Ashcraft, 2012). For this reason, participants recommended the employment of ABC tools in order to revitalise the IPD cost structure to enable better cost savings, including distribution among IPD team members. In this respect, overhead costs represent a significant proportion of total project costs, typically averaging 15% for some construction projects (Assaf et al., 2001). The corollary of this is that any misallocation of overhead costs in IPD has the potential of seriously impacting the profitability and performance of affected parties.

The final category was ABC and EVM integration, which scored a relatively high average of 3.58. This confirmed a strong interest in utilising EVM in the cost control tasks in the industry, and further suggests that a mandate was needed for the development of applications integrating these tools into IPD cost management processes. Moreover, BIM was also favourably assessed, especially as this is now increasingly being adopted at level 3 and above. It can therefore be proffered that the ongoing integration of BIM and IPD can be expected.

Finally, all of the proposed ten features for enhancing the IPD-based cost management process received positive responses, ranging from 3.36 to 3.98. Of particular note, respondents with greater experience tended to rank these features even more favourably (between 4 and 5). Notwithstanding this, consensus was secured on the need for an integrated framework to foster the wider adoption of IPD.

8.6. Conclusion

This chapter presented an overview of IPD as an approach for integrating all delivery aspects, from project dimensions to people, organisations, and business structures. This was discussed in context with traditional methods of project delivery such as design-bid-build, construction management at-risk, and design-build. Whilst the benefits were explained, it was noted that IPD was not used and that industry uptake had yet to gain momentum, thereby minimising the realisation of these benefits. One of the reasons for this lower-than-expected uptake rests with extensive support systems and measures needed for IPD to be fully effective, especially as systems are not always commonly available to all projects. Indeed, IPD would more than likely underperform as a delivery mode without this support. Since the heart of the IPD model is the equitable allocation of profit-at-risk compensation percentages to all project participants, this can only be achieved where information is timely, accurate, and transparently shared between all parties. As such, stakeholder parties' needs and expectations are aligned. This creates a collaborative problem-solving environment, which ultimately culminates in cost minimisation and profit maximisation for all involved; however, it was advocated that cost management practices and systems need to be investigated, particularly how best these might be harnessed. This research gap was explored by identifying antecedents to the successful design of cost management practices in IPD projects.

This study engaged 50 qualified experts to investigate the main challenges and opportunities available through IPD. Responses were captured in rank order and found to b: 'early involvement of all participants from the design stage', 'open pricing technique' (as there is no tender stage in IPD), 'fair compensation approach', and 'equitable allocation of responsibilities and risks'. This research further confirmed the available strategies needed to enhance IPD-based cost management. These were: (i) integrating ABC and EVM to enhance cost management practices for IPD (such as developing an automated model to show payments due for all parties based on their achievements against planned value); (ii) integrating Monte Carlo simulation into 5D BIM as a means of providing continuous cost estimation feedback (to enhance the conceptual cost estimation for TVD within IPD pre-detailed design stages); and (iii) utilising information communication technology to enhance collaboration and trust among IPD team members. It was advocated that pursuing these strategies would not only help strengthen the robustness of cost management practices on which IPD is so reliant, but that this in turn would promote IPD as a viable and resilient construction project procurement model for wider uptake.

Whilst the findings and recommendations provide important guidance to practitioners seeking to reap the benefits of IPD, it is equally important to share the limitations of this study. This includes the sampling strategy (50 respondents) and context (UK). It is also important to acknowledge that internal reliability was validated exclusively through Cronbach's Alpha. Given this, in order to further strengthen findings (to support external consistency, reliability, and repeatability), additional work should engage IPD/BIM specialists from both cognate and non-cognate sectors, including international contexts.

References

Ahmad, I., Azhar, N., & Chowdhury, A. (2019). Enhancement of IPD characteristics as impelled by information and communication technology. *Journal of Management in Engineering*, 35(1), 04018055. https://doi.org/10.1061/(ASCE)ME.1943-5479.0000670

AIA. (2007). *Integrated project delivery: A guide*. Retrieved June 26, from www.aia.org/resources/64146-integrated-project-delivery-a-guide

Aibinu, A., & Venkatesh, S. (2013). Status of BIM adoption and the BIM experience of cost consultants in Australia. *Journal of Professional Issues in Engineering Education and Practice, 140*(3), 04013021. https://doi.org/10.1061/(ASCE)EI.1943-5541.0000193

Allison, M., Ashcraft, H., Cheng, R., Klawens, S., & Pease, J. (2018a). *Integrated project delivery: An action guide for leaders*. Charles Pankow Foundation.

Allison, M., Ashcraft, H., Cheng, R., Klawens, S., & Pease, J. (2018b). *Integrated project delivery: An action guide for leaders*. https://leanipd.com/integrated-project-delivery-an-action-guide-for-leaders/

Alves, T. D. C. L., Lichtig, W., & Rybkowski, Z. K. (2017). Implementing target value design: Tools and techniques to manage the process. *HERD: Health Environments Research & Design Journal, 10*(3), 18–29. https://doi.org/10.1177/1937586717690865

Ashcraft, H. W. (2012). *IPD framework 2018*. www.hansonbridgett.com/Publications/pdf/ipd-framework.aspx

Ashcraft, H. W. (2014). Integrated project delivery: A prescription for an ailing industry. *Construction Law Review, 9*, 21.

Asmar, M. E., Hanna, A. S., & Loh, W.-Y. (2016). Evaluating integrated project delivery using the project quarterback rating. *Journal of Construction Engineering and Management, 142*(1), 04015046. https://doi.org/10.1061/(ASCE)CO.1943-7862.0001015

Assaf, S. A., Bubshait, A. A., Atiyah, S., & Al-Shahri, M. (2001). The management of construction company overhead costs. *International Journal of Project Management, 19*(5), 295–303.

Ballard, G., Dilsworth, B., Do, D., Low, W., Mobley, J., Phillips, P., Reed, D., Sargent, Z., Tillmann, P., & Wood, N. (2015, July 29). *How to make shared risk and reward sustainable*. 23rd Annual Conference of the International Group for Lean Construction, 257–266. http://iglc.net/Papers/Details/1193

Bensalah, M., Elouadi, A., & Mharzi, H. (2019). Overview: The opportunity of BIM in railway. *Smart and Sustainable Built Environment, 8*(2), 103–116. https://doi.org/10.1108/SASBE-11-2017-0060

Chou, J.-S., Chen, H.-M., Hou, C.-C., & Lin, C.-W. (2010). Visualized EVM system for assessing project performance. *Automation in Construction, 19*(5), 596–607. https://doi.org/10.1016/j.autcon.2010.02.006

Dawood, N., Rahimian, F., Seyedzadeh, S., & Sheikhkhoshkar, M. (2020). *Enabling the development and implementation of digital twins*. Proceedings of the 20th International Conference on Construction Applications of Virtual Reality, Teesside University. ISBN:9780992716127

Durdyev, S., Hosseini, M. R., Martek, I., Ismail, S., & Arashpour, M. (2019). *Barriers to the use of integrated project delivery (IPD): A quantified model for Malaysia*. Engineering, Construction and Architectural Management.

Eastman, C., Teicholz, P., Sacks, R., & Liston, K. (2011). *BIM handbook: A guide to building information modeling for owners, managers, designers, engineers and contractors*. Wiley. ISBN:9781118021699

El-Omari, S., & Moselhi, O. (2011). Integrating automated data acquisition technologies for progress reporting of construction projects. *Automation in Construction, 20*(6), 699–705. https://doi.org/10.1016/j.autcon.2010.12.001

Elghaish, F., Abrishami, S., Abu Samra, S., Gaterell, M., Hosseini, M. R., & Wise, R. (2019a). Cash flow system development framework within integrated project delivery (IPD) using BIM tools. *International Journal of Construction Management*, 1–16.

Elghaish, F., Abrishami, S., & Hosseini, M. R. (2020a). Integrated project delivery with blockchain: An automated financial system. *Automation in Construction, 114*, 103182.

Elghaish, F., Abrishami, S., Hosseini, M. R., Abu-Samra, S., & Gaterell, M. (2019b). Integrated project delivery with BIM: An automated EVM-based approach. *Automation in Construction, 106*, 102907.

Elghaish, F. A. K., Abrishami, S., Hosseini, M. R., & Abu-Samra, S. (2020b). *Revolutionising cost structure for integrated project delivery: A BIM-based solution*. Engineering, Construction and Architectural Management.

Etikan, I., Musa, S. A., & Alkassim, R. S. (2016). Comparison of convenience sampling and purposive sampling. *American journal of theoretical and applied statistics, 5*(1), 1–4.

Field, A. (2013). *Discovering statistics using IBM SPSS statistics*. Sage. ISBN:1446274586

Fischer, M. J. A., Khanzode, A., Reed, D. P., & Ashcraft, H. W., Jr. (2017). *Integrating project delivery*. John Wiley & Sons Inc. ISBN:978-0470587355

Ghassemi, R., & Becerik-Gerber, B. (2011). Transitioning to integrated project delivery: Potential barriers and lessons learned. *Lean Construction Journal*, 32–52.

Hamledari, H., McCabe, B., Davari, S., & Shahi, A. (2017). Automated schedule and progress updating of IFC-based 4D BIMs. *Journal of Computing in Civil Engineering*, 31(4), 04017012. https://doi.org/10.1061/(ASCE)CP.1943-5487.0000660

Hamzeh, F., Rached, F., Hraoui, Y., Karam, A. J., Malaeb, Z., El Asmar, M., & Abbas, Y. (2019). Integrated project delivery as an enabler for collaboration: A Middle East perspective. *Built Environment Project and Asset Management*, 0(0), null. http://doi.org/10.1108/BEPAM-05-2018-0084

Hartmann, T., Gao, J., & Fischer, M. (2008). Areas of application for 3D and 4D models on construction projects. *Journal of Construction Engineering and Management*, 134(10), 776–785. https://doi.org/10.1061/(ASCE)0733-9364(2008)134:10(776)

Hosseini, M. R., Maghrebi, M., Akbarnezhad, A., Martek, I., & Arashpour, M. (2018). Analysis of citation networks in building information modeling research. *Journal of Construction Engineering and Management*, 144(8), 04018064. https://doi.org/10.1061/(ASCE)CO.1943-7862.0001492

Kahvandi, Z., Saghatforoush, E., Ravasan, A. Z., & Mansouri, T. (2018). An FCM-based dynamic modelling of integrated project delivery implementation challenges in construction projects. *Lean Construction Journal*, 63–87.

Kent, D. C., & Becerik-Gerber, B. (2010). Understanding construction industry experience and attitudes toward integrated project delivery. *Journal of Construction Engineering and Management*, 136(8), 815–825. https://doi.org/10.1061/(ASCE)CO.1943-7862.0000188

Khamooshi, H., & Abdi, A. (2016). Project duration forecasting using earned duration management with exponential smoothing techniques. *Journal of Management in Engineering*, 33(1), 04016032. https://doi.org/10.1061/(ASCE)ME.1943-5479.0000475

Kim, Y. W., & Ballard, G. (2001, August). *Activity-based costing and its application to lean construction*. Proceedings of the 9th Annual Conference of the International Group for Lean Construction, National University of Singapore.

Kim, Y.-W., Han, S.-H., Yi, J.-S., & Chang, S. (2016). Supply chain cost model for prefabricated building material based on time-driven activity-based costing. *Canadian Journal of Civil Engineering*, 43(4), 287–293. https://doi.org/10.1139/cjce-2015-0010

Kreuze, J. G., & Newell, G. E. (1994). ABC and life-cycle costing for environmental expenditures. *Strategic Finance*, 75(8), 38.

Kumar, N., & Mahto, D. (2013). Current trends of application of activity based costing (abc): A review. *Global Journal of Management and Business Research Accounting and Auditing*, 13(3).

Lee, S.-K., Kim, K.-R., & Yu, J.-H. (2014). BIM and ontology-based approach for building cost estimation. *Automation in Construction*, 41, 96–105. https://doi.org/10.1016/j.autcon.2013.10.020

Lipke, W., Zwikael, O., Henderson, K., & Anbari, F. (2009). Prediction of project outcome: The application of statistical methods to earned value management and earned schedule performance indexes. *International Journal of Project Management*, 27(4), 400–407. https://doi.org/10.1016/j.ijproman.2008.02.009

Liu, M. M., & Bates, A. J. (2013a). Compensation structure and contingency allocation in integrated project delivery. *Age*, 23, 1.

Liu, M. M., & Bates, A. J. (2013b). *Compensation structure and contingency allocation in integrated project delivery*. Paper presented at the 120th ASEE Annual Conference and Exposition. Retrieved January 8, 2019, from https://nyuscholars.nyu.edu/en/publications/compensation-structure-and-contingency-allocation-in-integrated-p

Ma, Z., Zhang, D., & Li, J. (2018). A dedicated collaboration platform for integrated project delivery. *Automation in Construction*, 86, 199–209. https://doi.org/10.1016/j.autcon.2017.10.024

Manata, B., Miller, V., Mollaoglu, S., & Garcia, A. J. (2018). Measuring key communication behaviors in integrated project delivery teams. *Journal of Management in Engineering*, 34(4), 06018001. https://doi.org/10.1061/(ASCE)ME.1943-5479.0000622

Mesa, H. A., Molenaar, K. R., & Alarcón, L. F. (2016). Exploring performance of the integrated project delivery process on complex building projects. *International Journal of Project Management, 34*(7), 1089–1101.

Naeni, L. M., Shadrokh, S., & Salehipour, A. (2011). A fuzzy approach for the earned value management. *International Journal of Project Management, 29*(6), 764–772. https://doi.org/10.1016/j.ijproman.2010.07.012

Newman, C., Edwards, D., Martek, I., Lai, J., Thwala, W. D., & Rillie, I. (2020). Industry 4.0 deployment in the construction industry: A bibliometric literature review and UK-based case study. *Smart and Sustainable Built Environment, ahead-of-print*(ahead-of-print). https://doi.org/10.1108/SASBE-02-2020-0016

Oliver, S., Seyedzadeh, S., Rahimian, F., Dawood, N., & Rodriguez, S. (2020). Cost-effective as-built BIM modelling using 3D point-clouds and photogrammetry. *Current Trends in Civil & Structural Engineering-CTCSE, 4*(5), 000599. https://doi.org/10.33552/CTCSE.2020.04.000599

Omar, H., & Dulaimi, M. (2015). Using BIM to automate construction site activities. *Building Information Modelling (BIM) in Design, Construction and Operations, 149*, 45. http://doi.org/10.2495/BIM150051

Pajares, J., & López-Paredes, A. (2011). An extension of the EVM analysis for project monitoring: The cost control index and the schedule control index. *International Journal of Project Management, 29*(5), 615–621. https://doi.org/10.1016/j.ijproman.2010.04.005

Palinkas, L. A., Horwitz, S. M., Green, C. A., Wisdom, J. P., Duan, N., & Hoagwood, K. (2015). Purposeful sampling for qualitative data collection and analysis in mixed method implementation research. *Administration and policy in mental health and mental health services research, 42*(5), 533–544.

Pishdad-Bozorgi, P. (2017). Case studies on the role of integrated project delivery (IPD) approach on the establishment and promotion of trust. *International Journal of Construction Education and Research, 13*(2), 102–124. https://doi.org/10.1080/15578771.2016.1226213

Pishdad-Bozorgi, P., Moghaddam, E. H., & Karasulu, Y. (2013). *Advancing target price and target value design process in IPD using BIM and risk-sharing approaches.* Paper presented at the the 49th ASC Annual International Conference California Polytechnic State University.

Pishdad-Bozorgi, P., & Srivastava, D. (2018). *Assessment of Integrated Project Delivery (IPD) Risk and Reward Sharing Strategies from the Standpoint of Collaboration: A Game Theory Approach.* Paper presented at the Construction Research Congress 2018. New Orleans, Louisiana. https://doi.org/10.1061/9780784481271.020

PMI. (2013). *A guide to the project management body of knowledge (PMBOK® guide)* (5th ed.). Project Management Institute. ISBN:8925598620

Ross, J. (2003). *Introduction to project alliancing.* Alliance Contracting Conference.

Rowlinson, S. (2017). Building information modelling, integrated project delivery and all that. *Construction Innovation, 17*(1), 45–49. https://doi.org/10.1108/CI-05-2016-0025

Roy, D., Malsane, S., & Samanta, P. K. (2018). Identification of critical challenges for adoption of integrated project delivery. *Lean Construction Journal*, 1–15.

Seyedzadeh, S., Agapiou, A., Moghaddasi, M., Dado, M., & Glesk, I. (2021). WON-OCDMA system based on MW-ZCC codes for applications in optical wireless sensor networks. *Sensors, 21*. https://doi.org/10.3390/s21020539

Seyedzadeh, S., & Pour Rahimian, F. (2021). Machine learning for building energy forecasting. In *Data-driven modelling of non-domestic buildings energy performance* (pp. 41–76). Springer. https://doi.org/10.1007/978-3-030-64751-3_4

Singleton, M. S. (2010). *Implementing integrated project delivery on department of the navy construction projects.* https://digitalscholarship.unlv.edu/cgi/viewcontent.cgi?article=2941&context=thesesdissertations.

Smith, R. E., Mossman, A., & Emmitt, S. (2011). Editorial: Lean and integrated project delivery special issue. *Lean construction journal*, 1–16.

Sun, W., Mollaoglu, S., Miller, V., & Manata, B. (2015). Communication behaviors to implement innovations: How do AEC teams communicate in IPD projects? *Project Management Journal, 46*(1), 84–96. https://doi.org/10.1002/pmj.21478

Talebi, S., Koskela, L., Tzortzopoulos, P., & Kagioglou, M. (2020a). Tolerance management in construction: A conceptual framework. *Sustainability*, *12*(3), 1039. https://doi.org/10.3390/su12031039

Talebi, S., Koskela, L., Tzortzopoulos, P., Kagioglou, M., & Krulikowski, A. (2020b). Deploying geometric dimensioning and tolerancing in construction. *Buildings*, *10*(4), 62. https://doi.org/10.3390/buildings10040062

Tillmann, P. A., Do, D., & Ballard, G. (2017). A case study on the success factors of target value design. 25th Annual Conference of the International Group for Lean Construction, pp. 563–570. https://doi.org/10.24928/2017/0324

Tsai, W. H., Yang, C. H., Chang, J. C., & Lee, H. L. (2014). An activity-based costing decision model for life cycle assessment in green building projects. *European Journal of Operational Research*, *238*(2), 607–619. https://doi.org/10.1016/j.ejor.2014.03.024

Tsai, W. H., & Hung, S.-J. (2009). A fuzzy goal programming approach for green supply chain optimisation under activity-based costing and performance evaluation with a value-chain structure. *International Journal of Production Research*, *47*(18), 4991–5017. https://doi.org/10.1080/00207540801932498

Turkan, Y., Bosche, F., Haas, C. T., & Haas, R. (2012). Automated progress tracking using 4D schedule and 3D sensing technologies. *Automation in Construction*, *22*, 414–421. https://doi.org/10.1016/j.autcon.2011.10.003

Turkan, Y., Bosché, F., Haas, C. T., & Haas, R. (2013). Toward automated earned value tracking using 3D imaging tools. *Journal of Construction Engineering and Management*, *139*(4), 423–433. https://doi.org/10.1061/(ASCE)CO.1943-7862.0000629

Wang, K.-C., Wang, W.-C., Wang, H.-H., Hsu, P.-Y., Wu, W.-H., & Kung, C.-J. (2016). Applying building information modeling to integrate schedule and cost for establishing construction progress curves. *Automation in Construction*, *72*, 397–410.

Wang, P., Du, F., Lei, D., & Lin, T. W. (2010). The choice of cost drivers in activity-based costing: Application at a Chinese oil well cementing company. *International Journal of Management*, *27*(2), 367.

Zhang, L., & Chen, W. (2010). *The analysis of liability risk allocation for Integrated Project Delivery*. Paper presented at the the 2nd International Conference on Information Science and Engineering. https://doi.org/10.1109/ICISE.2010.5689527

Zhang, L., & Li, F. (2014). Risk/reward compensation model for integrated project delivery. *Engineering Economics*, *25*(5), 558–567.

9 4D BIM for structural design and construction integration

9.1 Introduction

It is argued that while seemingly unavoidable, inspections, monitoring, and comparing as-planned with as-built progress in construction projects do not readily add tangible intrinsic value to end users (other than perhaps some form of surety or confidence). That being said, in large-scale construction projects, the process of monitoring every single part of a building's progress is not only considered unfeasible, but also time consuming and challenging. Moreover, these inspections can be even more challenging if these comparisons and inspections relate to Building Information Models (BIM). Project complexity also amplifies this challenge, especially due to the vast amounts of data produced in the form of schedules, reports and photo logs etc. Take one instance of crack inspection as an example. Where concrete defects such as cracks can significantly reduce the structural integrity of buildings and structures (Ma et al., 2015). In this respect, meticulous attention to detail needs be given to these potential issues, particularly during the design and construction phases in order to prevent the occurrence of these defects (Ghodoosi et al., 2018). Several risk mitigation approaches can be used in concrete structures (Nawy, 2008). For instance, expansion joints can mitigate stresses from temperature changes on structural concrete, and contraction joints can accommodate drying shrinkage of the concrete without proliferating cracks (Merritt & Ricketts, 2001); however, is has been acknowledged that construction joints are somewhat unavoidable, and must therefore be controlled during both the design and construction phases (Issa et al., 2014). If not (for example, if placed incorrectly), then these joints can reduce overall structural integrity and lead to irreparable or costly damages to structures (Gerges et al., 2015). Conversely, the correct placement of joints can support greater structural health (Richardson, 2014).

In practice, various factors (beyond structural analysis considerations) typically control concrete pouring tasks, and therefore limit flexibility in the selection of joint positions. These include time considerations such as the speed of erection, human resources limitations such as the prerequisite skills and competence of workers, and temperature control between the concrete core and its surface (ACI Committee et al., 2008). Other major considerations include the accurate identification of locations to place concrete pouring, including the production of an efficient and effective concreting plan and schedule (CCAA, 2010; Ghasri et al., 2016). Evidence from the industry reveals that these items have presented major challenges, particularly for structures that require large pours (Bennink, 2006). These challenges could be overcome via the use of automation achieved through the digitisation of the design and construction process (Ghaffarianhoseini et al., 2019; Costin et al., 2018). An inherent benefit of Building Information Management (BIM) is its ability to produce information that supports insightful decision-making in structural analyses and designs (Hamledari

DOI: 10.1201/9781003106944-9

et al., 2017; Eleftheriadis et al., 2017). Such information could include the details of concrete pours, and more specifically, that of construction joints (Hardin, 2015; Pomares Torres et al., 2017).

Given these needs, research exploring the capabilities of BIM within the concrete supply chain has received scant academic attention (Aram et al., 2013; Hosseini et al., 2018). Hitherto, pertinent studies on the applications of BIM for concrete work has been limited to improving supply chain management, enhancing the quality of precast components on projects (Kim et al., 2015, 2016), or estimating the costs of production and reducing the carbon footprint of concrete structures (Eleftheriadis et al., 2018; Nizam et al., 2018). Hyun et al. (2018) presented a study with the closest alignment to the topic, highlighting the use of BIM for designing cast-in-place concrete formwork. That being said, there is a paucity of research and development that explores BIM's capabilities for controlling joints. Initial attempts to provide automated procedures still remain in their infancy stages (Ozumba & Shakantu, 2018; Stanton & Javadi, 2014). In order to address this identified knowledge gap, this study developed an automated concrete schedule programme using BIM. This work focuses on planning construction joints, given common limitations affecting concrete pouring activities on construction projects.

This chapter presents a framework and proof-of-concept prototype for on-demand automated simulation of construction projects. This integrates cutting-edge IT solutions, particularly in image processing, machine learning, BIM, and Virtual Reality (VR). This study presents the Unity game engine for integrating data from original BIM models with as-built images, the results of which are then processed via various computer vision techniques. These methods include object recognition and semantic segmentation to identify different structural elements through supervised training in order to superimpose real-world images on the as-planned model. This proof-of-concept prototype also generates an automated update of the 3D virtual environment with the current situation on the construction site. In doing so, it provides a range of stakeholders (such as project managers, clients, contractors, etc.) with vital information needed for decision-making, highlighting inconsistencies and defects in real time. This chapter contributes to the wider body knowledge of technical integration, particularly the concepts of Machine Learning (ML) and image processing approaches with immersive and interactive BIM interfaces, algorithms, and programme codes, all of which can help future product development.

9.2 The theoretical background

9.2.1. OpenBIM standards and the IFC format

Uninterrupted concrete pouring is often impractical due to a myriad of reasons, including size and/or complexity of structures, material supply limitations, allowable working times/conditions, and availability of labour (Gerges et al., 2015; CCAA, 2004). Consequently, it is usually necessary to place fresh concrete on concrete that has already hardened, where the 'contact surface' is termed as a 'construction joint' (Halvorsen et al., 2002). Whilst construction joints can be eliminated through increased reinforcement, the volume of reinforcement needed can sometimes make it unfeasible for ordinary construction projects (Nawy, 2008). According to Issa et al. (2014), "no concrete structure is built without the use of construction joints, whether planned or unplanned". Yet, construction joints require optimisation to reduce unfavourable impacts such as increased permeability. In addition, construction joints also reduce the loading capacity of respective structural elements by up to 20% below the computed value (Kirillov, 1969). The superlative option is to meticulously plan construction joints to coincide with contraction joints prior to concrete pouring – hence, to minimise the number of joints in structures (Gerges et al., 2016; CCAA, 2004). In this respect it is also important to recognise that joints are also formed due to unplanned interruptions of concrete supply, where if this interruption is long enough, then initial setting of the concrete takes place (CCAA, 2004; Gerges et al., 2016). Thus, it is important to appreciate that designers must specify each joint's location from the outset, thereby creating a concrete pouring schedule that accounts for determinatives such as the given daily batching volume (Richardson, 2014; Suprenant, 1988). Moreover, joint locations must be determined in conjunction with the contractor to incorporate the maximum volume of concrete placement and to mitigate any potential operational constraints applicable to the project (CCAA, 2004). Table 9.1 presents a summary of recommendations for concrete joint location derived from several guidelines and specifications in different countries or regions of economic collaboration.

From Table 9.1, given the importance of these recommendations, it is apparent that various guidelines might result in different solutions. For instance, both British and European standards recommend a special design of joint, whilst other specifications allocate these joints where the shear force is minimum – based on each designer's judgment. Moreover, despite the availability of these clear guidelines, a wide range of variables may affect the planning of joints, e.g. appearance, strength, and cost (Suprenant, 1988). Cumulatively, a diverse set of standards and variables serve to illustrate that planning the location of joints is a complex

Table 9.1 Various guidelines for construction joint placement in concrete

Country (Source)	Recommendation Summary	Type of Recommendation
Australia (Standards Australia, 2001)	Construction joints are located to facilitate the placement of concrete; unless otherwise specified, a construction joint shall be made between the soffits of slabs or beams and their supporting columns or walls.	Position of the joint based on the distribution of shear force
Canada (Canadian Standards Association, 2014)	Provision shall be made for the transfer of shear and other forces through construction joints.	Position of the joint based on designers' judgment
Hong Kong (Hong Kong Government, 2006)	Construction joints in concrete shall be formed only at the specified positions and by the specified method unless otherwise approved by the engineer.	Position of the joint based on the distribution of shear force
Japan (JSCE Concrete Committee, 2007)	Joints should be located in portions where the shear force is less and, at right angles to the direction of compressive force, according to the requirements specified herein.	Position of the joint based on the distribution of shear force
The EU (Technical Committee CEN/ TC 250, 2004)	Where tensile stresses are expected to occur in concrete, reinforcement should be detailed to control cracking.	The special design of joint
The UK (B/525, 1997)	Construction joint location should be carefully considered and agreed upon before concrete is placed.	The special design of joint
The USA (Halvorsen et al., 2002)	Desirable locations for joints: perpendicular to the main reinforcement, at points with minimum shear or points of contra-flexure.	Position of the joint based on the distribution of shear force

task that is prone to human acts, errors, or omissions (Bennink, 2006). Therefore, effective planning relies extensively upon the availability of proficient personnel (Heesom & Mahdjoubi, 2004), which can also be affected by the judgements, competence, and perceptions of staff undertaking these (Gledson & Greenwood, 2017). Given this situation, traditional forms of planning for complex circumstances can often produce poor or defective quality structures, mainly because of the human element. In this respect, the work presented in this chapter is founded upon the premise that the intrinsic capabilities of BIM for planning and scheduling concrete pouring should represent viable solutions (Crotty, 2013; Hosseini et al., 2018; Gledson & Greenwood, 2016).

9.2.2. 4D BIM for planning and scheduling

BIM is equipped with multiple dimensions for information delivery and data integration (Elghaish et al., 2019), which is capable of transforming existing practices across the construction sector (Ahmed & Kassem, 2018; Eastman et al., 2018; Chen & Nguyen, 2019; Pour Rahimian et al., 2019). Integration of BIM with other applications is often defined as nD modelling (Ghaffarianhoseini et al., 2019), where supplementary information is added to three-dimensional (3D) models to create additional aspects and visualisation opportunities (Pärn et al., 2017; Boton, 2018). There is consensus within academic literature that linking the time dimension to 3D models (colloquially termed as 4D BIM) is an innovative addition and remedial solution for overcoming the deficiencies of current planning practices (Gledson & Greenwood, 2017; Charef et al., 2018; Hosseini et al., 2018). This tends to entail adding a temporal dimension to 3D models – specifically, linking units of work (based on geometric graphical 3D models) to scheduling details (Heesom & Mahdjoubi, 2004; Park et al., 2017; Pärn et al., 2017).

Koo and Fischer (2000) and later Heesom and Mahdjoubi (2004) argued that the fourth dimension of BIM provides construction stakeholders (i.e. designers and contractors) with a useful alternative to traditional project scheduling tools such as critical path method (CPM) approaches. For example, the use of 4D BIM provides greater control over time and cost variances – estimated to be 40% more efficient than traditional planning procedures (Candelario-Garrido et al., 2017; Meister, 2011). Various 4D applications cover both activity and operations levels alike, including temporary components such as equipment movement, resource availability and congestion, operational problems, and the layout and dynamic analysis of construction sites (Zhou et al., 2015; Huang et al., 2007; Antwi-Afari et al., 2018). In addition, 4D applications can also improve the quality of the planning process in various ways by providing augmented vehicle tracking and transportation route planning (Chen & Nguyen, 2019); improved logistics management, spatial conflict detection, and workspace congestion avoidance (Bortolini et al., 2019); enhanced health and safety management (Golizadeh et al., 2018); and improved monitoring of construction progress and site layout designs with better resource utilisation (Hosseini et al., 2018; Costin et al., 2018).

In terms of use, project teams can be uniquely supported by 4D BIM's inherent ability to identify activities through model interrogation, using accurate durations and estimations of needed resources via automated quantity estimation processes (Gledson & Greenwood, 2016). The visualisation element provided by 4D results in higher productivity, better training, and enhanced communications and collaboration in undertaking scheduling and constructability analysis (Boton, 2018; Eastman et al., 2018). With these factors in mind, 4D can be a useful alternative for traditional methods of joint planning for concrete structures (Charef et al., 2018).

9.2.3. Research gap and methodological approach

Despite 4D BIM's potential, research into exploring its various applications (such as creating and validating new practices to perform project tasks for the benefit of practitioners) has been limited (Gledson & Greenwood, 2016; Hosseini et al., 2018; Boton, 2018). Table 9.2 reports upon major studies that applied 4D BIM for concrete structures, highlighting core areas of research and subsequent findings.

These studies (and other similar studies) are predominantly based upon optimisation objectives regarding formwork required for concrete structures with a view of reducing the workload of designers (Singh et al., 2017; Lee & Ham, 2018).

Arguably, current 4D activities on construction projects are the most labour-intensive parts. For example, most construction projects rely on human resources for manually

Table 9.2 Studies on the use of 4D BIM for concrete structures

Publication	Focus of study	Main method	Findings
Boton (2018)	Use of 4D and VR in constructability assessments	Integration of 4D and VR applications	Presented a procedure for transferring a 4D model into VR for constructability analysis
Wang et al. (2018)	Precast concrete structure	Integration of BIM and Genetic Algorithm	An optimal assembly sequence was presented to reduce the assembly difficulty of a precast concrete building
Lee and Ham (2018)	Formwork systems	Cost optimisation	An automated procedure to optimise the design and layout of formwork, to reduce costs
Wang et al. (2018)	Temporary structures (formwork)	Automation of temporary structures estimation	An automated procedure to estimate temporary structures requirements
Mansuri et al. (2017)	Formwork systems	BIM integration with a cascading algorithm	Generated a scheduled formwork reuse plan including calculating the minimum quantity of formwork required for a project
Singh et al. (2017)	Formwork design	Application Programming Interface (API) of BIM tools	A streamlined formwork design process in the BIM environment
Jiang and Leicht (2016)	Constructability checking for formwork	Pursuing automated constructability reasoning	Established constructability ontology
Stanton and Javadi (2014)	Cost optimisation of a reinforced concrete structure	Cost optimisation with Genetic Algorithm	Cost of reinforced concrete is optimised based on site-based variables like height limitations

Publication	Focus of study	Main method	Findings
Aram et al. (2013)	Exploring BIM capabilities for concrete structures supply chain	Conceptual study	Recommendation proposed to align BIM tools with the supply chain of concrete structures
Porwal and Hewage (2012)	Reducing the waste of rebar in concrete structures	Use of BIM models to simulate the architectural and structural design	Significant cost saving increases the diameter of rebar
Barak et al. (2009)	Defining BIM requirements for concrete structures' production modelling	A qualitative study based on experts' views	Providing a set of object schemas, defining relations, methods, and attributes needed for modelling the production of concrete structures

planning activities, including the inspection and control progress. The average share of these activities within a project budget typically lies up to 40%, as argued by Kropp et al. (2018). Therefore, automating 4D BIM application to reduce the workload of field personnel is of great importance; albeit, this is still an underexplored potential of BIM (Kropp et al., 2018).

From an analysis perspective, investigating typical damaged concrete structures after earthquakes show that the failure of joints was a major contributor to the collapse of concrete structures due to earthquake excitation (Khashi et al., 2018). Given this finding, one such area of analysis is 4D activities associated with optimising the layout of joints, which contemporary literature fails to address (despite its importance to structural integrity). In recognition of this fact, the work presented in this chapter posits that this might be due to the multidisciplinary nature of the issue, including the engagement of parochial construction practices – the corollary of which is that it is time to change established routines in order to evaluate these issues from a wider perspective.

9.3 Research design

To address these theoretical and technical gaps, this study presents a proof-of-concept prototype for 4D automated concrete joint layout planning. This study adopted a two-stage process to cover both technical and theoretical aspects at the same time:

Stage one: prototype development – this consisted of a three-stage iterative process: (i) identification of the factors and variables that affected the concrete pouring schedule; (ii) integration of selected software tools and data exchange procedures needed to automate the model; and (iii) creation of logic and analytical considerations for concrete pouring.

Stage two: application of the proof-of-concept – development of a case study to demonstrate and validate the approach developed in stage one.

Figure 9.1 illustrates the two stages of the research design. Full details of these stages are presented later in the chapter.

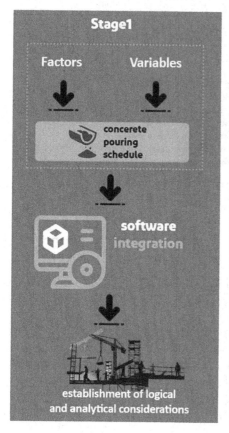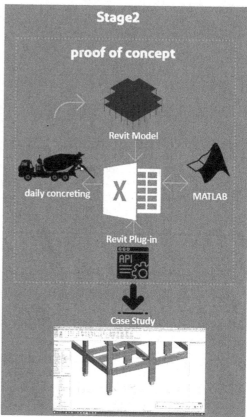

Figure 9.1 The two-stage research design

9.4 Stage one: prototype development

9.4.1. *Assumptions, factors, and variables*

For the initial proof-of-concept, several assumptions were considered: (i) the daily allowed concrete remains constant on various working days; (ii) structures used have rectangular plans; and (iii) the structural plan is similar across all floors of a building. Holding these assumptions constant enabled the proof-of-concept's basic design to be developed and tested; however, it was acknowledged from the outset that future work would be required to increase the application's scope (in terms of different production schedules, material availability, and building designs).

Many variables and factors affect the pouring technical requirements and also the scheduling and planning of the pour (ACI Committee et al., 2008). These variables are both context-dependent and unique to a bespoke project. Cumulatively, they place limitations on the pouring procedure, and therefore, must be incorporated into the proof-of-concept (refer to Table 9.3).

Table 9.3 Factors and variables affecting the concrete scheduling procedure

Factor/Variables	Associated considerations
Daily available concrete	Clarifying the limitation for daily concreting available on the project.
Concrete waste percentage	Estimating the amount of concrete waste.
Pour starting point	Clarifying the point of start for the pour.
Pour direction	Clarifying the direction to which pouring is heading.
Floor thickness,	Estimating the concreting volume.
Points: details of the floors	
The perimeter of the model	
Beams: start and end points	Locating midspan points.
Beams: midspan points	Locating construction joints and stop concreting.
Maximum length and width of the model	Clarifying the point of start for the pour and clarifying the direction to which pouring is heading.

9.4.2. *Project framework and data exchange procedure*

This study employed four software applications to develop the automated procedure within this proof-of-concept prototype, namely: (i) Autodesk Revit© 2018; (ii) Dynamo 1.3.2; (iii) Microsoft (MS) Excel 2016; and (iv) MATLAB 2014. The combination of Revit–Dynamo provided a convenient and automated data exchange procedure for importing data extracted in various forms from a BIM model in Revit (Ninić et al., 2017). Dynamo was recognised as a user-friendly input-output data interface, enabling users of visual programming to establish bilateral integration between Revit and MS Excel in order to store and manipulate BIM data in spreadsheets (Bueno et al., 2018). Dynamo's architecture of subroutine definitions and communication protocols provide access to the Revit API (Application Programming Interface) (Autodesk, 2008). This enables Dynamo users to interact with a Revit model, query and change element properties, and also add and modify some elements directly from the Dynamo environment (Pärn & Edwards, 2017). Data exchange structure and flowchart of how these four applications were integrated into the developed prototype are presented in Figure 9.2.

In the Revit model, the first step was to insert data into the model (refer to arrow 1). The next step was to identify relevant data to export from the 3D model into the Dynamo spreadsheet. To prevent unplanned joints and develop the concreting schedule, a pour volume was needed together with the identification of vulnerable points in beams and the perimeter of the model. For beams, the vulnerable points were located in one-third of midspans, and for floors, the vulnerable points were located at the building perimeter. Accordingly, the coordination details of beams and floors were extracted from the model and submitted to the spreadsheet (arrow 2). Data is exchanged between the Revit model and one of the two MS Excel spreadsheets. This data exchange provided the exact location coordinates for beams and floors (arrow 3). Figures 9.3 and 9.4 illustrate the data flow (of perimeter data) between the Revit model and MS Excel for floors in seven steps.

The next step was to identify the existing floors as structural elements using Dynamo. This involved separating and inserting the locations of the concrete pour into MS Excel. By specifying the start row and column in the MS Excel sheet, and the names for various types of points, the file path to store the data was defined in the Matlab algorithm. Extracting X, Y, and Z coordinates of the floor from the Revit model was the next step. By including the

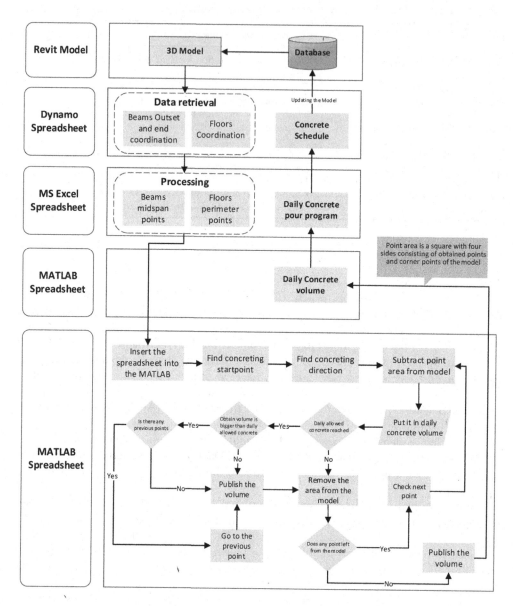

Figure 9.2 The structure of data exchange

headlines for every sheet of the extracted data in MS Excel, results were visually seen as X, Y, and Z for points (where every X, Y, and Z combination had its own headline in the spreadsheet). A similar procedure for the data flow between BIM and MS Excel was applicable for structural beams. The aim of data extraction for floors was to find the perimeter of the model for the whole building, whereas for beams, the aim was to find the midspans. This approach served to identify the points with the minimum negative effects on the structure, where the concreting activity could be ceased.

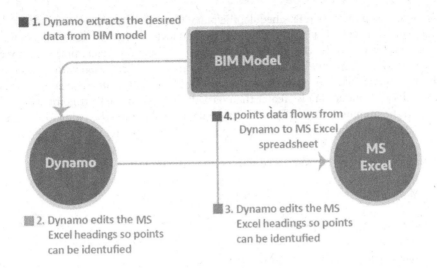

■ **1.** Dynamo extracts the desired data from BIM model

■ **4.** points data flows from Dynamo to MS Excel spreadsheet

■ **2.** Dynamo edits the MS Excel headings so points can be identufied

■ **3.** Dynamo edits the MS Excel headings so points can be identufied

Figure 9.3 Data flow from the 3D Revit model to MS Excel, using Dynamo

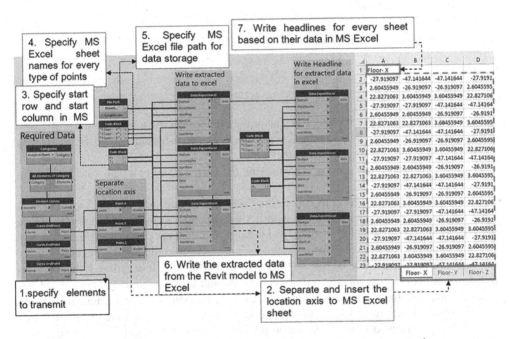

Figure 9.4 Exporting floors location from the 3D Revit model to MS Excel, using visual programming

In order to locate the beam's midspans as well as the floor's perimeter, coding was performed in MS Excel using macros and the programming language Visual Basic for Applications (VBA). Macros and VBAs are coding spreadsheets, where mathematical actions and formula insertions can be conducted. The developed coding had the capability of taking up to 10,000 data points in every Excel column for beams in order to locate their midspan.

9.4.3. Logic and analytical considerations

In order to develop the concrete schedule, the maximum width and length of the model was calculated in MS Excel – first to find the starting point location for the pour, and second, the direction towards which the concreting was headed. A common industry practice is for contractors to start and continue the concrete pouring process in a direction with the minimum length, thereby to make concreting cease controllably. Logical operator 1 (below) illustrates: if the length of the model (Y) is greater than its width (X), then the start point is one of the points with the minimum of (Y). From the aforementioned points, the point with the minimum (X) value would be selected as the start point. If (X) was greater than (Y), then the start point would be one of the points with the minimum (X) and from those points, the point with the minimum (Y) will be chosen as the start point of the project.

$$A \subseteq \text{Model points, } A = \left\{ x,y | x,y \subseteq \text{Model points, } \left(x,y_{min} \right) \right\},$$

$$B = \left\{ x,y | x,y \subseteq A, \left(x_{min},y \right) \right\}$$

$$C \subseteq \text{Model points, } C = \left\{ x,y | x,y \subseteq \text{Model points, } \left(x_{min},y \right) \right\},$$

$$D = \left\{ x,y | x,y \subseteq C, \left(x,y_{min} \right) \right\}$$

$$\text{Start Point and Direction} = \begin{cases} \text{Start Point : Set } B, \text{Directon : Augmented } X, \ X < Y \\ \text{Start Point : Set } D, \text{Directon : Augmented } Y, \ X \geq Y \end{cases}$$

Logical Operator 1

Where A is a subset of B, which includes all of the points with the minimum value of Y, and B is a start point if $X < Y$. In addition, C is a subset of D, which includes all of the points with the minimum value of X, and D is a start point if $X \geq Y$.

Figure 9.5 illustrates the process of identifying the corner point of a plan view building as the start of the project.

To choose the right direction in concreting (as shown in Logical Operator 1), in case the length of the model was greater than the width, the concrete pouring would start at the point with the minimum value of (X), with (Y) value fixed, and the next points would head on new (X) and same (Y) respectively. Conversely, concrete pouring would head on new (Y) direction with the fixed (X) value. While on the equal width and length situation, if there was no difference in the direction of concreting, then the default was set on fixed (Y).

By finding the start point, a rectangular area including this start point and the model's corner points (as the rectangle corners) were identified. This collision area was calculated and removed at the next step, where afterwards, the next point is checked. The arrangement strategy for this selection was based on the direction of concrete pouring as if the width of the model was greater than the length. The next point is an augmented Y with the fixed X value. Where the value of the model's length was greater than the width value, the next point is augmented X with the fixed Y value. This method continues up until the complete row is eliminated. The new fixed X or Y value was then selected as the start point, and the loop continues until there is no model left on the program.

Give this approach, it was acknowledged that available concrete is an important variable that differs across different projects (Dunlop & Smith, 2004), where volume depends on

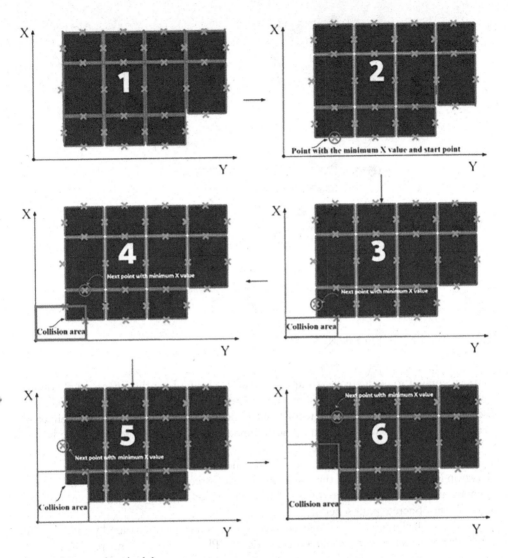

Figure 9.5 Logic of methodology

resource limitations such as financial resources, access restrictions to the site, and human resources restrictions, etc. (ACI Committee et al., 2008). Hence, each bespoke project has its own unique available concrete (Dunlop & Smith, 2004). Therefore, a variable representing the available concrete was inserted into the proof-of-concept as a default concrete volume number, which was used as the limitation volume in daily concrete operations. This value was defined as a variable, where its value was left (to be defined by individual users working on projects with various conditions affecting them).

Another variable for consideration was concrete waste. During on-site activities, some unavoidable factors (such as material transportation and human resource activities) traditionally affect the amount of concrete waste generated. Therefore, a coefficient was included

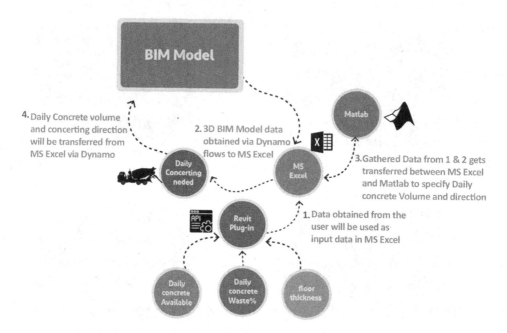

Figure 9.6 The developed Autodesk Revit plug-in

(as a percentage) to estimate concrete waste and determine how this affects total volume. Determining waste is a complex phenomenon that is almost entirely context-specific and based upon on-site experiences and records of technicians (Kazaz et al., 2015; Nikmehr et al., 2017). This coefficient was included in the automated schedule designed for this study.

The pour schedule data was linked with the Revit model for visualisation purposes. From Figure 9.2 (arrows 6, 7, and 8), the pour programme generated through the proposed methodology in MATLAB presented the results in MS Excel spreadsheet, which was inserted into the model using Dynamo. The extracted data was attached to the floor element in the Revit model for the concrete pour schedule.

In order to facilitate the automation process and create a user-friendly interface, an Autodesk Revit plug-in was developed using Revit API (Figure 9.6). This plug-in was able to access the daily concrete volume, as well as daily concrete waste, based on experimental and documented records of previous projects and floor thickness when calculating the surface of the pour. By using the structural design of the project and based on the information mentioned earlier, ceasing concrete pouring in critical points could be properly managed and structural problems avoided.

9.5 Stage two: application of the proof-of-concept

The developed proof-of-concept prototype was tested and validated via adopting a real-life case study. All factors and variables (mentioned in Table 9.3) were used in different stages of this case study. In this regard, the available daily concrete, concrete waste percentage, and floor thickness were obtained via a plug-in and used for concreting schedule calculations. The floor perimeter points and beams start and end points were obtained from the 3D model and were used to calculate concrete volume as well as beam midspan points. The maximum length and width of the model were also calculated based on floor perimeter points (to help calculate concrete pouring directions and starting points).

The case study presented in this chapter was a three-storey educational building (with a uniform design) based in Tehran, Iran. The total project budget was USD$345,210, which was developed over an 18-month period from February 2017 to August 2018. Using the developed plug-in in the Revit environment, the daily concrete volume and the starting point of the pour were defined (Figure 9.6.b). Data required for running the plug-in was the available daily concrete volume (as the first limitation to start the project), floor thickness, and the percentile concrete waste factor (Figure 9.7.a). Based on previous studies on the composition of construction waste, concrete has been seen to be the second largest contributor to waste generation. According to Poon's investigation, where 80% of the work is made from ready-mix concrete, 3–5% of concrete waste is mainly caused by excessive material ordering, broken formwork, or redoing due to poor concrete placement quality (Poon et al., 2004). Given this, in this research, the amount of concrete waste was therefore considered as 5%.

As illustrated in Figure 9.7.b, the generated concreting schedule volume is shown in the analytical floor schedule. The first column in this schedule shows the daily concrete volume required during construction. The daily allowed concrete volume in this case was 400 ft³. Yet, the available amount was automatically reduced by 5% to factor in the impact of waste. The second column shows the start point. The direction of the concrete is then calculated automatically using the logic discussed earlier.

The outcomes of applying the plug-in on scheduling the pour was visualised against a scenario in which the pour was planned based on the daily available concrete volume of 400 ft³. The total areas that need concrete are presented in Figure 9.8.

Figure 9.9 presents the pour plan that includes 5% of concrete waste and 400 ft³ daily concrete available on-site (assuming that the total available concrete volume could be poured).

(a)

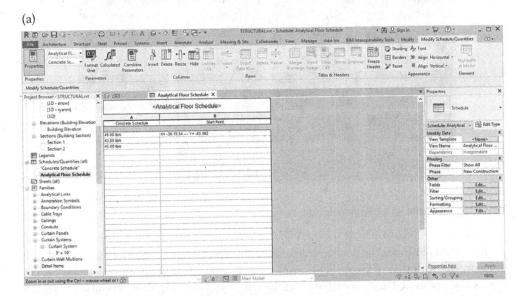

Figure 9.7 Snapshot of (a) the automated concreting schedule plug-in and (b) the developed concreting schedule

(b)

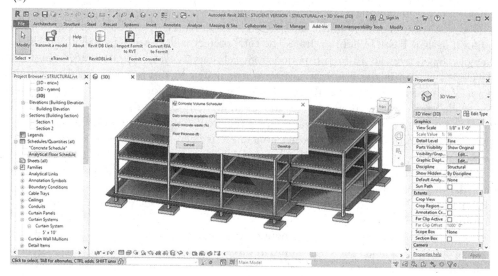

Figure 9.7 (Continued)

Figure 9.8 Total areas in the Revit model that need concreting

Figure 9.10 illustrates the pour schedule using the developed plug-in. Comparing the plan in Figure 9.9 with that of Figure 9.10 reveals that in Figure 9.10, the pour was ceased prior to reaching the available amount, based on the criterion: "where is the best position to stop", with the aim of having the least possible impact on the structural strength. As a result, daily pour activities must cease at 381.612 ft^3 (with 5% waste included), where the last day volume

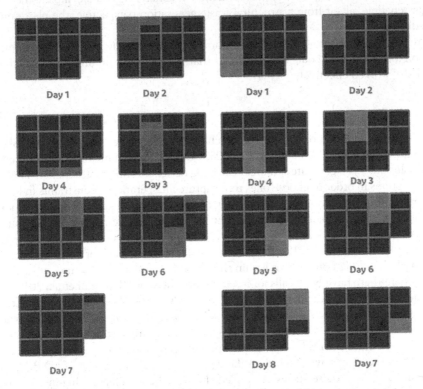

Figure 9.9 Concrete pouring plan with maximum daily concrete available without using the automated procedure (left) and via the proposed automated methodology (right)

Point1			POINT2			POINT3			MIDSPAN1			MIDSPAN2			MIDSPAN		
X	Y	Z	X	Y	Z	X	Y	Z	X	Y	Z	X	Y	Z	X	Y	Z
-45.1416	23.6305	0	-29.9191	23.6305	0	-47.6416	23.6305	0	-37.5304	23.6305	0	-38.7804	23.6305	0	-38.155	23.6305	0
-24.9191	23.6305	0	0.60456	23.6305	0	-27.4191	23.6305	0	-12.1573	23.6305	0	-13.4073	23.6305	0	-12.782	23.6305	0
17.8271	23.6305	0	5.60456	23.6305	0	20.3271	23.6305	0	11.7158	23.6305	0	12.9658	23.6305	0	12.3408	23.6305	0
-45.1416	-3.03619	0	-29.9191	-3.03619	0	-47.6416	-3.03619	0	-37.5304	-3.03619	0	-38.7804	-3.03619	0	-38.155	-3.0362	0
-24.9191	-3.03619	0	0.60456	-3.03619	0	-27.4191	-3.03619	0	-12.1573	-3.03619	0	-13.4073	-3.03619	0	-12.782	-3.0362	0
20.8271	-3.03619	0	5.60456	-3.03619	0	23.3271	-3.03619	0	13.2158	-3.03619	0	14.4658	-3.03619	0	13.8408	-3.0362	0
-45.1416	-29.7029	0	-29.9191	-29.7029	0	-47.6416	-29.7029	0	-37.5304	-29.7029	0	-38.7804	-29.7029	0	-38.155	-29.703	0
-24.9191	-29.7029	0	0.60456	-29.7029	0	-27.4191	-29.7029	0	-12.1573	-29.7029	0	-13.4073	-29.7029	0	-12.782	-29.703	0
20.8271	-29.7029	0	5.60456	-29.7029	0	23.3271	-29.7029	0	13.2158	-29.7029	0	14.4658	-29.7029	0	13.8408	-29.703	0
-45.1416	-56.3695	0	-29.9191	-56.3695	0	-47.6416	-56.3695	0	-37.5304	-56.3695	0	-38.7804	-56.3695	0	-38.155	-56.37	0
-24.9191	-56.3695	0	0.60456	-56.3695	0	-27.4191	-56.3695	0	-12.1573	-56.3695	0	-13.4073	-56.3695	0	-12.782	-56.37	0
20.8271	-56.3695	0	5.60456	-56.3695	0	23.3271	-56.3695	0	13.2158	-56.3695	0	14.4658	-56.3695	0	13.8408	-56.37	0
-45.1416	-83.0362	0	-29.9191	-83.0362	0	-47.6416	-83.0362	0	-37.5304	-83.0362	0	-38.7804	-83.0362	0	-38.155	-83.036	0
-24.9191	-83.0362	0	0.60456	-83.0362	0	-27.4191	-83.0362	0	-12.1573	-83.0362	0	-13.4073	-83.0362	0	-12.782	-83.036	0
20.8271	-83.0362	0	5.60456	-83.0362	0	23.3271	-83.0362	0	13.2158	-83.0362	0	14.4658	-83.0362	0	13.8408	-83.036	0
-47.6416	21.1305	0	-47.6416	-0.53619	0	-47.6416	23.6305	0	-47.6416	10.2971	0	-47.6416	11.5471	0	-47.642	10.9221	0
-47.6416	-5.53619	0	-47.6416	-27.2029	0	-47.6416	-3.03619	0	-47.6416	-16.3695	0	-47.6416	-15.1195	0	-47.642	-15.745	0
-47.6416	-32.2029	0	-47.6416	-53.8695	0	-47.6416	-29.7029	0	-47.6416	-43.0362	0	-47.6416	-41.7862	0	-47.642	-42.411	0
-47.6416	-58.8695	0	-47.6416	-80.5362	0	-47.6416	-56.3695	0	-47.6416	-69.7029	0	-47.6416	-68.4529	0	-47.642	-69.078	0
-27.4191	-58.8695	0	-27.4191	-80.5362	0	-27.4191	-56.3695	0	-27.4191	-69.7029	0	-27.4191	-68.4529	0	-27.419	-69.078	0
-27.4191	-32.2029	0	-27.4191	-53.8695	0	-27.4191	-29.7029	0	-27.4191	-43.0362	0	-27.4191	-41.7862	0	-27.419	-42.411	0
-27.4191	-5.53619	0	-27.4191	-27.2029	0	-27.4191	-3.03619	0	-27.4191	-16.3695	0	-27.4191	-15.1195	0	-27.419	-15.745	0
-27.4191	21.1305	0	-27.4191	-0.53619	0	-27.4191	23.6305	0	-27.4191	10.2971	0	-27.4191	11.5471	0	-27.419	10.9221	0
3.10456	21.1305	0	3.10456	-0.53619	0	3.10456	23.6305	0	3.10456	10.2971	0	3.10456	11.5471	0	3.10456	10.9221	0
3.10456	-5.53619	0	3.10456	-27.2029	0	3.10456	-3.03619	0	3.10456	-16.3695	0	3.10456	-15.1195	0	3.10456	-15.745	0
3.10456	-32.2029	0	3.10456	-53.8695	0	3.10456	-29.7029	0	3.10456	-43.0362	0	3.10456	-41.7862	0	3.10456	-42.411	0

P1 MIN		P2 MIN		P3 MIN	
X	Y	X	Y	X	Y
-47.6416	-83.0362	-47.6416	-83.0362	-47.6416	-83.0362

MIN X	MIN Y
-47.6416	-83.03

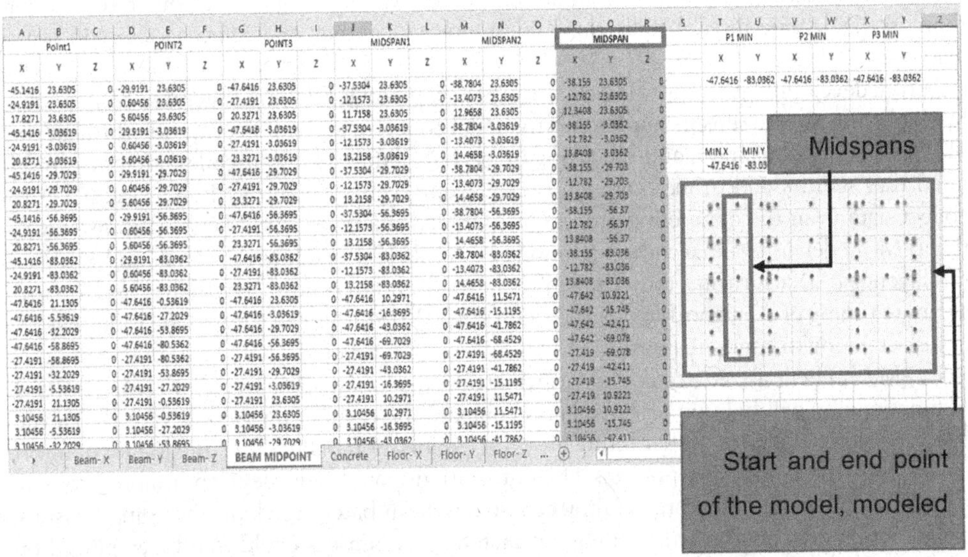

Figure 9.10 Midspan and start floor points

was 165.76 ft³. While using the traditional method of concreting could complete the task on a tighter schedule, albeit from a structural waste generation perspective, the plan would be clearly different.

Figure 9.10 illustrates how the extracted data from the model fulfilled the structural consideration for ceasing the pour. This illustrates that the midspans finding process from Revit to MS Excel worked effectively.

9.6 Discussion

Efforts on the automation of various construction activities have often been cited as being required from the outset (Rahimian et al., 2008; Goulding & Rahimian, 2012). This enables optimal solutions to be generated, thereby enabling designers and practitioners to analyse all available options in order to choose ones that capture all operational constraints in line with design principles. In this respect, a large body of BIM literature suggests drawing upon the potential of 4D BIM to tackle operational considerations and design principles simultaneously (Eastman et al., 2018; Zhou et al., 2015). Nevertheless, except for a few existing studies such as that by Hyun et al. (2018), practical applications of 4D BIM for bringing operational aspects back into design procedures within the context of concrete structures are somewhat limited, where there have been calls for research into the topic (Hosseini et al., 2018).

The developed Revit plug-in presented in this study, along with the proposed analytical considerations, are the first steps in this process. For example, bringing 4D BIM applications into design procedures of concrete structures can directly support connections along with operational constraints. In this respect, the findings presented here are considered complementary to studies that have focused on the design or post-design stages of 4D BIM use. In this instance, both of these stages are linked through the developed plug-in for concrete structures, making this work somewhat unique. That being said, it is also important to note at this juncture that certain limitations exist that have yet to be resolved. For example, the application's inability to accommodate circular building structures (or other different types of building plan layouts). Other limitations include varying degrees of pour rates, rather than assumed static rates. Whilst these are minor issues, they may affect usability in some instances.

Future development works will look at the possibility of developing calibrated BIM models by employing on-site sensors and drones to model concrete activities constantly through the 4D BIM model – mapping and specifying items with different colour codes. Here it is proffered that sensors and drones can act as intelligent agents (Asgari & Rahimian, 2017) to collect data from the concreting process; it is argued that data could be synchronised with the BIM model using linked-data (Pauwels et al., 2018), machine learning, and artificial intelligence (AI) algorithms (Seyedzadeh et al., 2018). It has been shown that by undertaking accurate tuning of AI algorithms (Seyedzadeh et al., 2019), this can support decision-making in tasks relating to concrete pouring and forecasting (Ghasri et al., 2016), estimating productivity (Maghrebi et al., 2016), and identifying real-time hazards (Kim et al., 2018). This type of functionality will help project managers compare the concreting process with the developed concrete schedules. Other future areas for exploitation include the integration of mixer trucks with the concreting process. This integration would help develop a more accurate concreting schedule based on the project distance from batching plant, including logistics, traffic conditions, on-site truck limitation factors, etc. Sensors could also be employed on trucks to facilitate this integration – with on-site drones and sensors directly supporting real-time concrete scheduling programs.

9.7 Conclusion

This study contributes to the field of digital construction and project management in several ways. Firstly, in terms of research, this work presents a real-life application of 4D BIM capabilities, which combines operational data with design considerations to increase the efficiency of on-site activities (which historically is affected by many different variables). Secondly, it proposes a novel method of using BIM in structural engineering activities – an area that needs more detailed attention. The study, therefore, bridges these theoretical and technical gaps by providing a bespoke integrated platform that brings together the capabilities of various software applications through a specific use case of 4D BIM for structural engineering activities.

From an application development point of view, the work presented here offers a readily available plug-in and a cost-effective solution that takes into account construction managers' requirements (e.g. the need to order and schedule the correct volumes of concrete) on the one hand and structural engineers' concerns (e.g. structural integrity – particularly on joints) on the other hand. It is argued that this endeavour not only expedites the design and planning stages of concrete pouring, but also helps improve the delivery of construction projects using realistic concrete schedules and work orders. In doing so, it has the capacity of being able to reduce waste (material and time), compared to traditional projects. Particularly, where unplanned joints are imposed (e.g. running out of materials, or added extra layers of complexity concerning rectification measures) or excessive concrete waste generated by fluctuating patterns of consumption or working days. This proof-of-concept prototype application is able to overlay these findings against traditional manual methods.

The findings and innovative approaches presented in this chapter provide fertile grounds for further discussion. In particular, development of software applications such as this as a baseline for developing a robust commercial plug-in – acting as an industry standard for future BIM design tools. Such integrated design software applications would be invaluable for practitioners. Not just for time/cost savings *per se*, but as an approach that embraces a more accurate methodology for addressing the limitations and inefficiencies of traditional methods associated with planning concrete pours and designing construction joints. This approach also takes into consideration structural principles, constructional procedures, and operational constraints.

Finally, and upon reflection, despite the positive progress made and findings articulated, it is important to note that this is only the beginning. Notwithstanding the limitations highlighted, future opportunities are seemingly endless – in particular, the incorporation of ML and AI supported by live data. Systems can be trained based on data collected from on-site sensors and additional information conduits (e.g. temperature, productivity, logistics, traffic conditions, equipment availability, etc.), making future pouring activities much more controllable, predictable, and dynamic.

Acknowledgments

The authors would like to acknowledge the contributions of Moslem Sheikhkhoshkar, Mohammad Hossein Kaveh, M. Reza Hosseini, and David J. Edwards to the underpinning research presented in this chapter.

References

ACI CommitteeAmerican Concrete Institute and International Organization for Standardization. (2008). *Building code requirements for structural concrete (ACI 318–08) and commentary.* American Concrete Institute. ISBN: 978087031264-9

Ahmed, A. L., & Kassem, M. (2018). A unified BIM adoption taxonomy: Conceptual development, empirical validation and application. *Automation in Construction*, 96, 103–127. https://doi.org/10.1016/j.autcon.2018.08.017

Antwi-Afari, M. F.Li, H.Pärn, E. A., & Edwards, D. J. (2018). Critical success factors for implementing building information modelling (BIM): A longitudinal review. *Automation in Construction*, 91, 100–110. https://doi.org/10.1016/j.autcon.2018.03.010

Aram, S.Eastman, C., & Sacks, R. (2013). Requirements for BIM platforms in the concrete reinforcement supply chain. *Automation in Construction*, 35, 1–17. https://doi.org/10.1016/j.autcon.2013.01.013

Asgari, Z., & Rahimian, F. P. (2017). Advanced virtual reality applications and intelligent agents for construction process optimisation and defect prevention. *Procedia Engineering*, 196, 1130–1137. https://doi.org/10.1016/j.proeng.2017.08.070

Autodesk. (2008). *BIM and API extensions*. Autodesk, Inc. Retrieved July 11, 2021, from http://images.autodesk.com/latin_am_main/files/revit_bim_and_api_extensions_mar08.pdf

B/525, T. C. (1997). *Structural use of concrete, part 1: Code of practice for design and construction*. The British Standards Institution. ISBN:0580262081

Barak, R.Jeong, Y.-S.Sacks, R., & Eastman, C. M. (2009). Unique requirements of building information modeling for cast-in-place reinforced concrete. *Journal of Computing in Civil Engineering*, 23 (2), 64–74. https://ascelibrary.org/doi/abs/10.1061/(ASCE)0887-3801(2009)23:2(64)

Bennink, C. (2006). *Plan for the large pours for construction pros*. Retrieved July 11, 2019, from www.forconstructionpros.com/concrete/article/10305422/plan-for-the-large-pours

Bortolini, R.Formoso, C. T., & Viana, D. D. (2019). Site logistics planning and control for engineer-to-order prefabricated building systems using BIM 4D modeling. *Automation in Construction*, 98, 248–264. https://doi.org/10.1016/j.autcon.2018.11.031

Boton, C. (2018). Supporting constructability analysis meetings with immersive virtual reality-based collaborative BIM 4D simulation. *Automation in Construction*, 96, 1–15. https://doi.org/10.1016/j.autcon.2018.08.020

Bueno, C., Pereira, L. M., & Fabricio, M. M. (2018). Life cycle assessment and environmental-based choices at the early design stages: An application using building information modelling. *Architectural Engineering and Design Management*, 14(5), 332–346. https://doi.org/10.1080/17452007.2018.1458593

Canadian Standards Association. (2014). *A23.3–14: Design of concrete structures*. Canadian Standards Association. ISBN:1553975596

Candelario-Garrido, A.García-Sanz-Calcedo, J., & Reyes Rodríguez, A. M. (2017). A quantitative analysis on the feasibility of 4D planning graphic systems versus conventional systems in building projects. *Sustainable Cities and Society*, 35, 378–384. https://doi.org/10.1016/j.scs.2017.08.024

CCAA. (2004). *Joints in concrete buildings*. Cement Concrete & Aggregates Australia. Retrieved July 11, 2019, from www.ccaa.com.au/imis_prod/documents/Library%20Documents/CCAA%20Datasheets/TN63-2004JointsTBR.pdf

CCAA. (2010). *Concrete basics a guide to concrete practice*. Cement Concrete & Aggregates Australia. Retrieved July 11, 2019, from www.elvingroup.com.au/wp-content/uploads/2016/01/concrete_basics.pdf

Charef, R.Alaka, H., & Emmitt, S. (2018). Beyond the third dimension of BIM: A systematic review of literature and assessment of professional views. *Journal of Building Engineering*, 19, 242–257. https://doi.org/10.1016/j.jobe.2018.04.028

Chen, P.-H., & Nguyen, T. C. (2019). A BIM-WMS integrated decision support tool for supply chain management in construction. *Automation in Construction*, 98, 289–301. https://doi.org/10.1016/j.autcon.2018.11.019

Costin, A.Adibfar, A.Hu, H., & Chen, S. S. (2018). Building information modeling (BIM) for transportation infrastructure – literature review, applications, challenges, and recommendations. *Automation in Construction*, 94, 257–281. https://doi.org/10.1016/j.autcon.2018.07.001

Crotty, R. (2013). *The impact of building information modelling: Transforming construction*. Routledge. ISBN:1136860576

Dunlop, P., & Smith, S. D. (2004). Planning, estimation and productivity in the lean concrete pour. *Engineering, Construction and Architectural Management, 11*(1), 55–64. https://doi.org/10.1108/09699980410512665

Eastman, C. M. A., Lee, G., Sacks, R., & Teicholz, P. M. (2018). *BIM handbook: A guide to building information modeling for owners, managers, designers, engineers and contractors.* Wiley. ISBN:9780470541371

Eleftheriadis, S.Duffour, P.Greening, P.James, J.Stephenson, B., & Mumovic, D. (2018). Investigating relationships between cost and CO2 emissions in reinforced concrete structures using a BIM-based design optimisation approach. *Energy and Buildings, 166,* 330–346. https://doi.org/10.1016/j.enbuild.2018.01.059

Eleftheriadis, S.Mumovic, D., & Greening, P. (2017). Life cycle energy efficiency in building structures: A review of current developments and future outlooks based on BIM capabilities. *Renewable and Sustainable Energy Reviews, 67,* 811–825. https://doi.org/10.1016/j.rser.2016.09.028

Elghaish, F.Abrishami, S.Abu Samra, S.Gaterell, M.Hosseini, M. R., & Wise, R. (2019). Cash flow system development framework within integrated project delivery (IPD) using BIM tools. *International Journal of Construction Management.* 1–16. https://doi.org/10.1080/15623599.2019.1573477

Gerges, N. N.Issa, C. A., & Fawaz, S. (2015). Effect of construction joints on the splitting tensile strength of concrete. *Case Studies in Construction Materials, 3,* 83–91. https://doi.org/10.1016/j.cscm.2015.07.001

Gerges, N. N.Issa, C. A., & Fawaz, S. (2016). The effect of construction joints on the flexural bending capacity of singly reinforced beams. *Case Studies in Construction Materials, 5,* 112–123. https://doi.org/10.1016/j.cscm.2016.09.004

Ghaffarianhoseini, A.Zhang, T.Naismith, N.Ghaffarianhoseini, A.Doan, D. T.Rehman, A. U.Nwadigo, O., & Tookey, J. (2019). ND BIM-integrated knowledge-based building management: Inspecting post-construction energy efficiency. *Automation in Construction, 97,* 13–28. https://doi.org/10.1016/j.autcon.2018.10.003

Ghasri, M.Maghrebi, M.Rashidi, T. H., & Waller, S. T. (2016). Hazard-based model for concrete pouring duration using construction site and supply chain parameters. *Automation in Construction, 71,* 283–293. https://doi.org/10.1016/j.autcon.2016.08.012

Ghodoosi, F.Bagchi, A.Zayed, T., & Hosseini, M. R. (2018). Method for developing and updating deterioration models for concrete bridge decks using GPR data. *Automation in Construction, 91,* 133–141. https://doi.org/10.1016/j.autcon.2018.03.014

Gledson, B., & Greenwood, D. (2016). Surveying the extent and use of 4D BIM in the UK. *Journal of Information Technology in Construction (ITcon), 21,* 57–71. Retrieved July 11, 2019, from www.itcon.org/paper/2016/4

Gledson, B. J., & Greenwood, D. (2017). The adoption of 4D BIM in the UK construction industry: An innovation diffusion approach. *Engineering, Construction and Architectural Management, 24*(6), 950–967. https://doi.org/10.1108/ECAM-03-2016-0066

Golizadeh, H.Hon, C. K. H.Drogemuller, R., & Reza Hosseini, M. (2018). Digital engineering potential in addressing causes of construction accidents. *Automation in Construction, 95,* 284–295. https://doi.org/10.1016/j.autcon.2018.08.013

Goulding, J. S., & Rahimian, F. P. (2012). Industry preparedness: Advanced learning paradigms for exploitation. *Construction Innovation and Process Improvement,* 409–433. https://doi.org/10.1002/9781118280294.ch18

Halvorsen, G. T., Poston, R. W., Barlow, P., Fowler, D. W., Palmbaum, H. M., Barth, F. G., Gergely, P., Pashina, K. A., Bishara, A. G., & Hansen, W. (2002). *224.3R-95: Joints in concrete construction (reapproved 2013).* American Concrete Institute, ACI Committee 224. ISBN:9780870313288

Hamledari, H.McCabe, B.Davari, S., & Shahi, A. (2017). Automated schedule and progress updating of IFC-based 4D BIMs. *Journal of Computing in Civil Engineering, 31*(4), 04017012. https://ascelibrary.org/doi/10.1061/%28ASCE%29CP.1943-5487.0000660

Hardin, B. (2015). *BIM and construction management: Proven tools, methods, and workflows.* John Wiley & Sons. ISBN:9781118942765

Heesom, D., & Mahdjoubi, L. (2004). Trends of 4D CAD applications for construction planning. *Construction Management and Economics, 22*(2), 171–182. https://doi.org/10.1080/0144619042000201376

Hong Kong Government. (2006). *General specification for civil engineering works*. Civil Engineering and Development Department (CEDD). Retrieved July 11, 2019, from www.cedd.gov.hk/filemanager/eng/content_71/gsvol1upto2_02.pdf

Hosseini, M. R.Maghrebi, M.Akbarnezhad, A.Martek, I., & Arashpour, M. (2018). Analysis of citation networks in building information modeling research. *Journal of Construction Engineering and Management, 144*(8), 04018064. https://doi.org/10.1061/(ASCE)CO.1943-7862.0001492

Huang, T.Kong, C. W.Guo, H. L.Baldwin, A., & Li, H. (2007). A virtual prototyping system for simulating construction processes. *Automation in Construction, 16*(5), 576–585. https://doi.org/10.1016/j.autcon.2006.09.007

Hyun, C.Jin, C.Shen, Z., & Kim, H. (2018). Automated optimization of formwork design through spatial analysis in building information modeling. *Automation in Construction, 95*, 193–205. https://doi.org/10.1016/j.autcon.2018.07.023

Issa, C. A.Gerges, N. N., & Fawaz, S. (2014). The effect of concrete vertical construction joints on the modulus of rupture. *Case Studies in Construction Materials, 1*, 25–32. https://doi.org/10.1016/j.cscm.2013.12.001

Jiang, L., & Leicht, R. M. (2016). Supporting automated constructability checking for formwork construction: An ontology. *Journal of Information Technology in Construction (ITcon), 21*(28), 456–478. Retrieved July 11, 2019, from www.itcon.org/paper/2016/28

Jsce Concrete Committee. (2007). *Standard specifications for concrete structures 2007, design*. Japan Society of Civil Engineers. ISBN:9784810607529

Kazaz, A.Ulubeyli, S.Er, B.Arslan, V.Arslan, A., & Atici, M. (2015). Fresh ready-mixed concrete waste in construction projects: A planning approach. *Procedia Engineering, 123*, 268–275. https://doi.org/10.1016/j.proeng.2015.10.088

Khashi, K.Dehghani, H., & Jahanara, A. A. (2018). An optimization procedure for concrete beam-column joints strengthened with FRP. *International Journal of Optimization in Civil Engineering, 8*(4), 675–687. http://ijoce.iust.ac.ir/article-1-369-en.html

Kim, I.Chin, S., & Ko, J. (2018). *An accident notification system in concrete pouring using sound analysis*. ISARC 2018, 35th International Symposium on Automation and Robotics in Construction and International AEC/FM Hackathon: The Future of Building Things, 2018. https://doi.org/10.22260/ISARC2018/0018

Kim, M.-K.Cheng, J. C.Sohn, H., & Chang, C.-C. (2015). A framework for dimensional and surface quality assessment of precast concrete elements using BIM and 3D laser scanning. *Automation in Construction, 49*, 225–238. https://doi.org/10.1016/j.autcon.2014.07.010

Kim, M.-K.Wang, Q.Park, J.-W.Cheng, J. C.Sohn, H., & Chang, C.-C. (2016). Automated dimensional quality assurance of full-scale precast concrete elements using laser scanning and BIM. *Automation in Construction, 72*, 102–114. https://doi.org/10.1016/j.autcon.2016.08.035

Kirillov, A. P. (1969). Effect of construction joints on the performance of reinforced concrete structures. *Hydrotechnical Construction, 3*(3), 214–222. https://doi.org/10.1007/BF02377208

Koo, B., & Fischer, M. (2000). Feasibility study of 4D CAD in commercial construction. *Journal of Construction Engineering and Management, 126*(4), 251–260. https://doi.org/10.1061/(ASCE)0733-9364(2000)126:4(251)

Kropp, C.Koch, C., & König, M. (2018). Interior construction state recognition with 4D BIM registered image sequences. *Automation in Construction, 86*, 11–32. https://doi.org/10.1016/j.autcon.2017.10.027

Lee, C., & Ham, S. (2018). Automated system for form layout to increase the proportion of standard forms and improve work efficiency. *Automation in Construction, 87*, 273–286. https://doi.org/10.1016/j.autcon.2017.12.028

Ma, L.Sacks, R., & Zeibak-Shini, R. (2015). Information modeling of earthquake-damaged reinforced concrete structures. *Advanced Engineering Informatics, 29*(3), 396–407. https://doi.org/10.1016/j.aei.2015.01.007

Maghrebi, M.Shamsoddini, A., & Waller, S. T. (2016). Fusion based learning approach for predicting concrete pouring productivity based on construction and supply parameters. *Construction Innovation, 16*(2), 185–202. https://doi.org/10.1108/CI-05-2015-0025

Mansuri, D.Chakraborty, D.Elzarka, H.Deshpande, A., & Grönseth, T. (2017). Building information modeling enabled cascading formwork management tool. *Automation in Construction, 83,* 259–272. https://doi.org/10.1016/j.autcon.2017.08.016

Meister, S. B. (2011). *Commercial real estate restructuring revolution: Strategies, tranche warfare, and prospects for recovery.* Wiley. ISBN: 9780470626832

Merritt, F. S., & Ricketts, J. T. (2001). *Building design and construction handbook.* McGraw-Hill. ISBN: 9780070419995

Nawy, E. G. (2008). *Concrete construction engineering handbook.* CRC Press. ISBN: 9781420007657

Nikmehr, B.Hosseini, M. R.Rameezdeen, R.Chileshe, N.Ghoddousi, P., & Arashpour, M. (2017). An integrated model for factors affecting construction and demolition waste management in Iran. *Engineering, Construction and Architectural Management, 24*(6), 1246–1268. https://doi.org/10.1108/ECAM-01-2016-0015

Ninić, J.Koch, C., & Stascheit, J. (2017). An integrated platform for design and numerical analysis of shield tunnelling processes on different levels of detail. *Advances in Engineering Software, 112,* 165–179. https://doi.org/10.1016/j.advengsoft.2017.05.012

Nizam, R. S.Zhang, C., & Tian, L. (2018). A BIM based tool for assessing embodied energy for buildings. *Energy and Buildings, 170,* 1–14. https://doi.org/10.1016/j.enbuild.2018.03.067

Ozumba, A. O. U., & Shakantu, W. (2018). Exploring challenges to ICT utilisation in construction site management. *Construction Innovation, 18*(3), 321–349. https://doi.org/10.1108/CI-03-2017-0027

Park, J.Cai, H.Dunston, P. S., & Ghasemkhani, H. (2017). Database-supported and web-based visualization for daily 4D BIM. *Journal of Construction Engineering and Management, 143*(10), 04017078. https://doi.org/10.1061/(ASCE)CO.1943-7862.0001392

Pärn, E. A., & Edwards, D. J. (2017). Conceptualising the FinDD API plug-in: A study of BIM-FM integration. *Automation in Construction, 80,* 11–21. https://doi.org/10.1016/j.autcon.2017.03.015

Pärn, E. A.Edwards, D. J., & Sing, M. C. P. (2017). The building information modelling trajectory in facilities management: A review. *Automation in Construction, 75,* 45–55. https://doi.org/10.1016/j.autcon.2016.12.003

Pauwels, P.Mcglinn, K.Törmä, S., & Beetz, J. 2018. Linked data. *Building Information Modeling: Technology Foundations and Industry Practice,* 181–197. https://doi.org/10.1007/978-3-319-92862-3_10

Pomares Torres, J. C.Baeza, F. J.Varona Moya, F. D. B., & Bru Orts, D. 2017. BIM implementation for structural design courses in civil engineering. In A. G. Garigos, L. Mahdjoubi, & C. A. Brebbia (Eds.), *Building information modelling (BIM) in desing, constructrion and operations II* (pp. 79–86). WIT Press. https://doi.org/10.2495/bim170081

Poon, C. S., Yu, A. T., & Jaillon, L. (2004). Reducing building waste at construction sites in Hong Kong. *Construction Management and Economics, 22*(5), 461–470. https://doi.org/10.1080/0144619042000202816

Porwal, A., & Hewage, K. N. (2012). Building information modeling – based analysis to minimize waste rate of structural reinforcement. *Journal of Construction Engineering and Management, 138*(8), 943–954. https://doi.org/10.1061/(ASCE)CO.1943-7862.0000508

Pour Rahimian, F.Chavdarova, V.Oliver, S., & Chamo, F. (2019). OpenBIM-Tango integrated virtual showroom for offsite manufactured production of self-build housing. *Automation in Construction, 102,* 1–16. https://doi.org/10.1016/j.autcon.2019.02.009

Rahimian, F. P.Ibrahim, R., & Baharudin, M. N. (2008). *Using IT/ICT as a new medium toward implementation of interactive architectural communication cultures.* 2008 International Symposium on Information Technology. https://doi.org/10.1109/ITSIM.2008.4631984

Richardson, J. (2014). *Supervision of Concrete construction 1.* CRC Press. ISBN: 1482275627

Seyedzadeh, S., Pour Rahimian, F., Rastogi, P., & Glesk, I. (2019). Tuning machine learning models for prediction of building energy loads. *Sustainable Cities and Society, 47,* 101484. https://doi.org/10.1016/j.scs.2019.101484

Seyedzadeh, S.Rahimian, F. P.Glesk, I., & Roper, M. (2018). Machine learning for estimation of build-ing energy consumption and performance: A review. *Visualization in Engineering, 6*(1). https://doi.org/10.1186/s40327-018-0064-7

Singh, M. M.Sawhney, A., & Sharma, V. (2017). Utilising building component data from BIM for formwork planning. *Construction Economics and Building, 17*(4), 20. https://doi.org/10.5130/AJCEB.v17i4.5546

Standards Australia. (2001). *AS 3600-concrete structures.* Standards Australia International. ISBN: 9771760721466

Stanton, A., & Javadi, A. A. (2014). An automated approach for an optimised least cost solution of reinforced concrete reservoirs using site parameters. *Engineering Structures, 60,* 32–40. https://doi.org/10.1016/j.engstruct.2013.12.020

Suprenant, B. (1988). *Construction joints for multistory structures.* The Aberdeen Group. Retrieved July 11, 2019, from www.concreteconstruction.net/_view-object?id=00000153-96b1-dbf3-a177-96b93fa70000

Technical Committee Cen/Tc 250. (2004). *Eurocode 2: Design of concrete structures – Part 1-1: General rules and rules for buildings.* Comite Europeen de Normalisation. Retrieved July 11, 2019, from www.phd.eng.br/wp-content/uploads/2015/12/en.1992.1.1.2004.pdf

Wang, Y.Yuan, Z., & Sun, C. (2018). Research on assembly sequence planning and optimization of precast concrete buildings. *Journal of Civil Engineering and Management, 24*(2), 106–115. https://doi.org/10.3846/jcem.2018.458

Zhou, Y.Ding, L.Wang, X.Truijens, M., & Luo, H. (2015). Applicability of 4D modeling for resource allocation in mega liquefied natural gas plant construction. *Automation in Construction, 50,* 50–63. https://doi.org/10.1016/j.autcon.2014.10.016

10 BIM integrated project delivery

An automated earned value management–based approach

10.1. Introduction

Traditional project delivery systems have often proven inefficient in improving overall performance (Matthews & Howell, 2005). In response to this, Integrated Project Delivery (IPD) has been proffered for use in projects across the construction industry (Zhang & Li, 2014), where it is argued that IPD provides a new contractual, behavioural, and organisational context for delivering construction projects through enhancing integrated and collaborative practices (Pishdad-Bozorgi & Beliveau, 2016). This is seen as an innovative project delivery method, characterised by early collaborative and collective engagement of key stakeholders across all phases of project delivery (Ahmad et al., 2018; AIA, 2007).

Successful delivery of a project through IPD is not easy, however, especially as this requires fulfilling a wide range of requirements (Fischer et al., 2017). Of these requirements, the IPD compensation model, also called risk/reward compensation, is of cardinal importance (Ma et al., 2018). This is described as a key principle of IPD (Zhang et al., 2018), which plays a pivotal role in stimulating creativity, motivating collaboration, and sustaining performance (Liu & Bates, 2013a; Zhang & Li, 2014). That being said, the risk and reward must be shared and allocated to all participants in core project teams, necessitating joint project control (Ashcraft, 2011; Fischer et al., 2017). For designing the risk and reward model (hereafter referred to as the compensation approach), economic models provide a sound foundation based on project costs (AIA, 2007; Zhang & Chen, 2010). The compensation approach typically depends on achievements throughout the project, as well as two cost lines; namely: target cost, which defines the cost baseline, and agreed percentage for profit-at-risk (Kent & Becerik-Gerber, 2010; Seyedzadeh & Pour Rahimian, 2021b). For example, if a project achieved below its target cost, this means the cost saving percentages are shared among key participants; however, if performance indicated that the level exceeded the profit-at-risk line, then the client has sole responsibility for paying the direct costs (AIA, 2007; Ashcraft, 2011). As such, any cost management system under IPD must be rigorous, dynamically integrated, and capable of avoiding erroneous cost information throughout all project stages (Ma et al., 2018; Zhang & Li, 2014).

With this goal in mind, IPD as a delivery method has largely been promoted for its potential in facilitating Building Information Modelling (BIM) implementation on construction projects (Dawood et al., 2020; Fischer et al., 2017). Coupling BIM with IPD has proven to improve efficiency, reduce errors, enable exploring alternative approaches, as well as expanding market opportunities on projects (Kent & Becerik-Gerber, 2010). In fact, "the full potential benefits of IPD and BIM are achieved only when they are used together" (Ashcraft, 2008, p. 15). Acknowledging this, construction IPD projects rely on data-rich BIM models

DOI: 10.1201/9781003106944-10

that focus on exploiting BIM in integrating information flows (Ma et al., 2017; Turkan et al., 2012). Such combined use of IPD and BIM also makes sense from a theoretical perspective, but in reality this faces substantial roadblocks (Holzer, 2011).

BIM-based project control activities have to date largely relied on automated site data collection tools that use various methods such as spatial sensing technologies and exploitation of links between the 3D BIM model and performed works, etc. (Hosseini et al., 2018; Jaselskis et al., 2005). These methods and approaches almost entirely measure physical items and components on construction sites, but tend to overlook the value of wider activities (Turkan et al., 2012, 2013). Problems therefore accrue regarding the sharing of acquired control information across projects, which tend to exacerbate matters given that project team members are predominantly dominated by silo thinking (Merschbrock et al., 2018; Mignone et al., 2016). Moreover, information systems are typically loosely coupled (Hosseini et al., 2018; Shen et al., 2012), where automated processes that integrate information of physical components with managerial attributes (such as allocated resources and values) to facilitate controlling cost-time integrated progress are needed for a more complete solution (Lee et al., 2017; Pishdad-Bozorgi et al., 2013).

From this analysis, it is acknowledged that such automated cost structures should be capable of differentiating overhead costs from profit, ensuring that no profit items remain hidden in overhead costs and labour rates (Allison et al., 2018; Teng et al., 2017). This is particularly important, as all parties in IPD are held equally responsible for the entire project performance (AIA, 2007; Allison et al., 2018). As such, the automated cost structure must provide a financial tracking system that (i) aggregates all cost data, (ii) presents data in a clear format, (iii) is readily accessible by all parties, and (iv) shows saved costs and by whom (Allison et al., 2018). With this in mind, the study presented in this chapter provides a clear solution for enhancing the cost structure and compensation mechanism, where an automated cost structure for risk/reward sharing is presented for discussion. This solution was designed to enhance collaborative interaction among project participants using a web-based platform for instant sharing of risk/reward mechanism outcomes. This automated model of a cost control system of IPD projects is discussed in the following sections. This incorporates the capabilities of Earned Value Management (EVM) – an approach used for effectively tracking, analysing, and controlling project costs, where EVM has been identified as an effective tool for cost management in projects (PMI, 2013). Moreover, EVM can also provide performance metrics for both cost and schedule alike through an integrated pool (de Andrade et al., 2019). The model, therefore, is designed to draw upon EVM.

The following sections present the theoretical background on the key concepts surrounding cost control and EVM. From this, the research gap is established, followed by the methodology, logic, and framework development. Results and analysis are presented, concluding with a discussion on findings, limitations, and future research opportunities.

10.2. Theoretical background

The following theoretical background is divided into five sub-sections, each of which covers one major area associated with the development of this work.

10.2.1. BIM and cost management

In recognising the need to move towards more efficient project delivery, the ultimate goal is to have a comprehensive database of information available to all project participants, with confidence in its accuracy, universal utility, and clarity of deliverables (Ashcraft, 2014; Oraee et al., 2017). The main drive for adopting BIM is to manage all project documents and stages (i.e. design, planning, and costing) within a single/dynamic context to secure the proper exploitation of all available information (Abrishami et al., 2015; Merschbrock et al., 2018; Redmond et al., 2012). In this respect, it is important to observe that BIM design elements must contain the required information in various formats (including design/management) (Banihashemi et al., 2018), in order to secure smartly designed elements – as opposed to traditional 3D components (Fu et al., 2006; Pärn & Edwards, 2017). This requires BIM users to acquire all the required information from single BIM elements – a prerequisite for making informed decisions (Abrishami et al., 2014; Motamedi & Hammad, 2009; Shen et al., 2012). It is also important to note that four-dimensional modelling (4D BIM) can embed progress data into 3D model objects by adjusting the task-object relationship (Hamledari et al., 2017). This is particularly useful, as the application of 4D BIM leads to more easily operated workflows, greater efficiency with on-site management, and wider ability to assess constructability (Hartmann et al., 2008). As for cost management, BIM has been acknowledged as being one of the most efficient Architecture, Engineering, and Construction (AEC) tools for increasing productivity (Aibinu & Venkatesh, 2013; Lee et al., 2014; Pour Rahimian et al., 2020; Wang et al., 2016). Finally, colloquially termed as 5D BIM (Aibinu & Venkatesh, 2013), this enhanced capability offers techniques for extracting quantities from 3D models, whilst also allowing cost consultants to incorporate productivity allowances and pricing values (Eastman et al., 2011; Lee et al., 2014), where the cost estimating process starts with exporting data from 3D models to BIM-based cost estimating software (e.g. CostX®) to prepare quantity take-off. Following this, Bills of Quantities (BoQs) are generated and exported to an external database (Aibinu & Venkatesh, 2013). From this, prices and productivity allowances can be added to the project schedule (Eastman et al., 2011; Lee et al., 2014). Such automated quantification procedures shorten the quantity take-off processing time, which automatically considers changes in design, thereby enabling projects to be fast-tracked (Popov et al., 2010; Wang et al., 2016).

10.2.2. IPD-based cost estimation

IPD is a project delivery system designed to deliver value, where 'value' includes considerations other than pure cost (Pishdad-Bozorgi & Srivastava, 2018). In this respect, it is important to note that the determination, projection, and tracking of costs are critical to IPD success (Ashcraft, 2012; Love et al., 2011). Cost estimation hence has a vital role in the application of IPD (AIA, 2007; Zhang & Chen, 2010). Therefore, this must be tracked using some form of scrutinising method, typically engaging all core team members, the process of which not only determines profit (and shared benefits/risks), but also identifies deviations between actual and target costs (AIA, 2007; Zhang & Li, 2014). Therefore, the compensation approach structure must be capable of drawing upon effective methods for determining several factors, not least of which are cost overrun proportions, cost underruns, and saving potential against the agreed cost (Thomsen et al., 2009). This is particularly important as the risk/reward proportion relies on the degree of achievement made throughout the project stages (Love et al., 2011); however, the compensation approach has two limits (Figure 10.1). Firstly, the direct, indirect, and overhead costs (which can be nominated as agreed costs), and secondly, the profit-at-risk percentage after estimating the agreed cost (AIA, 2007; Zhang & Li, 2014).

The precise determination of risk perception is critical for ensuring the agreed compensation structure is implemented correctly throughout the project, particularly so that the

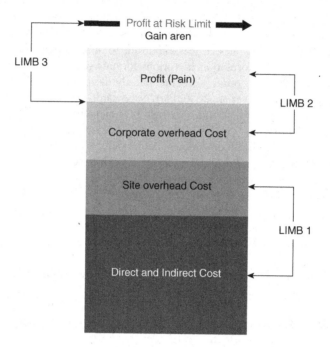

Figure 10.1 Compensation structure (adapted from Zhang and Li (2014))

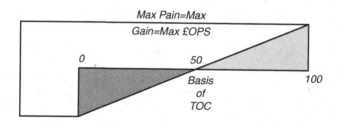

Figure 10.2 OPS ranges for risk/reward ratio (adapted from Ross (2003))

risk/reward ratio can be fairly allocated among project participants. In some instances, participants that carry higher proportions of uncertain works can be compensated with higher profit-at-risk percentages (Das & Teng, 2001). Another aspect of this is alliancing agreements, where these can be used to reduce risk impacts through sharing information and risks depending upon data availability (Delerue & Simon, 2009).

Figure 10.1 presented a typical compensation structure. This identified three IPD limbs, where Limb-1 represents the reimbursement of project costs, Limb-2 indicates the overhead costs for all participants, and Limb-3 covers the profit-at-risk ratio. Limb-3 in particular represents the risk/reward sharing scheme, the details of which must be specified at the beginning of the project (Ross, 2003). According to Ross (2003), the risk/reward ratio is measured by the Overall Performance Score (OPS), which is a scale between 0 and 100, where 0 to 50 represents the 'pain' scope, and 50 to 100 represents the 'gain' range (see Figure 10.2). After determining the risk/reward ratio using OPS, project participants generally share this ratio in contract correspondence.

10.2.3. Earned value management

The concept and approach of EVM is a quantitative project management technique used for measuring project progress, the underlying rationale of which aims to provide project participants with early warnings where the project is running 'over budget' or 'behind schedule'. In this respect, EVM has been successfully applied on several real-life projects to deliver accurate cost/schedule metrics (Pajares & López-Paredes, 2011; PMI, 2013). Khamooshi and Abdi (2016). Moreover, the "earned value technique is a crucial technique in analysing and controlling the performance of a project"(Naeni et al., 2011, p. 764).

From this, the Actual Cost of Work Performed (ACWP) is seen as one of the fundamental inputs of EVM. Similarly, the Budgeted Cost of Work Scheduled (BCWS) is the term used to ascribe the time-phased budget in line with the scheduled work. Achievement values are determined in accordance with several parameters. For example, Cost Performance Index (CPI), where $CPI < 1$ indicates that the cost performance is poor; $CPI = 1$ indicates that the cost performance is efficient, and $CPI > 1$ indicates that cost performance is excellent. Using EVM, achievements can be measured as a variance (not performance), such as Cost Variance (CV) and Schedule Variance (SV). For example, $CV < 0$ indicates a project is over budget, $CV = 0$ indicates the project is on budget, and $CV > 0$ indicates that a project is under budget (Pajares & López-Paredes, 2011).

The granularity of project scheduling can be represented through the Work Breakdown Structure (WBS), which can include detail such as expenditure, but this in itself can be a problem for the accurate implementation of EVM (Pajares & López-Paredes, 2011). An EVM system, therefore, not only has to be functional, but it also needs to have advanced capabilities in order to correlate data from multiple sources to generate cost control reports (Lipke et al., 2009; Seyedzadeh et al., 2021). In addition, the concept of interoperability (among various data sources) needs to be addressed in order to build federated project cost control sheets, which it is argued can be best resolved through advanced technologies and visualisation techniques (Chou et al., 2010).

10.2.4. Activity-based costing

Construction projects are more often than not highly fragmented, engaging multiple stakeholders and project participants from diverse backgrounds and organisations. Acknowledging this situation, this fragmentation can also lead to increases in overhead activities and overhead costs (Ashcraft, 2008; Mignone et al., 2016). Several traditional cost accounting methods are used in the industry. These include Resource-Based Costing (RBC), which relies on resources' cost, and Volume-Based Allocation (VBA), which is based on allocating the cost of resources directly to the objects, regardless of the cost structure (Holland & Jr, 1999). Additional issues to mention here include cost distortion, which is often embroiled with using these traditional methods, due in part to the conflation of all indirect costs into one, which can distort the pricing of company products (Miller, 1996). As a solution to this issue, Activity-Based Costing (ABC) aims to reduce or minimise such distortion, by allocating costs through multi-pools and determining the overhead activities and the associated costs needed to transform resources into activities that deliver the final product (Kim & Ballard, 2001; Kim et al., 2011).

10.2.5. *BIM 4D/5D automation*

Integrating BIM into daily construction activities can facilitate automatic updating of site information, and as such, can enhance productivity, strengthen relationships amongst stakeholders, and increase trust in site-collected data (Omar & Dulaimi, 2015). As such, El-Omari and Moselhi (2011) asserted that using unsystematic procedures for collecting site data could lead to a significant loss of information, whilst also generating unreliable results. That being said, BIM 4D automation can enhance the quality of collected data and reduce the human interference from the data collection process (Hamledari et al., 2017; Hartmann et al., 2008). Similarly, the use of 5D BIM provides an effective methodology for cost data collection and analysis of construction projects (Aibinu & Venkatesh, 2013; Lee et al., 2014; Oliver et al., 2020; Popov et al., 2010; Wang et al., 2016). Further studies also recommended that BIM cost systems should be actively engaged with decision-making, rather than merely generating the BoQ (Lee et al., 2014).

Over recent years, automated data collection methods have significantly improved, particularly through various kinds of technology such as bar coding, radio frequency identification, 3D laser scanning, photogrammetry, multimedia, and pen-based computers (El-Omari & Moselhi, 2011; Turkan et al., 2012, 2013). Eastman et al. (2011). Whilst these are significant improvements, arguably there do not seem to be many fully comprehensive BIM-based cost management platforms in the market that can perform all cost-related processes, namely: estimation, budgeting, and control. There is therefore a need for the construction industry to capitalise on this need, especially to explore the means towards analysing data in much more efficient ways (Hosseini et al., 2018; Wang et al., 2016).

10.3. Research gap

Literature in IPD, BIM, and costing shows several trends and opportunities. Of these, the research gap presented here focusses on the potential of available tools and techniques (i.e. EVM and ABC within IPD) (Holzer, 2011; Hosseini et al., 2018) to move industry challenges forward. Whilst studies in this area (for the most part) stop at providing an outline of how these methods and techniques actually add value to the risk/reward sharing mechanism in IPD (c.f. Ilozor & Kelly, 2012; Pishdad-Bozorgi et al., 2013; Pishdad-Bozorgi & Srivastava, 2018), this is seen as an ideal point of departure for further investigation.

BIM integration with IPD practices are discussed in several research studies (Allison et al., 2018; Ashcraft, 2012; Fischer et al., 2017; Ilozor & Kelly, 2012; Nawi et al., 2014; Rowlinson, 2017). Integration is seen as a real issue here, where the challenges of such integration are explored in a stream of studies, including financial challenges, differences in cost accounting, and the risk/reward sharing mechanism (Holzer, 2011; Roy et al., 2018; Kahvandi et al., 2018). Of note, it seems that there are limited workable methodologies demonstrating the interrelationship of BIM tools/dimensions and IPD stages in practical terms (Roy et al., 2018; Kahvandi et al., 2018).

Acknowledging this paucity in knowledge and applications, some research has attempted to provide additional information on IPD compensation structures and frameworks. For example, Zhang and Li (2014) developed a risk/reward compensation mechanism by combining risk perception with the Nash Bargaining Solution (NBS) technique. Whilst this model did not consider the method of sharing actual risk/reward amongst participants, nor evaluated the impact of IPD compensation structures, this was nevertheless a start in the right direction. Other developments in this area include Liu and Bates (2013b), who articulated a probabilistic contingency calculation model for predicting contingency in order to minimise cost overruns.

In recognition of these challenges, there is significant potential to explore the integration of BIM, ABCm and EVM into IPD cost structure practices. Moreover, whilst acknowledging that a workable and theoretically based solution that supports such integration is missing (Allison et al., 2018; Ballard et al., 2015), the research gap presented in this chapter addresses this challenge through the development of a practice-based solution.

10.4. Methodology

The work presented hereafter presents the methodological processes and structures engaged for the development of a practical and feasible solution to the problems identified herein. This research examines the applicability and validity of assumptions derived from literature (qualitative evidence), and places this within a real-life setting (Figure 10.3). This approach follows an amalgamation of *exploratory case study* and *experiment* research, both of which were deemed suitable for accomplishing issues such as this – following arguments from Banihashemi et al. (2018). Real-life context is an essential part of case study research, where many variables can often affect the outcome, particularly on inference between cause and effects (Yin, 1981). Assessing the impacts of proposed workflows in real-life scenarios are acknowledged

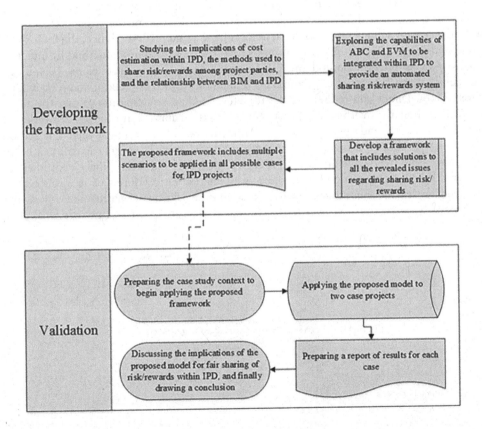

Figure 10.3 Research methodology

to be affected by many factors and procedures. Therefore, a case study approach using observational techniques was seen to offer reliable outcomes for research purposes, particularly to reveal whether real data can support or refute any proposed procedures (Zellmer-Bruhn et al., 2016) to demonstrate a match (if any) between data and theory.

10.4.1. Logic

Following the methodology, a framework was developed to address the identified challenges and deficiencies highlighted in literature. This framework highlights the concepts, scope, model development, and validation procedure (see Figure 10.3).

10.5. Framework development

The framework presented in Figure 10.3 relies on estimating the costs within the IPD approach based on ABC, given the proven capability of ABC in assigning different costs: direct, indirect, and overhead. In this respect, the cost estimator is able to distinguish between direct, indirect, and overhead costs in order to apply costs to tasks (which is vital for ensuring the appropriate application of IPD). EVM is also used to measure project progress. This framework therefore adopted EVM integration with an IPD approach using an ABC-based estimation method.

The compensation structure in IPD depends heavily on distinguishing direct and overhead costs, such that owners and non-owner parties can manage their activities in accordance with achievements ascribed in each Limb (Figure 10.4). Hence, the framework presented in this chapter involves an innovative risk/reward sharing method by integrating the ABC estimation method into EVM controlling technique. From Figure 10.4, it can be seen that the compensation structure formulation of the proposed framework follows three Limbs. Direct and indirect costs are determined as a summation of costs of direct activities. Similarly, overhead costs are estimated as a summation of costs of overheard activities for each trade package (derived from the ABC estimation sheet). The reason behind using ABC for articulating the compensation approach was its capability of being able to measure the degree of savings for each participant. This accordingly leads to effective and precise computation of the risk/reward sharing ratio. Moreover, it was accepted that the cost-saving share for the owners differed from non-owner participants, given the difference between the cost overhead saving at the organisation and project level. The

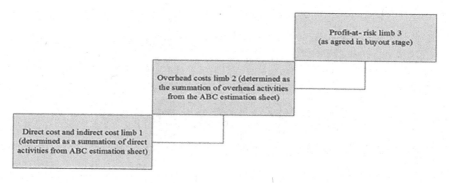

Figure 10.4 Compensation under the IPD approach using ABC estimation

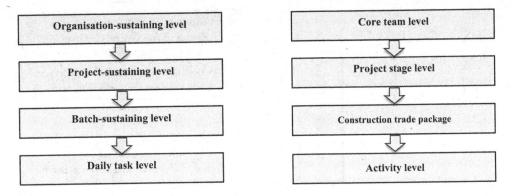

Figure 10.5 ABC functional level: comparison between the traditional delivery approach (left) and the IPD approach (right)

overall approach was therefore to determine the participants' risk/reward sharing ratio for transparency and equity.

An EVM-grid output was used to measure project progress (since this represents cost and schedule progress in a single index). In this respect, the framework needed to integrate EVM and ABC, to articulate three models for dealing with differing scenarios, where the cost-savings sharing approach adopted a different scheme for distinguishing between the overhead costs for each participant (Figure 10.5).

10.5.1. *Framework implementation process*

After determining the BCWS, ACWP, and Budgeted Cost of Work Performed (BCWP) costs for controlling milestones, it is normal for quantity surveyors to determine the values of Cost Performance Ratio (CPR) and Schedule Performance Ratio (SPR). These are then entered into the grid as positive or negative percentages in order to determine the current project situation. The EVM-grid divides the project into four areas, where each area represents the project situation, typically distinguished by a specific colour. Whilst considering the X-axis as the schedule and the Y-axis as the cost, each area was then divided into small squares around the planned point. In this respect, users determine the value of the CPR and SPR and enter these values into the grid as positive or negative percentages to determine the project situation at each milestone (or for each package). Quantity surveyors then mark the square in accordance with CPR and SPR percentages to determine the cumulative progress throughout the project execution stages. Thereafter, the 'profit-at-risk' percentage is shared in accordance with the output of the developed EVM-based IPD grid. For instance, if the output was 63%, this means that the project is running over cost and is also behind schedule. Thus, the profit-at-risk percentage is used to determine if the project is still within Limb-2 or exceeds it. This relationship of EVO limbs can be seen in Figure 10.6.

Hence, the proportion for each participant is determined based on the limb location. After determining project progress (in accordance with the IPD compensation approach), the model applies the risk/reward proportion among the core team members (Table 10.1), where EVO represents the EVM-grid output.

The following narrative discusses the cases from Table 10.1.

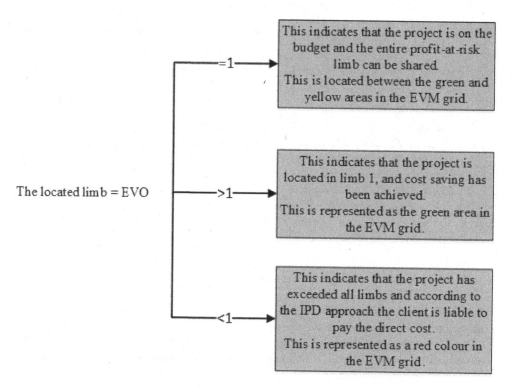

The located limb = EVO

=1→ This indicates that the project is on the budget and the entire profit-at-risk limb can be shared. This is located between the green and yellow areas in the EVM grid.

>1→ This indicates that the project is located in limb 1, and cost saving has been achieved. This is represented as the green area in the EVM grid.

<1→ This indicates that the project has exceeded all limbs and according to the IPD approach the client is liable to pay the direct cost. This is represented as a red colour in the EVM grid.

Figure 10.6 Function for assessing overall project situation and limb location

Table 10.1 Cases summary and required models

No of cases	Case	EVM output values	Equation(s)
1	On cost/schedule	EVO =1	Equations 5 and 6 (Case 1)
2	Ahead of schedule and/or cost underrun	EVO >1	Equations 7 and 8 (Case 2)
3	Behind schedule and/or cost overrun	EVO <1	Equations 9 and 10 (Case 3)

Case 1

The first case occurs when a project is progressing on schedule and within budget. Equations 5 and 6 are used to determine the risk/reward sharing among the project participants.

$$\text{Rewards value} = \big((\text{EVO}) \times P@R\,\text{Per}\big) \times \text{MVoLIMB2}\big) \tag{5}$$

$$\text{MV for R or RD for each party} = \text{Rewards value} \times \text{PoO or PoNO} \tag{6}$$

Where: P or G Per represents the risk/reward (pain/gain) percentage of all project participants (%); P@R Per represents the profit@Risk percentage (%); MVoLIMB2 represents the monetary value of LIMB2 (£); MV for R/RD for each party represents the monetary value for risk or rewards for each participant (£); and PoO/PoNO represents the proportion of owner or non-owner party (%).

Case 2

The second case is when the project is progressing ahead of schedule and below the budget. Equations 7 and 8 are used to determine the cost-savings sharing among project participants.

$$\text{CSoOC for NO} = \Sigma\text{CSoOOA from ABC estimation sheet} +$$
$$\Sigma\text{CSoOPA from ABC estimation sheet} \times \text{NOARP} \qquad (7)$$

$$\text{CSoOC for O} = \Sigma\text{CSoOPA from ABC estimation sheet} \times \text{OARP} \qquad (8)$$

Where: CSoOC for NO represents the overhead cost saving for non-owner participants (£); CSoOOA from ABC estimation sheet represents the overhead organisation activities costs' saving from the ABC estimation sheet (£); CSoOPA from the ABC estimation sheet represents the overhead project activities costs' saving from the ABC estimation sheet (£); NOARP represents the Non-Owner Agreed Rewards Proportion (%); CSoOC for O represents the overhead cost saving for owner participants (£); and OARP represents the owner agreed rewards proportion (%).

Case 3

The third case is where the project is behind schedule and over the budget. This implies that the project is in the crisis area (red zone). In this case, the owner is liable for paying the direct costs only to the non-owner (i.e. constructor and trade contractors), as shown in equation 9. In case the P@R< EVO<1, the project progress is at risk/crisis area; however, the profit-at-risk percentage will cover the determined deviation.

$$\text{DC} = \Sigma\text{DAC from ABC estimation sheets} \qquad (9)$$

In case the deviation is less than the profit-at-risk percentage, there will be a remaining rewards value, which can be determined as equation 10.

$$\text{Rewards value} = \left((\text{OoEVMG} - 1) + \text{P@R Per}\right) \times \text{MVoP@Rper} \qquad (10)$$

Where: DC represents the direct cost (£); DAC from the ABC estimation sheets represents the direct activities' costs from the ABC sheet as BCWS (£); RV represents the Rewards Value (£); MVoP@Rper represents the monetary value of Profit@Risk percentage (£).

From these cases, an EVM-IPD grid (Figure 10.7) is presented, which considers a range of positive and negative CPR and SPR values (depending on the project's degree of complexity and other factors including potential risks and mitigation). ON implies that the project is on the schedule and budget; OC implies that the project is on budget; OS implies that the project is on schedule; AS represents ahead of schedule; BS represents behind schedule; VS represents cost overrun; and UC represents cost underrun.

Figure 10.8 summarises the framework in the form of a flowchart. This provides a comprehensive solution for structuring the IPD's compensation approach, offering an easy method

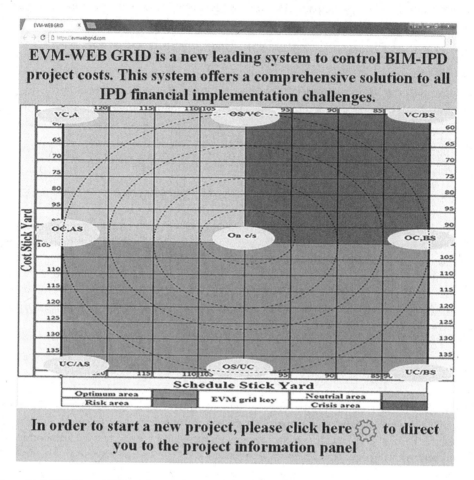

Figure 10.7 EVM-based IPD with considering ABC estimating approach

of managing the IPD compensation structure under different circumstances, whilst also considering different participants' organisational needs in terms of risk and reward sharing. This new approach integrates ABC into IPD using the EVM technique.

10.6. Model integration and flow of data

The flow of data in the proposed model reflects the documentation and buyout stage, through to the close-out stage. This also highlights BIM integration at each stage – the details of which are discussed as follows.

During the documentation stage, core team members conduct cost estimation based on ABC and then load costs to each corresponding activity – whether the activity is direct, indirect, or an overhead. This is implemented by estimating costs using a 5D BIM platform (i.e. Navisworks) after configuring its layers in accordance with ABC levels. Subsequently, BCWS values are prepared by exporting data created through 4D/5D BIM platform to another software package such as Microsoft Project. Hence, the buyout stage takes place

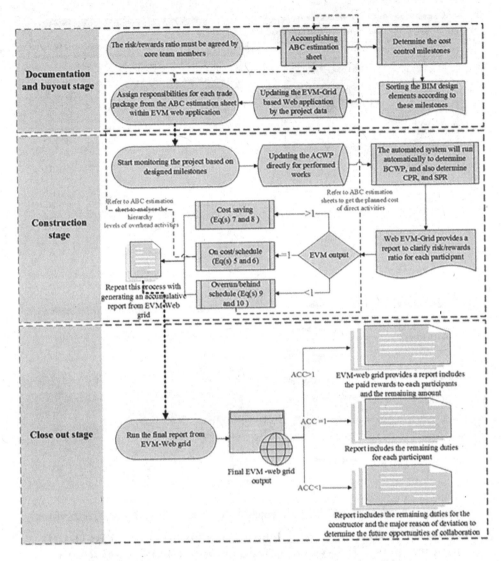

Figure 10.8 EVM-based IPD approach implementation flowchart

to agree on the percentage of profit-at-risk (P@R%), as well as risk/reward among owner/ non-owner parties. Subsequently, the agreed-upon P@R% is added to BCWS to develop the project compensation approach, where all project data (i.e. BCWS for each package, P@R %, risk/reward sharing %) is recorded to enable the determination of actual percentages within the construction stage. Once the construction stage begins, the project manager starts loading the project information (CPR and SPR) into the EVM-Web grid, as shown in Figure 10.9. In this respect, the steps presented in Figure 10.9 must be followed during the construction stage in order to generate the report at each milestone. All of the previously mentioned equations for three cases are coded to receive the input of equations, through which the outcomes are then displayed automatically. From a data perspective,

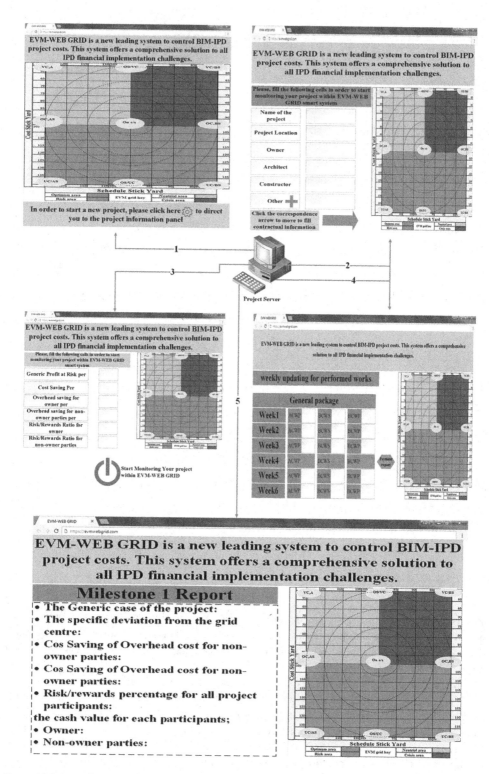

Figure 10.9 Data flow for the construction stages

this is centred in the project server, where the project manager attaches the initial documents, including the budgeted cost of work schedule (BCWS). Afterwards, progress data is updated on the server and used to generate a milestone report (see Figure 10.9, step 5). For the close-out stage, the report includes accumulative monetary profit and risk values for each party (and all participants), as all parties are responsible for profits and risks regardless of the causes of profits/risks. The profit/risk outcome for each milestone is kept in the profit/risk pool, which is shared during close-out stage.

From Figure 10.9, the system begins by providing context and background information to operational issues. Following this, the system provides empty cells for all project parties to complete (Figure 10.9, section 2). This includes the compensation structure terms. At each payment milestone, the system is linked to a spreadsheet (Microsoft Excel) in order to generate outcomes based on risk/reward monetary values.

10.7. Results and analysis

In order to validate the proposed methodology, this model was applied to a case study – namely, a property development company, whose managers decided to build a new house. The client decided to use the IPD delivery approach for the project and commenced by appointing the architect and the main contractor from the conceptualisation stage (in accordance with IPD principles). Thereafter, the client formulated the core project team from the architectural firm (Company X) and the contractor (Company Y), including trade contractors to obtain the required information during start-up meetings. The project works were categorised into five trade packages: general works, ceiling, lighting fixture, finishing, and doors/windows packages. Since IPD depends on sharing the benefits and risks, it was necessary to assign the expenses and costs to specific activities. Using the ABC technique (after adapting it to be consistent with IPD levels), all parties were gathered into one unified cost pool under joint venture cooperation (Kim et al., 2011). The costs of implementing IPD was then determined from the conceptualisation stage to buyout stages. The compensation structure was agreed upon as follows: (1) the agreed profit-at-risk percentage was 20%, (2) the saving cost allocation percentage for overhead project level cost was 70% for non-owner participants and 30% for the owner, (3) the risk/reward ratio was 80% for non-owners and 20% for the owner. Although, within the existing IPD model, the owner does not attract any proportion from P@R% (AIA, 2007; Allison et al., 2018), it was assumed that the owner would secure a proportion from P@R% for two reasons: providing any service such as participating in managing project workflow, and showing capabilities of the presented framework to work on various scenarios; (4) the direct and indirect cost limit (Limb 1) was £118,484.9; (5) Limb 2, which involved direct, indirect, and overhead costs was £190,484.9; and (6) Limb 3, which comprised the total cost and the profit-at-risk percentage was £228,581.9. The project was packaged by trade, in accordance with the ABC method, and articulated across the three limbs.

Finally, the detailed cost estimate was prepared by package for the three limbs, as shown in Table 10.2, where Limb 1 represents the direct and indirect costs, Limb 2 represents the summation of overhead activities, and Limb 3 represents the profit-at-risk percentage after estimating the entire project cost.

The model was specifically designed to manage progress, whether positive or negative, and moreover, to share the risk/reward in accordance with the agreed percentage of IPD. The case study considered two different scenarios (Scenario 1 and Scenario 2) to display the model's flexibility in capturing different circumstances. These two scenarios are discussed as follows.

Table 10.2 Compensation structure components

Construction Packages	General (£)	Ceiling (£)	Lighting fixtures (£)	Finishings (£)	Doors and windows (£)
Total material costs	38,038.9	2,140.2	17,037.9	3,553.8	31,919.1
Total labour costs	21,318.9	1,715	296.5	1,334.4	763
Total equipment costs	366.8	0	0	0	0
Total direct and indirect costs (Limb 1)	59,724.7	3,855.2	17,334.4	4,888.3	32,682.2
Overhead costs (Limb 2)	27,557.6	11,519.2	7,134.6	15,403.8	7,134.6
Total costs (ABC) (starting point of profit-at-risk percentage)	89,014.7	15,474.1	24,916.7	20,418.4	40,660.9
Profit-at-risk limit (Limb 3)	106,817.6	18,568.9	29,900.1	24,502.1	48,793.1

Scenario 1

Scenario 1 highlights how the risk/reward was shared among all project participants. In this respect, project payments were assumed to be vired monthly, with collected data from the project cost centre shown in Table 10.3.

The following steps were applied to determine the risk/reward sharing proportion for each participant. Project CPR and SPR were determined using the EVM-grid (see Figure 10.8). The EVM output was 82.5%. This is located in the red area, implying a crisis situation due to the considerable deviation from the planned values. The EVM output was 0.825, which is less than 1, with a 17.5% P@R deviation, less than 20% deviation, then the third case (B) in the framework, model 11, should be applied. By applying equation 10, the total reward was valued at £341.5. Thereafter, in order to split the reward between the owner and non-owner, equation 6 was applied. The reward outcome displayed £68.3 and £273.2 for the owners and non-owners, respectively.

Scenario 2

Scenario 2 shows how the cost saving was shared among all project participants without cost distortion. Project payments were assumed to be vired monthly with collected data from the project cost centre shown in Table 10.4.

The project CPR and SPR values were determined using the EVM-grid (see Figure 10.8). The EVM output was 104%, located in the green area, implying an optimum situation due to the positive deviation from the planned values. Since the EVM output was 1.04, which is greater than 1 with a 4% P@R deviation, equation 7 should be applied to calculate the entire savings cost and the proportion for each participant. The entire saving cost was valued at £2,732.4. Thereafter, applying equation 7 and equation 8, the cost savings for the owners and non-owners was calculated. The cost savings were estimated at £304 and £709.1 for the owners and non-owners, respectively, noting that the savings from the overhead daily activities was estimated at £1,013 from the estimation sheet. Furthermore, the total planned overhead cost was valued at £17,038 for the first month and the actual project overheads was £16,025 (showing a £1,013 saving). The direct and indirect cost value for the owners and non-owners was estimated at £859.7. Finally, the profit-at-risk percentage was calculated using equation 6 and equation 7 for the owners and non-owners, respectively. The gain ratio was valued at £13,662, split into £2,732 and £10,929 for the owners and non-owners, respectively (see Figure 10.10).

Table 10.3 Monthly cost data – Scenario 1

Activities	Feb W2	Feb W3	Feb W4	Mar W1
CPR	0.91	0.91	1.25	0.92
SPR	0.79	0.83	1.39	0.90

Table 10.4 Monthly cost data – Scenario 2

Activities	Feb W2	Feb W3	Feb W4	Mar W1
CPR	1.05	1.00	1.25	1.02
SPR	1.17	1.00	1.39	1.02

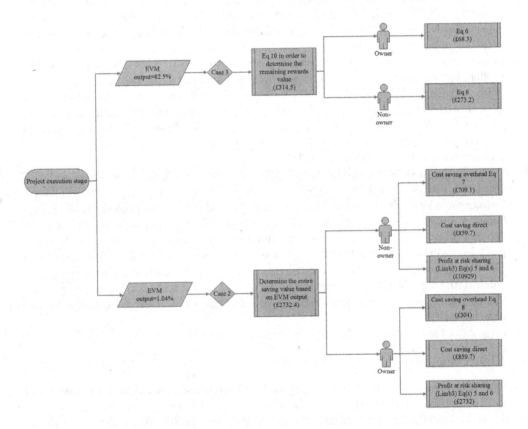

Figure 10.10 Results analysis flowchart

Figure 10.10 presents the above-mentioned scenario steps and corresponding results of implementing the framework for both owner and non-owner parties. The scenarios display two scenarios for an EVM output less than 1 and greater than 1 (where the sharing proportion was calculated accordingly).

10.8. Discussion

These results present an opportunity to reflect on the findings and process of developing this model. Using ABC enabled practitioners to identify the source of cost savings and monetary sharing value (both owner and non-owner parties). The model was able to distinguish between overhead cost sources and determine the proportion of sharing. On reflection (and in this instance), findings highlighted that non-owner parties received approximately twice the percentage of owners. Previous studies, c.f. Zhang and Li (2014), developed models that were capable of differentiating overhead cost levels such as corporate and project levels, but did not commute accurate overhead costs to project progress. Whilst the model presented in this chapter addressed this issue, there are still a number of wider issues to meet.

A novel feature of using EVM with tailored mathematical equations for an IPD's characteristics is seen as a significant step forward. The proposed model supports the automation of sharing risk/reward process, which is seen as an extension of integrated models proposed in previous studies (c.f.Pishdad-Bozorgi et al., 2013). One of the main contributions arising from this study is its ability to integrate BIM with IPD. This is particularly useful as several calls have advocated for the need for workable solutions that incorporate and integrate BIM with IPD, especially as cost-oriented tools have proven their potential for cost estimation purposes such as EVM and ABC (Pishdad-Bozorgi et al., 2013). Such integration has been proffered as a solution to cost distortion problems that occur in applying existing methods (Kim & Ballard, 2001; Kim et al., 2011; Miller, 1996). In this respect, the model presented here enhances the cost structure of BIM for IPD – the aspects of which have been recommended from literature (AIA, 2007; Allison et al., 2018; Rowlinson, 2017), particularly, an integrated cost control system that determines the proportion of risk/reward automatically for sustainable risk/reward sharing (Ballard et al., 2015).

Finally, the model presented here also addresses some of the deficiencies associated with EVM (c.f. Kim & Ballard, 2002, 2010) – in particular, the EVM concept, which relies on 'management by results' thinking. This quantities method tends to overlook the relationships between activities at an operational level, which by default does not take into account interdependencies and workflows of resources within project packages. The corollary of which can result in unfair or skewed results. This was one of the main reasons for integrating ABC into EVM for this model to mitigate this challenge. It is argued that there is an intrinsic need to enhance the capability of analysing unit costs, resources, and activities, including the need to tracking the source of resources and activities (Morgan et al., 1998; Seyedzadeh & Pour Rahimian, 2021a).

10.9. Conclusion

The work presented in this chapter proposed and presented a model for exploiting EVM to structure the compensation approach in IPD (using ABC to optimise the cost structure for IPD projects). Due to the complexity of structuring a compensation system fairly within the IPD approach for BIM projects, this model was specifically designed and aligned to BIM under the IPD approach. The model helps the apportionment of cost sharing and savings, which historically represented a significant barrier in the implementation of IPD. In doing so, it manages this issue by adopting the ABC estimation method to distinguish among the different types of activities within the project organisation hierarchy. Thus, it is able to differentiate between the overhead sustaining level and project level. On the issue of sharing overhead cost savings and overhead resources, these sources are determined from the outset (to minimise conflicts amongst stakeholders). In summary, the research also presented an EVM-Web grid for enhancing stakeholder collaboration in order to increase trust among project participants. This is achieved automatically, with minimal human interaction. The case study was applied to several scenarios for analysis and wider reflection.

The model is therefore novel in several ways, as it purposefully uses an innovative grid that locates the CPR and SPR into one integrated system. This provides a 'picture' of a project's position in terms of both cost and schedule. In addition, the EVM-Grid is integrated with the ABC estimating method to optimise the cost structure, which is positively reflected in the compensation structure. In doing so, the findings support risk/reward sharing and also open up new directions for wider discourse on fair sharing using ABC sheets and allocation of costs. For example, for overhead costs, the framework is able to distinguish between the sustaining/organisation level and the project level. Finally, the EVM-Grid has been developed as a web-based system, the rationalisation of which allows participants to easily track progress.

In practical terms, it is hoped that these findings will help AEC reflect upon the potential of this system to inform decision-making. It was designed specifically to meet the needs of both experienced and novice professionals alike, particularly as all tasks are aligned to implementation stages. That being said, the work presented here is just seen as a starting point, as there are many other avenues to explore. For example, the findings here could be used to develop a prototype using an Application Programming Interface (API) coded by C#.NET or on any BIM 4D and 5D platform, such as Navisworks. This might focus on the concept of open BIM to develop a vendor-free Industry Foundation Class (IFC)–based platform that is compatible with various other BIM packages. Whilst the developed EVM-Web grid presented in this chapter acts as a 'smart tool', providing recommendations for the optimal corrective actions that need to be taken, this is just seen as the starting point.

Finally, despite the contributions noted here, it is important to highlight a number of limitations arising from the work from a generalisability and repeatability perspective. The first issue concerns the appropriation of the proposed sharing risk/reward equations. These rely on the application of identical weights to cost/schedule to determine performance. Outcomes of the case study were also based on this assumption. That being said, users are able to change the degree of importance, including an ability to multiply the CPR and SPR by any agreed decimal value (to give preference to one parameter over another). This work also did not embed project participant performance into any of the outcomes. From experience, risk registers could be used here alongside tacit knowledge. It is also important to acknowledge that IPD with construction projects ultimately relies upon the principles of trust (to both engender and enable sustainable relationships).

References

Abrishami, S., Goulding, J., Pour Rahimian, F., & Ganah, A. (2015). Virtual generative BIM workspace for maximising AEC conceptual design innovation: A paradigm of future opportunities. *Construction Innovation*, 15(1), 24–41. https://doi.org/10.1108/CI-07-2014-0036

Abrishami, S., Goulding, J., Rahimian, F. P., & Ganah, A. (2014). Integration of BIM and generative design to exploit AEC conceptual design innovation. *Information Technology in Construction*, 19, 350–359. Retrieved June 1, 2021, from www.itcon.org/paper/2014/21

Ahmad, I., Azhar, N., & Chowdhury, A. (2018). Enhancement of IPD characteristics as impelled by information and communication technology. *Journal of Management in Engineering*, 35(1), 04018055. https://doi.org/10.1061/(ASCE)ME.1943-5479.0000670

AIA. (2007). *Integrated project delivery: A guide*. Retrieved June 26, 2020, from www.aia.org/resources/64146-integrated-project-delivery-a-guide

Aibinu, A., & Venkatesh, S. (2013). Status of BIM adoption and the BIM experience of cost consultants in Australia. *Journal of Professional Issues in Engineering Education and Practice*, 140(3), 04013021. https://doi.org/10.1061/(ASCE)EI.1943-5541.0000193

Allison, M., Ashcraft, H., Cheng, R., Klawens, S., & Pease, J. (2018). *Integrated project delivery: An action guide for leaders*. IPDA. Retrieved June 1, 2021.

Ashcraft, H. W. (2008). Building information modeling: A framework for collaboration. *Construction Law*, 28(3), 5–18. Retrieved June 1, 2021.

Ashcraft, H. W. (2011). Negotiating an integrated project delivery agreement. *Construction Law*, 31(3), 17–50. Retrieved June 1, 2021.

Ashcraft, H. W. (2012). *IPD framework*. Retrieved June 26, 2020, from www.hansonbridgett.com/Publications/pdf/ipd-framework.aspx

Ashcraft, H. W. (2014). The transformation of project delivery. *Construction Law*, 34(4), 35–58. Retrieved June 1, 2021.

Ballard, G., Dilsworth, B., Do, D., Low, W., Mobley, J., Phillips, P., Reed, D., Sargent, Z., Tillmann, P., & Wood, N. (2015). *How to make shared risk and reward sustainable*. Proceedings 23rd Annual Conference of the International Group for Lean Construction, pp. 257–266. Retrieved June 1, 2021.

Banihashemi, S., Tabadkani, A., & Hosseini, M. R. (2018). Integration of parametric design into modular coordination: A construction waste reduction workflow. *Automation in Construction*, 88, 1–12. https://doi.org/10.1016/j.autcon.2017.12.026

Chou, J.-S., Chen, H.-M., Hou, C.-C., & Lin, C.-W. (2010). Visualized EVM system for assessing project performance. *Automation in Construction*, 19(5), 596–607. https://doi.org/10.1016/j.autcon.2010.02.006

Das, T., & Teng, B.-S. (2001). A risk perception model of alliance structuring. *Journal of International Management*, 7(1), 1–29. https://doi.org/10.1016/S1075-4253(00)00037-5

Dawood, N., Rahimian, F., Seyedzadeh, S., & Sheikhkhoshkar, M. (2020). *Enabling the development and implementation of digital twins*. Proceedings of the 20th International Conference on Construction Applications of Virtual Reality, Teesside University. ISBN:9780992716127

de Andrade, P. A., Martens, A., & Vanhoucke, M. (2019). Using real project schedule data to compare earned schedule and earned duration management project time forecasting capabilities. *Automation in Construction*, 99, 68–78. https://doi.org/10.1016/j.autcon.2018.11.030

Delerue, H., & Simon, E. (2009). National cultural values and the perceived relational risks in biotechnology alliance relationships. *International Business Review*, 18(1), 14–25. https://doi.org/10.1016/j.ibusrev.2008.11.003

Eastman, C., Teicholz, P., Sacks, R., & Liston, K. (2011). *BIM handbook: A guide to building information modeling for owners, managers, designers, engineers and contractors*. Wiley. ISBN: 9781118021699

El-Omari, S., & Moselhi, O. (2011). Integrating automated data acquisition technologies for progress reporting of construction projects. *Automation in Construction*, 20(6), 699–705. https://doi.org/10.1016/j.autcon.2010.12.001

Fischer, M. J. A., Khanzode, A., Reed, D. P., & Ashcraft, H. W., Jr. (2017). *Integrating project delivery*. John Wiley & Sons Inc. ISBN: 0470587350

Fu, C., Aouad, G., Lee, A., Mashall-Ponting, A., & Wu, S. (2006). IFC model viewer to support nD model application. *Automation in Construction*, 15(2), 178–185. https://doi.org/10.1016/j.autcon.2005.04.002

Hamledari, H., McCabe, B., Davari, S., & Shahi, A. (2017). Automated schedule and progress updating of IFC-based 4D BIMs. *Journal of Computing in Civil Engineering*, 31(4), 04017012. https://doi.org/10.1061/(ASCE)CP.1943-5487.0000660

Hartmann, T., Gao, J., & Fischer, M. (2008). Areas of application for 3D and 4D models on construction projects. *Journal of Construction Engineering and Management*, 134(10), 776–785. https://doi.org/10.1061/(ASCE)0733-9364(2008)134:10(776)

Holland, N. L., & Jr, D. H. (1999). Indirect cost categorization and allocation by construction contractors. *Journal of Architectural Engineering*, 5(2), 49–56. https://doi.org/10.1061/(ASCE)1076-0431(1999)5:2(49)

Holzer, D. (2011). BIM's seven deadly sins. *International Journal of Architectural Computing*, 9(4), 463–480. https://doi.org/10.1260/1478-0771.9.4.463

Hosseini, M. R., Maghrebi, M., Akbarnezhad, A., Martek, I., & Arashpour, M. (2018). Analysis of citation networks in building information modeling research. *Journal of Construction Engineering and Management*, 144(8), 04018064. https://doi.org/10.1061/(ASCE)CO.1943-7862.0001492

Ilozor, B. D., & Kelly, D. J. (2012). Building information modeling and integrated project delivery in the commercial construction industry: A conceptual study. *Journal of Engineering, Project, and Production Management*, 2(1), 23–36. 10.32738/JEPPM.201201.0004

Jaselskis, E. J., Gao, Z., & Walters, R. C. (2005). Improving transportation projects using laser scanning. *Journal of Construction Engineering and Management*, 131(3), 377–384. https://doi.org/10.1061/(ASCE)0733-9364(2005)131:3(377)

Kahvandi, Z., Saghatforoush, E., Ravasan, A. Z., & Mansouri, T. (2018). An FCM-based dynamic modelling of integrated project delivery implementation challenges in construction projects. *Lean Construction Journal*, 2018, 63–87.

Kent, D. C., & Becerik-Gerber, B. (2010). Understanding construction industry experience and attitudes toward integrated project delivery. *Journal of Construction Engineering and Management*, 136(8), 815–825. https://doi.org/10.1061/(ASCE)CO.1943-7862.0000188

Khamooshi, H., & Abdi, A. (2016). Project duration forecasting using earned duration management with exponential smoothing techniques. *Journal of Management in Engineering*, 33(1), 04016032. https://doi.org/10.1061/(ASCE)ME.1943-5479.0000475

Kim, Y. W., & Ballard, G. (2001, August). *Activity-based costing and its application to lean construction*. Proceedings of the 9th Annual Conference of the International Group for Lean Construction, National University of Singapore.

Kim, Y.-W., & Ballard, G. (2002). Earned value method and customer earned value. *Journal of Construction Research*, 3(01), 55–66. https://doi.org/10.1142/S1609945102000096

Kim, Y.-W., & Ballard, G. (2010). Management thinking in the earned value method system and the last planner system. *Journal of Management in Engineering*, 26(4), 223–228. https://doi.org/10.1061/(ASCE)ME.1943-5479.0000026

Kim, Y. W., Han, S., Shin, S., & Choi, K. (2011). A case study of activity-based costing in allocating rebar fabrication costs to projects. *Construction Management and Economics*, 29(5), 449–461. https://doi.org/10.1080/01446193.2011.570354

Lee, J., Park, Y.-J., Choi, C.-H., & Han, C.-H. (2017). BIM-assisted labor productivity measurement method for structural formwork. *Automation in Construction*, 84, 121–132. https://doi.org/10.1016/j.autcon.2017.08.009

Lee, S.-K., Kim, K.-R., & Yu, J.-H. (2014). BIM and ontology-based approach for building cost estimation. *Automation in Construction*, 41, 96–105. https://doi.org/10.1016/j.autcon.2013.10.020

Lipke, W., Zwikael, O., Henderson, K., & Anbari, F. (2009). Prediction of project outcome: The application of statistical methods to earned value management and earned schedule performance indexes. *International Journal of Project Management*, 27(4), 400–407. https://doi.org/10.1016/j.ijproman.2008.02.009

Liu, M. M., & Bates, A. J. (2013a). *Compensation structure and contingency allocation in integrated project delivery*. Paper presented at the 120th ASEE Annual Conference and Exposition. Retrieved June 1, 2021, from https://nyuscholars.nyu.edu/en/publications/compensation-structure-and-contingency-allocation-in-integrated-p

Liu, M. M., & Bates, A. J. (2013b). Compensation structure and contingency allocation in integrated project delivery. *Age*, 23, 1.

Love, P. E., Davis, P. R., Chevis, R., & Edwards, D. J. (2011). Risk/reward compensation model for civil engineering infrastructure alliance projects. *Journal of Construction Engineering and Management*, 137(2), 127–136. https://doi.org/10.1061/(ASCE)CO.1943-7862.0000263

Ma, J., Ma, Z., & Li, J. (2017). An IPD-based incentive mechanism to eliminate change orders in construction projects in China. *KSCE Journal of Civil Engineering*, 21(7), 2538–2550. https://doi.org/10.1007/s12205-017-0957-3

Ma, Z., Zhang, D., & Li, J. (2018). A dedicated collaboration platform for integrated project delivery. *Automation in Construction, 86*, 199–209. https://doi.org/10.1016/j.autcon.2017.10.024

Matthews, O., & Howell, G. A. (2005). Integrated project delivery an example of relational contracting. *Lean construction journal, 2*(1), 46–61.

Merschbrock, C., Hosseini, M. R., Martek, I., Arashpour, M., & Mignone, G. (2018). Collaborative role of sociotechnical components in BIM-based construction networks in two hospitals. *Journal of Management in Engineering, 34*(4), 05018006. https://doi.org/10.1061/(ASCE)ME.1943-5479.0000605

Mignone, G., Hosseini, M. R., Chileshe, N., & Arashpour, M. (2016). Enhancing collaboration in BIM-based construction networks through organisational discontinuity theory: A case study of the new Royal Adelaide Hospital. *Architectural Engineering and Design Management, 12*(5), 333–352. https://doi.org/10.1080/17452007.2016.1169987

Miller, J. A. (1996). *Implementing activity-based management in daily operations.* The University of Michigan. John Wiley & Sons. ISBN: 9780471040033

Morgan, J. J., Johnson, T. L., Keefer, L. E., Smith, P. A., Bradford, W. H., Wells, K. L., & Mason, E. T. (1998). *Automated activity-based management system.* Google Patents. Retrieved June 1, 2021.

Motamedi, A., & Hammad, A. (2009). Lifecycle management of facilities components using radio frequency identification and building information model. *Journal of Information Technology in Construction (ITCON), 14*(18), 238–262.

Naeni, L. M., Shadrokh, S., & Salehipour, A. (2011). A fuzzy approach for the earned value management. *International Journal of Project Management, 29*(6), 764–772. https://doi.org/10.1016/j.ijproman.2010.07.012

Nawi, M. N. M., Haron, A. T., Hamid, Z. A., Kamar, K. A. M., & Baharuddin, Y. (2014). Improving integrated practice through building information modeling-integrated project delivery (BIM-IPD) for Malaysian industrialised building system (IBS) construction projects. *Malaysia Construction Research Journal (MCRJ), 15*(2), 29–38.

Oliver, S., Seyedzadeh, S., Rahimian, F., Dawood, N., & Rodriguez, S. (2020). Cost-effective as-built BIM modelling using 3D point-clouds and photogrammetry. *Current Trends in Civil & Structural Engineering-CTCSE, 4*(5), 000599. https://doi.org/10.33552/CTCSE.2020.04.000599

Omar, H., & Dulaimi, M. (2015). Using BIM to automate construction site activities. *Building Information Modelling (BIM) in Design, Construction and Operations, 149*, 45. https://doi.org/10.2495/BIM150051

Oraee, M., Hosseini, M. R., Papadonikolaki, E., Palliyaguru, R., & Arashpour, M. (2017). Collaboration in BIM-based construction networks: A bibliometric-qualitative literature review. *International Journal of Project Management, 35*(7), 1288–1301. https://doi.org/10.1016/j.ijproman.2017.07.001

Pajares, J., & López-Paredes, A. (2011). An extension of the EVM analysis for project monitoring: The cost control index and the schedule control index. *International Journal of Project Management, 29*(5), 615–621. https://doi.org/10.1016/j.ijproman.2010.04.005

Pärn, E. A., & Edwards, D. J. (2017). Conceptualising the FinDD API plug-in: A study of BIM-FM integration. *Automation in Construction, 80*, 11–21. https://doi.org/10.1016/j.autcon.2017.03.015

Pishdad-Bozorgi, P., & Beliveau, Y. J. (2016). Symbiotic relationships between integrated project delivery (IPD) and trust. *International Journal of Construction Education and Research, 12*(3), 179–192. https://doi.org/10.1080/15578771.2015.1118170

Pishdad-Bozorgi, P., Moghaddam, E. H., & Karasulu, Y. (2013). *Advancing target price and target value design process in IPD using BIM and risk-sharing approaches.* Paper presented at the the 49th ASC Annual International Conference California Polytechnic State University.

Pishdad-Bozorgi, P., & Srivastava, D. (2018). *Assessment of Integrated Project Delivery (IPD) Risk and Reward Sharing Strategies from the Standpoint of Collaboration: A Game Theory Approach.* Paper presented at the Construction Research Congress 2018. New Orleans, Louisiana. https://doi.org/10.1061/9780784481271.020

PMI. (2013). *A guide to the project management body of knowledge (PMBOK® guide)* (5th ed.). Project Management Institute. ISBN: 978-1-935589-67-9

Popov, V., Juocevicius, V., Migilinskas, D., Ustinovichius, L., & Mikalauskas, S. (2010). The use of a virtual building design and construction model for developing an effective project concept in 5D environment. *Automation in Construction, 19*(3), 357–367. https://doi.org/10.1016/j.autcon.2009.12.005

Pour Rahimian, F., Seyedzadeh, S., Oliver, S., Rodriguez, S., & Dawood, N. (2020). On-demand monitoring of construction projects through a game-like hybrid application of BIM and machine learning. *Automation in Construction, 110*, 103012. https://doi.org/10.1016/j.autcon.2019.103012

Redmond, A., Hore, A., Alshawi, M., & West, R. (2012). Exploring how information exchanges can be enhanced through cloud BIM. *Automation in Construction, 24*, 175–183. https://doi.org/10.1016/j.autcon.2012.02.003

Ross, J. (2003). *Introduction to project alliancing*. Alliance Contracting Conference.

Rowlinson, S. (2017). Building information modelling, integrated project delivery and all that. *Construction Innovation, 17*(1), 45–49. https://doi.org/10.1108/CI-05-2016-0025

Roy, D., Malsane, S., & Samanta, P. K. (2018). Identification of critical challenges for adoption of integrated project delivery. *Lean Construction Journal*, 1–15.

Seyedzadeh, S., Agapiou, A., Moghaddasi, M., Dado, M., & Glesk, I. (2021). WON-OCDMA system based on MW-ZCC codes for applications in optical wireless sensor networks. *Sensors, 21*. https://doi.org/10.3390/s21020539

Seyedzadeh, S., & Pour Rahimian, F. (2021a). Building energy data-driven model improved by multi-objective optimisation. In *Data-driven modelling of non-domestic buildings energy performance* (pp. 99–109). Springer. https://doi.org/10.1007/978-3-030-64751-3_6

Seyedzadeh, S., & Pour Rahimian, F. (2021b). Modelling energy performance of non-domestic buildings. In *Data-driven modelling of non-domestic buildings energy performance* (pp. 111–133). Springer. https://doi.org/10.1007/978-3-030-64751-3_7

Shen, W., Hao, Q., & Xue, Y. (2012). A loosely coupled system integration approach for decision support in facility management and maintenance. *Automation in Construction, 25*, 41–48. https://doi.org/10.1016/j.autcon.2012.04.003

Teng, Y., Li, X., Wu, P., & Wang, X. (2017). Using cooperative game theory to determine profit distribution in IPD projects. *International Journal of Construction Management*, 1–14. https://doi.org/10.1080/15623599.2017.1358075

Thomsen, C., Darrington, J., Dunne, D., & Lichtig, W. (2009). Managing integrated project delivery. *Construction Management Association of America (CMAA), McLean, VA, 105*. www.leanconstruction.org/wp-content/uploads/2016/02/CMAA_Managing_Integrated_Project_Delivery_1.pdf

Turkan, Y., Bosche, F., Haas, C. T., & Haas, R. (2012). Automated progress tracking using 4D schedule and 3D sensing technologies. *Automation in Construction, 22*, 414–421. https://doi.org/10.1016/j.autcon.2011.10.003

Turkan, Y., Bosché, F., Haas, C. T., & Haas, R. (2013). Toward automated earned value tracking using 3D imaging tools. *Journal of construction engineering and management, 139*(4), 423–433. https://doi.org/10.1061/(ASCE)CO.1943-7862.0000629

Wang, K.-C., Wang, W.-C., Wang, H.-H., Hsu, P.-Y., Wu, W.-H., & Kung, C.-J. (2016). Applying building information modeling to integrate schedule and cost for establishing construction progress curves. *Automation in Construction, 72*, 397–410. https://doi.org/10.1016/j.autcon.2016.10.005

Yin, R. K. (1981). The case study crisis: Some answers. *Administrative science quarterly, 26*(1), 58–65.

Zellmer-Bruhn, M., Caligiuri, P., & Thomas, D. C. (2016). From the editors: Experimental designs in international business research. *Journal of International Business Studies, 47*(4), 399–407. https://doi.org/10.1057/jibs.2016.12

Zhang, L., Cao, T., & Wang, Y. (2018). The mediation role of leadership styles in integrated project collaboration: An emotional intelligence perspective. *International Journal of Project Management, 36*(2), 317–330. https://doi.org/10.1016/j.ijproman.2017.08.014

Zhang, L., & Chen, W. (2010). *The analysis of liability risk allocation for Integrated Project Delivery*. Paper presented at the the 2nd International Conference on Information Science and Engineering, Hangzhou, China https://doi.org/10.1109/ICISE.2010.5689527

Zhang, L., & Li, F. (2014). Risk/reward compensation model for integrated project delivery. *Engineering Economics, 25*(5), 558–567.

11 Revolutionising cost structure within integrated project delivery

A BIM-based solution

11.1. Introduction

Integrated Project Delivery (IPD) is characterised by the early, collaborative, and collective engagement of key stakeholders throughout all delivery phases of a project (Ahmad et al., 2019; Ashcraft, 2014). Compared to common methods of project delivery, such as design-bid-build, construction management at-risk, and design-build, IPD is proving distinctly superior in performance (Asmar et al., 2016; Manata et al., 2018). In addition, evidence shows that IPD can improve project performance over 14 metrics, covering areas such as quality, scheduling, communication management, and cost performance (Ahmad et al., 2019; Asmar et al., 2016).

Traditional forms of IPD, such as alliancing, can be implemented without Building Information Modelling (BIM); however, new forms of IPD are defined in relation to their integration with BIM (Dawood et al., 2020; Fischer et al., 2017; Rowlinson, 2017). This facilitates smooth data exchange between a project's packages and parties in line with IPD's aims and objectives (AIA, 2007; Niemann, 2017). Therefore, the integration of BIM and IPD improves the likely outcomes of the design and construction process, including cost/profit, scheduling, return on investment, safety, productivity, and relationships (Azhar et al., 2015; Ilozor & Kelly, 2012); however, IPD relies on open pricing techniques and fiscal transparency among participants (Ahmad et al., 2019). For example, project stakeholders such as designers and contractors typically assess and determine their profit and shared risks, according to the deviation between actual and target costs (AIA, 2007).

Cost estimation is essential for any compensation arrangement, the nuances of which defines accurate risk/reward proportions (Love et al., 2011). Hence, accurate cost estimation is vital for the successful delivery of IPD-based projects (AIA, 2007; Allison et al., 2017; Ebrahimi & Dowlatabadi, 2018). In this respect, Target Value Design (TVD) is treated as part of the IPD approach, with TVD requiring rapid cycles of suggestions and analyses of costs (Alves et al., 2017). Therefore, continuous estimation feedback is essential for accomplishing the pre-construction IPD stages in order to make informed decisions (Allison et al., 2017; Zimina et al., 2012). With these facts in mind, a precise semi-automated, agile estimation technique that is interoperable with BIM data has been proffered as an ideal solution (Pishdad-Bozorgi et al., 2013).

Other developments in this area note that cost management practices in IPD are not yet fully established (Chen et al., 2012). This warrants further expansion of the capacity of BIM and cost estimation to support the TVD process (Alves et al., 2017; Hall et al., 2018; Pishdad-Bozorgi et al., 2013).

DOI: 10.1201/9781003106944-11

Given this knowledge gap, this chapter uncovers some of these challenges. In doing so, it contributes to the field by addressing the need for a TVD-based solution for IPD (based on BIM's capabilities). In broader terms, it provides a background of the principal challenges and need for accurate cost estimation at the planning stages of IPD projects. Little research currently exists in this area (Andersen et al., 2016; Welde & Odeck, 2017). This study addresses this challenge through the development of a BIM-based IPD solution. This solution was tested and compared against traditional methods in order to evaluate performance and practicality.

11.2. Theoretical background

11.2.1. Cost estimation of construction projects

Cost performance (meeting cost requirements), although frequently criticised, is still considered the gold standard for measuring project success (Berssaneti & Carvalho, 2015; Kim et al., 2004). Thus, cost estimation is an important element of project planning (Torp & Klakegg, 2016). According to the Project Management Institute (PMI, 2017), cost estimation is the iterative process of estimating the project resources required for project activities. In this respect, it is important to link project resources and activities in order to ensure successful cost management. Given this need, major cost estimation activities typically occur at the very early stages of a project, where minimal project information is available (Kim et al., 2004; Welde & Odeck, 2017). This is a significant challenge, as uncertainty remains a major cause of poor cost estimation across the construction industry (Andersen et al., 2016; Johansen et al., 2014; Torp & Klakegg, 2016). Uncertainty is identified as "controllable and non-controllable factors that may occur, and variation and foreseeable events that occur during a project execution, and that has a significant impact on the project objective" (Johansen et al., 2014, p. 592).

Several research studies provide evidence that the greatest level of uncertainty in cost estimation belongs to the feasibility study stages of projects, colloquially termed the 'front end' of projects (Andersen et al., 2016; Caffieri et al., 2018; Welde & Odeck, 2017), where uncertainty levels ranging from –30% to +50% can be expected (Johansen et al., 2014). In IPD, the overall risk is equal to that of traditional methods, where the owner has a responsibility to guarantee the direct cost of projects (Ghassemi & Becerik-Gerber, 2011). As a result, IPD relies heavily on cost estimation at the project feasibility study phase in order to (i) develop a reliable business case for the client and (ii) help the decision-making process (Allison et al., 2017; Pishdad-Bozorgi et al., 2013).

11.2.2. Cost estimation in integrated project delivery

A major tenet of the IPD approach is its compensation system for allocating the gain/pain ratio among project participants (Ashcraft, 2014; Fischer et al., 2017). This necessitates developing a truly cooperative contracting relationship with parties, which intrinsically ties the individual success of participants to the wider success of achieving the project objectives (Ahmad et al., 2019; AIA, 2007; Pishdad-Bozorgi, 2017). In this respect, all project participants must agree upon a suitable compensation scheme (Pishdad-Bozorgi et al., 2013), where this scheme determines the proportions of cost overrun, cost underrun, and any other fees in the total budget at the agreed cost (Fischer et al., 2017; Pishdad-Bozorgi et al., 2013).

Thus, the cost scheme contains direct, indirect, and overhead costs, including risk/reward proportions based on the degree of achievement during project delivery (Love et al., 2011; Pishdad-Bozorgi et al., 2013; Zhang & Li, 2014). In IPD, three components (or limbs) are defined, where Limb 1 represents the reimbursement of project costs and captures all project implementation costs (guaranteed); Limb 2 represents the overhead costs for all participants in addition to the profit (at-risk); and Limb 3 identifies the pain/gain ratios (the contractual agreement) (Raisbeck et al., 2010; Zhang & Li, 2014). Therefore, according to Das and Teng (2001), a precise determination of risk is critical, through which project participants who are exposed to more uncertainty can be compensated for the risks against a higher profit-at-risk percentage. In this respect, TVD has been proven to offer a highly reliable route to successful project cost estimation for IPD arrangements (Do et al., 2014; Zimina et al., 2012).

11.2.3. Target value design

The use and application of TVD within the construction industry can be seen as a management strategy that aims to eliminate waste and deliver value using a 'design-to-cost' method (Meijon Morêda Neto et al., 2019). The aim of TVD is to position a client's 'value' (e.g. cost, schedule, etc.) as a key driver of design in order to reduce waste and satisfy the client's expectations (Zimina et al., 2012). TVD therefore introduces a philosophy towards design based on the budget, in contrast to the idea of budgeting for the design – a traditional design concept; therefore, cost estimating becomes a crucial part of design development (Allison et al., 2017). Empirical research shows that TVD projects can achieve cost reductions of 15–20% and contingency costs of approximately 3.5% compared to 7.9% for non-TVD projects (Meijon Morêda Neto et al., 2019; Silveira & Alves, 2018).

Given these benefits, TVD is often recommended as an effective solution for IPD projects (de Melo et al., 2016; Pishdad-Bozorgi et al., 2013); however, successful TVD requires extensive collaboration among designers, builders, quantity surveyors, and trade partners (Alwisy et al., 2018). This is particularly important as all parties must be at the table and offer continuous feedback to influence the design and achieve the client's goals within the set budget (Pishdad-Bozorgi et al., 2013; Allison et al., 2017). This collaboration is based on multiple interactions and rapid circles of suggestions, analysis, and feedback to allow continuous improvements and to find the solutions that meet the client's (or multiple stakeholders) definition of value (Alves et al., 2017; Silveira & Alves, 2018). TVD can therefore be seen to support lean management tools in order to facilitate effective collaboration and make possible these rapid circles of conceptualisation, analysis, and estimation (Allison et al., 2017; Alves et al., 2017; Alwisy et al., 2018; Meijon Morêda Neto et al., 2019). A reflection of some of these tools are described as follows.

11.2.4. Five-dimensional (5D) building information modelling

The notion and concept of BIM has been proffered as a primary tool best suited to facilitate TVD in IPD projects (Allison et al., 2017; Alves et al., 2017; Meijon Morêda Neto et al., 2019; Pishdad-Bozorgi et al., 2013), where this has the potential to facilitate a comprehensive and accurate design from the early stages of a project (Eastman et al., 2018; Lu et al., 2016; Nassar, 2012). In particular, the five-dimensional cost model (5D BIM) has been promoted as the preferred method for extracting quantities and cost estimations from 3D models (Aibinu & Venkatesh, 2013; Nassar, 2012; Oliver et al., 2020; Zheng et al., 2019). BIM is able to provide the project team with enhanced levels of capability, particularly being able to accommodate changes in the design development process and subsequent impact on value

(Eastman et al., 2018; Hannon, 2007; Lu et al., 2018; Nassar, 2012; Zheng et al., 2019); however, in order to maximise this capability, a schematic is needed to organise the design and estimation of value – linking this BIM model and various external databases to efficiently extract cost items (Lu et al., 2018). Other advantages of 5D BIM over traditional methods are well documented – for example, increased efficiency, improved visualisation of construction details, and early risk identification (Lu et al., 2016; Stanley & Thurnell, 2014). In doing so, this enables cost estimates to reach optimum solutions and reduce errors and misleading estimates (Shen & Issa, 2010). In addition, Alves et al. (2017, p. 25) noted that BIM was "an important part of the TVD process. BIM allows project participants to quickly develop solutions, visualize them in three- and four-dimensional (time added) environments, while also understanding the impact of their decisions on the cost of the project". In summary, there is significant scope and potential for integrating BIM into IPD delivery (Alves et al., 2017; Porwal & Hewage, 2013; Zheng et al., 2019).

11.2.5. Capabilities of Monte Carlo simulation

Monte Carlo simulation can be implemented using several computer programs that engage a large number of trials to analyse cost data in order to develop a probabilistic model for early cost estimation (Alashwal & Chew, 2017; Loizou & French, 2012). The use of Monte Carlo simulation in costing provides a range of cost values against specific degrees of certainty, whilst also offering wide flexibility in cost predictions (Potts & Ankrah, 2014). The process of simulation entails cost data collection; formulation of a statistical model by choosing one available distribution (beta, normal, etc.); analysis of the cost data by running the model; result visualisation of the whole-cost estimates; and finally, the provision of a sensitivity analysis chart (Chou, 2011; Seyedzadeh & Pour Rahimian, 2021; Seyedzadeh et al., 2020). As a result, Monte Carlo simulation has been described as an important method for evaluation through probabilistic cost estimation (Khedr, 2006), the most used technique in the literature for early cost estimation of construction projects (Alashwal & Chew, 2017), and one of the earliest methods available for property evaluation purposes (Jahangirian et al., 2010).

11.2.6. Activity-based costing

Traditional costing methods such as resource-based costing (RBC) rely on the cost of the required resources (Kim & Ballard, 2001). This type of costing is frequently applied in the industry; however, cost distortion can sometimes occur when using these traditional methods as the methods combine and allocate all indirect costs to a single pool of costs, based on the resources common to all products of an organisation (Kim et al., 2011; Miller, 1996; Wang et al., 2010). In other words, traditional methods tend to fail to find the key decision variables that affect the total cost, particularly overhead costs (Kim et al., 2016). In this respect, activity-based costing (ABC) prevents this distortion by allocating costs through multiple pools, which in turn determines the overhead activities and costs needed to transform resources into activities that can deliver the final product (Kim & Ballard, 2001; Wang et al., 2010). ABC measures costs based on activities – linking cost drivers to impact measures of a certain product or service (Tsai & Hung, 2009). The use of this approach has been acknowledged as being able to improve the efficiency and accuracy of cost-related information, whilst also being able to provide a greater level of sophistication in monitoring and controlling project costs (Tsai et al., 2014). This is particularly important in collaborative working environments such as IPD, where multiple stakeholders are engaged with cost drivers (Kim et al., 2016).

11.3. Costing studies and research gaps

A review of costing literature reveals several research trends, where emphasis seems to have been placed on the need to inform practitioners of the potential approaches, available tools, and techniques (such as TVD and BIM). Literature also highlights approaches for developing better IPD solutions. For example, Pishdad-Bozorgi et al. (2013) discussed the potential of integrating TVD, BIM, and IPD cost estimation, while Alves et al. (2017) presented various techniques commonly used for TVD applicable to IPD. Similarly, Zimina et al. (2012) and later de Melo et al. (2016) showed how systematic TVD could achieve noticeable improvements in project performance. Several studies have also mentioned the potential of BIM to add value to a project's objectives through IPD implementation (c.f.Ahmad et al., 2019; Azhar et al., 2015; Chang et al., 2017; Fischer et al., 2017; Hosseini et al., 2018; Succar, 2009).

Another stream of studies discussed the challenges and barriers of using TVD or BIM for IPD cost estimation tasks. Examples include Ghassemi and Becerik-Gerber (2011), Manata et al. (2018), Pishdad-Bozorgi (2017), and Kahvandi et al. (2018), who focused on various key critical success factors, largely from a managerial perspective, albeit limiting their attention to cost estimation issues.

However, Tillmann et al. (2017) discussed the underlying mechanisms of TVD cost estimation within IPD-oriented projects, exploring the factors that influenced success when TVD was applied to these projects. In addition, Ballard et al. (2015) explored the relationship between IPD and TVD, and they recommended a set of procedures for enhancing the chance of success in applying TVD to IPD cost management processes. Although the authors acknowledged that following TVD principles was a critical success factor, no explicit technique or procedure was recommended. On this theme, Roy et al. (2018) identified the challenges and cost structure of implementing IPD, which included profit pooling, misunderstandings in risk contingency accounting, and hard pricing.

Extant literature highlights a number of models and frameworks to address IPD cost estimation issues. For example, Zhang and Li (2014) combined risk perception and Nash Bargaining Solution (NBS) techniques to formulate a risk/reward compensation model. Similarly, Pishdad-Bozorgi and Srivastava (2018) developed a model for sharing risks and rewards using a game theory approach, particularly for cases in which project cost exceeded the profit-at-risk percentage.

In summary, literature reveals that IPD, TVD, and BIM are regarded as a winning combination to improve project delivery success (Pishdad-Bozorgi et al., 2013). That being said, limited research is available in this field. This is especially so on the validation of these relationships, or indeed on workable practitioner-based solutions (Azhar et al., 2015; Kahvandi et al., 2017). The following sections present the research methodological approach and development strategy employed to meet this challenge.

11.4. Research methodology

11.4.1. *Research approach*

Given the challenges highlighted here, this chapter presents the structure adopted for developing a workable solution to meet these challenges, particularly to explore its practical validity within real-life settings. Adopting this position as a starting point, exploratory case study and experiment methods are typically the most applicable methods used to accomplish tasks of this nature. To be specific, context is an essential part of case study research, in which many variables can affect the outcome and produce inferences about the causes and effects of the methodologies and procedures under question (Yin, 1981). As a result, it was acknowledged that assessing the impact of these types of workflows in a real-life case would not only be affected by many factors, but also through various procedures. Consequently, running a case study and applying observational techniques was deemed inappropriate.

As a research approach, experiments can often be effective in revealing practical solutions. According to Zellmer-Bruhn et al. (2016, p. 400) "experiments isolate causal variables and enable a strong test of the robustness of a theory: they provide convincing evidence for theories". In this respect, the validity of an assumption about causes and effects in which a match between data and theory is observed can be demonstrated through experiments (Shadish et al., 2002; Yin, 1981). The work presented in this chapter therefore selected the use of an experiment as the principal method for testing the assumptions about the positive effects of the proposed cost estimation methodology.

11.4.2. Requirements and techniques

The first part of this work commenced with a critical review of the existing best practice in IPD, particularly to identify the gaps in IPD cost structures. In this respect, the high level of technology needed in IPD projects was recognised (Ahmad et al., 2019), including the correlation between BIM and IPD. The research design schema is illustrated in Figure 11.1.

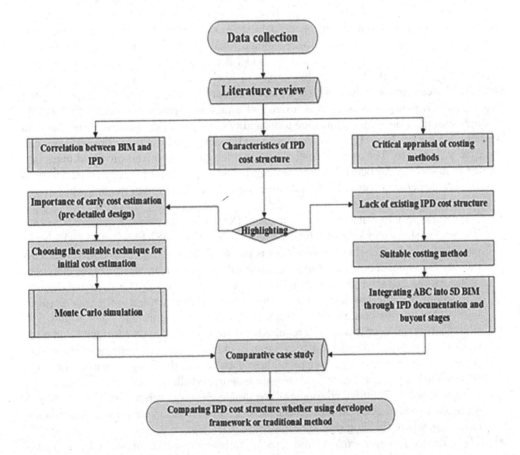

Figure 11.1 Research design process

Given the earlier discussion on the need to integrate BIM tools into a Monte Carlo simulation to provide precise cost estimation throughout the IPD pre-construction stages, it was also important to appreciate that a suitable approach was needed to support all parties engaged in IPD in order to obtain and deliver supported decisions. To achieve the optimal IPD cost structure, ABC was used to appropriately allocate costs to each activity in order to justify that the right package prices were obtained. The following experiment compares the case project results using the traditional costing method against the proposed framework to highlight deviations (if any) between the two methods in terms of the monetary value of the profit-at-risk percentage for each trade's package (Figure 11.1).

11.5. Framework development

IPD has five pre-construction stages: conceptualisation, outline design, detailed design, documentation, and buyout (AIA, 2007). The proposed framework therefore involves tools to manage cost estimation at each of these stages, each of which are discussed in the following sections.

11.5.1. Conceptualisation and outline design stages

The conceptualisation and outline design stages begin by forming the core project team with its members, normally comprising the owner, the architect, the main contractor, and trade contractors. The essential project decision is reached by evaluating the owner's requirements for the design criteria. From this, a conceptual BIM model of the architectural and engineering intentions is created by using a three-dimensional (3D) BIM platform (i.e. Autodesk Revit or Graphisoft ArchiCAD). This BIM model is used to obtain indications of the proposed quantities and identities. At this stage, it is likely that some project information will not be complete (or available), which invites considerable uncertainty. Consequently, the cost estimation model is presented as a range of total costs against the degree of certainty through a Monte Carlo simulation (due to its ability to deal with different types of cost data distribution). Once the architect and design team has developed the BIM conceptual model, the quantity surveyor begins to extract the quantities and type of the proposed materials. A BIM tool, such as Autodesk Navisworks, can be used for this purpose, given its ability to simultaneously perform cost and schedule processes – typically through the following steps:

- Navisworks in Extensible Markup Language (XML) format extracts quantities so that the pricing sheet can be built (typically using Microsoft [MS] Excel), from which the proposed initial price sheet of materials is prepared. Given that BIM design elements include the same amount of information, regardless of the entire degree of development of design, the extracted Bills of Quantities (BoQs) should be valid for the same element throughout the different design stages. This is highly recommended in the case of adopting TVD to provide continuous feedback of cost estimation during the design stages;
- The quantity surveyor collects the required cost data from the main contractor and trade contractors in order to build statistical samples of the labour and equipment required to perform the proposed design elements. These data include the range of material prices to draw reliable samples for each BIM design element, and allowances of labour and equipment that will be required to execute BIM design elements (preferred using analogous estimation (Amos, 2004) as most of the project parties in IPD join at the conceptualisation stage, where the data should be easily accessible);
- The quantity surveyor explores the type of statistical distribution needed, compatible with the collected data (normal, beta, etc.);

- The quantity surveyor identifies each proposed cost element to estimate the total value of all distributed elements to enable the simulation to run;
- When the simulation starts to run, the extracted graphs show the total costs for the project, corresponding to the percentage of certainty of the input data.

Formulation of statistical model:

To determine the proposed total cost, the following equations are applied. Equation 1 represents the total cost that must be collected for each design element D to assign the package cost for contractor j:

$$PC_{Dj} = IQP + \forall L_{BM} \,\&\, E_{BM} \tag{1}$$

where PC_{Dj} is the proposed cost for the design element D that is proposed to be assigned to contractor j; IQP is the initial quantity prices for Dj; and $L_{BM} \,\&\, E_{BM}$ are the labour and equipment prices for the best scenario B for the specific material M.

The statistical model requires a wide range of proposed values to enable a reliable total cost to be obtained. Equations 2.1, 2.2, and 2.3 show how the BIM data are integrated into the Monte Carlo simulation. These equations rely on using beta distribution; however, if a wider range of prices is used, these equations are extended to provide a more accurate material cost:

$$IQP_{AVM} = IQ_{BIM\ conceptual\ model} \times RP_M \tag{2.1}$$

$$IQP_{OPM} = IQ_{BIM\ conceptual\ model} \times OP_M \tag{2.2}$$

$$IQP_{PM} = IQ_{BIM\ conceptual\ model} \times PP_M \tag{2.3}$$

where IQP represents the initial quantity prices for average, optimistic, and pessimistic values; IQ is the initial quantities extracted using BIM tools; while RP_M is the recent price for material M; OP_M is the optimistic price for material M; and PP_M is the pessimistic price for material M. Other costs such as labour and equipment can be easily collected using IPD core team members, drawing upon their early involvement.

Equation 2.4 shows the formula for calculations:

$$\forall L_{BM} \,\&\, E_{BM} = UP_M \times TU_M \tag{2.4}$$

where $L_{BM} \,\&\, E_{BM}$ are labour and equipment prices for the best scenario B for specific material M; UP_M is the unit price for material M; and TU_M is the total units for material M. Equation 2.5) is another version of Equation (2.4) to capture the worst-case scenario, as follows:

$$\forall L_{WM} \,\&\, E_{WM} = UP_M \times TU_M \tag{2.5}$$

where $L_{WM} \,\&\, E_{WM}$ are the labour and equipment price for the worst-case scenario W for specific material M. To complete the beta distribution, the average value is determined as in equation 2.6:

$$\forall L_{AVM} \,\&\, E_{AVM} = \frac{\left(\forall L_{BM} \,\&\, E_{BM} + \forall L_{WM} \,\&\, E_{WM}\right)}{2} \tag{2.6}$$

Figure 11.2 shows the interoperability and process of integrating BIM data into a Monte Carlo simulation to obtain the proposed material cost. Based on the data and using analogous cost estimation or expert judgement from core team members, the cost range of the statistical model was determined. For example, if core team members agreed that three values for each cost element were reliable, the distribution was loaded for three probable costs. Based on the pre-identified range of costs, the distribution system was then selected. Given this result, the three values mentioned previously were consistent with both the beta and normal distribution.

Obtaining proposed entire cost against certainty percentages: At this stage, the model is ready to run. The Monte Carlo simulation has two important features, the first of which is the total cost, corresponding to the degree of certainty. This cost range is necessary to develop the business case for the client, based on the TVD system, before moving to the detailed design stage, as recommended by Allison et al. (2017). The second feature is the sensitivity analysis chart that presents the degree of importance of each project design element. This is vital to support decisions regarding the use of sensitive elements in the design. Through these features, the necessary data are available for making the right decision. Therefore, the client can decide whether the proposed whole cost is located within the allowable budget. Once the client has approved the proposed cost, the project moves to the detailed design stage, in which an additional cost estimation strategy is used.

If the client does not approve the proposed cost, an ongoing negotiation is necessary to fulfil the client's requirements. The sensitivity analysis chart plays a key role here, identifying

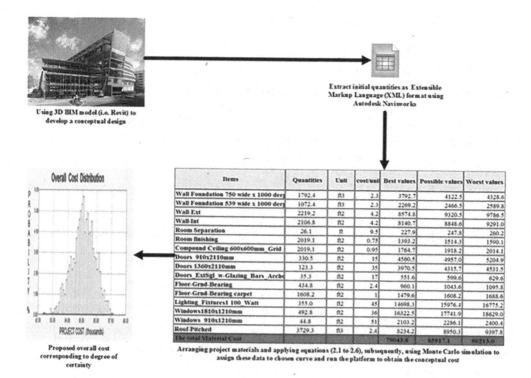

Using 3D BIM model (i.e. Revit) to develop a conceptual design

Extract initial quantities as Extensible Markup Language (XML) format using Autodesk Navisworks

Overall Cost Distribution

Proposed overall cost corresponding to degree of certainty

Items	Quantities	Unit	cost/unit	Best values	Possible values	Worst values
Wall Foundation 750 wide x 1000 deep	1792.4	ft3	2.3	3792.7	4122.5	4328.6
Wall Foundation 539 wide x 1000 deep	1072.4	ft3	2.3	2269.2	2466.5	2589.8
Wall-Ext	2219.2	ft2	4.2	8574.8	9320.5	9786.5
Wall-Int	2106.8	ft2	4.2	8140.7	8848.6	9291.0
Room Separation	26.1	ft	9.5	227.9	247.8	260.2
Room finishing	2019.1	ft2	0.75	1393.2	1514.3	1590.1
Compound Ceiling 600x600mm_Grid	2019.1	ft2	0.95	1764.7	1918.2	2014.1
Doors 910x2110mm	330.5	ft2	15	4560.5	4957.0	5204.9
Doors 1360x2110mm	123.3	ft2	35	3970.5	4315.7	4531.5
Doors ExtSgl_w-Glazing Bars_Arche	35.3	ft2	17	551.6	599.6	629.6
Floor-Grnd-Bearing	434.8	ft2	2.4	960.1	1043.6	1095.8
Floor-Grnd-Bearing carpet	1608.2	ft2	1	1479.6	1608.2	1688.6
Lighting Fixtures1 100 Watt	355.0	ft2	45	14698.3	15976.4	16775.2
Windows1810x1210mm	492.8	ft2	36	16322.5	17741.9	18629.0
Windows 910x1210mm	44.8	ft2	51	2103.2	2286.1	2400.4
Roof-Pitched	3729.3	ft3	2.4	8234.2	8950.3	9397.8
The total Material Cost				79043.8	85917.1	90213.0

Arranging project materials and applying equations (2.1 to 2.6), subsequently, using Monte Carlo simulation to assign these data to chosen curve and run the platform to obtain the conceptual cost

Figure 11.2 Cost estimation within conceptualisation and outline design stages

the elements that are sensitive to increased costs, and consequently, seeks to minimise costs by targeting these elements.

11.5.2. Detailed design stage

The detailed design stage is traditionally seen as the most significant part of the project, where information is formulated, and is therefore a vital stage of the IPD approach (Allison et al., 2017). In this stage, the 3D BIM model is enhanced by adding other dimensions such as scheduling (4D BIM) and cost (5D BIM). The BoQs are extracted using Navisworks in XML, including various data for each element, such as dimensions, the exact place in the project hierarchy, etc. The quantity surveyor collects the corresponding unit price of each element used in order to move to the documentation stage. With adequate information, a reliable cost structure is then prepared.

11.5.3. Documentation and buyout stages

The IPD cost structure relies on distinguishing all cost elements, such as direct, indirect, and overhead costs (given that risks/rewards are determined based on the rate of achievement of each individual element). To extend this, according to the AIA (2007), the overhead cost represents a separate limb after the direct and indirect limbs, where the final limb is the profit-at-risk percentage. The risks/rewards are determined based on the progress of each individual limb (i.e. whether the progress indicates a cost saving or is located as a profit-at-risk percentage); however, if progress indicates that the expanded cost exceeds the three limbs, the client is responsible only for the direct cost. Therefore, as discussed, having a scrutinising costing system is vital for successful IPD delivery.

Using the example presented here, adopting the ABC approach provides a solution, with each stakeholder involved from the conceptualisation stage. Throughout the first three IPD stages, all stakeholders participate in determining the cost of the project. In this respect, overhead costs represent a significant proportion of the whole project cost, with costs for each construction package obtained from the activities required to proceed with that package. Therefore, the ABC system is able to allocate overhead costs to relevant activities in order to determine the overhead resources for each package. Figure 11.3 illustrates the comparison between the traditional ABC hierarchy levels and the proposed IPD based on ABC adapted levels – following overhead resources within the defined/specific levels. As a result, overhead resources required at different levels can be evaluated.

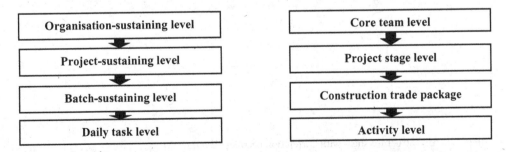

Figure 11.3 ABC functional level: comparison between traditional delivery approach (left) and IPD approach (right)

Overhead costs such as inspection and quality control (including cost control reports) should be converted into a unit allocated by its proportion of the cost driver. This process generates an accurate cost estimation value for each trade package (i.e. civil package, mechanical package, electrical package, etc.). The target cost in the IPD payment method should be fair for each package/party in the IPD project (as some packages require a low consumption rate of overhead resources, whereas for others, a higher consumption rate of overhead resources is required regardless of their proportion of the entire project). Notwithstanding this issue, the consumption of this significant proportion of overhead costs is needed, and it is therefore imperative that these costs are allocated to overhead activity consumption.

Activity-based costing estimation sheet: At this stage, the proposed coding system was developed to work as a bridge between ABC and BIM tools. This included digital numbers as well as alphabetical letters. Digital numbers refer to the type of operation. For example, a cost control activity had a 010 code. Alphabetical letters indicate the trade package; for example, the windows and doors package code was WD. Therefore, the cost control activity code for the windows and doors trade package was 010WD. If the project requires a sub-task, then codes can be extended to include another two letters derived from the first two letters of the sub-task. For example, the set of activities for door frame erection had the code 010WDDF. For interoperability between ABC and 4D/5D BIM, the task type and code correlation was interconnected. Various colours can also be used to identify the task types, such as the colour red for the package level (see Figure 11.3). Therefore, the code was designated as 010WDDF/red. Figure 11.4 illustrates the proposed ABC structure sheet.

Formulation of the IPD-based ABC model: At the buyout stage, each party should know the cost structure of the proposed works, where equations (3), (4), and (5) are used to determine the total cost of each limb. Extracting the BoQs using Navisworks was followed by pricing the extracted quantities and adding productivity allowances (labour and equipment) to complete the project pricing. Equations 3, 4, and 5 were used to categorise the estimated costs into three limbs for each package using the proposed coding system, as discussed hereafter.

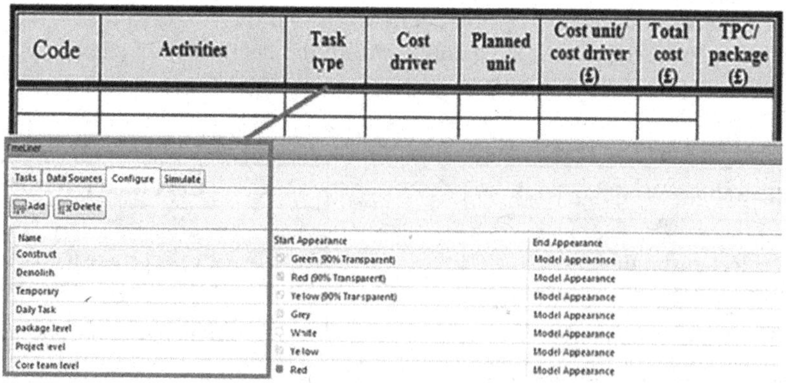

Figure 11.4 ABC structure sheet with correlation between 4D/5D BIM

Note: TPC = total project cost

Equation 3 shows the structure of Limb 1, including direct and indirect costs, with these two terms being able to be automatically estimated for each package (participant) by extracting costs using the coding system from the ABC sheet.

$$LIMB1_{ij} = \sum_{i=1}^{n}(CoDA_{Kj} + CoIA_{Kj}) \tag{3}$$

where $LIMB1_{ij}$ represents the direct and indirect costs for trade contractor i to perform trade package j; $CoDA_{Kj}$ represents the cost of direct activity for design element k and trade package j; and $CoIA_{Kj}$ represents the cost of indirect activity for design element k and trade package j.

Equation 4 shows the structure of Limb 2, representing overhead costs as the summation of the number of overhead activities for each package multiplied by the cost driver's estimated costs. For the purpose of automation, all costs can be automatically extracted from the ABC sheet.

$$LIMB2_{OA} = \sum_{i=1}^{n}(NOA_{OA} \times MVoCD_{DA}) \tag{4}$$

where $LIMB2_{OA}$ represents the overhead costs of specific operation O, such as cost control to perform overhead activity A; NOA_{OA} represents the summation of the number of operations O needs to perform in overhead activity A; and $MVoCD_{DA}$. reflects the monetary value of cost driver D performing overhead activity A.

Equation 5 represents the structure of Limb 3, which can be estimated by adding the profit-at-risk percentage (P@R%) to the pre-estimated Limbs 1 and 2.

$$LIMB3_{ij} = \sum_{i=1}^{n}\left(LMB2 \,\&\, 3_{ij}\right) \times P@R\,\%_{ij} \tag{5}$$

where $LIMB3_{ij}$. is the profit-at-risk percentage for trade contractor i to implement specific trade package j; $LMB2 \,\&\, 3_{ij}$. reflects the total costs for each package assigned to a specific party in the buyout stage; and $P@R\,\%_{ij}$ represents the profit-at-risk percentage for trade contractor i to implement trade package j.

According to Allison et al. (2017), splitting all overhead resources in a single pool can help to avoid waste when some project members implement more work than is required; however, determining overhead resources for a separate limb minimises the opportunity to hide a proportion of profit in the overhead percentage (Allison et al., 2017). Figure 11.5 illustrates the structure of the IPD approach based on ABC estimation. As all non-owner parties carry the same level of responsibility, the relationships between contractors and other parties are at the same level of inference.

Given the level of compensation presented in Figure 11.5, Figure 11.6 presents the structure of IPD cost estimation for each party.

The nature and structure of the IPD approach requires the completion of several tasks prior to the construction stage. In this respect, Figure 11.7 illustrates these tasks: the cost estimation process within conceptualisation, outline and detailed design, and documentation stages;

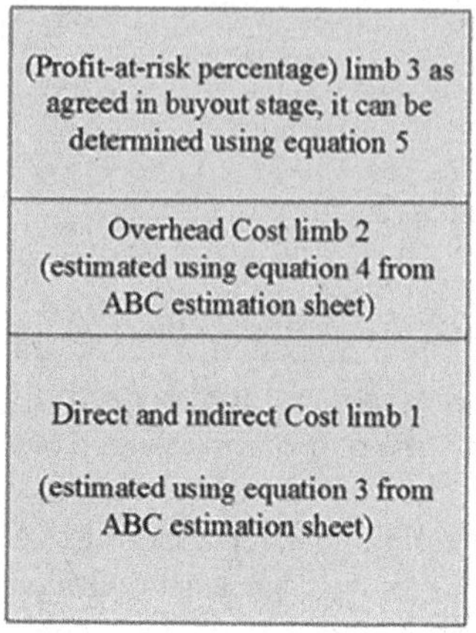

(Profit-at-risk percentage) limb 3 as agreed in buyout stage, it can be determined using equation 5
Overhead Cost limb 2 (estimated using equation 4 from ABC estimation sheet)
Direct and indirect Cost limb 1
(estimated using equation 3 from ABC estimation sheet)

Figure 11.5 Compensation under the IPD approach using ABC estimation

Package cost structure
(Profit-at-risk percentage)
Overhead cost
Direct and indirect cost
(Profit-at-risk percentage)
Overhead cost
Direct and indirect cost
(Profit-at-risk percentage)
Overhead cost
Direct and indirect cost
Project package

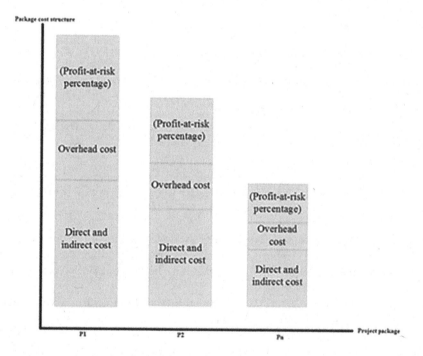

Figure 11.6 Structure of IPD cost estimation for each party

methods and tools to deal with various types of data; the amount of cost data to be analysed; the input and output of each stage; and the proposed tool to analyse the available data.

Case Study

The following discussion presents a single property development company. This company was wanting to build 100 identical houses. The specification of each house was as follows: (1)

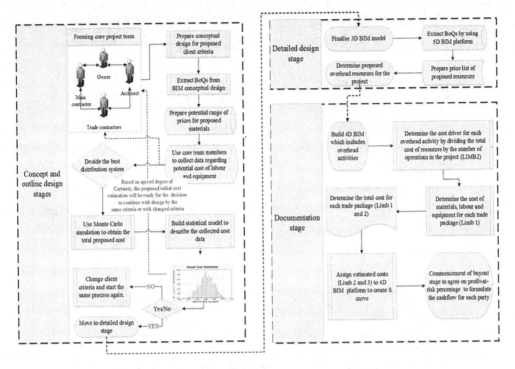

Figure 11.7 Cost estimation data flow within IPD pre-construction stages

gross floor area approximately 192 m²; (2) each house had a single floor; (3) from reviewing the Revit architectural plan, the spaces included a master bedroom with its own facilities of a bathroom and a robe room, which included three bedrooms, a large living room, kitchen, dining room, another bathroom, family room, and utility room.

From this, project works were categorised into five trade packages (general works and ceiling, lighting fixtures, finishings, and doors and windows packages). The client intended to use IPD as a means of delivering the project. In forming the core project team, an architectural firm and five trade contractors were appointed to build the project's core group, which also involved trade contractors to obtain the required information during kick-off meetings. As previously discussed, the IPD approach relies on sharing the benefits and risks from the outset. Hence, it was important to determine all expenses and costs in order to assign them to specific activities. Using the ABC estimation technique (after adapting it to be consistent with IPD levels) allowed all parties to create one cost pool under a cooperative joint venture (Kim et al., 2011). The costs from the conceptualisation stage to the buyout stage are discussed in the next section.

11.5.4. Initial cost estimation at outline design stage

Figure 11.8 illustrates the total material and labour costs for this project, prepared by a Monte Carlo simulation using cost data collected by the IPD core team's quantity surveyor, with beta distribution used to distribute these cost elements. The output from this process presents the total costs graph, showing how the total cost corresponds to a specific certainty percentage. The sensitivity analysis charts reveal the impacts of each cost element in the

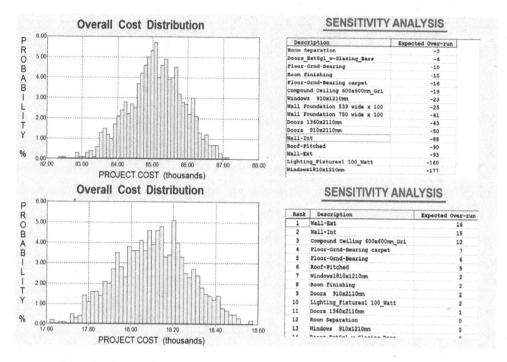

Figure 11.8 Total material and labour costs prepared by Monte Carlo simulation

project, thus determining each element's importance in the detailed design stage and the execution process. From this, the client makes decisions based on these outputs. If the client accepts the solution, the project progresses to the detailed design stage. If not, the client/ quantity surveyor can alter the requirements by changing the cost elements in order to repeat the process.

In the case project, the decision-making scope revealed that the cost would be almost £103,000, while the actual case study identified that the direct and indirect costs totalled £118,484. The deviation between the decision-making scope and the precise cost estimation is therefore approximately 12%. This level of deviation is generally more acceptable at the feasibility study and budget authorisation stages – in accordance with class 3 of the cost estimate classification matrix developed by Amos (2004), with this class accepting a deviation below detailed estimation of from –10% to –20%.

11.5.5. Cost estimation during detailed design stage

After finalising the 3D BIM model, the estimator uses this model for detailed cost estimation by importing it into a 5D BIM platform to extract quantities for the pricing stage. Based on the agreed length of the contract, overhead resources are determined to enable the costing process. The proposed resources needed to perform each activity are presented in Table 11.1.

From this, the cost drivers can be determined as the total cost of each operation divided by the number of operations (activities) in the project (Table 11.2).

Table 11.1 Organisation overhead costs (ABC estimation)

No.	Function	Salary (£)
1	Quantity Surveyor	12,000
2	Quality Control Engineer	9,000
3	Quality Assurance Engineer	9,000
4	Accountant	7,500
5	Project Manager	12,000
6	Site Engineer	9,000
7	Supervisor	7,500
8	Warehouse Manager	6,000
9	Total overhead cost	**72,000**

Table 11.2 Cost drivers of overhead operations

No.	Task/Unit	Value of cost driver (£)
1	Inspection	2,884.69
2	Mobilising	1,000
3	Cost control	3,250
4	Setting out	1,500

11.5.6. Calculations of cost drivers/cost units

The inspection process required a quality control engineer, quality assurance engineer, supervisor, and project manager. In total, 13 inspection activities were needed for this project. The mobilisation process occurred six times during the project, with the warehouse manager assigned this responsibility. Cost control needed a quantity surveyor and an accountant, which was run six times during project execution. Setting out was also run six times during the project, with the site engineer having responsibility for its implementation.

11.5.7. Integrated project delivery cost structure

With the extracted quantities priced, material costs are ready, and the summary of each trade package's materials can be presented (Table 11.3). Additional labour and equipment resources can then be determined using the same MS Excel spreadsheet (Table 11.3). Limb 1 is thus ready, and the estimator can then move to Limb 2, which relates to overhead costs. Table 11.3 summarises both cost estimation approaches; namely, the traditional costing system and the use of ABC estimation to validate the significance of the developed framework.

From Table 11.3, the total project costs were calculated as £190,484, where overhead costs represented about 37.8% of the total costs. This required a very precise allocation so the actual target cost for each package could be determined. When the project was completed, project parties were then able to assess whether each package achieved cost savings or not. This allowed them to be able to determine the percentage of cost savings made so that the rewards could be allocated fairly between project parties, especially where packages have different expenditure and overhead costs. For instance, the concrete package needed to be inspected three times: after the formwork, the rebar, and concreting. In contrast, the doors

Table 11.3 Compensation structure components

Table sections	Construction packages	General £	Ceiling £	Lighting fixture £	Finishing £	Doors and windows £
Limb 1 for traditional and proposed estimation methods	Total material costs	38,038.9	2,140.2	17,037.9	3,553.8	31,919.1
	Total labour costs	21,318.9	1,715	296.5	1,334.4	763
	Total equipment costs	366.8	0	0	0	0
	Total direct and indirect costs (Limb 1)	59,724.7	3,855.2	17,334.4	4,888.3	32,682.2
Limbs 2 and 3 for traditional cost estimation	Proportion of overhead costs	0.533	0.031	0.138	0.039	0.260
	Total overhead/package (Limb 2)	38,377	2,206	9,919	2,797	18,701
	Total costs	98,102	6,061	27,253	7,685	51,383
	Profit-at-risk limit (Limb 3)	19,620.4	1212.2	5450.6	1537	10276.6
Limbs 2 and 3 for the proposed ABC estimation method	Overhead costs (Limb 2)	27,557.6	11,519.2	7,134.6	15,403.8	7,134.6
	Total costs (ABC) (starting point of profit-at-risk percentage)	89,014.7	15,474.1	24,916.7	20,418.4	40,660.9
	Profit-at-risk limit (Limb 3)	106,817.6	18,568.9	29,900.1	24,502.1	48,793.1

and windows package only needed one inspection to ensure that the installation was in compliance with requirements.

11.6. Discussion of findings

From Figure 11.9, Limb 1 can be seen to be similar for both the traditional method and ABC estimation; however, the overhead cost differs between these two methods. The fluctuation percentage between ABC estimation and traditional cost estimation was higher than 100% in the finishing package, given that the case study project was relatively small with a limited number of activities, with the lowest level being 8% fluctuation in the lighting fixture package.

Figure 11.10 illustrates all deviations between the ABC estimation and traditional cost estimation for each package. To validate the significance of integrating ABC into IPD using BIM capabilities, Figure 11.10 revealed that the deviation for Limb 3 values (the profit-at-risk percentage) were elevated by £2521.42 for the finishing package, which was more than twice the value in the traditional method; however, other packages decreased in value, such as the doors and windows package, which was 22% lower than the traditional cost estimation approach.

To provide some additional context here, Ballard et al. (2015) set out to identify the factors that led to the failure of risks/rewards sharing, with case research comprising a 250,000 ft² patient care pavilion. The findings referred to cost overrun as the main reason, with the completed project having a cost overrun of almost 6.4% more than what had been planned. Subsequently, the risk pool firms did not receive any profit. Placing these results in context with the work presented in this chapter, it seems that a number of parallels can be drawn – in particular, the importance of continuous scrutiny of cost estimation, as this is seen as a vital component for revealing any potential cost overrun at an early stage. If this is undertaken, then the source of any overruns can be rectified with appropriate corrective action. It is also

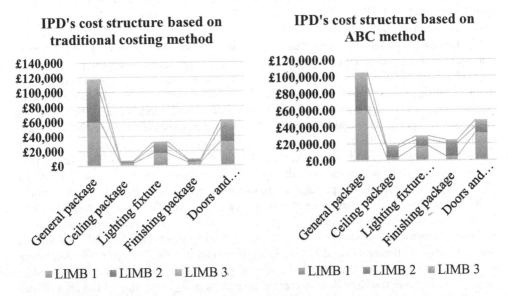

Figure 11.9 IPD cost structure using two costing methods

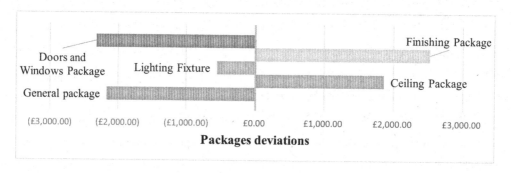

Figure 11.10 Deviations between ABC estimation and traditional estimation for each package

important to acknowledge that accurate cost estimation from the outset is also considered a prerequisite, as this not only helps secure greater accuracy, but also provides high-level evidence for future reduction of costs.

11.7. Conclusion

Exploiting the full potential of BIM, IPD, and non-traditional cost estimation approaches such as TVD requires solutions that draw upon each approach's strengths. In doing so, capabilities and advantages can be harvested from the synergy between them. As research in this field is still relatively in its infancy, this chapter provided a number of insights for future reflection. Firstly, the theoretical foundations and details of an innovative framework were presented, along with analytical considerations for integrating these methodologies. Secondly, the findings from this study highlight the need to promote the integration of various solutions to provide a workable, practical solution based on the integration of Monte Carlo simulation, TVD, and ABC with BIM-enabled IPD. Such integrative approaches are considered important for improvements.

The framework presented in this chapter provides a workable solution for BIM–IPD integration, producing reliable cost data from different sources applicable to various project delivery modes. Using BIM as a means of developing a conceptual model to address client criteria enables costing professionals (such as estimators) to build statistical models with a range of proposed costs to predict higher levels of cost certainty. These models can all be translated into innovative practices in real-life projects. In this respect, the case study used Monte Carlo simulation integrated with BIM to develop a conceptual cost estimation, where the deviation between the conceptual and detailed cost estimation did not exceed 12%. This was considered acceptable, especially at the early design stage. Moreover, using ABC provided a better IPD cost structure, minimising package fluctuations due to better overhead cost allocation.

Whilst recognising the advantages of this approach, it is equally important to place these findings into context. For example, this study presents only a small part of ongoing research into the development of an automated cost system for IPD projects based on BIM. Additional work and data are therefore needed to support the veracity of findings. In particular, future studies may wish to focus on developing several proposals at the buyout stage to enable participants to make the right decision regarding acceptance or rejection of the offer. Other improvements include innovative ways for cost control using Earned Value Management.

One of the main challenges when looking into issues like this one is that of interoperability. This is typically an issue when using different types of platforms, but the study presented in this chapter demonstrated interoperability with Revit, Navisworks, and Excel. A number of Navisworks plug-ins could be used to develop further cost management system with IPD, using application programming interface coded by C#.NET. Finally, it is important to recognise that the cost estimation approach used in this work was the expected cost. Therefore, market and allowable costs were not considered.

References

Ahmad, I., Azhar, N., & Chowdhury, A. (2019). Enhancement of IPD characteristics as impelled by information and communication technology. *Journal of Management in Engineering*, 35(1), 04018055. https://doi.org/10.1061/(ASCE)ME.1943-5479.0000670

AIA. (2007). *Integrated project delivery: A guide*. Retrieved June 26, from www.aia.org/resources/64146-integrated-project-delivery-a-guide

Aibinu, A., & Venkatesh, S. (2013). Status of BIM adoption and the BIM experience of cost consultants in Australia. *Journal of Professional Issues in Engineering Education and Practice*, 140(3), 04013021. https://doi.org/10.1061/(ASCE)EI.1943-5541.0000193

Alashwal, A. M., & Chew, M. Y. (2017). Simulation techniques for cost management and performance in construction projects in Malaysia. *Built Environment Project and Asset Management*, 7(5), 534–545. https://doi.org/10.1108/BEPAM-11-2016-0058

Allison, M., Ashcraft, H., Cheng, R., Klawans, S., & Pease, J. (2017). *Integrated project delivery an action guide for leaders*. Retrieved December 20, from www.leanconstruction.org/lci-news/integrated-project-delivery-an-action-guide-for-leaders/

Alves, T. d. C. L., Lichtig, W., & Rybkowski, Z. K. (2017). Implementing target value design: Tools and techniques to manage the process. *HERD: Health Environments Research & Design Journal*, 10(3), 18–29. https://doi.org/10.1177/1937586717690865

Alwisy, A., Bouferguene, A., & Al-Hussein, M. (2018). Framework for target cost modelling in construction projects. *International Journal of Construction Management*, 1–16. https://doi.org/10.1080/15623599.2018.1462446

Amos, S. J. (2004). *Skills & knowledge of cost engineering*: AACE International. ISBN:1885517491

Andersen, B., Samset, K., & Welde, M. (2016). Low estimates – high stakes: Underestimation of costs at the front-end of projects. *International Journal of Managing Projects in Business*, 9(1), 171–193. https://doi.org/10.1108/IJMPB-01-2015-0008

Ashcraft, H. W. (2014). Integrated project delivery: A prescription for an ailing industry. *International Construction Law Review*, 9, 21.

Asmar, M. E., Hanna, A. S., & Loh, W.-Y. (2016). Evaluating integrated project delivery using the project quarterback rating. *Journal of Construction Engineering and Management*, 142(1), 04015046. https://doi.org/10.1061/(ASCE)CO.1943-7862.0001015

Azhar, N., Kang, Y., & Ahmad, I. (2015). Critical look into the relationship between information and communication technology and integrated project delivery in public sector construction. *Journal of Management in Engineering*, 31(5), 04014091. https://doi.org/10.1061/(ASCE)ME.1943-5479.0000334

Ballard, G., Dilsworth, B., Do, D., Low, W., Mobley, J., Phillips, P., Reed, D., Sargent, Z., Tillmann, P., & Wood, N. (2015, July). How to make shared risk and reward sustainable. In O. Seppänen, V. A. González, & P. Arroyo (Eds.), *23rd Annual Conference of the International Group for Lean Construction* (pp. 257–266). Retrieved June 7, 2021, from http://iglc.net/Papers/Details/1193

Berssaneti, F. T., & Carvalho, M. M. (2015). Identification of variables that impact project success in Brazilian companies. *International Journal of Project Management*, 33(3), 638–649. https://doi.org/10.1016/j.ijproman.2014.07.002

Caffieri, J. J., Love, P. E. D., Whyte, A., & Ahiaga-Dagbui, D. D. (2018). Planning for production in construction: Controlling costs in major capital projects. *Production Planning & Control*, 29(1), 41–50. https://doi.org/10.1080/09537287.2017.1376258

Chang, C.-Y., Pan, W., & Howard, R. (2017). Impact of building information modeling implementation on the acceptance of integrated delivery systems: Structural equation modeling analysis. *Journal of Construction Engineering and Management, 143*(8), 04017044. https://doi.org/10.1061/(ASCE)CO.1943-7862.0001335

Chen, G., Zhang, G., Xie, Y.-M., & Jin, X.-H. (2012). Overview of alliancing research and practice in the construction industry. *Architectural Engineering and Design Management, 8*(2), 103–119. https://doi.org/10.1080/17452007.2012.659505

Chou, J.-S. (2011). Cost simulation in an item-based project involving construction engineering and management. *International Journal of Project Management, 29*(6), 706–717. https://doi.org/10.1016/j.ijproman.2010.07.010

Das, T. K., & Teng, B. S. (2001). Trust, control, and risk in strategic alliances: An integrated framework. *Organization Studies, 22*(2), 251–283. https://doi.org/10.1177/0170840601222004

Dawood, N., Rahimian, F., Seyedzadeh, S., & Sheikhkhoshkar, M. (2020). *Enabling the development and implementation of digital twins*. Proceedings of the 20th International Conference on Construction Applications of Virtual Reality, Teesside University. ISBN:9780992716127

de Melo, R. S. S., Do, D., Tillmann, P., Ballard, G., & Granja, A. D. (2016). Target value design in the public sector: Evidence from a hospital project in San Francisco, CA. *Architectural Engineering and Design Management, 12*(2), 125–137. https://doi.org/10.1080/17452007.2015.1106398

Do, D., Chen, C., Ballard, G., & Tommelein, I. (2014). Target value design as a method for controlling project cost overruns. *International Group for Lean Construction, 22*.

Eastman, C. M. A., Lee, G., Sacks, R., & Teicholz, P. M. (2018). *BIM handbook: A guide to building information modeling for owners, managers, designers, engineers and contractors* (3rd ed.). Wiley. ISBN:1119287545

Ebrahimi, G., & Dowlatabadi, H. (2018). Perceived challenges in implementing integrated project delivery (IPD): Insights from stakeholders in the U.S. and Canada for a path forward. *International Journal of Construction Education and Research*, 1–24. https://doi.org/10.1080/15578771.2018.1525446

Fischer, M., Khanzode, A., Reed, D., & Ashcraft, H. W. (2017). *Integrated project delivery*. John Wiley & Sons. ISBN: 9781118415382

Ghassemi, R., & Becerik-Gerber, B. (2011). Transitioning to integrated project delivery: Potential barriers and lessons learned. *Lean Construction Journal (Lean and Integrated Project Delivery Special issue)*, 32–52.

Hall, D. M., Algiers, A., & Levitt, R. E. (2018). Identifying the role of supply chain integration practices in the adoption of systemic innovations. *Journal of Management in Engineering, 34*(6), 04018030. https://doi.org/10.1061/(ASCE)ME.1943-5479.0000640

Hannon, J. J. (2007). Estimators' functional role change with BIM. *AACE International Transactions*, IT31.

Hosseini, M. R., Maghrebi, M., Akbarnezhad, A., Martek, I., & Arashpour, M. (2018). Analysis of citation networks in building information modeling research. *Journal of Construction Engineering and Management, 144*(8), 04018064. https://doi.org/10.1061/(ASCE)CO.1943-7862.0001492

Ilozor, B. D., & Kelly, D. J. (2012). Building information modeling and integrated project delivery in the commercial construction industry: A conceptual study. *Journal of Engineering, Project, and Production Management, 2*(1), 23–36. 10.32738/JEPPM.201201.0004

Jahangirian, M., Eldabi, T., Naseer, A., Stergioulas, L. K., & Young, T. (2010). Simulation in manufacturing and business: A review. *European Journal of Operational Research, 203*(1), 1–13. https://doi.org/10.1016/j.ejor.2009.06.004

Johansen, A., Sandvin, B., Torp, O., & Økland, A. (2014). Uncertainty analysis – 5 challenges with today's practice. *Procedia-Social and Behavioral Sciences, 119*, 591–600. 10.1016/j.sbspro.2014.03.066

Kahvandi, Z., Saghatforoush, E., Alinezhad, M., & Noghli, F. (2017). Integrated project delivery (IPD) research trends. *Journal of Engineering, Project, and Production Management, 7*(2), 99–114. https://doi.org/10.32738/JEPPM.201707.0006

Kahvandi, Z., Saghatforoush, E., Ravasan, A. Z., & Mansouri, T. (2018). An FCM-based dynamic modelling of integrated project delivery implementation challenges in construction projects. *Lean Construction Journal*, 63–87.

Khedr, M. K. (2006). Project risk management using Monte Carlo simulation. *AACE International Transactions, RI21*.

Kim, G.-H., An, S.-H., & Kang, K.-I. (2004). Comparison of construction cost estimating models based on regression analysis, neural networks, and case-based reasoning. *Building and Environment, 39*(10), 1235–1242. https://doi.org/10.1016/j.buildenv.2004.02.013

Kim, Y. W., & Ballard, G. (2001, August). *Activity-based costing and its application to lean construction.* Proceedings of the 9th Annual Conference of the International Group for Lean Construction, National University of Singapore.

Kim, Y.-W., Han, S.-H., Yi, J.-S., & Chang, S. (2016). Supply chain cost model for prefabricated building material based on time-driven activity-based costing. *Canadian Journal of Civil Engineering, 43*(4), 287–293. https://doi.org/10.1139/cjce-2015-0010

Kim, Y. W., Han, S., Shin, S., & Choi, K. (2011). A case study of activity-based costing in allocating rebar fabrication costs to projects. *Construction management and Economics, 29*(5), 449–461. https://doi.org/10.1080/01446193.2011.570354

Loizou, P., & French, N. (2012). Risk and uncertainty in development: A critical evaluation of using the Monte Carlo simulation method as a decision tool in real estate development projects. *Journal of Property Investment & Finance, 30*(2), 198–210. https://doi.org/10.1108/14635781211206922

Love, P. E., Davis, P. R., Chevis, R., & Edwards, D. J. (2011). Risk/reward compensation model for civil engineering infrastructure alliance projects. *Journal of Construction Engineering and Management, 137*(2), 127–136. https://doi.org/10.1061/(ASCE)CO.1943-7862.0000263

Lu, Q., Won, J., & Cheng, J. C. P. (2016). A financial decision making framework for construction projects based on 5D building information modeling (BIM). *International Journal of Project Management, 34*(1), 3–21. https://doi.org/10.1016/j.ijproman.2015.09.004

Lu, W., Lai, C. C., & Tse, T. (2018). *BIM and big data for construction cost management.* Routledge. ISBN:9781351172318

Manata, B., Miller, V., Mollaoglu, S., & Garcia, A. J. (2018). Measuring key communication behaviors in integrated project delivery teams. *Journal of Management in Engineering, 34*(4), 06018001. https://doi.org/10.1061/(ASCE)ME.1943-5479.0000622

Meijon Morêda Neto, H., Bastos Costa, D., & Coelho Ravazzano, T. (2019). Recommendations for target value design implementation for real estate development in Brazil. *Architectural Engineering and Design Management, 15*(1), 48–65. https://doi.org/10.1080/17452007.2018.1509054

Miller, J. A. (1996). *Implementing activity-based management in daily operations:* Wiley. ISBN:0471040037

Nassar, K. (2012). Assessing building information modeling estimating techniques using data from the classroom. *Journal of Professional Issues in Engineering Education and Practice, 138*(3), 171–180. https://doi.org/10.1061/(ASCE)EI.1943-5541.0000101

Niemann, R. (2017). *IPD and BIM – the future of project delivery in Australia.* Retrieved January 9, from www.mccullough.com.au/2017/10/05/ipd-bim-future-project-delivery-australia/

Oliver, S., Seyedzadeh, S., Rahimian, F., Dawood, N., & Rodriguez, S. (2020). Cost-effective as-built BIM modelling using 3D point-clouds and photogrammetry. *Current Trends in Civil & Structural Engineering-CTCSE, 4*(5), 000599. https://doi.org/10.33552/CTCSE.2020.04.000599

Pishdad-Bozorgi, P. (2017). Case studies on the role of integrated project delivery (IPD) approach on the establishment and promotion of trust. *International Journal of Construction Education and Research, 13*(2), 102–124. https://doi.org/10.1080/15578771.2016.1226213

Pishdad-Bozorgi, P., Moghaddam, E. H., & Karasulu, Y. (2013). *Advancing target price and target value design process in IPD using BIM and risk-sharing approaches.* Paper presented at the the 49th ASC Annual International Conference California Polytechnic State University.

Pishdad-Bozorgi, P., & Srivastava, D. (2018). *Assessment of Integrated Project Delivery (IPD) Risk and Reward Sharing Strategies from the Standpoint of Collaboration: A Game Theory Approach.* Paper presented at the Construction Research Congress 2018. New Orleans, Louisiana. https://doi. org/10.1061/9780784481271.020

PMI. (2017). *A guide to the project management body of knowledge (PMBOK Guide)* (6th ed.). Project Management Institute Standards Committee. ISBN: 9781628251845

Porwal, A., & Hewage, K. N. (2013). Building information modeling (BIM) partnering framework for public construction projects. *Automation in Construction, 31,* 204–214. https://doi.org/10.1016/j. autcon.2012.12.004

Potts, K., & Ankrah, N. (2014). *Construction cost management: Learning from case studies* (2nd ed.). Taylor and Francis. ISBN:1135013454, 9781135013455

Raisbeck, P., Millie, R., & Maher, A. (2010). Assessing integrated project delivery: A comparative analysis of IPD and alliance contracting procurement routes. *Management, 1019,* 1028.

Rowlinson, S. (2017). Building information modelling, integrated project delivery and all that. *Construction Innovation, 17*(1), 45–49. https://doi.org/10.1108/CI-05-2016-0025

Roy, D., Malsane, S., & Samanta, P. K. (2018). Identification of critical challenges for adoption of integrated project delivery. *Lean Construction Journal,* 1–15. Retrieved June 7, 2021, from www.lean-construction.org/media/docs/lcj/2018/LCJ_17_007.pdf

Seyedzadeh, S., & Pour Rahimian, F. (2021). Building energy data-driven model improved by multi-objective optimisation. In *Data-driven modelling of non-domestic buildings energy performance* (pp. 99–109). Springer. https://doi.org/10.1007/978-3-030-64751-3_6

Seyedzadeh, S., Rahimian, F. P., Oliver, S., Glesk, I., & Kumar, B. (2020). Data driven model improved by multi-objective optimisation for prediction of building energy loads. *Automation in Construction, 116,* 103188. https://doi.org/10.1016/j.autcon.2020.103188

Shadish, W. R., Cook, T. D., & Campbell, D. T. (2002). *Experimental and quasi-experimental designs for generalized causal inference:* Wadsworth Cengage learning. ISBN:0395615569

Shen, Z., & Issa, R. R. (2010). Quantitative evaluation of the BIM-assisted construction detailed cost estimates. *Journal of Information Technology in Construction, 15,* 234–257.

Silveira, S., & Alves, T. (2018). Target value design inspired practices to deliver sustainable buildings. *Buildings, 8*(9), 116. https://doi.org/10.3390/buildings8090116

Stanley, R., & Thurnell, D. (2014). The benefits of, and barriers to, implementation of 5D BIM for quantity surveying in New Zealand. *Construction Economics and Building, 14*(1), 105–117. https://doi. org/10.5130/AJCEB.v14i1.3786

Succar, B. (2009). Building information modelling framework: A research and delivery foundation for industry stakeholders. *Automation in Construction, 18*(3), 357–375. https://doi.org/10.1016/j. autcon.2008.10.003

Tillmann, P. A., Do, D., & Ballard, G. (2017). *A case study on the success factors of target value design.* 25th Annual Conference of the International Group for Lean Construction, pp. 563–570. https:// doi.org/10.24928/2017/0324

Torp, O., & Klakegg, O. J. (2016). Challenges in cost estimation under uncertainty – a case study of the decommissioning of Barsebäck nuclear power plant. *Administrative Sciences, 6*(4), 14. https://doi. org/10.3390/admsci6040014

Tsai, W. H., Yang, C. H., Chang, J. C., & Lee, H. L. (2014). An activity-based costing decision model for life cycle assessment in green building projects. *European Journal of Operational Research, 238*(2), 607–619. https://doi.org/10.1016/j.ejor.2014.03.024

Tsai, W. H., & Hung, S.-J. (2009). A fuzzy goal programming approach for green supply chain optimisation under activity-based costing and performance evaluation with a value-chain structure. *International Journal of Production Research, 47*(18), 4991–5017. https://doi.org/10.1080/00207540801932498

Wang, P., Du, F., Lei, D., & Lin, T. W. (2010). The choice of cost drivers in activity-based costing: Application at a Chinese oil well cementing company. *International Journal of Management, 27*(2), 367.

Welde, M., & Odeck, J. (2017). Cost escalations in the front-end of projects – empirical evidence from Norwegian road projects. *Transport Reviews, 37*(5), 612–630. https://doi.org/10.1080/01441647.201 6.1278285

Yin, R. K. (1981). The case study crisis: Some answers. *Administrative science quarterly, 26*(1), 58–65.

Zellmer-Bruhn, M., Caligiuri, P., & Thomas, D. C. (2016). From the Editors: Experimental designs in international business research. *Journal of International Business Studies, 47*(4), 399–407. https://doi.org/10.1057/jibs.2016.12

Zhang, L., & Li, F. (2014). Risk/reward compensation model for integrated project delivery. *Engineering Economics, 25*(5), 558–567.

Zheng, X., Lu, Y., Li, Y., Le, Y., & Xiao, J. (2019). Quantifying and visualizing value exchanges in building information modeling (BIM) projects. *Automation in Construction, 99*, 91–108. https://doi.org/10.1016/j.autcon.2018.12.001

Zimina, D., Ballard, G., & Pasquire, C. (2012). Target value design: Using collaboration and a lean approach to reduce construction cost. *Construction Management and Economics, 30*(5), 383–398. https://doi.org/10.1080/01446193.2012.676658

12 Dynamic sustainable success prediction model for infrastructure projects

A rough set–based fuzzy inference system

12.1. Introduction

Over the last decade (2010–2020), there has been a growing emphasis on the successful implementation of construction projects (Bensalah et al., 2019; Moshtaghian et al., 2020; Pour Rahimian et al., 2019, 2020; Zarghami et al., 2018). In this respect, the primary challenge has always been the ambiguities associated with assessing and anticipating success on such projects, since predicting several probable issues from a vast set of data is historically problematic (Cheng et al., 2013; Moshtaghian et al., 2020); however, there is still no consensus in project management research as to how project success is measured or predicted (Pinto & Slevin, 1988) since the critical characteristics of construction projects is their unpredictability, especially in comparison with static production industries (Loosemore et al., 2003).

Traditional approaches used in the evaluation of project success have tended to focus on project time, quality, and cost. This (arguably) has been somewhat restrictive, insofar as there is a wider need to reflect on other issues, including construction investment, as well as social, environmental, and sustainable dimensions. In this respect, any such evaluation therefore needs to embrace prediction models (Kibert, 2016; Kolo et al., 2014). This reflection should also embrace the different types of approaches and methodologies associated with monitoring and control strategies, especially where project managers and project team have to make judgments on project success (Ding & Banihashemi, 2017; Rahimian et al., 2019; Seyedzadeh et al., 2021).

On this theme, and in an attempt to capture these issues, the study presented in this chapter presents a dynamic Decision Support System (DSS) capable of predicting project success from a sustainability perspective across the whole project lifecycle. This DSS model employs Rough Set Theory (RST; Pawlak, 1982) in conjunction with the Fuzzy Inference System (FIS) to help develop robust rulesets for better prediction. Where given the technical challenges and level of complexity, the work presented here considers sustainability criteria – Sustainable Success Factors (SSF) and Sustainable Success Criteria (SSC) – as part of this process. In this respect, RST was engaged for rules generation, where a decision table was used as the input for the rough set, which returned a set of rules (as the output). The generated rulesets were then filtered using FIS, which served as the basis for the DSS. The developed prediction tool was then tested and validated by applying data from a real infrastructure project.

This tool is of particular benefit to decision-makers. Findings from this study identified that the developed rough set fuzzy method was able to evaluate and predict project success. In this respect, the efficacy of the DSS and application of RST was able to generate rules filtered

DOI: 10.1201/9781003106944-12

through the FIS to predict and deliver sustainable success. This approach is now gaining significant momentum, supported by developments and integration of RST and FIS, the results of which are now able to deliver robust rulesets for enhanced prediction. Of particular benefit is its ability to dynamically evaluate and predict project success based on different sustainability criteria throughout the project lifecycle.

12.2. Literature review

The definition of project success has undergone several transformations over the years. Traditionally, a construction project was deemed successful if it met criteria related to time, cost, and quality (Atkinson, 1999), and in particular, success measures, factors, and criteria underpinning this Belassi and Tukel (1996). \Cooke-Davies (2002) highlighted the difference between success criteria and success factors. Given this, extant literature increasingly uses Critical Success Factors (CSFs) to measure success, cf. independent variables of a project that contribute to project success (Müller & Turner, 2007; Rockart, 1982).

When considering the term 'success', it is often accepted that project success criteria are predominantly seen as dependent variables by which the success or failure of a project is judged and measured by its stakeholders (Belassi & Tukel, 1996; Pinto & Slevin, 1988; Rockart, 1982). Factors underpinning success criteria are commonly referred to CSFs or key performance indicators (KPIs). Cox et al. (2003) argued that CSFs were seen as the efforts made or strategies adopted to achieve the success of a project, whereas KPIs are typically seen as the compilation of data measures used to assess the performance of a project. In other words, KPIs are essential for assessing the effectiveness, efficiency, and quality of actual versus estimated performance.

Literature on project success reveals that various authors have identified many success determinants, either from experience or research. For example, Westerveld (2003) revealed that along with the conventional measures of cost, time, quality, and scope, an additional five KPIs were most frequently used, namely: (i) client's appreciation, (ii) project personnel appreciation, (iii) users' appreciation, (iv) contracting partners' appreciation, and (v) stakeholders' appreciation. Similarly, Belout and Gauvreau (2004) emphasised the importance of the project team's ability to manage project risks (and resolution thereof), which naturally includes the need to evaluate project success through risk factors (Moshtaghian et al., 2020). Moreover, a study by Cserhati and Szabo (2014), analysis of correlations, revealed that relationship-oriented success factors such as communication, cooperation, and project leadership played a vital role in the successful implementation of projects.

Given that the nature and scope of projects undertaken in Architecture, Engineering, and Construction (AEC) vary significantly, the determinants of 'success' are invariably scaled accordingly, especially with large infrastructure projects and concomitant capital investment and large delivery schedules. In this respect, several attempts have been made to address project success in these projects, where Brundtland et al. (1987) identified success factors for large projects using the factor analysis method. These factors were grouped into four major categories: (i) incompetent designers and contractors, (ii) poor estimation and change management, (iii) social and technological issues, and (iv) improper techniques and tools. Similarly, Ogunlana (2010) argued that traditional criteria for success were not sufficient to determine whether or not a project was successful in isolation, requiring other issues to be incorporated (quantitative and qualitative), using criteria such as environmental regulations, building performance, and client satisfaction.

Literature on sustainability has covered a range of areas. For example, the work by Al-Tmeemy et al. (2011) measured the success of projects for the provision of sustainable social housing in Nigeria. This work identified several CSFs that influenced success. In a more recent study, Krajangsri and Pongpeng (2016) addressed the impacts of sustainable infrastructure assessments on construction project success using the Structural Equation Modelling (SEM) approach. Construction project success was measured using six criteria: time, cost, quality, client satisfaction, safety, and environment. In other works, Banihashemi et al. (2017) looked at CSFs that affected the integration of sustainability in construction projects, particularly on project management practices in developing countries. This work incorporated innovation diffusion theory as the theoretical point of departure, which identified 59 CSFs pertaining to the triple bottom line of sustainability (i.e. environmental, social, and economic). A number of related studies with different perspectives on success factors (Table 12.1) and success criteria (Table 12.2) are presented as examples of work undertaken in these areas.

Table 12.1 Sustainable success factor studies

Group	Code	Success Factors	Sources
Economic	1	Time management	(Yuan et al., 2011)
	2	Cost management	(Yuan et al., 2011)
	3	Quality management	(Krajangsri & Pongpeng, 2016; Aquilani et al., 2017; Skibniewski & Ghosh, 2009)
	4	Feasibility study	(Li et al., 2005; Yuan et al., 2011)
	5	Risk management	Frödell et al., 2008; Bakar et al., 2009; Yuan et al., 2011; Khang & Moe, 2008; Ihuah et al., 2014)
	6	Adequate project fund and resources	(Shen et al., 2010; Ihuah et al., 2014)
	7	Level of local economy	(Bennett et al., 1999)
	8	Safety/implementation of HSE	(Yuan et al., 2011; Bennett et al., 1999; Park, 2009; Elbarkouky, 2012; Skibniewski & Ghosh, 2009)
	9	Effective communication	(Aquilani et al., 2017; Yeung et al., 2007; Frödell et al., 2008)
	10	Teamwork	(Frödell et al., 2008; Park 2009; Yang et al., 2011; Chileshe & John Kikwasi, 2014; Aquilani et al., 2017)
Social	11	Job satisfaction	(Yeung et al., 2007; Abidin, 2010; Loosemore et al., 2003; Lai & Lam, 2010)
	12	Leadership	(Frödell et al. 2008; Bakar et al., 2009; Park, 2009; Ihuah et al., 2014; Aquilani et al., 2017)
	13	Competent project team	(Du Plessis, 2007; Ihuah et al., 2014; Bakar et al., 2009; Cooke-Davies, 2002)
	14	Motivation	(Schianetz & Kavanagh, 2008; Park, 2009; Ashley, 1986)
	15	Attempt to preserve environment/ environmental protection	(Yuan et al., 2011; Yeung et al., 2007)
	16	Waste management	(Chen et al., 2008; Ross et al., 2010; Fernández-Sánchez & Rodríguez-López, 2010; Elbarkouky, 2012; Shane et al., 2013)

Group	Code	Success Factors	Sources
Environmental	17	Utilising clean and renewable energies	(Elbarkouky, 2012), (Fernández-Sánchez & Rodríguez-López, 2010; Bennett et al., 1999; Manoliadis et al., 2006)
	18	Environment protection measures in project design	(Shen et al., 2010; Chen et al., 2008)
	19	Cleaning up contaminated water and land	(Bourdeau, 1999; Huang & Hsu, 2011)
	20	Using clean technologies and materials	(Bennett et al., 1999; Chen et al., 2008; Ross et al., 2010)

Table 12.2 Sustainable success criteria studies

Group	Code	Success Factors	Sources
	1	Project completion within time	(Adinyira et al., 2012; Atkinson, 1999; Elattar, 2009; Ahadzie et al., 2008)
Economic	2	Project completion within budget	(Adinyira et al., 2012), (Atkinson, 1999; Elattar, 2009; Ahadzie et al., 2008)
	3	Project quality	(Adinyira et al., 2012; Atkinson, 1999; Elattar, 2009; Ahadzie et al., 2008)
	4	Internal return ratio (IRR)/ Return on investment (ROI)	(Elattar, 2009; Shen et al., 2010)
	5	Respond to project risks/ overall risk containment	(Adinyira et al., 2012; Ahadzie et al., 2008)
	6	Employer satisfaction/ client satisfaction/owner satisfaction	(Yeung et al., 2007; Pheng & Chuan, 2006; Ashley, 1986; Krajangsri & Pongpeng, 2016; Skibniewski & Ghosh, 2009; Ahadzie et al., 2008; Chan & Chan, 2004)
Social	7	Satisfaction of people in project neighbourhood/ end user satisfaction/ customer satisfaction	(Adinyira et al., 2012; Elattar, 2009; Yuan et al., 2011; Yeung et al., 2007; Müller & Turner, 2007; Ahadzie et al., 2008; Chan & Chan, 2004)
	8	Provision of employment opportunities	(Bennett et al., 1999; Shen et al., 2010; Chen et al., 2010; Alnaser et al., 2008)
	9	Overall health and safety measures/accident rates	(Adinyira et al., 2012; Shen et al., 2010; Fernández-Sánchez & Rodríguez-López, 2010; Ahadzie et al., 2008; Chan & Chan, 2004)
	10	Satisfaction of staff/team satisfaction	(Müller & Turner, 2007; Chan & Chan, 2004)
	11	Environmental degradation	(Adinyira et al., 2012; Ahadzie et al., 2008; Chan & Chan, 2004)
	12	Noise pollution	(Shen et al., 2010; Fernández-Sánchez & Rodríguez-López, 2010)
	13	Effect on air and land quality	(Chen et al., 2010; Shen et al., 2010)
Environmental	14	Adverse impact on historical sites and cultural heritage	(Krajangsri & Pongpeng, 2016; Shen et al., 2010)
	15	Energy consumption	(Alnaser et al., 2008; Bennett et al., 1999)

Literature on prediction models has also highlighted significant developments, including studies focussing on different prediction models to evaluate project success. For example, Kim et al. (2009) developed an SEM to predict project success of uncertain international construction projects. This study engaged a comparative analysis of SEM with multiple regression analysis and artificial neural network, where SEM showed a more accurate prediction of performance. Similarly, Cheng et al. (2010) proposed an Evolutionary Support Vector Machine Inference Model (ESIM) (Seyedzadeh & Pour Rahimian, 2021a, 2021b) for dynamically predicting project success. This model integrated the process of continuous assessment of project performance (CAPP) to select factors that dynamically influenced project success. Furthermore, Cheng et al. (2013) proposed an evolutionary Gaussian process inference model for dynamic success prediction. This work produced an Evolutionary Gaussian Process Inference Model (EGPIM), using a Gaussian process, along with Bayesian inference and particle swarm optimisation. The model was trained using the EGPIM, which proved quite precise at predicting project success, delivering exceptional performance in time-series applications. Other models for predicting construction cost and schedule success have used artificial neural networks and support vector machines classification (Wang et al. 2012). Findings to date show that early planning status can be efficiently adapted to predict project success, where proposed artificial intelligence (AI) models can deliver acceptable prediction results.

From a slightly different perspective, Khosravi and Afshari (2011) developed a success measurement model for construction projects that established a benchmark for measuring future construction projects based on success comparisons with other completed projects. This work used a two-round Delphi study complemented with a questionnaire survey. Khang and Moe (2008) also presented a conceptual framework for international development project success. This delineated a link between success factors and criteria to assess project status and forecasting capabilities throughout the project lifecycle. Success predictors were also developed by Abd-Hamid et al. (2015) through a conceptual framework, which evaluated the success of construction industry entrepreneurs. Other approaches include work developed by Kim et al. (2009), which developed an analytic network process model to predict the Performance of International Construction Joint Ventures (ICJV). Through this study, the relationship between the important determinants of ICJV success were identified, underpinned by findings from eight construction projects.

12.3. Research methodology

From a research development perspective, the underpinning philosophical positioning of this work (often termed the research lens) aimed to develop sustainable success measurement criteria. In this respect, it follows Akbari et al. (2018), who established a list of sustainable success indicators, namely: Sustainable Success Factors (SSF) and Sustainable Success Criteria (SSC). The rationale behind this was to be grounded in replication. Akbari et al. (2018) designed a questionnaire survey using 20 SSFs and 15 SSCs, which were classified into the triple bottom line of economic, social, and environmental groups. From this, 11 factors and 9 criteria from these three groups are discussed as follows.

The identified SSFs: Time management (F1), Cost management (F2), Quality management (F3), Risk management (F5), Leadership (F12), Competent project team (F13), Motivation (F14), Teamwork (F10), Attempt to preserve environment/environmental protection (F15), Waste management (F16), and Environment protection measures in project design (F18).

The selected SSCs: Project completion within time (C1), Project completion within budget (C2), Project quality (C3), Employer satisfaction/Client satisfaction/owner satisfaction (C6), Provision of employment opportunities (C8), Overall health and safety measures (C9), Environmental degradation (C11), Energy consumption (C15), and Effect on air and land pollution (C13).

The next process required the identification of sustainable success indicators, where Analytic Hierarchy Process (AHP) was utilised to quantify and calculate success based on the data collected from 20 experts. With respect to the weight factors, the success indices of the three categories of sustainability were calculated using the following equations:

$$ECSI = 0.31\ C3 + 0.38C1 + 0.31C2 \tag{1}$$
$$SOSI = 0.43C11 + 0.34C15 + 0.23C13 \tag{2}$$
$$ENSI = 0.48C6 + 0.37C9 + 0.15C8 \tag{3}$$

Where: ECSI = Economical Success Index, SOSI = Social Success Index, and ENSI = Environmental Success Index. From this, the economic category was considered most important, with 62% of the weighting. This was greater than that of the social category (24%) and environmental category (14%), respectively.

Therefore, it was concluded that most attention should be paid to the economic point of view, whilst the social and environmental aspects should be taken into account as well. Given this, the sustainable success index (SSI) was calculated as follows:

$$SSI = 0.62\ ECEI + 0.24\ SOSI + 0.14\ ENSI \tag{4}$$

Subsequently, respondents were asked to allocate a score based on the status of the selected success factors and criteria in the aforementioned projects (for which they were responsible). These numbers were between 1 and 5, where 1 represented "very bad", 2 – "slightly bad", 3 – "moderate", 4 – "good", and 5 – "excellent". Finally, based on equation 4, the SSI of these projects was calculated, and a decision table used in RST was created.

From a research methodological design perspective, several studies have also focussed on AI models, where, for example, AI has been deployed as a powerful tool to support decision-making in construction projects (Martínez Magaña & Fernández-Rodríguez, 2015). In addition, machine learning techniques have been used for modelling building energy-related neural networks through support vector machine and decision trees (Seyedzadeh et al., 2020a; Seyedzadeh et al., 2019). These accurate and fast predictors enable exploration of a vast combination of building characteristics – both in the design stage and retrofit planning (Seyedzadeh et al., 2020a, 2020b). To deal with the uncertainty in building and construction data (e.g. weather), rough set can be coupled with machine learning techniques (Shi & Li, 2008). Fuzzy logic is another AI method that has also received significant attention in this area due to its ability to handle subjectivity and variability during the construction process (Lam et al., 2007). In this respect, fuzzy logic provides an accurate approach for dealing with 'vagueness'. For example, its rule-based nature allows the use of information expressed in the form of natural language statements. In essence, fuzzy inference is an expert system that interprets the values in the input vector; then based on a set of "if-then" rules, values are assigned to the output vector.

RST does not need a membership function, and compared to other statistical methods, it does not require a large number of data points to find a rule (Pheng & Hongbin, 2006).

The most important advantage of this theory is its capability of estimating the significance of specific attributes (Liu & Yu, 2009). The fundamental concept of the rough set algorithm for the proposed application is presented as follows.

Definition 1: Information systems

Information systems are the set of objects described by their attributes and attribute values. The information system is defined as follows:

$$IS = (U, A) \tag{5}$$

Where U is the universe, a finite non-empty set of objects, $U = \{x_1, x_2, ..., x_m\}$, and A is the set of attributes. Each attribute $a \in A$ (attribute a, belonging to the considered set of attributes A) defines an information function:

$$f_a : U \to V_a \tag{6}$$

Where V_a is the set of values of a, called the domain of attribute a. In all attributes, there are decision attributes and condition attributes.

Definition 2: Core and reduct of attributes

Core and reduct are two fundamental concepts of RST. A reduct is the minimal subset of attributes that enables the same classification of elements of the universe as the entire set of attributes. In other words, properties that do not belong to a reduct are superfluous with regard to classification of elements of the universe. The core is the necessary element for rules and is the common portion of all reducts. For example, let B be a subset of A. The core of B is the set of all indispensable attributes of B. The following is an important property, linking the concept of the core and reducts:

$$Core(B) = \bigcap Red(B) \tag{7}$$

Where Red (B) is the set of all reducts of B.

The significance of an attribute can be measured by comparing the degree of partial dependency (γ) of a set, which includes the attribute with the degree of a set without the attribute. This idea can be formally described as follows:

$$\sigma_{(C,D)}(a) = \frac{(\gamma(C,D) - \gamma(C - \{a\}, D))}{\gamma(C,D)} \tag{8}$$

Where a $\in$ C, and $\sigma(a)$ is the significance of attribute a ($0 \leq \sigma(a) \leq 1$). The significance of a set of attributes can be calculated in the same way as follows:

$$\sigma(a) = \frac{(\gamma(C,D) - \gamma(C - B, D))}{\gamma(C,D)} = 1 - \frac{\gamma(C - B, D)}{\gamma(C,D)} \tag{9}$$

Where B is a subset of C. The significance of a set B, i.e. $\sigma(B)$, represents the effect of eliminating the set. Thus, the set of decision attributes D will not be properly classified to the same extent as the degree of $\sigma(B)$ when taking out set B from C, the set of condition attributes. Thus, we can determine an approximate reduct, the best subset for explaining a decision, by determining the significance of all possible sets.

12.3.1. *RST attribute reduction and rule generation*

As described earlier, a unique feature of the RST method is its generation of rules, which has great importance in the prediction of outputs. For this purpose, the Rosetta toolkit was employed to induce rough-based models. Rosetta tool lists the rules and provides some statistics for filtering the rules. The basic concepts of RST and its ability for rule generation are recognised through literature. RST can be used to identify the most significant features by computing subsets and cores. In this respect, in order to generate reducts, Genetic Algorithm (GA) can be applied as this provides a more exhaustive exploration of the search space (Wroblewski, 1995). This reduction produces a set of decision rules and minimal attributes subset that distinguishes on a per-object basis, whilst reduct with full object reduction creates a set of minimal attributes subset that designates functional dependencies (Sulaiman et al., 2008).

In this study, a full object reduction approach was adopted. The reducts used to generate rules in the economic, social, and environmental categories are [F15, F16, F18], [F12, F10, F13, F14], [F15, F16, F18], respectively. Rosetta tool lists the rules and provides some statistics for the rules, which are support, accuracy, coverage, stability, and length. This study adopted this definition of the rule statistics adopted from Sulaiman et al. (2008).

The number of primary decision rules generated based on produced reducts in the economic, social, and environmental categories are 17, 18, and 16, respectively. To raise effectiveness, this study filtered the decision rules according to the principle LHS support 2. Using Rosetta, rules with the highest LHS Support in all sustainable groups are extracted. These rules are sorted based on LHS Support in Table 12.3. It is worth mentioning that the LHS Support indicates the number of projects satisfying the condition of the rule, while the RHS Support indicates the number of projects satisfying the decision of the rule. In this study, based on the set of generated rules, projects were classified into four categories.

Table 12.3 Filtered rules extracted from Rosetta software

Success Perspective	Rule	LHS Support	RHS Support	RHS Accuracy	LHS Coverage	RHS Coverage	Class
Economical	F1(4) AND F2(4) AND F3(4) AND F4(3) => Economical Success(4)	4	4	1	0.21	0.57	3
	F1(3) AND F2(3) AND F3(4) AND F4(4) => Economical Success(3)	3	3	1	0.16	0.50	2
	F1(3) AND F2(4) AND F3(4) AND F4(2) => Economical Success(4)	2	2	1	0.11	0.29	3
	F1(1) AND F2(2) AND F3(1) AND F4(3) => Economical Success(2)	2	2	1	0.11	0.50	1
	F1(2) AND F2(1) AND F3(4) AND F4(1) => Economical Success(2)	2	2	1	0.11	0.50	1
	F1(1) AND F2(3) AND F3(4) AND F4(2) => Economical Success(3)	2	2	1	0.11	0.33	2

(Continued)

Table 12.3 (Continued)

Success Perspective	Rule	LHS Support	RHS Support	RHS Accuracy	LHS Coverage	RHS Coverage	Class
	F1(4) AND F2(5) AND F3(5) AND F4(3) => Economical Success(5)	2	2	1	0.11	1.00	4
	F1(5) AND F2(3) AND F3(4) AND F4(3) => Economical Success(3) OR (4)	2	1, 1	0.5, 0.5	0.11	0.17, 0.14	2,3
Social	F12(4) AND F10(3) AND F13(3) AND F14(4) => social success(3) OR (4)	4	3, 1	0.75, 0.25	0.22	0.5, 0.14	2,3
	F12(3) AND F10(4) AND F13(4) AND F14(4) => social success(4)	2	2	1	0.11	0.29	3
	F12(4) AND F10(5) AND F13(4) AND F14(5) => social success(5)	2	2	1	0.11	1.00	4
	F12(5) AND F10(4) AND F13(3) AND F14(5) => social success(4)	2	2	1	0.11	0.29	3
	F12(5) AND F10(4) AND F13(4) AND F14(4) => social success(4)	2	2	1	0.11	0.29	3
	F12(2) AND F10(2) AND F13(3) AND F14(2) => social success(2)	2	2	1	0.11	0.67	1
	F12(2) AND F10(3) AND F13(2) AND F14(4) => social success(2) OR (3)	2	1, 1	0.5, 0.5	0.11	0.34, 0.17	1,2
	F12(3) AND F10(2) AND F13(2) AND F14(4) => social success(3)	2	2	1	0.11	0.33	2
Environmental	F15(4) AND F16(3) AND F18(2) => environmental success(4) OR (3)	3	1, 2	0.34, 0.67	0.16	0.14, 0.5	2,3
	F15(1) AND F16(1) AND F18(3) => environmental success(2)	3	3	1	0.16	0.60	1
	F15(4) AND F16(3) AND F18(3) => environmental success(3)	2	2	1	0.11	0.50	2
	F15(5) AND F16(4) AND F18(4) => environmental success(5) OR (4)	5	3, 2	0.6, 0.4	0.26	1.0, 0.29	3,4
	F15(4) AND F16(4) AND F18(3) => environmental success(4)	2	2	1	0.11	0.29	3
	F15(4) AND F16(4) AND F18(5) => environmental success(4)	2	2	1	0.11	0.29	3
	F15(3) AND F16(2) AND F18(2) => environmental success(2)	2	2	1	0.11	0.40	1

12.4. Results and analysis

This section presents the development of a success prediction model using Mamdani and Takagi–Sugeno fuzzy inference expert systems. A set of rules obtained from RST was utilised to implement the proposed DSS. The description of the proposed model is presented through two stages, as shown in Figure 12.1. The first step was to generate the 'if-then' rules deriving from Table 12.3 and define membership functions, then build the fuzzy inference engine. In order to implement this model, in the first stage, decision-makers need to give a score to inputs in each sustainability category of economic, social, and environmental. The output of this stage is success scores defuzzified to obtain crisp values. There are three output variables in the first stage, which are considered as input variables in the second stage. In the second stage, these three input variables represent the status of economic, environmental, and social success. The relative importance weightage of each category is used to obtain the sustainable success index in total. For this purpose, to design this DSS, the basic concepts of the FIS are discussed.

12.4.1. Proposed model fuzzy membership functions

In this study, membership functions were applied in the trapezoidal and triangular forms. A trapezoidal membership function was defined as equation 10. According to equation 10, if b = c, then the number is called a triangular fuzzy number.

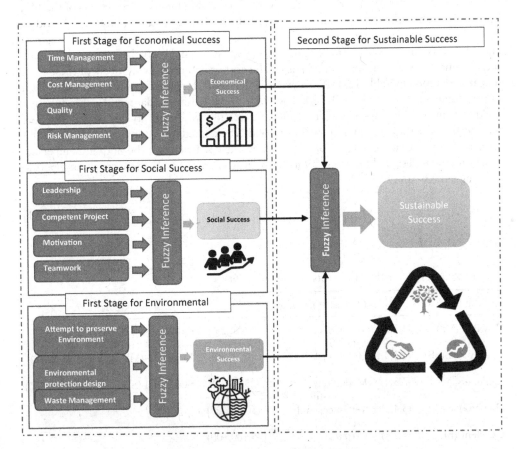

Figure 12.1 Overview of the proposed model development

$$\mu_{\tilde{A}}(x) = \begin{cases} 0 & x < a \\ \dfrac{1}{b-a}(x-a) & a \le x \le b \\ \dfrac{1}{c-d}(x-d) & c \le x \le d \\ 0 & x > d \end{cases} \tag{10}$$

In the first stage of the model, five fuzzy sets of membership functions were applied for both inputs and outputs of the FIS system. The fuzzy sets in the form of linguistic variables for inputs of stage 1 include "very bad", "slightly bad", "moderate", "good", and "excellent". These variables are equivalent to fuzzy numbers on a numeric range of 1–5, as shown in Table 12.4. As for outputs of this stage, linguistic variables are considered as "very unsuccessful", "unsuccessful", "moderately successful", "very successful", and "extremely successful". These variables are equivalent to fuzzy numbers on the numeric range of 0–100 (see Table 12.4).

The corresponding triangular membership functions for the inputs and outputs of stage 1 are shown in Figure 12.2. The DSS in this stage was developed using Mamdani's FIS, whereas the FIS in the second stage was conducted based on the Takagi—Sugeno method. To support clarity, this study only applied three fuzzy sets of membership functions for inputs of the FIS; however, a higher number of qualifiers could be used in the conceptual model to provide a better assessment. The fuzzy input sets in this stage were arranged in the form of linguistic variables, including "unsuccessful", "moderate", and "successful", which are used to evaluate the sustainable success in the second stage. The corresponding fuzzy numbers of these fuzzy sets are presented in Table 12.4. As shown in Table 12–4, linguistic variables are equivalent to fuzzy numbers on a scale of 0–100. The corresponding trapezoidal membership function for the inputs in this stage is shown in Figure 12.2. Yet, for output in this stage, the linear membership function of Sugeno-type was applied. The related fuzzy number for output of this stage is obtained using equation 10. In this study, the coefficients of the output membership functions of the designed FIS were obtained using AHP method. As there are three inputs in this stage, the output would be:

$$y = a_1 x_1 + a_2 x_2 + a_3 x_3 \tag{11}$$

Table 12.4 Linguistic variables for inputs and outputs at first stage

Inputs at first stage		Outputs at first stage		Inputs at second stage	
Very Bad (VB)	(1,1,2)	Very unsuccessful (VUS)	(0,0,25)	Unsuccessful (U)	(0,0,20,40)
Slightly Bad (SB)	(1,2,3)	Unsuccessful (US)	(0,25,50)	Moderate (M)	(30,50,50,70)
Moderate (M)	(2,3,4)	Moderately successful (MS)	(25,50,75)	Successful (S)	(60,80,100,100)
Good (G)	(3,4,5)	Very successful (VS)	(50,75,100)		
Excellent (E)	(4,5,5)	Extremely successful (ES)	(75,100,100)		

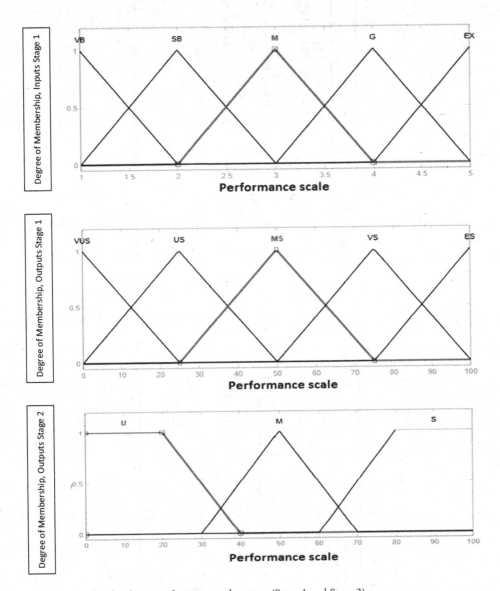

Figure 12.2 Membership functions for inputs and outputs (Stage 1 and Stage 2)

where a_1, a_2, a_3 denote weights resulted from the pairwise comparison matrices of the sustainability categories presented earlier. Finally, the fuzzy rule-based matrix in the second stage can be seen in Table 12.5.

12.4.2. *Defuzzification and calculation of outputs in each stage*

Defuzzification refers to the way in which the fuzzy number is converted into a crisp value. In this study, in the first stage, in order to calculate the project success in the sustainability categories, the centre of area method (COM) was used, as shown in equation 12.

Table 12.5 Fuzzy rule-based matrix (Stage 2)

First Input	Second Input	Third Input	Output	First Input	Second Input	Third Input	Output
U	U	U	U	M	M	S	0.86*U+0.14*M
U	U	M	0.86*U+0.14*M	M	S	U	0.62*U+0.24*M+0.14*S
U	U	S	0.86*U+0.14*S	M	S	M	0.74*M+0.24*S
U	M	U	0.76*U+0.24*M	M	S	S	0.62*M+038*S
U	M	M	0.62*U+0.38*M	S	U	U	0.62*S+0.38*U
U	M	S	0.62*U+0.24*M+0.14*S	S	U	M	0.62*U+0.24*M+0.14*S
U	S	U	0.76*U+0.24*S	S	U	S	0.76*S+0.24*U
U	S	M	0.62*U+0.24*M+0.14*S	S	M	U	0.62*U+0.24*M+0.14*S
U	S	S	0.62*U+0.38*S	S	M	M	0.62*S+0.38*M
M	U	U	0.62*M+0.38*U	S	M	S	0.76*S+0.24*M
M	U	M	0.76*M+0.24*U	S	S	U	0.86*S+0.14*U
M	U	S	0.62*U+0.24*M+0.14*S	S	S	M	0.86*S+0.14*M
M	M	U	0.86*M+0.14*U	S	S	S	S
M	M	M	M				

$$X_{COM} = \frac{\sum\limits_{i=1}^{n} x_i \cdot \mu_i(x_i)}{\sum\limits_{i=1}^{n} \mu_i(x_i)} \qquad (12)$$

In Sugeno FIS, the conclusion of a fuzzy rule is constituted by a weighted linear combination of the crisp inputs rather than by a fuzzy set. For example, in this study, during the second stage, the weighted average defuzzification method was used to calculate the output of DSS (known as the sustainability success index).

12.5. Case study model validation

12.5.1. Case study outline

A case study (Figure 12.3) was chosen to test the validity of the model. This case study was an urban tunnel project (two lines – 3993m each) based in Tehran (Iran), with a corresponding budget of approximately USD$64 million. This project commenced in 2014, and was completed at the end of 2019. Hereafter, this project will be referred to as project B.

The validation team consisted of a three-member assessment team, engaging project managers who were also the head of their departments in this project. Decision makers were asked for their perception on the status of success factors for this project (Table 12.6). Two virtual projects were defined as extremely successful (A) and unsuccessful (C). Finally, the sustainable success indexes of these projects were computed by applying this model.

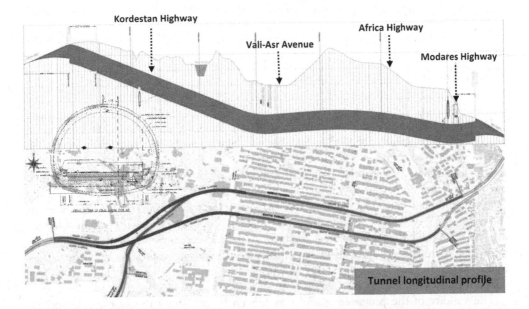

Figure 12.3 Case study details

Table 12.6 Decision makers' perception on project B success factors

Category	Success Factors	Decision Makers' Perception
Economical Perspective	Time management	4
	Cost management	4
	Quality management	3
	Risk management	2
Social Perspective	Leadership	4
	Competent project team	2
	Teamwork	3
	Motivation	3
Environmental Perspective	Attempt to preserve the environment	3
	Waste management	3
	Environment protection measures in project design	2

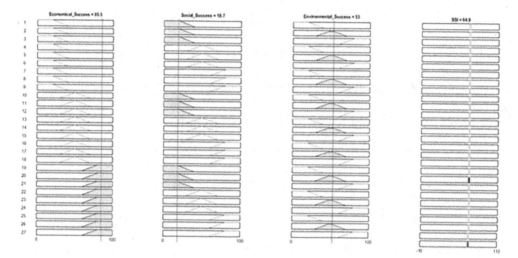

Figure 12.4 Rule viewer of the FIS for project B

During the implementation and result extraction, input values from the data collection process were transferred to the FIS system. For visualisation of the structure of the model, the rule viewer of the second stage of project B can be seen in Figure 12.4. The second stage of the model contained three input variables (economical, social, and environmental) and three membership functions. Therefore, the rule base consisted of 27 (3^3) if-then rules. To verify these rules, inputs were increased and outputs were examined. Inputs and outputs were in the range of [0–100]. The output surface of the second stage of FIS for the sustainable success index on the basis of economical and environmental success can be seen in Figure 12.5, where the more the input values increase, the more the output values (sustainable success index) surges. In addition, as linear Takagi–Sugeno was applied in this stage, inputs therefore varied linearly with outputs.

The validity of the proposed model was proven by obtaining the sustainable success index of project B, which varied between sustainable success index of projects A and C.

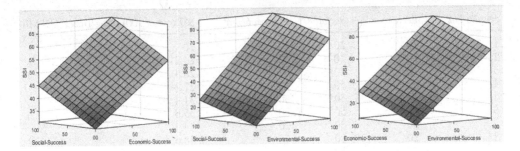

Figure 12.5 Output surface of the FIS for project B

Table 12.7 Validation of proposed model (first stage)

Project	Output	Assessment results using different defuzzification methods				
		COM	MOM	SOM	LOM	BOM
	Economical success score	85	87.5	90	100	87
A	Social success score	91	91	82	100	82
	Environmental success score	82	82	82	87	82
	Economical success score	76	76	65	96	69
B	Social success score	67	67	53	76	67
	Environmental success score	52	52	48	63	55
	Economical success score	20	20	4	10	20
C	Social success score	10	10	2	15	20
	Environmental success score	15	15	7	20	20

This model was also tested by applying different defuzzification methods to reinforce the validity of the model. Different approaches such as the centre of area method (COM), the bisector of area method (BOM), mean of the maximum method (MOM), smallest of the maximum method (SOM), and largest of the maximum method (LOM) were applied. As shown in Table 12.7, the assessment results for A, B, and C all had the same trend in all the defuzzification methods (which further reinforces the validity of this model). Finally, the obtained final value of stage 2 (70.5) provides a sustainable success index calculated using equation 4.

12.5.2. Prediction of sustainable success index based on scenario analysis

This section addresses different scenarios for evaluating the prediction of the sustainable success index. Given this, it was assumed that each of the 27 rules in stage 2 could be considered as a potential scenario in case three numbers in each membership function range – considered as inputs. These numbers were determined as follows: 20 for unsuccessful range, 50 for moderate range, and 80 for successful range. Hence the sustainable success status of projects could be predicted in these developed scenarios. For example, in the case of scenario 27, all parameter values were considered excellent, resulting in the highest score of 80. By reducing the economic values to the lowest value of 20, the SSI score was

Table 12.8 Prediction of SSI based on scenario analysis

Scenario No.	Economic	Social	Environmental	SSI (out of 100) using Trapezoidal MF
1	20	20	20	20
2	20	20	50	24.2
3	20	20	80	28.4
4	20	50	20	27.2
5	20	50	50	31.4
6	20	50	80	35.6
7	20	80	20	34.4
8	20	80	50	38.6
9	20	80	80	42.8
10	50	20	20	38.6
11	50	20	50	42.8
12	50	20	80	47
13	50	50	20	45.8
14	50	50	50	50
15	50	50	80	54.2
16	50	80	20	53
17	50	80	50	57.2
18	50	80	80	61.4
19	80	20	20	57.2
20	80	20	50	61.4
21	80	20	80	65.6
22	80	50	20	64.4
23	80	50	50	68.6
24	80	50	80	72.8
25	80	80	20	71.6
26	80	80	50	75.8
27	80	80	80	80

reduced to 42.8 (scenario 9), notwithstanding other parameters at maximum. Table 12.8 presents the results from these 27 scenarios.

12.5.3. Sensitivity analysis using FIS

To evaluate the impact of each of the economic, social, and environmental areas on the sustainable success index, a sensitivity analysis was performed. In this respect, two areas were chosen to replicate identical conditions in order to evaluate the impact of the third area on the SSI. These results are presented in Figure 12.6.

From Figure 12.6, the X-axis presents the sustainability areas and the Y-axis represents the sustainable success index. Accordingly, the equation related to each graph with a different slope shows the link between sustainability areas and the sustainable success index. There is a striking difference in the coefficients of the equations. This can be expected among the sustainability areas, as the economic area had the most impact on the sustainable success index compared to environmental and social areas.

Based on these scenarios, a comparison study was undertaken across the sustainability areas. The bar chart presented in Figure 12.7 confirms the results, noting that special attention should be paid to the economic sustainability area; albeit equally, not forgetting the importance of the social and environmental aspects.

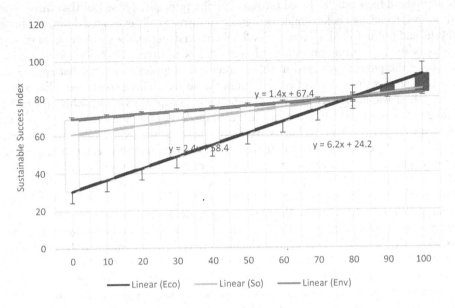

Figure 12.6 Sensitivity analysis scenarios

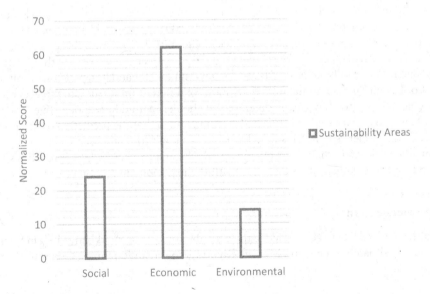

Figure 12.7 Sustainability areas score

12.6. Conclusion

Infrastructure projects can make a significant contribution to economic growth, social development, and environmental activities, especially in developing countries. With respect to heavy
resource consumption and concomitant investment in these types of infrastructure projects,
the determination of subsequent 'success' is paramount. Moreover, it is argued that success

indicators should not only be based on sustainability principles *per se*, but that these should be (i) properly predicted before implementation, (ii) dynamically evaluated during the project life cycle, and (iii) reflected upon post-development. In doing so, the measures from these are more likely to shape future projects (and decisions) of this type. On reflection, the original rationale of the project presented in this chapter aimed to design a DSS that applied RST and FIS to predict the success of projects from the sustainability perspective. This goal was achieved. RST was implemented for rules generation, as explained earlier, where the input of the rough set was a decision table, which included success factors and criteria, and the output presented a set of rules. Filtered rules served as the basis for the DSS. Drawing on the results and discussion, it can be inferred that the rough set fuzzy method demonstrated a strong capability of being able to evaluate and predict project success. That being said, the developed model presents several other extended opportunities for further improvements. In doing so, this will enhance the evaluation and prediction of sustainable success endeavours of this type.

Analysis of results derived from the rough-fuzzy model suggests that economic success can be seen as the main triggering fact for strengthening the social and environmental dimensions. Similarly, the social aspect needs to be taken into account as much as the environmental aspect in order to enhance the wider inclusion of supportive sustainable success criteria. In this respect, the model presented in this chapter provides a unique starting trajectory for stakeholders. For example, it presents decision-makers such as project managers with a tool to help them maintain and balance economic, social, and environmental success performance indicators throughout a project's lifecycle.

As a closing caveat and acknowledgement, it should be noted that this research was limited to infrastructure projects with budgets around USD$30 million. Hence, from a reliability and generalisability perspective, any inference or causality derived from this model should be treated with caution, especially for smaller projects. Moreover, it is advocated that contextual drivers such as country-specific constraints or governmental mandates, policies, or initiatives may also skew findings. This includes derivations of different infrastructure projects because large projects can often have many different stakeholders with different objectives and mandates. This is to be expected. It is therefore recommended that the prediction of success from different stakeholders should be studied separately. Finally, the proposed technique calls for further studies to strengthen this work – in particular, applying different approaches such as ANFIS to achieve improved results, along with the inclusion of additional rules.

Acknowledgements

The authors would like to acknowledge the contributions of Saeed Akbari, Moslem Sheikhkhoshkar, Saeed Banihashemi, and Mostafa Khanzadi to the underpinning research presented in this chapter.

References

Abd-Hamid, Z., Azizan, N. A., & Sorooshian, S. (2015). Predictors for the success and survival of entrepreneurs in the construction industry. *International Journal of Engineering Business Management, 7*.

Abidin, N. Z. (2010). Investigating the awareness and application of sustainable construction concept by Malaysian developers. *Habitat International, 34*(4), 421–426.

Adinyira, E., Botchway, E., & Kwofie, T. E. (2012). Determining critical project success criteria for public housing building projects (PHBPS) in Ghana. *Engineering Management Research, 1*(2). https://doi.org/10.5539/emr.v1n2p122

Ahadzie, D. K., Proverbs, D. G., & Olomolaiye, P. O. (2008). Critical success criteria for mass house building projects in developing countries. *International Journal of Project Management, 26*(6), 675–687. https://doi.org/10.1016/j.ijproman.2007.09.006

Akbari, S., Khanzadi, M., & Gholamian, M. R. (2018). Building a rough sets-based prediction model for classifying large-scale construction projects based on sustainable success index. *Engineering, Construction and Architectural Management, 25*(4), 534–558. https://doi.org/10.1108/ECAM-05-2016-0110

Al-Tmeemy, S. M. H. M., Abdul-Rahman, H., & Harun, Z. (2011). Future criteria for success of building projects in Malaysia. *International Journal of Project Management, 29*(3), 337–348.

Alnaser, N., Flanagan, R., & Alnaser, W. (2008). Model for calculating the sustainable building index (SBI) in the kingdom of Bahrain. *Energy and Buildings, 40*(11), 2037–2043.

Aquilani, B., Silvestri, C., Ruggieri, A., & Gatti, C. (2017). A systematic literature review on total quality management critical success factors and the identification of new avenues of research. *The TQM Journal, 29*(1), 184–213. https://doi.org/10.1108/tqm-01-2016-0003

Ashley, D. (1986). *New trends in risk management.* Internet's 10th International Expert Seminar on New Approaches in Project Management.

Atkinson, R. (1999). Project management: Cost, time and quality, two best guesses and a phenomenon, its time to accept other success criteria. *International Journal of Project Management, 17*(6), 337–342.

Bakar, A., Hassan, A., Abd Razak, A., Abdullah, S., & Awang, A. (2009). Project management success factors for sustainable housing: A framework. *Asian Journal of Management Research, 66.*

Banihashemi, S., Hosseini, M. R., Golizadeh, H., & Sankaran, S. (2017). Critical success factors (CSFs) for integration of sustainability into construction project management practices in developing countries. *International Journal of Project Management, 35*(6), 1103–1119.

Belassi, W., & Tukel, O. I. (1996). A new framework for determining critical success/failure factors in projects. *International Journal of Project Management, 14*(3), 141–151.

Belout, A., & Gauvreau, C. (2004). Factors influencing project success: The impact of human resource management. *International Journal of Project Management, 22*(1), 1–11.

Bennett, M., James, P., & Klinkers, L. (1999). *Sustainable measures: Evaluation and reporting of environmental and social performance.* Greenleaf Publishing. ISBN: 1907643192

Bensalah, M., Elouadi, A., & Mharzi, H. (2019). Overview: The opportunity of BIM in railway. *Smart and Sustainable Built Environment, 8*(2), 103–116. https://doi.org/10.1108/SASBE-11-2017-0060

Bourdeau, L. (1999). Sustainable development and the future of construction: A comparison of visions from various countries. *Building Research & Information, 27*(6), 354–366.

Brundtland, G., Khalid, M., Agnelli, S., Al-Athel, S., Chidzero, B., Fadika, L., Hauff, V., Lang, I., Shijun, M., & de Botero, M. M. (1987). *Our common future (\'brundtland report\').* Report of the World Commission.

Chan, A. P., & Chan, A. P. (2004). Key performance indicators for measuring construction success. *Benchmarking: An International Journal, 11*(2), 203–221.

Chen, Y., Okudan, G. E., & Riley, D. R. (2010). Sustainable performance criteria for construction method selection in concrete buildings. *Automation in Construction, 19*(2), 235–244. https://doi.org/10.1016/j.autcon.2009.10.004

Chen, Z., Li, H., Ross, A., Khalfan, M. M., & Kong, S. C. (2008). Knowledge-driven ANP approach to vendors evaluation for sustainable construction. *Journal of Construction Engineering and Management, 134*(12), 928–941.

Cheng, M. Y., Huang, C. C., & Roy, A. F. V. (2013). Predicting project success in construction using an evolutionary Gaussian process inference model. *Journal of Civil Engineering and Management, 19*(Sup1), S202–S211.

Cheng, M. Y., Wu, Y. W., & Wu, C. F. (2010). Project success prediction using an evolutionary support vector machine inference model. *Automation in Construction, 19*(3), 302–307. https://doi.org/10.1016/j.autcon.2009.12.003

Chileshe, N., & John Kikwasi, G. (2014). Critical success factors for implementation of risk assessment and management practices within the Tanzanian construction industry. *Engineering, Construction and Architectural Management, 21*(3), 291–319.

Cooke-Davies, T. (2002). The "real" success factors on projects. *International Journal of Project Management*, *20*(3), 185–190.

Cox, R. F., Issa, R. R., & Ahrens, D. (2003). Management's perception of key performance indicators for construction. *Journal of Construction Engineering and Management*, *129*(2), 142–151.

Cserhati, G., & Szabo, L. (2014). The relationship between success criteria and success factors in organisational event projects. *International Journal of Project Management*, *32*(4), 613–624.

Ding, G., & Banihashemi, S. (2017). Ecological and carbon footprints – the future for city sustainability. In *Encyclopedia of Sustainable Technologies* (pp. 43–51). Elsevier.

Du Plessis, C. (2007). A strategic framework for sustainable construction in developing countries. *Construction Management and Economics*, *25*(1), 67–76.

Elattar, S. M. S. (2009). Towards developing an improved methodology for evaluating performance and achieving success in construction projects. *Scientific Research and Essays*, *4*(6), 549–554.

Elbarkouky, M. M. G. (2012). A multi-criteria prioritization framework (MCPF) to assess infrastructure sustainability objectives. *Journal of Sustainable Development*, *5*(9). https://doi.org/10.5539/jsd.v5n9p1

Fernández-Sánchez, G., & Rodríguez-López, F. (2010). A methodology to identify sustainability indicators in construction project management – application to infrastructure projects in Spain. *Ecological Indicators*, *10*(6), 1193–1201. https://doi.org/10.1016/j.ecolind.2010.04.009

Frödell, M., Josephson, P. E., & Lindahl, G. (2008). Swedish construction clients' views on project success and measuring performance. *Journal of Engineering, Design and Technology*, *6*(1), 21–32.

Huang, R. Y., & Hsu, W. T. (2011). Framework development for state-level appraisal indicators of sustainable construction. *Civil Engineering and Environmental Systems*, *28*(2), 143–164.

Ihuah, P. W., Kakulu, I. I., & Eaton, D. (2014). A review of critical project management Success Factors (CPMSF) for sustainable social housing in Nigeria. *International Journal of Sustainable Built Environment*, *3*(1), 62–71. https://doi.org/10.1016/j.ijsbe.2014.08.001

Khang, D. B., & Moe, T. L. (2008). Success criteria and factors for international development projects: A life-cycle-based framework. *Project Management Journal*, *39*(1), 72–84. https://doi.org/10.1002/pmj.20034

Khosravi, S., & Afshari, H. (2011). *A success measurement model for construction projects*. International Conference on Financial Management and Economics IPEDR, IACSIT Press, pp. 186–190.

Kibert, C. J. (2016). *Sustainable construction: Green building design and delivery*. John Wiley & Sons. ISBN: 1119055172

Kim, D. Y., Han, S. H., Kim, H., & Park, H. (2009). Structuring the prediction model of project performance for international construction projects: A comparative analysis. *Expert Systems with Applications*, *36*(2), 1961–1971.

Kolo, S. J., Rahimian, F. P., & Goulding, J. S. (2014). Offsite manufacturing construction: A big opportunity for housing delivery in Nigeria. *Procedia Engineering*, *85*, 319–327. https://doi.org/10.1016/j.proeng.2014.10.557

Krajangsri, T., & Pongpeng, J. (2016). Effect of sustainable infrastructure assessments on construction project success using structural equation modeling. *Journal of Management in Engineering*, 04016056.

Lai, I. K. W., & Lam, F. K. S. (2010). Perception of various performance criteria by stakeholders in the construction sector in Hong Kong. *Construction Management and Economics*, *28*(4), 377–391. https://doi.org/10.1080/01446190903521515

Lam, K. C., Wang, D., Lee, P. T. K., & Tsang, Y. T. (2007). Modelling risk allocation decision in construction contracts. *International Journal of Project Management*, *25*(5), 485–493. https://doi.org/10.1016/j.ijproman.2006.11.005

Li, B., Akintoye, A., Edwards, P. J., & Hardcastle, C. (2005). Critical success factors for PPP/PFI projects in the UK construction industry. *Construction Management and Economics*, *23*(5), 459–471.

Liu, G., & Yu, W. (2009). Smart case-based indexing in worsted roving process: Combination of rough set and case-based reasoning. *Applied Mathematics and Computation*, *214*(1), 280–286.

Loosemore, M., Dainty, A., & Lingard, H. (2003). *Human resource management in construction projects: Strategic and operational approaches*. Taylor & Francis. ISBN: 0415261643

Manoliadis, O., Tsolas, I., & Nakou, A. (2006). Sustainable construction and drivers of change in Greece: A Delphi study. *Construction Management and Economics, 24*(2), 113–120.

Martínez Magaña, D., & Fernández-Rodríguez, J. C. (2015). Artificial intelligence applied to project success: A literature review. *International Journal of Interactive Multimedia and Artificial Intelligence, 3*(5), 77–84.

Moshtaghian, F., Golabchi, M., & Noorzai, E. (2020). A framework to dynamic identification of project risks. *Smart and Sustainable Built Environment*, ahead-of-print. https://doi.org/10.1108/sasbe-09-2019-0123

Müller, R., & Turner, R. (2007). The influence of project managers on project success criteria and project success by type of project. *European Management Journal, 25*(4), 298–309. https://doi.org/10.1016/j.emj.2007.06.003

Ogunlana, S. O. (2010). Beyond the "iron triangle": Stakeholder perception of key performance indicators (KPIs) for large-scale public sector development projects. *International Journal of Project Management, 28*(3), 228–236.

Park, S. H. (2009). Whole life performance assessment: Critical success factors. *Journal of Construction Engineering and Management, 135*(11), 1146–1161.

Pawlak, Z. (1982). Rough sets. *International Journal of Computer & Information Sciences, 11*(5), 341–356.

Pheng, L. S., & Chuan, Q. T. (2006). Environmental factors and work performance of project managers in the construction industry. *International Journal of Project Management, 24*(1), 24–37. https://doi.org/10.1016/j.ijproman.2005.06.001

Pheng, L. S., & Hongbin, J. (2006). Analysing ownership, locational and internalization advantages of Chinese construction MNCs using rough sets analysis. *Construction Management and Economics, 24*(11), 1149–1165.

Pinto, J. K., & Slevin, D. P. (1998). *Project success: Definitions and measurement techniques.* Project Management Institute.

Pour Rahimian, F., Chavdarova, V., Oliver, S., & Chamo, F. (2019). OpenBIM-Tango integrated virtual showroom for offsite manufactured production of self-build housing. *Automation in Construction, 102*, 1–16. https://doi.org/10.1016/j.autcon.2019.02.009

Pour Rahimian, F., Seyedzadeh, S., Oliver, S., Rodriguez, S., & Dawood, N. (2020). On-demand monitoring of construction projects through a game-like hybrid application of BIM and machine learning. *Automation in Construction, 110*, 103012. https://doi.org/10.1016/j.autcon.2019.103012

Rahimian, F. P., Seyedzadeh, S., & Glesk, I. (2019). OCDMA-based sensor network for monitoring construction sites affected by vibrations. *Journal of Information Technology in Construction, 24*, 299–317.

Rockart, J. F. (1982). *The changing role of the information systems executive: A critical success factors perspective.* MIT.

Ross, N., Bowen, P. A., & Lincoln, D. (2010). Sustainable housing for low-income communities: Lessons for South Africa in local and other developing world cases. *Construction Management and Economics, 28*(5), 433–449.

Schianetz, K., & Kavanagh, L. (2008). Sustainability indicators for tourism destinations: A complex adaptive systems approach using systemic indicator systems. *Journal of Sustainable Tourism, 16*(6), 601–628.

Seyedzadeh, S., Agapiou, A., Moghaddasi, M., Dado, M., & Glesk, I. (2021). WON-OCDMA system based on MW-ZCC codes for applications in optical wireless sensor networks. *Sensors, 21*. https://doi.org/10.3390/s21020539

Seyedzadeh, S., & Pour Rahimian, F. (2021a). Machine learning for building energy forecasting. In *Data-driven modelling of non-domestic buildings energy performance* (pp. 41–76). Springer. https://doi.org/10.1007/978-3-030-64751-3_4

Seyedzadeh, S., & Pour Rahimian, F. (2021b). Machine learning models for prediction of building energy performance. In *Data-driven modelling of non-domestic buildings energy performance* (pp. 77–98). Springer. https://doi.org/10.1007/978-3-030-64751-3_5

Seyedzadeh, S., Pour Rahimian, F., Oliver, S., Rodriguez, S., & Glesk, I. (2020a). Machine learning modelling for predicting non-domestic buildings energy performance: A model to support deep energy retrofit decision-making. *Applied Energy, 279*, 115908. https://doi.org/10.1016/j.apenergy.2020.115908

Seyedzadeh, S., Pour Rahimian, F., Rastogi, P., & Glesk, I. (2019). Tuning machine learning models for prediction of building energy loads. *Sustainable Cities and Society, 47*, 101484. https://doi.org/10.1016/j.scs.2019.101484

Seyedzadeh, S., Rahimian, F. P., Oliver, S., Glesk, I., & Kumar, B. (2020b). Data driven model improved by multi-objective optimisation for prediction of building energy loads. *Automation in Construction, 116*, 103188. https://doi.org/10.1016/j.autcon.2020.103188

Shane, J. S., Bogu, S. M., & Molenaar, K. R. (2013). Municipal water/wastewater project delivery performance comparison. *Journal of Management in Engineering, 29*(3). https://doi.org/10.1061/(asce)me.1943-5479

Shen, L., Wu, Y., & Zhang, X. (2010). Key assessment indicators for the sustainability of infrastructure projects. *Journal of Construction Engineering and Management, 137*(6), 441–451.

Shi, H., & Li, W. (2008, December 20–22). *The integrated methodology of rough set theory and artificial neural-network for construction project cost prediction.* 2008 Second International Symposium on Intelligent Information Technology Application, pp. 60–64. https://doi.org/10.1109/IITA.2008.238

Skibniewski, M. J., & Ghosh, S. (2009). Determination of key performance indicators with enterprise resource planning systems in engineering construction firms. *Journal of Construction Engineering and Management, 135*(10), 965–978.

Sulaiman, S., Shamsuddin, S. M., & Abraham, A. (2008). *An implementation of rough set in optimizing mobile Web caching performance.* Computer Modeling and Simulation, UKSIM 2008, Tenth International Conference, IEEE, pp. 655–660.

Wang, Y.-R., Yu, C.-Y., & Chan, H.-H. (2012). Predicting construction cost and schedule success using artificial neural networks ensemble and support vector machines classification models. *International Journal of Project Management, 30*(4), 470–478. https://doi.org/10.1016/j.ijproman.2011.09.002

Westerveld, E. (2003). The project excellence model®: Linking success criteria and critical success factors. *International Journal of Project Management, 21*(6), 411–418.

Wroblewski, J. (1995). *Finding minimal reducts using genetic algorithms.* Proceedings of the Second Annual Join Conference on Infomation Science, pp. 186–189.

Yang, L.-R., Huang, C.-F., & Wu, K.-S. (2011). The association among project manager's leadership style, teamwork and project success. *International Journal of Project Management, 29*(3), 258–267. https://doi.org/10.1016/j.ijproman.2010.03.006

Yeung, J. F. Y., Chan, A. P. C., Chan, D. W. M., & Li, L. K. (2007). Development of a partnering performance index (PPI) for construction projects in Hong Kong: A Delphi study. *Construction Management and Economics, 25*(12), 1219–1237. https://doi.org/10.1080/01446190701598673

Yuan, J., Wang, C., Skibniewski, M. J., & Li, Q. (2011). Developing key performance indicators for public-private partnership projects: Questionnaire survey and analysis. *Journal of Management in Engineering, 28*(3), 252–264.

Zarghami, E., Azemati, H., Fatourehchi, D., & Karamloo, M. (2018). Customizing well-known sustainability assessment tools for Iranian residential buildings using fuzzy analytic hierarchy process. *Building and Environment, 128*, 107–128.

13 Multi-objective optimisation to support building window design

13.1 Introduction

This chapter presents a multi-objective method for analysing and optimising the energy processes associated with window system design in office buildings. The rationale behind this is that simultaneous consideration of multiple and conflicting design objectives can often make the architectural design process more complicated. This study is based on the fundamental recognition that optimising parameters on the building energy loads via window system design tend to reduce the quality of the view to outside and the received daylight – both qualities highly valued by building occupants. The chapter proposes an approach for quantifying Quality of View in office buildings in balance with energy performance and daylighting, thus enabling an optimisation framework for office window design. The study builds on previous research by developing a multi-objective method of assessment for a reference room, which was parametrically modelled using actual climate data. A method of Pareto frontier and a weighting sum was applied for multi-objective optimisation to determine best outcomes that balance design requirements. The method identifies the maximum possible window-to-wall ratio for the reference room. The optimisation model also indicates that the room geometry should be altered to achieve the lighting and view requirements set out in building performance standards. This exploratory approach to a methodology and framework considers both building parameters and the local climate condition. It has the potential to be adopted and further refined by other researchers and designers to support complex, multi-factorial design decision-making.

13.2 Background challenge

Artificial lighting contributes to a large proportion of electricity consumption in commercial buildings across the globe. For example, in the US, artificial lighting contributes to one-third of electricity consumption in commercial buildings (Energy, 2013). In the UK, this sector accounts for almost 24 million tons of carbon dioxide (CO_2) per year, equal to 47% of the CO_2 emission of the UK (Mhalas et al., 2013). In Iran, artificial lighting is responsible for 25% of electricity usage in office buildings (Mahdavinejad et al., 2012). This level remains relatively high despite Iran having a high daylight availability during working hours (n.b. Tehran has an average of 8.5 sunshine hours per day). In light of global recognition of the importance of more sustainable and efficient building performance, it was therefore considered important to develop methods for minimising electricity usage for lighting through best practice design decisions. One efficient method is to utilise the natural daylight in indoor areas more

DOI: 10.1201/9781003106944-13

effectively. To achieve this goal, a considered design approach to the placement and size of windows in office buildings is imperative.

Several studies have examined the effect of daylight on occupant behaviour (Al Horr et al., 2016), increasing productivity (Alrubaih et al., 2013), improving job satisfaction of employees (Edwards & Torcellini, 2002) and their health conditions (Beute & de Kort, 2014). Studying the lighting conditions in office types shows strong relationships between the illuminance at eye level and the health parameters, namely fatigue and sleep quality (Aries, 2005). For example, even the colour temperature of light has significant correlation with the performance and alertness of office workers (Krüger et al., 2018). In addition, it has also been shown that daylight has direct physical effects on occupants and is an efficient energiser to human visual and circadian systems (Boyce et al., 2003); however, daylight can also cause visual discomfort through glare and distraction (Alrubaih et al., 2013). Hence, the productiveness of daylight for visual efficiency depends on how it is delivered; it is recommended that direct sunlight should be avoided in areas in which visual activities are required (Ne'Eman, 1974). In the context of this study, the views and perspectives provided by windows have been shown to impact both the visual performance and comfort of workers (Farley & Veitch, 2001). The positive effects of a pleasant or attractive view in the workplace include reduced physical and psychological discomfort, enhanced sleep quality (Aries et al., 2010), eye health (Ko et al., 2017), and increased job satisfaction (Altomonte, 2008). As such, the optimisation of window design, especially in commercial buildings, typically involves the careful balancing of three main objectives: (i) maximising energy efficiency through natural lighting, (ii) providing the best possible view, and (iii) delivering optimum visual comfort.

Several researchers have considered window type, size, and glazing in their calculations to optimise various objectives, including lifecycle cost and lifecycle environmental impact (Wang et al., 2005), energy efficiency (Seyedzadeh & Pour Rahimian, 2021c, 2021d) and occupant comfort (Diakaki et al., 2008), and retrofitting actions (Asadi et al., 2012); however, some of these approaches provide architects with restricted information, thereby compromising design objectives. To address this issue, several researchers have focused on daylighting and thermal efficiency considering window glazing (Hee et al., 2015), external shading (Manzan & Padovan, 2015), or window size, orientation, and wall reflectance (Mangkuto et al., 2016). Others have investigated visual comfort and energy performance together, focusing on the orientation of windows (Tagliabue et al., 2012), window size (Ochoa et al., 2012), exterior components (Uribe et al., 2017), and whole building characteristics (Carlucci et al., 2015a). In addition, Iommi (Iommi, 2019) evaluated daylighting performance and visual comfort in specific buildings, but these interdependent factors have not been simultaneously considered in the actual process of building design. One reason for this is the subjectivity of assessing 'visual comfort', which is typically a case-based quality assessment based on a person's individual experience of architecture (Carlucci et al., 2015b). Another barrier has been the need to consider all objectives, with their concomitant increase in complexity throughout the design decision-making process.

To address the challenges mentioned here, this study presents a framework for the evaluation of Quality of View (QV) in office buildings, applying a multi-objective optimisation method that minimises energy consumption and maximises daylight and visual comfort (absence of glare). The objectives of minimisation of energy consumption and Annual Sun Exposure (ASE), including maximisation of daylight, are assessed using simulation software, whilst the target values of the objective of QV are input using the proposed framework. This research applies the framework to a case study of a typical office building located in Tehran (Iran), to determine the most appropriate window dimensions and positions.

This chapter first presents a review of previous studies and issues associated with building windows design, the optimisation of energy efficiency, daylighting, and visual parameters, followed by an explanation of the framework methodology. The results from the application of the multi-objective optimisation method are then demonstrated, followed by a discussion of the criteria for selection of the best solution from the generated Pareto front. The final section summarises the proposed method and findings, highlighting the main knowledge contribution and recommendations for future works.

13.3 Critical challenge

The background and motivation for this study stemmed from the need to optimise office window design (position and area) – specifically, to minimise energy usage, whilst also optimising daylight and visual performance. Criteria for assessing these objectives are discussed throughout this chapter; for example, the first part of the literature review focusses on studies that have developed evaluation indices for the optimisation process to enhance integrated building design. This is followed by a discussion on the body of evidence that addresses the optimisation of window design and the challenges faced within the field. The critical challenge rests with the analysis of performance issues, design criteria, quality of view, energy use, and application of algorithms to deliver viable evidenced-based solutions.

13.4 Performance goals and building design indices

Building openings, windows (Hee et al., 2015), and doors or generally all key elements of building façades (Halawa et al., 2018), allow for daylighting, visual connection to the outside, and also heat penetration. Windows contribute to building energy usage through two main ways: (i) heating from direct sunlight through windows (imposing high cooling loads in warm seasons); and (ii) in cold seasons (heat loss). This is predominantly down to the high thermal transmittance of glazing (represented as a higher U-value) when compared to (non-glazed) walls (Seyedzadeh & Pour Rahimian, 2021b). Hence, a critical factor in the design of energy-efficient buildings is the design of a window system that takes into consideration the impact of high thermal transmittance of glazing (Omrany et al., 2016). User comfort is another factor to be considered in estimating energy performance, using various indexes and prognostic methods (Djongyang et al., 2010; Yang et al., 2014). Ochoa et al, (2012) investigated the suitability of combined optimisation criteria on window sizing methods for low energy consumption, focusing on user visual comfort and performance. Similarly, Ghaffarian-hoseini et al. (2015) investigated the ability of vegetation, including unshaded courtyards, on various design configurations and scenarios. In addition, Rupp et al. (2015) revealed a gap in thermal comfort studies in relation to interdisciplinary research and a connection with other related fields such as psychology, physiology, and sociology, noting the need for an integrated research approach to support a better understanding of the perception of thermal comfort, its physiological and psychological dimensions. Similarly, Khatami and Hashemi (2017) investigated the influence of decreasing internal heat gain by introducing automated ventilation strategies into lightweight open-plan offices to improve energy performance thermal comfort and indoor air quality.

Several methodological approaches can be used to predict and model the thermal behaviour of buildings. One way is by using a simplified physical model (Ahmad et al., 2006;

Rahimian et al., 2019; Seyedzadeh et al., 2021) based on thermodynamic laws, heat transfer, and thermodynamic variables. The 'degree days' approach is another method, which uses a measure of local outside temperature over time to calculate energy consumption required to heat or cool buildings (De Rosa et al., 2014).

A number of sustainable certification systems are also available. These include the Building Research Establishment Environmental Assessment Method (Neto & Fiorelli, 2008) and Leadership in Energy and Environmental Design (LEED), both of which have been developed to continually assess energy performance across all life stages of a building. They also incorporate measures to incentivise better design and analysis of low energy-consuming building systems, whether by classical (Ghaffarianhoseini et al., 2016) or even machine learning (Pour Rahimian et al., 2020; Seyedzadeh et al., 2020a, 2019, 2018, 2020b) methods. The analysis of indoor daylight performance is generally performed using software simulation, which calculates a range of metrics, e.g. Daylight Illuminance (DI), Daylight Factor (DF), Daylight Coefficient (DC), Daylight Autonomy (DA) (Reinhart & Walkenhorst, 2001), and ASE (Heschong et al., 2012). The metrics of Spatial Daylight Autonomy (sDA) and ASE are the most common daylight indices used in the LEED 4 rating.

Daylight Illuminance (DI) is the most common daylight performance measure, which designates the brightness of the daylight for illuminating the indoor environment. The unit for DI is lux (*cf.* lx, the SI unit of illuminance). Based on the application of each environment, the recommended illuminance level is often very different. For instance, an illumination level higher than 500 lux is suggested for an office (Dubois, 2003), where a range between 200 to 500 lux is often required for a classroom (Krüger & Dorigo, 2008).

DF is determined as the ratio of daylight illuminance at an indoor point to the illuminance at the same outdoor point under the sky. DF is a traditional approach to evaluate the daylight illuminance inside a building and mostly used due to simplicity; however, using DF metrics can sometimes lead to inaccurate calculations as the ratio of internal to external illuminance can differ considerably in real-life situations. Moreover, the impact of direct sunlight is sometimes ignored in this method (Mardaljevic, 2000). Given this, the DC approach was developed as a more practical measure in comparison with the DF. It considers dynamic changes in the luminance of the sky elements under various sky conditions and solar positions (Li et al., 2004). The assumption here is that the sky is divided into several patches, contributing to the internal illuminance level at a given point (Hensen & Lamberts, 2012). This often means that calculations are predominantly time-consuming and complicated (Yu et al., 2014).

DA can also be referred to as a dynamic daylight metric or climate-based metric, determined as a percentage of annual daytime hours over the year, when a specific illuminance threshold is achieved by daylight alone (Walsh, 1951). The continuous DA (cDA) and DA_{max} are two modified versions of DA. The former metric assigns the partial participation of daylight to illuminance when it is lower than the minimum threshold, whereas the second one indicates the percentage of the time when daylight illuminance is ten times the recommended illuminance – beyond which condition the risk of glare from direct sunlight patch would rise (Reinhart & Weissman, 2012).

sDA describes the annual amount of self-sufficiency in the environment, in terms of the amount of daylight received in the interior. It is the ratio of the analysed space, with the minimum received brightness defined for the desired activity during the working hours of a year. Analysis typically only includes the working area, but usually the total space is also considered. This assumes a grid of N points and assigns a function ST(*i*) *whose value becomes one* for every point i in the grid, achieving the minimum required illuminance for more than the given fraction of total occupancy time. sDA can be represented as:

$$sDA = \frac{\sum_{i=1}^{N} ST(i)}{N} \text{ with } ST(i) = \begin{cases} 1 : st_i \geq \tau t_y \\ 0 : st_i \leq \tau t_y \end{cases} \quad (1)$$

Where st_i. denotes the occurrence count of exceeding the sDA illuminance threshold at point i, t_y is the annual timestamp count, and τ. represents the temporal fraction threshold.

The Illuminating Engineering Society (IES) introduced the idea of the sDA (Heschong et al., 2012) to set the minimum illumination of 300 lux for 50% of the year when the zone is occupied (8 am to 6 pm), which is written as sDA$_{300/50\%}$. The IES also provided two metrics for sDA (Heschong et al., 2012). In order to have preferred daylight sufficiency in space, at least 75% of the analysed points should receive more than 300 lux in at least 50% of the year when it is occupied [3 points]. If nominally accepted, the brightness of more than 300 lux in 50% of the time of the year is considered for at least 55% of points [2 points] (Heschong et al., 2012). These recommendations are based on the comparison of occupants' opinions with the results of daylight simulations (Paule et al., 2018). The most important advantage of the DA family compared to the other daylighting indices is that the annual daylight performance assessment is achieved with regards to the sun and sky conditions based on meteorological climatic data. These indices help designers understand the overall conditions of daylight in buildings over a period of one year. Given that these data are recorded continuously (n.b. prediction is based on continuous daytime measurements), it is often difficult to assess the instantaneous light situation with the DAs. In addition, DAs are based on the percentage of occupied space per year and do not take into account the changes in hourly light (which it is argued here as being one of the most important aspects of building design). Where the sDA index is estimated to be the same for different models, a more accurate analysis can be made using the DA$_{ave}$.

ASE is the ratio of analysed space that receives more than a certain amount of direct sunlight in a number of specific hours of the year. Both factors determination (number of hours and amount of radiation) in the index definition is required. The IES Recommendation is ASE$_{1000, 250h}$, which is the percentage of space in which in more than 250 hours of the year, the direct exposure of the sun is more than 1,000 lux. Similar to sDA, ASE can be mathematically represented as:

$$ASE = \frac{\sum_{i=1}^{N} AT(i)}{N} \text{ with } AT(i) = \begin{cases} 1 : at_i \geq T_i \\ 0 : at_i \leq T_i \end{cases} \quad (2)$$

Where at_i is the occurrence count of exceeding the ASE illuminance threshold at the point i, and T_i represents the annual absolute hour threshold.

LEED (v4) requires the attention of designers to the ASE and sDA to score 2 to 3, by reaching 55–75% respectively of the area of the occupied spaces. To achieve this goal, designers using annual computer simulations need to show that the annual direct solar radiation of ASE$_{1000, 250h}$ is received in less than 10% of the space. It is therefore necessary that simulation run is based on sDA$_{300/50\%}$.

Among all window-related properties, the most appealing and challenging one is the view to outdoor. In the property market, there is a direct relationship between the value of the high-rise building (Yu et al., 2007) or a neighbourhood (Lake et al., 1998) and a pleasant outdoor view. There is also evidence that views can positively impact on eye health (Knecht, 2004), well-being (Heerwagen & Orians, 1986), and comfort (Aries et al., 2007). Being able to see natural landscapes from inside a workplace building has a significant impact on reducing stress and increasing individual attention. Studies have demonstrated that the relationship between view and daylight in contemporary façades is less perceptible, and the indices and studies related to view are very limited (Andersen, 2015). Interestingly, only the LEED v4, the Chartered Institution of Building Services Engineers (CIBSE), as well as the New European Daylighting Standard EN 17037, have introduced design guidelines for achieving a desired view to outdoors.

The basis of LEED v4 (Mitrache, 2012) for meeting QV needs is to provide a direct view through glazing for 75% of the occupied space and seeks to enhance the connection between the perimeter environment and the building occupants. In addition to having 75% of the occupied space in the index, these parts must also provide at least two of the following four view types. View type 1 is the horizontal and vertical viewing angles of at least 90 degrees to the view glazing. View type 2 is the viewing content of two of the following three objects: the visibility of flora, fauna, or sky or movement. View type 3 is the distance from the window, which is less than three times the height of the window from the floor. View type 4 is a view factor (VF) of 3 or more. VF is categorised from 0 to 5 based on the minimum horizontal and vertical angle of each point of occupied space than the window (where the user is sitting on a chair and the height of the eye is 1.2 m). If this view angle is more than 50 degrees, the rate of 5 is granted. VF of 3 and higher means a view angle of at least 11 degrees (Heschong et al., 2012).

The UK-based CIBSE provides methods for assessing view based on the position of the observer in space (Chartered Institution, Daylighting – a guide for designers: Lighting for the Built Environment 2014). This standard evaluates the quality of view based on four factors: window width, the distance of the view, view layers (sky, landscape, and foreground), and environmental information (contents). The view quality is rated based on four levels: unacceptable, acceptable, good, or excellent. The New European Daylighting standard EN 17037 recently introduced three general principles of horizontal viewing angles, distances, and visibility levels (which is similar to CIBSE's standards). According to its categorisation, the horizontal angle of at least 14 degrees is minimum, more than 28 degrees is average, and an angle more than 54 degrees is maximum. In addition, the window distance from the obstacle of more than 6 m is minimum, more than 20 m is medium, and more than 50 m is maximum. Visible layers could be the sky, landscape, or earth; if at least one of them is visible, the minimum score is achieved. The average score is for the view to two layers, and if all layers can be covered, a maximum score will be obtained (Standard, 2018). Therefore, view angles, view contents, and distance of the view and observer position are some indices considered in recent regulations. Among these indices, the process of simulating some of these can be very difficult (or impossible) due to differences in the models, such as view content simulation or view layouts and also environmental information. As such, the approach adopted in this study focuses on measurable indices.

To date, no readily discernible tools can simulate views based on these indicators. For example, recent versions of DIAL+ software, built to assess the new daylight requirements of EN-17037 (Paule et al., 2018), assessed only one of the three conditions, thereby reducing the possibility of full optimisation. In this respect, there is an opportunity for developing a tool that can simulate the view based on LEED v4 guidelines.

13.5 Related studies

There is an established body of evidence built on optimising the design of window systems for a range of outcomes and objectives – some isolated single objectives, others combined objective optimisation. Lee et al. (2013) investigated different types and characteristics of window systems to optimise energy usage. This was limited to assessing energy performance in isolation; however, it considered a specific climate in the proposed approach. Futrell et al. (2015) examined a process for optimising the thermal and lighting performance of a typical classroom by investigating the orientation of windows. Results revealed that thermal and lighting performance was strongly conflicted in a north orientation (n.b. this research was based in the US northern hemisphere). Mangkuto et al. (2016) also investigated different characteristics of window systems including a Window-to-Wall ratio (WWR), wall reflectance, and window orientation on daylight metrics and lighting electricity consumption. This research studied an office room in a tropical climate and was one of the first to demonstrate the possibility of incorporating the view/visual aspects of window systems into an optimisation process.

Ochoa et al. (2012) also employed an optimisation approach to identify the window size while optimising energy consumption and visual comfort. Vanhoutteghem et al. (2015) discovered suitable window solutions for various room sizes by assessing the impact of the window design variables on the thermal performance and comfort as well as daylighting using a contour plot. Similarly, Zhang et al. (2019) studied the feasibility of an optimisation workflow for the cooling and heating load of a residential building with changes in spatial form and building envelope parameters. This verified the optimisation results achieved by Octopus, comparing average objective values and correlation analysis between design parameters and performance. Fang and Cho (2019) presented an optimised solution based on simple building geometry, window and skylight size, and placement for energy and daylighting performance. A genetic algorithm was utilised to increase 28% in daylight performance and decrease 17% in energy consumption in different climates. Dino Ipek and Üçoluk (2017) developed a design optimisation tool to support high-performance building design, employing buildings' energy and daylighting performance optimisation in different layout designs and building openings.

Zhai et al. (2019) applied a multi-objective optimisation algorithm to minimise energy use, whilst also improving the visual and thermal comfort of a reference office room by finding the most appropriate parameters for the window system. These factors included WWR, outer and inner glass metrical, and filling gas. Although the recommendations presented remarkable improvements in the target values, results were constrained by lack of climate and orientation. Hiyama and Wen (2015) proposed a method for reducing the number of required simulations for optimising window geometries and electric energy usage by creating response surfaces between those targets and likening DF and cDA. The outside view has also been considered, for example, in the early design phases of adaptive façades (Kasinalis et al., 2014) and automated shading control (Tzempelikos & Shen, 2013); however, these only considered quantitative measures (i.e. achieving maximised WWR). In summary, there is a real need to incorporate view performance into simulation tools, particularly to optimise window system designs in office environments.

13.6 Research gap

It is often acknowledged that the process of designing modern buildings tends to require consideration of many different trade-off factors. Whilst there is an expectation that buildings should provide comfort and support the well-being of users, buildings must also perform against a raft of sustainably issues throughout their lifecycle, including the minimisation of energy use in addition to compliance with prescribed building regulations and requirements.

Architects and decision-makers, therefore, need decision-making tools to help evaluate and balancing competing factors. Window design is a particularly complex optimisation task due to its contribution to building energy performance, daylighting, and QV (particularly in office spaces); however, as presented so far, QV is notably absent in this decision-making process. The challenge from literature on comprehensive optimisation approach faces several challenges, not least is the lack of a method for properly evaluating QV, along with the interoperability of assessment tools (Pour Rahimian et al., 2019). This chapter, therefore, presents a framework for assessing the QV in an office environment, along with an approach for considering three main factors of window design.

13.7 Research methodology

This study applies a multi-objective optimisation method for maximising energy performance, daylighting, and the quality of view for an office environment, across different window system scenarios. The following sections outline the research methodology, optimisation algorithm, and tools used.

13.7.1 Multi-objective optimisation

Multi-objective optimisation differs from a single objective enhancement primarily in its increased complexity (Seyedzadeh & Pour Rahimian, 2021e). This is a direct result of the complicated nature of simultaneously satisfying several goals, often with competing outcomes. In order to accurately optimise multiple objectives, a set of circumstances that define optimal solutions must be defined, and a Pareto frontier generated. Within this set of circumstances, all points within this set (also called non-dominated or feasible solutions) are logically valid and result in various values of the objectives. Generally, in most applications, including building design, only one best solution is required by decision-makers.

The criteria of selecting the final point from the non-dominated points differs for each application. The representations for describing different objectives under the investigation are often related to the maximum or minimum functions; however, the two extreme points can be transferred to each other, by the following formula (Seyedzadeh & Pour Rahimian, 2021a):

$$\max\{f(x)\} \Leftrightarrow \min\{-f(x)\} \tag{3}$$

Then, mathematically a multi-objective optimisation problem can be expressed as:
    Minimise:

$$F(\vec{x}) = \left[f_1(\vec{x}), f_2(\vec{x}), \ldots, f_m(\vec{x}) \right]^T \tag{4}$$

Subject to

$$g(\vec{x}) \le 0$$

$$h(\vec{x}) = 0$$

where:

$$y = [y_1, y_2, ..., y_m]^T \in \psi$$

Here m is the number of objective functions, which is three in the case of the problem investigated in this study. Φ is the search space with n dimensions and identified by upper and low bounds of the decision variable x_i $(i = 1, 2, .., n)$.

$$x^{max} = \left[x_1^{max}, x_2^{max}, ..., x_n^{max} \right]^T \tag{5}$$

$$x^{min} = \left[x_1^{min}, x_2^{min}, ..., x_n^{min} \right]^T \tag{6}$$

Ψ is the m-dimensional vector space of objective functions and is defined by Θ and the objective function $f(x)$. $g_j(\bar{x}) \leq 0 (j = 1, 2, .., p)$ and $h(\bar{x}) = 0 (j = 1, 2, .., q)$ denotes p and q, which are respectively the number of inequality and equality constraints. If both p and q are equal to zero, then the problem is simplified as an unconstrained optimisation problem. Pareto solutions incorporate a vector of an ideal solution and a vector of dominated solutions, determining the upper and the lower bounds of optimal solutions. An ideal or utopia point is a hypothetical concept with reference to a perfect target in which each objective is optimised without paying attention to the satisfaction of the others.

Multi-objective optimisation algorithms attempt to generate solutions that are as close to the Pareto optimal front with a possible uniform distribution (Figure 13.1). When non-dominated solutions are identified, decision-makers can choose from this set a final resolution according to the particular problem or personal preference. For this study, a hypervolume-based evolutionary optimisation (HypE) algorithm was utilised due to its effectiveness compared to other multi-objective optimisation techniques (Bader & Zitzler, 2011). HypE is an

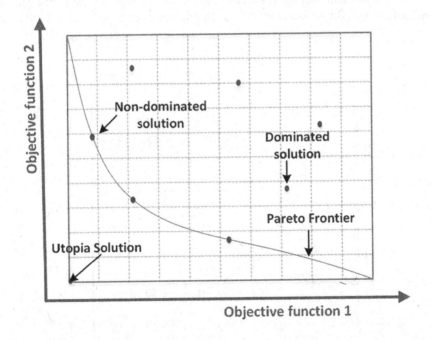

Figure 13.1 Example Pareto frontier multi-objective optimisation

evolutionary multi-objective algorithm that features hypervolume indicator (Beume et al., 2009), non-dominated sorting, and a fast search method based on Monte Carlo simulation (Bader et al., 2010).

The optimisation method used Octopus (Neira, 2019), a plug-in of Grasshopper (Domínguez-Amarillo et al., 2018), which is used for creating models for energy performance simulation analysis. The hypervolume indicator of a point set is determined as the volume of the region dominated by the point set and bounded by a reference point. Hence, it is crucial to carefully define this reference point. If this point is too close to the Pareto front, it will invariably cause incomplete cover of the non-dominated set. If it is too far from the Pareto front, it will lead to low accuracy in Monte Carlo simulation. The HypE algorithm developed in Octopus uses a dynamic reference point based on normalisation and slight changes in objective values.

13.7.2 Case study application

The reference office room for the case study was adapted from a standardised specification defined in a previous research (Reinhart et al., 2013), which consists of single-zone working space, with dimensions of 3.9m × 8.5m × 2.8m (Figure 13.2). The room is located in the middle floor of a multi-storey building, with an approximate area of 2300m². It is enclosed by other office rooms, except for the façade that is faced to the south.

The façade was considered diabatic and the rest of the wall as adiabatic. The input parameters for this case study are summarised in Table 13.1.

The Heating, Ventilation, and Air Conditioning (HVAC) system for such a building is considered packaged rooftop variable air volume (VAV), with reheat based on ASHRAE 90.1 (ASHRAE, 2007) recommendations. In this system, fan control is VAV, the cooling type is chilled water, and heating type is hot-water fossil fuel (natural gas) boiler. The annual mean Coefficient of Performance (COP) for this system is considered 3.02 for the cooling

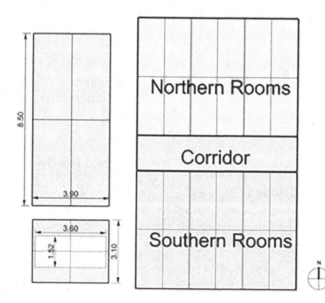

Figure 13.2 Research model (including multiple stacked reference offices)

Table 13.1 Input parameters of the research base model

Parameter	Value
interior wall thickness	0.15 m
floor to floor distance	3.10 m
occupied period	8 am to 6 pm
heating and cooling setpoints	20 and 26°C
heating and cooling Setback	15 and 30°C
peak occupant load	7.38 m²/ppl
lighting power density per area	10.1 W/m²
peak plug loads*	8 W/m²
infiltration rate per area**	0.0006 m³/s-m²
ventilation per area	0.00045 m³/s-m²

*: one Energy Star–rated LCD monitor and laptop per occupant present
**: according to ASHRAE recommendation for Leaky building and poor construction details in the research context

Table 13.2 Characteristics of construction materials

Material Name	Roughness	Conductivity	Density	Specific heat	Thermal emittance	Solar absorptance	Visible absorptance
		W/m.k	kg/m³	J/kg-K			
Mortar	Medium rough	1.0	1800	1840	0.9	0.6	0.6
Hollow brick	Medium rough	0.5	1300	840	0.9	0.7	0.7
Plaster	Smooth	0.4	900	1100	0.7	0.6	0.6
EPS	Medium rough	–	15	1340	0.9	0.6	0.6

system, 0.8 for the natural gas boiler, and 1.0 for conventional electric resistance for > 70kW and < 223kW cooling capacity based on ASHRAE/USGBC/ANSI 189.1 (ASHRAE, 2011).

The walls are finished in white plaster, the floor is covered with grey tiles, and ceilings are white. Exterior walls layers are (out to in) mortar 0.03 m, hollow brick 0.15 m, expanded polystyrene (EPS) 0.10 m, and plaster 0.03 m. Other materials characteristics are shown in Table 13.2. These materials are considered common construction materials in buildings in Tehran's official buildings.

The total U-value for the exterior wall is 0.329 W/m²K, and the reflectance is 50% inside and 35% outside. The interior wall reflectance is considered 50%. The reflectance of the ceiling and floor is 80% and 20%, respectively. All surfaces (except on one façade) were considered adiabatic. Glazing consists of double clear glass with air in the middle based on ASHRAE 169 (ASHRAE, 2013) for cities in climate zone 3B. The SHGC, U-value, and VT for such glazing was 0.25, 0.65 W/m²K, and 0.45, respectively.

Description of the climate regions

Geographically, B category (dry) in the Köppen climate classification accounts for 82.28% of Iran. In this research, Tehran was selected as an example of this climatic range. It has a dry climate with little precipitation throughout the year. This climate receives precipitation from the inter-tropical convergence zone (ITCZ) or from mid-latitude cyclones. The weather files used in this research are from the Mehrabad International Airport, with latitude of 35.683 and longitude of 51.317, and an elevation of 1190.0 m. The file is available to download from the EnergyPlus website (Najafi & Shariff, 2011). The climate parameters are summarised in Table 13.3.

Table 13.3 Tehran climate parameters (data sourced from weather file provided by the EnergyPlus database for Tehran)

Weather Data	Unit	Hourly			Average monthly	
		Average	Max	Min	Max	Min
Dry-bulb temperature	C	17.27	40	−5	30.07	3.88
Relative humidity	%	40.57	99	3	62.99	21.92
Dew point temperature	C	1.61	18.5	20.0	6.78	−3.5
Wind speed	m/s	2.71	16.3	0	4.25	1.67
Direct normal radiation	Wh/m^2	206.98	775	0	299.97	120.21
Diffuse horizontal radiation	Wh/m^2	121.15	540	0	177.11	64.73
Global horizontal radiation	Wh/m^2	244.25	1069	0	364.24	117.26
Horizontal infrared radiation	Wh/m^2	340.58	489.0	229.0	409.04	274.93
Total sky cover	tenth	4.44	9.0	0.0	4.60	4.24
Barometric pressure	Pa	87943.21	98300.0	86900.0	88416.26	87419.58

13.8 Optimisation objectives and simulation tools

The objective functions for the window system design problem are building energy loads for lighting, daylighting, and view to the outside. For the simulation of the office room, the 3D graphics software used Rhinoceros (Akhmetkal et al., 2015) and the Grasshopper plug-in was employed to control the parameters. In this respect, parametric models are useful for design exploration in complex and dynamic design settings (Dino, 2012), which are window location and dimensions in this study.

To quantify and evaluate annual daylighting performance, sDA and ASE metrics were utilised. These indices are somewhat contradictory to each other, and it is not therefore possible to calculate one metric for representing daylight. The Energy Use Intensity (EUI) metric was used to assess electricity usage, which represents the office energy consumption as a function of its conditioned floor area. So, EUI in this study was determined as the sum of normalised heating, cooling, electric equipment, and electric lighting load in a year (Kwh/m^2/y).

The view to the outside was assessed using the proposed QV metric discussed earlier, and the optimisation process considered four different functions. Daylight and energy metrics were calculated using Grasshopper plug-ins, namely Ladybug and Honeybee (Crawley et al., 2001). These simulation tools use EnergyPlus (Engstrom, 2016) and Open Studio (Lotte et al., 2018) engines for energy simulations. To simulate the integrated daylight and energy simulation, the lighting schedule was updated according to annual daylight luminance. This schedule was imported into the energy model to incorporate the electrical lighting energy requirement differences due to daylighting. For the calculation of ASE, an extra algorithm was developed in Grasshopper, to use the EnergyPlus weather file of Tehran and determine the direct illuminance in the horizontal plane (which is recorded at the end of each hour). The average illuminance for each hour is then calculated. In the next phase, sun vectors were plotted for the hours, which is more than 1000 lux. Using these sun vectors, the hours in which the sunlight hits the test surface (similar to the one used in daylighting analysis) are then simulated. With the number of hours of direct sunlight received by each of the test points in the test surface, the portion of the space below 250 annual hours is then calculated. Given the need to include view performance, this study is the first of its type to incorporate this into a framework for quantification (elaborated in the next subsection).

The following optimisation objective function extended fitness functions introduced (Konis et al., 2016) and applied (Toutou, 2018), using the weighting method to accurately find the optimum solution in Pareto front solutions.

$$FF_i = \left(sDA_i - sDA_{min}\right)C_1 - \left(ASE_i - ASE_{min}\right)C_2 - \left(EUI_i - EUI_{min}\right)C_3$$
$$+ \left(QV_i - QV_{min}\right)C_4 \tag{7}$$

Where:

i = result of iteration

Min = minimum value of optimisation set

Max = maximum value of optimisation set

$$C_1 = \frac{100}{sDA_{max} - sDA_{min}} \tag{8}$$

$$C_2 = \frac{100}{ASE_{max} - ASE_{min}} \tag{9}$$

$$C_3 = \frac{100}{EUI_{max} - EUI_{min}} \tag{10}$$

$$C_4 = \frac{100}{QV_{max} - QV_{min}} \tag{11}$$

Here, min and max values present each objective's minimum and maximum values that appeared in the solutions generated by the optimisation algorithm. The fitness function was calculated for all Pareto front solutions, which results in diversity in EUI, daylight, and view values. It should be noted that equation 7 is different from the weighting method that converts multi-objective optimisation to single-objective one. The difference is that the latter algorithm only optimises one function, but here, the first four different objectives are optimised, and then the fitness function is calculated over Pareto front to rank the solutions to determine the best one (Marler & Arora, 2010).

13.9 Metrics for the outside view

To simulate QV in this research, a Python-based (Zejnilovic & Husukic, 2018) plug-in for Grasshopper was used. The developed plug-in is able to evaluate and visualise five view types based on observer positions throughout the space. To define the viewer positions, a user-defined grid on a view analysis surface was constructed.

The evaluated view types are view access, view angles, VF, and view depth. With the results of these evaluations for the viewpoints, view quality can be assessed in two steps as per the LEED approach. For N points on the view analysis surface, in the first step, view access is determined. In the second step, other view types are evaluated for each point passing the last step threshold. Finally, the viewpoints, which passed the first step and also two out of three other view type thresholds, are considered as points with the QV. Figure 13.3 shows the QV assessment process for each point (i) on the view analysis surface. Therefore, the QV value in this research is the percentage of viewpoints that pass these two steps (j).

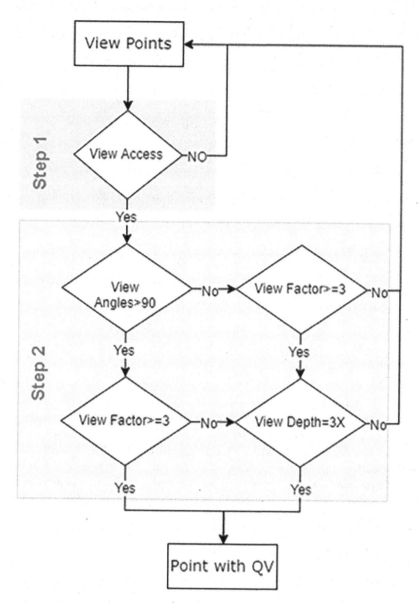

Figure 13.3 QV assessment flowchart for viewpoints (N is the number of viewpoints)

Each view type must pass its own threshold, where the minimum acceptable rate for view access is 75% of all viewpoints. To pass the vertical and horizontal view angle evaluation, it should be more than 90 degrees for both angles. To achieve the VF of 3 or greater, both horizontal and vertical view angles should be more than 11 degrees. To pass the effective view depth evaluation, viewpoints should be located in a specific area near the window. In this area, the maximum distance of the viewpoints from the window can be three times the window head height. These thresholds are based on LEED v.4 parameters to evaluate the outdoor view.

The view analysis surface was located at eye height of a seated user and is shown with the blue line in perspective and also in room section (Figure 13.4), where the view access is the percentage of the 360-degree horizontal view band visible from each viewpoint in the room. This view type acknowledges the finding that a view to the outdoor is a highly valued quality of a window (Reinhart et al., 2006) and demonstrates the amount of regularly occupied spaces that have a direct line of sight to the outside.

View angles can be affected by the viewer's location, eye height, and also the size and location of the window on the façade, which affects users' judgments of minimum acceptable window size (Ne'Eman & Hopkinson, 1970). In this study, the eye height of the viewer was set at 1.2 m above the floor as a seated observer (U.S. Department of Energy, Human factors/ergonomics handbook for the design for ease of maintenance, 2001), and horizontal and vertical angles for each point were defined (Figure 13.4).

The VF was based on a technical report of Windows and Offices Report (Heschong & Saxena, 2003), which focussed on productivity in interior environments. The report presented the results of a statistical study into the relationship between the indoor office environment and worker performance. Having a high VF is strongly and positively correlated to having a 'large size window view', 'interesting', and/or a 'relaxing' view. The VF for each viewpoint is rated from 0 to 5 based on both view angles (Table 13.4). The minimum value of both vertical angles (α) and horizontal angles (β) for each viewpoint was considered to modify VF; however, it was assumed that the views assessed in this study had no vegetation content.

The largest VF rating of 5 was defined as filling the seated observer's field of view. This was empirically determined to be at least a 50-degree viewing angle for both the vertical and horizontal view angles, which almost completely filled the visual field. Each subsequent lower category represents about one half of the previous angle. Research shows that a view depth

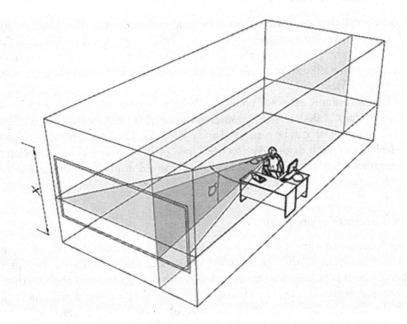

Figure 13.4 Studied parameters in QV evaluation (n.b. α and β are the vertical and horizontal view angles respectively; the minimum α and β values are used to calculate the VF rating for each viewpoint; X is used to determine the view depth)

Table 13.4 View factor rating based on both view angles (n.b. the smaller of the vertical and horizontal view angle values were used to define the VF for each viewpoint)

View Factor	Degrees
1	1–5
2	5–11
3	11–20
4	20–50
5	50–90

(or user's distance from the window) affects a viewer's judgment about the minimum width of an acceptable window (Ne'Eman, 1974), their satisfaction with the view (Markus, 1967), and also comfort perception (Aries et al., 2010). In this research, to have a QV according to LEED v4, viewer position should be located at a distance of three times the head height of the window. This distance actually defines the acceptable view depth in a room. From Figure 13.4, X is the distance to head height of window and was used to determine the view depth. View access is therefore equal to 3X. Moreover, a grid size of 0.75 m was overlaid onto the view analysis surface, where 119 points were defined. Each view type evaluation and QV result (the area enclosed with a black line on view access analysis figure) was visualised in the upper part of each model diagram.

13.10 Optimisation criteria

In the next step, the algorithm of location and dimensions of the window opening is defined, considering the limitations of the sill height and head height of 0.76 m and 2.28 m, respectively. The window parameters applied in this research were window width and height, window sill and head height, and also distance of window edges from façade edges (Figure 13.5).

These window parameters could change with an increase of 20 cm (Table 13.5). With this method, more than 12,000 diverse openings in the range from 0.2 m to 1.52 m in height and 0.2 m to 3.60 m in width can be generated and evaluated. The largest possible window is also used as the base research model. An electrical load of 20.88 kWh/m^2 was considered for the equipment in all models; however, this load was removed from energy usage objective function in order to achieve a better comparison.

13.11 Optimisation procedure

The procedure of optimising the window system followed a structured workflow (Figure 13.6). When the simulation of the optimisation objectives was completed, the optimisation phase begins. In this phase, two processes are performed. In the first process, the research parameters are evaluated, by a back-and-forth process, ensuring a reasonable trade-off between the objectives (using Octopus software) – see Octopus settings (Table 13.6). By producing different generations, the Pareto front is drawn and optimal solutions are extracted.

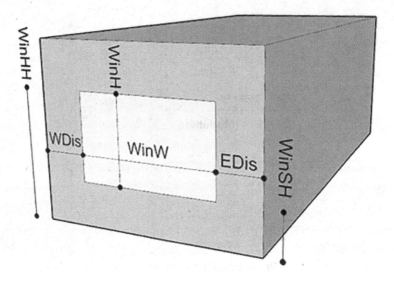

Figure 13.5 Core research parameters

Table 13.5 Research parameters: detail and range

Parameters	Description	Base model value	Lower limit	Upper limit	Increment
		metres			
WinW	Window width	3.60	0.6	3.60	0.20
WinH	Window height	1.52	0.20	1.52	0.20
WinHH	Window head height	2.28	0.95	2.28	0.20
WinSH	Window sill height	0.76	0.76	1.71	0.20
WDis	Distance from the western edge of the window to the western wall	0.15	0.15	1.95	0.20
EDis	Distance from the eastern edge of the window to the eastern wall	0.15	0.15	1.95	0.20

In the second process, the analysis of the research findings then determines the optimum result. This is done by importing the parameters and objectives into Microsoft Excel, extracting the smallest values of objectives along with the weighting fitness function. Using the fitness function (presented earlier), the function value for each model is then calculated, including the absolute optimum genome, as well as the fittest genomes in each objective. This optimisation process can take a considerable amount of time, as during this optimisation process, about 2900 simulations were separately conducted for each objective, and 28 generations of genomes were produced. About 1500 generated models were duplicated and removed, leaving 1400 unique genomes to be investigated.

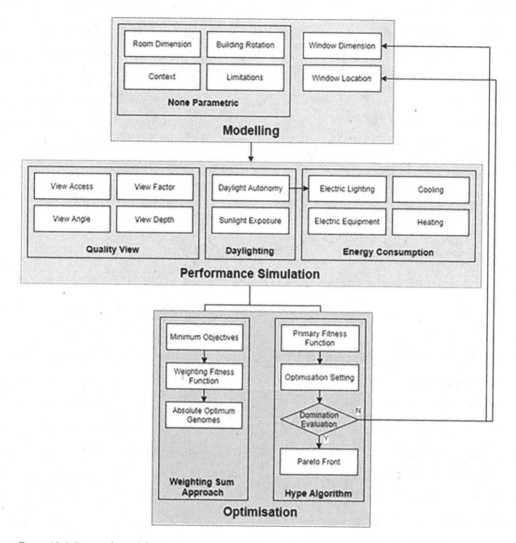

Figure 13.6 Research workflow

Table 13.6 Optimisation settings in Octopus

Elitism	Mutation Probability	Mutation Rate	Crossover Rate	Population Size	Max Generation	Max Evaluation Time
0.5	0.1	0.5	0.8	100	None	None

13.12 Optimisation deviation

A similar method of calculating the standard deviation was used to check the adequacy of the number of generations and iterations. In statistics, the standard deviation is a measure of the amount of variation or dispersion of a set of values. A low standard deviation indicates that the values tend to be close to the mean of the set, while a high standard deviation indicates that the values are spread out over a wider range.

In the proposed method of this research, the mean value was replaced by the value of the optimal absolute genome, and the distance between the parameters/objectives and the optimal genome for each iteration is then calculated.

The formula for the optimisation deviation is:

$$\sigma_{opt} = \sqrt{\frac{\sum_{i=1}^{N}\left(x_i - x_{opt}\right)^2}{N}} \tag{12}$$

Where $\{x_1, x_2, \ldots, x_n\}$ are the values of the parameters/objectives in a generation, x_{opt} is the value of the same parameter/objective in the absolute optimum genome, and N is the number of population in a studied generation.

The lower these values are, the greater the convergence of the optimisation process. The results of these optimisation deviations (σ_{opt}) for each parameter or objective are shown in Table 13.7. The highlighted numbers represent the least amount of distance from the optimal genome in different generations.

Table 13.7 Results of σ_{opt} for parameters and objectives of 28 generations (n.b. highlighted cells have the least values in each column)

Optimisation Generations	WinW	WinH	WWR	WinHH	WinSH	WDis	EDis	ASE	sDA	QV	EUI
G01	2.080		36.303	0.511	0.412	0.693	0.804	34.751	40.106	62.500	11.901
G02	2.046	0.866	35.134	0.420	0.446	0.760	0.810	33.475	38.630	59.493	11.534
G03	1.880	0.859	34.169	0.431	0.427	0.650	0.786	32.761	35.966	56.292	11.479
G04	1.774	0.807	32.940	0.334	0.473	0.714	0.736	32.383	33.788	53.233	11.317
G05	1.626	0.773	31.564	0.253	0.521	0.574	0.780	31.131	30.747	50.023	11.084
G06	1.618	0.771	31.821	0.249	0.522	0.630	0.820	31.727	31.326	50.107	11.149
G07	1.498	0.783	31.149	0.325	0.458	0.540	0.718	30.030	30.159	49.392	11.094
G08	1.310	0.768	29.140	0.285	0.483	0.542	0.536	29.408	26.108	47.393	10.323
G09	1.390	0.690	28.622	0.203	0.486	0.562	0.636	28.458	25.125	42.645	9.971
G10	1.516	0.693	29.722	0.260	0.433	0.518	0.742	27.988	27.579	44.661	10.572
G11	1.486	0.682	29.122	0.200	0.483	0.506	0.616	29.366	26.520	44.737	9.975
G12	1.344	0.695	28.166	0.182	0.513	0.530	0.602	28.307	23.999	43.939	9.815
G13	1.424	0.764	30.193	0.262	0.502	0.496	0.700	30.551	27.134	46.527	10.094
G14	1.362	0.610	27.223	0.173	0.437	0.478	0.716	26.946	23.654	39.292	9.271
G15	1.650	0.768	31.215	0.239	0.528	0.494	0.720	30.979	30.747	48.569	10.505
G16	1.492	0.747	29.992	0.217	0.530	0.500	0.696	30.231	27.453	45.997	9.963
G17	1.464	0.705	28.483	0.247	0.458	0.496	0.636	28.021	26.284	43.762	9.785
G18	1.464	0.617	28.003	0.160	0.458	0.586	0.602	27.601	24.806	39.149	9.699
G19	1.432	0.752	29.854	0.241	0.511	0.536	0.608	29.946	27.402	46.939	10.088
G20	1.426	0.598	27.214	0.198	0.401	0.518	0.716	26.475	23.873	38.351	9.478
G21	1.456	0.657	28.593	0.245	0.412	0.512	0.672	27.719	25.411	41.535	10.146
G22	1.452	0.680	28.546	0.184	0.496	0.540	0.688	26.853	26.015	42.275	9.653
G23	1.260	0.583	26.431	0.175	0.408	0.570	0.506	25.156	22.536	38.519	9.287
G24	1.514	0.610	28.638	0.182	0.427	0.568	0.606	27.996	25.402	40.233	9.857
G25	1.440	0.726	29.624	0.207	0.519	0.582	0.578	30.458	26.025	44.132	9.906
G26	1.376	0.673	27.996	0.228	0.445	0.574	0.602	26.820	25.470	41.653	9.683
G27	1.448	0.777	30.954	0.255	0.522	0.504	0.616	30.299	28.982	49.140	10.482
G28	1.456	0.709	29.411	0.236	0.473	0.460	0.600	28.299	27.696	46.653	9.929

The lowest values tended to be obtained in the 23rd generation and the WDis values were carefully examined in the 28th generation. The trend lines (Figure 13.7) shows the distance between the WDis in each population, highlighting the optimal WDis had generally decreased compared to previous generations. Trend lines of other parameters/objectives also behaved similarly to WDis. The lowest value for research optimisation deviation was reported in the 28th generation; therefore, the results of subsequent generations have not been reported.

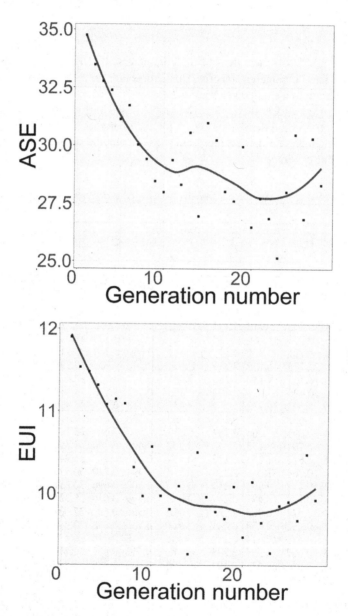

Figure 13.7 Results of σ_{opt} for research objectives in 28 generations (n.b. downward trend line shows the convergence of the optimisation process)

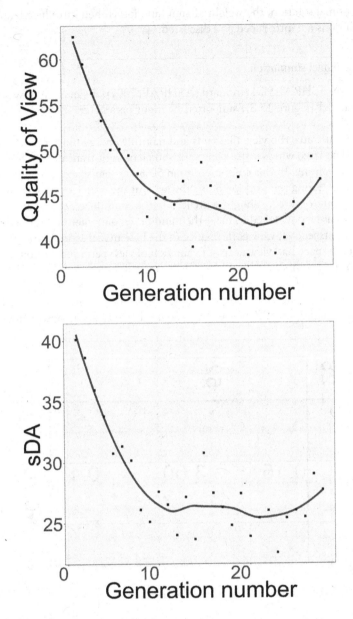

Figure 13.7 (Continued)

13.13 Results

This section presents the results of the parametric optimisation of daylight, energy, and view to outdoor. Initially, the base model for the optimisation algorithm was defined and the model simulation results were derived for the purpose of comparison, with the recommended solutions from the proposed algorithm. However, the HypE algorithm mutates the model to create new generations, which are then evaluated and returned to the optimisation algorithm.

To find the optimal solution, the weighted-sum function is then introduced, and ultimately, optimised solutions are introduced and discussed.

13.13.1 Base model simulation

According to the ASHRAE 90.1 standard (ASHRAE, 2007), the maximum WWR is selected for the base model (Figure 13.8), supported by input parameters and calculated objectives (Table 13.8).

Figure 13.9 illustrates the view diagrams and monthly energy usage chart generated from Grasshopper. The large window size allows for sufficient light transmission to the interior to provide natural lighting. In this case, more than 50% of room space received a mean of 300 lux, with daylight during the working hours throughout the year. Conversely, the aggregation of annual 250 hours of direct sunlight with 40% in the south face and the absence of a shading device, the base model failed to meet the minimum requirement prescribed by LEED v4.

From a view perspective, view performance in the base model satisfied the proposed conditions, as all of the space had view access. Evaluation of view performance conditions revealed that 82% of the grid points were located at the distance of three times the window head height (view depth), 15% had a viewing angle of more than 90 degrees, and the VF for 80% of them was more than thrée. Therefore, considering the majority of points that passed two out of three secondary conditions, 80% of the office room achieved satisfactory quality performance.

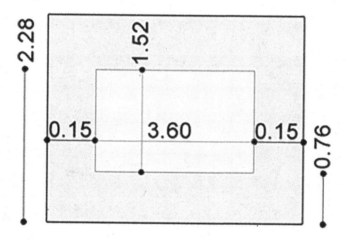

Figure 13.8 Base model elevation and parameter values

Table 13.8 Parameters and objective functions for the base case

Model in optimisation process	Parameters						WWR	Objectives					Fitness Function
	WinW	WinH	WinHH	WinSH	WDis	EDis		ASE	sDA	DA$_{ave}$	QV	EUI	
	m	m	m	m	m	m	%	%	%	%	%	kWh/m²	
Base	3.60	1.52	2.28	0.76	0.15	0.15	45.26	38.66	52.94	52.01	80.67	81.27	82.26

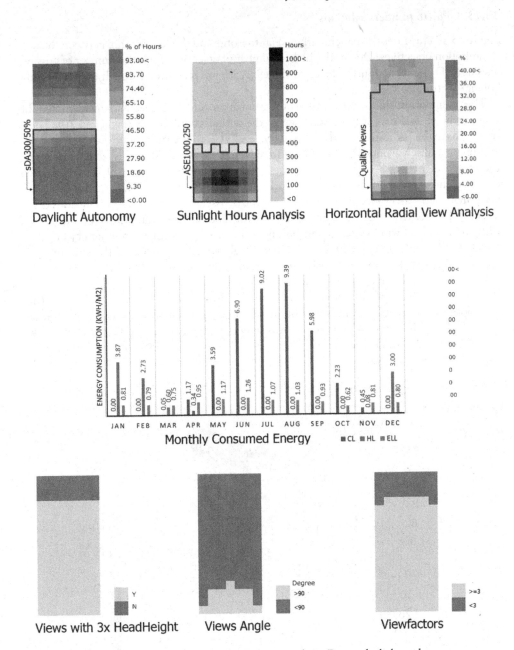

Figure 13.9 Base case simulation results (n.b. Up: view analysis; Down: daylight and energy consumption analysis)

The energy consumption for the base model obtained 81.27 kWh/m². Figure 13.9 (down right) highlights the monthly distribution of EUI separated by the cooling, heating, and electricity load. The cooling, heating, and electrical lighting loads account for 48%, 13%, and 14% of total EUI, respectively. As this case study was based in Tehran (with a warm and dry climate), the priority of design strategies would be to reduce the cooling load to prevent excessive sunlight penetration in summer.

13.13.2 Pareto frontiers solutions

Figure 13.10 illustrates the benchmarking Pareto fronts and the relationship between the four optimisation objectives. Due to the higher number of objectives than the number of possible axes in a 3D chart, the fourth objective values (ASE) are presented through the blue-yellow colour spectrum.

The optimisation algorithm and weighting sum approach (described earlier) identified the yellow spheres as the most fitted solutions in the optimisation process (Figure 13.10). These spheres have the highest QV and sDA and the least EUI, where their value of ASE is undesirably high. Whilst the optimisation was set to reduce the value of ASE, this led to considering the same weight for all objectives; however, other solutions (with an ASE less than 10%) could be chosen in the proposed framework, but models with low WWR often have lower performance in other objectives. In addition, in the base model, no devices were considered to control glare. It seems that using glare-control tools such as shadings can reduce ASE, with an increase in window size along with performance improvement in other objectives.

Due to difficulties in explaining four-dimensional models, the values of the objectives are presented through four two-dimensional representations (Figure 13.11).

Figure 13.11 presents the base model (shown as a green square), compared to other models, where the top-ten genomes based on the fitness function are highlighted. Since the coordination centre point represents the best theoretical solution, the best solutions should be near the origin of the coordinates. From Figure 13.11, each diagram has its own Pareto front and optimum genomes. It is notable with respect to Pareto front, although the

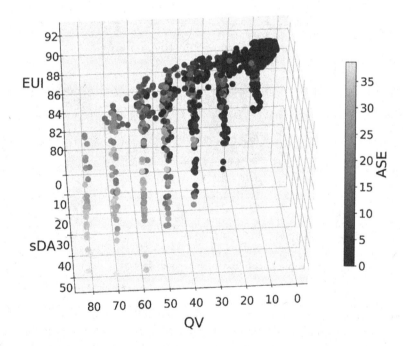

Figure 13.10 4D representation of simulated models from the optimisation process (n.b. ASE values illustrated by colour; yellow spheres are elite solutions)

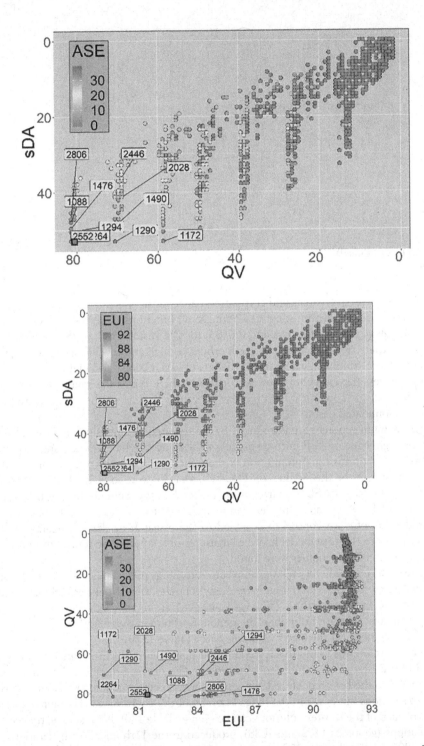

Figure 13.11 Genomes produced in Octopus (n.b. normally the most optimum genomes are closest to the origin coordinates; square point with a black border is the base model)

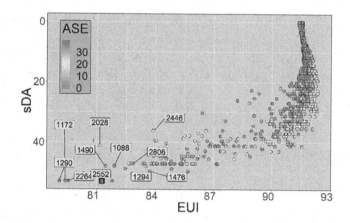

Figure 13.11 (Continued)

whole space of objective values and correlation between them is not the main focus of this research; the MOO algorithm sought to optimise all functions. Pareto solutions are found within intervals of (40, 53) for sDA and (60, 83) for QV. Inside this box, there is no correlation between these two objectives, nor with others, as high and low values of EUI and ASE exist in this space. Similarly, where inside the box of QV (60, 83) and EUI (79, 84), it is not possible to regress the variables; likewise, for sDA vs EUI. So, as the solutions get closer to the Pareto front, the conflict of the objective functions increases, where the algorithm has to find non-dominated cases.

As the optimum genomes are introduced by the research optimisation algorithm, the reason for the high value of ASE in these models is the same weight of the research objectives. Formula 7 also confirms the optimisation of these models. In the top ten optimised models, the algebraic sum of the three values of sDA, EUI, and QV were so high relative to the other models that the effect of ASE had virtually disappeared. In these circumstances, having models with acceptable ASE values, models can be restricted to ASE less than 10%, or through the introduction of glare-control tools added to the research models configuration. From Figure 13.11, since the base model had the largest possible window, no further improvements in sDA or QV in other genomes were found.

Because of the contradictory relationship between energy performance and view to outside (including energy and daylight), there were many genomes that evidenced high performance in one objective, but not in others. When genome generations are produced, each generation contains genomes that are fitter than the genomes of previous generations. Moreover, the fitness function value for the base model was 82.26 (Table 13.8).

The best solution was found in the 22nd generation, where its fitness function value represented the highest in this study, equal to 96.48. The fitness function value improved by 3% compared to the base model. In later generations, no better genome was found, and the density of solutions increased in the range of origin of the coordinates. The number of generations of each of the top-ten solutions can be seen in Table 13.9. For instance, the fifth and sixth solutions (genomes 1172 and 1088), produced in the 11th and 10th generations, and genome 2806 produced in the 28th generation. Finding genomes with better fitness function continued until the 28th generation, and no better genome was found in later generations.

Table 13.9 Optimised genomes: parameters and objective functions (n.b. the best single objective genomes among the top ten are highlighted and bordered with solid lines)

Model	Generation	Ranking	Parameters						WWR	Objectives				Fitness Function
			WinW	*WinH*	*WinHH*	*WinSH*	*WDis*	*EDis*		*ASE*	*sDA*	*QV*	*EUI*	
			m	m	m	m	m	m	%	%	%	%	kWh/m²	
2264	22	1	3.40	1.52	2.28	0.76	0.35	0.15	42.75	38.66	52.94	81.51	79.52	96.48
1290	13	2	3.60	1.33	2.28	0.95	0.15	0.15	39.60	38.66	52.94	70.59	79.05	86.32
2552	25	3	3.20	1.52	2.28	0.76	0.15	0.55	40.23	36.13	52.94	81.51	81.79	85.91
2028	21	4	2.00	1.33	2.28	0.95	0.55	1.35	22.00	26.05	41.18	68.91	81.15	78.81
1172	12	5	3.60	1.14	2.28	1.14	0.15	0.15	33.95	35.29	52.94	58.82	79.36	77.93
1088	11	6	3.00	1.52	2.28	0.76	0.35	0.55	37.72	35.29	47.90	81.51	81.90	77.71
1294	13	6	3.00	1.52	2.28	0.76	0.55	0.35	37.72	35.29	47.90	81.51	81.90	77.71
2806	28	7	2.80	1.52	2.28	0.76	0.15	0.95	35.20	34.45	47.06	81.51	82.84	71.28
2446	24	8	1.60	1.33	2.28	0.95	0.75	1.55	17.60	17.65	36.13	68.91	83.88	70.49
1476	14	9	3.00	1.52	2.28	0.76	0.15	0.75	37.72	34.45	49.58	81.51	83.70	69.54
1490	15	10	3.00	1.33	2.28	0.95	0.35	0.15	33.00	34.45	47.90	69.75	81.46	68.46

Table 13.10 Objectives, information, and range

	ASE (%)	sDA (%)	QV (%)	EUI (kWh/m²)
Min	0.84	6.72	5.88	79.05
Max	38.66	52.94	81.51	92.34
Average	5.80	17.13	24.14	90.91
Standard Deviation	9.89	13.91	21.19	1.96

As such, although the optimisation for the further six generations continued, just the first 28 generations' results are reported in this research.

In all, 2910 generated models were produced, where the minimum values, maximum values, average values, and standard deviations of the total objectives can be seen in Table 13.10. A lower standard deviation in EUI and ASE indicates that these objective values tend to be close to the average of their sets in the optimisation process, whereas a higher standard deviation in sDA and QV indicates that their values are spread out over a wider range. So, the fitness difference among the optimised objectives and the rest of the models are suitably large.

13.13.3 Optimised solutions

Table 13.9 presented the ten optimum solutions obtained as the Pareto front. Among all solutions, only the first three genomes had a fitness function value greater than the base model. Because of the antagonistic relation among sDA and EUI or QV and EUI, many solutions were found that consumed more energy, while receiving less daylight. These solutions shared similar parameters, as demonstrated in Table 13.11.

Table 13.11 Genomes with highest EUI and lowest sDA and QV: parameters, objectives, and fitness values

Model	Parameters						WWR	Objectives				Fitness Function
	WinW	WinH	WinHH	WinSH	WDis	EDis		ASE	sDA	QV	EUI	
	m	m	m	m	m	m	%	%	%	%	kWh/m²	
1524	3.00	0.57	1.33	0.76	0.35	0.55	14.14	21.01	23.53	27.73	92.34	−77.27
1771	3.00	0.57	1.33	0.76	0.15	0.75	14.14	21.01	22.69	26.05	92.33	−80.90
873	2.60	0.57	1.33	0.76	0.15	1.15	12.26	17.65	19.33	26.05	92.32	−78.49
1599	2.00	0.76	1.52	0.76	0.55	1.35	12.57	15.97	19.33	37.82	92.31	−59.28
1006	3.00	0.76	1.52	0.76	0.55	0.35	18.86	28.57	29.41	39.50	92.30	−70.69

Table 13.12 Energy consumption comparison: selected models vs. base model

Model	Cooling Load	Heating Load	Electrical Lighting Load	Electrical Equipment Load	Total Thermal Load
	kWh/m²				
1524	35.29 ↓	10.79 ↑	25.39 ↑	20.88	92.34 ↑
Base model	38.78	10.63	10.99	20.88	81.26
2264	36.37 ↑	10.86 ↑	11.41 ↑	20.88	79.52 ↓
2264-mirrored	39.20 ↑	10.60 ↓	12.37 ↑	20.88	83.04 ↑
1290	35.92 ↓	11.11 ↑	11.14 ↑	20.88	79.05 ↓

These similarities include low WWR, an aspect ratio of 3:1 to 6:1, and the lowest possible position of the window. Due to low WinHH in these models, on average only 22% of the room received enough light in 50% of the occupied hours in a year, and more than 20% of the room received more than 1,000 lux over 250 hours per annum. In addition, the low position of the window (in spite of the high aspect ratio) restricted the view to the sky compared to the base model. To investigate the cause of these increases in EUI, model 1524 was compared to the base model. Table 13.12 shows a significant increase of more than 130% in the electric lighting energy consumption, whilst its cooling decreased by 9%, and heating load increased by 1.5%. Thereby, the lack of sufficient daylight in this model led to an increase in electrical lighting load and higher total energy consumption.

Coincidentally, there were solutions that posed higher sDA and QV values with less energy usage, such as solutions 2264, 1290, and 2552 (Table 13.12). The main difference between models with high EUI and lower sDA and QV is the larger WWR, about 40% on average. Thus, WinHH is maximum so that the daylight penetration and view to the outside is increased, where the WinSH is enlarged to 1.5 m to reduce the possibility of sunlight penetration.

13.14 Selection of the best model

This section discusses the selection of the best models, based on the different objectives, presenting a comparison between the Pareto front solutions with the base research model to show how different parameters can affect building performance – particularly the effect this can have on improving the efficiency of the genomes.

13.14.1 *Best optimum genomes*

The best optimum solution at the Pareto front is defined as a balance between daylight performance, energy consumption, and QV. The best fitness value achieved for the model was 2264 (Figure 13.12), by 96.47. The energy consumption of the model was 79.52 Kwh/m² with 2% reduction (compared with the base model). The annual energy distribution is shown for the best optimum model, on the right-hand side of Figure 13.13. Because of the lower WWR than the base model, the penetration of the solar radiation and the adequate distribution of natural light decreased, which results in a 2.16 and 4% increase in heating and electrical light loads and a 6.62% reduction in cooling load (Table 13.12); however, due to the nature of Tehran's hot climate, the total energy consumption decreased.

The daylight simulation of the best optimum genome in the bottom left-hand side of Figure 13.13 shows that the sDA and ASE are the same, as the base model with values of 52.94 and 38.66, respectively. For a better comparison, DA_{ave} was calculated, given that the sDA is a spatial index and does not give information about light distribution (Table 13.13). In the optimum model, DA_{ave} fell 2% (50.88%), which means that, in less than 2% of occupancy hours, there was less daylighting. Due to the smaller WWR, this decrease seems natural and in line with an increase in electric lighting consumption.

In the best optimum genome, 81% of the points in the office room have quality views, which is 1% higher than the base model. By comparing the analysis of the three conditions of QV in Figure 13.12 and Figure 13.9, reveals that in both models there are limited points that had a view angles of 90 degrees or more. In addition, the points located in the distance of three times, the WHH were the same. Therefore, an effective factor, which led to a difference in QV between the two models is the VF.

To analyse the effect of EDis and WDis, on the best optimum genome, a new mirrored model was generated, with the reverse dimensions. This model was not among the Pareto front solutions; consequently, this was individually simulated. The result of this simulation (Table 13.13) indicates that these two models are similar in terms of ASE, sDA, and QV,

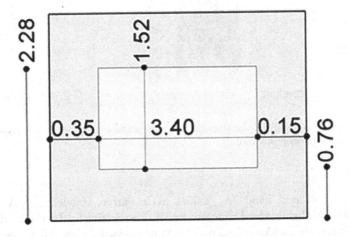

Figure 13.12 Absolute optimum genome elevation and parameter values

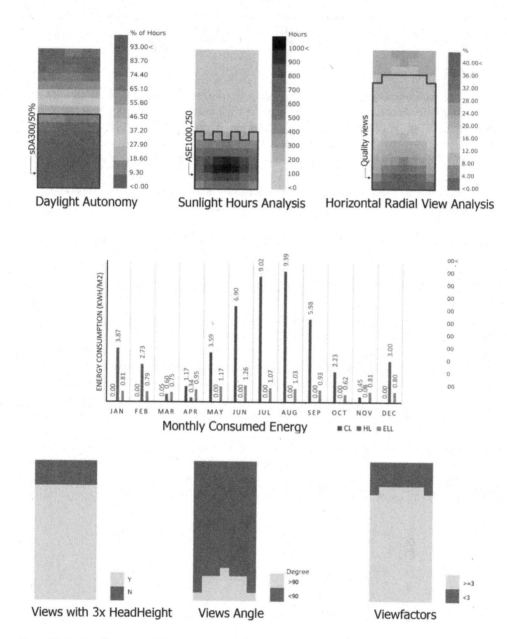

Figure 13.13 Simulation results for the absolute optimum genome (n.b. Up: view analysis. Down: Day light and energy analysis)

and they deviate only in EUI and DA_{ave} values. In the mirrored model, the DA_{ave} dropped by 0.08%, yet its electric lighting load increased by 8%. The heating load in this model decreased by 2% and the cooling load increased by 7%. As mentioned earlier, due to the greater sensitivity of the cooling load, the total energy consumption increased by about 4%. For this

Table 13.13 Parameters and objectives of the absolute optimum genome

Model in the optimisation process	Total optimum genome	Parameters						WWR	Objectives					Fitness Function
		WinW	WinH	WinHH	WinSH	WDis	EDis		ASE	sDA	DA$_{ave}$	QV	EUI	
		m	m	m	m	m	m	%	%	%	%	%	kWh/m²	
2264	1	3.40	1.52	2.28	0.76	0.35	0.15	42.75	38.66	52.94	50.88	81.51	79.52	96.47
2264-mirrored	–					0.15	0.35				50.80		83.04	69.94

reason, it was concluded that in a window of a specific size, it is better to place a window near the eastern wall in this research climate.

13.14.2 The Energy Optimum Model

Figure 13.14 shows the optimum energy consumption genome (model 1290). This model holds the same sDA and ASE as the best optimum model and the base model; however, in terms of DA$_{avg}$ value, it stands between those two models Table 13.14). Since the WWR is smaller, the DA$_{avg}$ is 0.83% lower than the base model. Nevertheless, as a result of the lower EDis as well as its symmetry, light distribution is more uniform and about 0.3% higher than the best optimum model.

QV in the optimum energy model was 2.7% and 4.8% less than the base and best optimum models, due to the reduced WinSH and VF. The energy optimum genome shows 7.4% and 1.2% decline in cooling load, in comparison to the base and the best models, respectively, and 4.5% and 2.3% increase in heating loads. Furthermore, as a result of the reduction in WWR (by 5.7% and 2.7%, against the base and the best optimal models), the possibility of solar radiation penetration in the summer and winter declined, which justifies those changes in thermal loads. At the same time, the interstitial state of this model in daylight, relative to the base and the best optimal models, had a consistent effect on the lighting energy consumption, and its electrical load was 1.4% higher than the base and 2.4% lower than the best optimal model (Table 13.14). Although this model had a higher heating load than the other two, and the corresponding electrical lighting load was larger than the base model, the impact of the cooling load on the energy consumption was higher; accordingly, the total energy consumption was lower.

13.14.3 The Best sDA genome

Amongst the 2910 models investigated in the optimisation process, four models were found fitter than the others (Figure 13.15). These four models had equally the highest sDA values (52.9%). The average sDA was 29.5% with a standard deviation of 13.91 (Table 13.10), so the sDA values were spread out over an acceptable range. Parameters and objectives of some randomly generated models are presented in Table 13.15.

In these models, possessing different parameters, almost 53% of the room had enough daylight at all occupied hours. These genomes were among the top five solutions, with the highest fitness function value. The models in which the sDA value was maximised represented the three highest possible values for WinH and WinW. DA$_{avg}$ was calculated for comparison

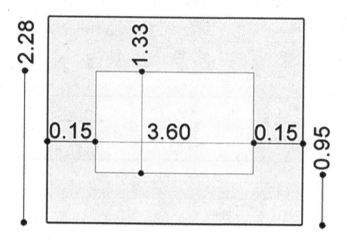

Figure 13.14 Energy optimum genome elevation and parameter values

Table 13.14 Parameters and objectives of the energy optimum genome

Model in the optimisation process	Total optimum genome	Parameters						WWR	Objectives					Fitness Function
		WinW	WinH	WinHH	WinSH	WDis	EDis		ASE	sDA	DA_{ave}	QV	EUI	
		m	*m*	*m*	*m*	*m*	*m*	%	%	%	%	%	*kWh/m²*	
1290	2	3.60	1.33	2.28	0.95	0.15	0.15	39.60	38.66	52.94	51.18	70.59	79.05	86.32

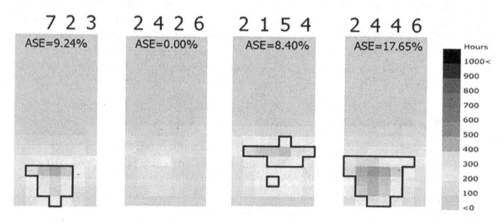

Figure 13.15 Daylight autonomy in optimum genomes of SDA (n.b. model 1290 had the highest DA_{avg}. Areas with $sDA_{300/50\%}$ demonstrated by black lines; the back of the room was brighter in model 1290 than others).

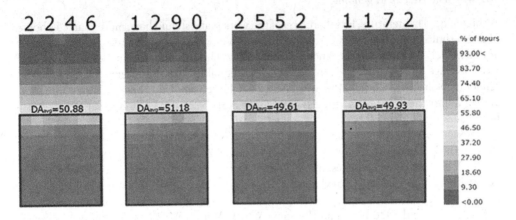

Figure 13.15 (Continued)

Table 13.15 Parameters and objectives of sDA optimum genome

Total optimum genome Model in the optimisation process		Parameters						WWR	Objectives					Fitness Function
		WinW	*WinH*	*WinHH*	*WinSH*	*WDis*	*EDis*		*ASE*	*sDA*	*DA$_{ave}$*	*QV*	*EUI*	
		m	m	m	m	m	m	%	%	%	%	%	kWh/m²	
2264	1	3.40	1.52	2.28	0.76	0.35	0.15	42.75	38.66	52.94	50.88	81.51	79.52	96.47
1290	2	3.60	1.33	2.28	0.95	0.15	0.15	39.60	38.66	52.94	51.18	70.59	79.05	86.32
2552	3	3.20	1.52	2.28	0.76	0.15	0.55	40.23	36.13	52.94	49.61	81.51	81.79	85.91
1172	5	3.60	1.14	2.28	1.14	0.15	0.15	33.95	35.29	52.94	49.93	58.82	79.36	77.93

of the options. As shown in Table 13.15, model 1290, which was the second optimum model, had the highest average annual gain (more than 300 lux), which is 0.83% lower than the base model.

It is perceived that the high WinW and the symmetry of the model result in gaining more daylight than the models 2264 and 2552. In addition, model 1290 had a better fitness function than model 1172, due to the lower WinSH and the higher WinH. Figure 13.15 illustrates the higher brightness of the back position of the room, in this model (in comparison with other models). In the last row of the grid, there are points that in 9–27% of the year had daylight more than 300 lux. Such points are rare in the best optimal model and do not exist in models 2552 and 1172.

None of the four fitting models having the best sDA as well as the base model (with the largest possible WWR and sDA = 52.94%) met the LEED needs. As the sDA has a spatial definition, one of the ways to qualify for its LEED point was reducing the depth of the room. Calculations show that if the simulated room had a maximum of 5.8 m depth, the maximum points in daylighting could be attainable.

13.14.4 The Best ASE genome

As highlighted in literature, less than 10% of the work plane should have more than 250 hours of direct sunlight in excess of 1000 lux in order to meet LEED requirements. Among the generated models, 1033 models met these specifications. This was reduced further as 107 models (with average WWR of 1.22%), in which sDA is equal to zero, were removed. Among the remaining solutions, model 2154 with ASE of 8.40 achieved the highest fitness function value. Whilst there were ten models with higher ASE (9.24), these had lower fitness function values.

Among these top ten models, model 723 scored the highest fitness function of 10.79. Among the models with the lowest ASE (zero), model 2426 had the highest fitness function of 28.99. In the top ten genomes, model 2446 with the fitness function of 70.49 had the best ASE with 17.65% (Table 13.16). Figure 13.16 presents a comparison among the ASE of all these models.

Models 723 and 2426 both had the same WinW, WinH, WWR, and location; however, higher WinSH and lower WinHH in model 723 caused a reduction in ASE. It was observed that the high WinW of model 2154, the decrease in the WinSH, and the increase in the

Table 13.16 Parameters and objectives of ASE optimum genomes

Model in the optimisation process	Total optimum genome	Parameters						WWR	Objectives					Fitness Function
		WinW	WinH	WinHH	WinSH	WDis	EDis		ASE	sDA	DA_ave	QV	EUI	
		m	m	m	m	m	m	%	%	%	%	%	kWh/m2	
2446	8	1.60	1.33	2.28	0.95	0.75	1.55	17.60	17.65	36.13	NA	68.91	83.88	70.49
2154	1250	3.20	0.76	2.28	1.52	0.55	0.15	20.12	8.40	44.54	NA	38.66	85.53	59.96
2426	22	1.40	0.95	2.28	1.33	0.95	1.55	11.00	0.00	29.41	27.43	49.58	90.55	28.99
723	92	1.40	1.14	2.09	0.95	1.15	1.35	13.20	9.24	30.25	NA	57.14	91.26	10.79

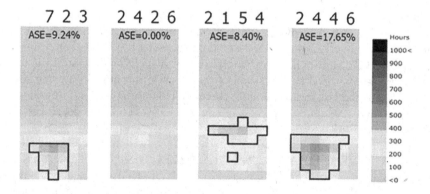

Figure 13.16 Sunlight hour analysis of the best ASE optimum genomes (n.b. areas with an ASE more than 250 hours with more than1000 lux annually demonstrated by black lines)

WinHH of model 2446 led to an increase in ASE. Comparison of model 2154 (the ASE optimum model with ASE less than 10% and the highest fitness function) with other models identified that if a high WinW is recommended (to improve other objective function values), an increased WinSH can reduce solar penetration and glare.

There are 707 models with zero ASE among the investigated models. The standard deviation for WinW, WinH, WinHH, and WinSH parameters of these models was 0.8, 0.3, 0.4, and 0.3, respectively. Given that the lower the standard deviation, the less dispersion of data, this indicates the high impact of architectural design parameters on the loss of glare.

Figure 13.17 presents the optimum ASE genome. Model 2426 had the rank of 22nd in the list of the best genome. Although this genome was less likely to have glare than the absolute optimal genome, it still did not receive enough daylight in 24% of the room, and 32% of the rooms had no QV (a major change occurred in VF), as well as 14% more energy consumption (Figure 13.18). Due to the smaller WWR, the cooling load decreased by 12% and 6%, relative to the basic model and the absolute optimum.

In winter, due to the decrease in received solar radiation, the heating load increased by 11% and 9% (relative to the basic model and absolute optimum). In this model, due to reduced daylight, the electrical lighting load doubled (116% and 112%, respectively, relative to the absolute optimal model and base model).

13.14.5 *The best quality views genomes*

In all solutions, 14 models had the highest QV of 82%, which were more than 1% higher than the base model. To find the QV optimum model, more accurate simulations were performed by having the grid size and doubling the grid points (Table 13.17). The comparison shows that among parameters, just the view angles were effective, and the other parameters had constant values. Although the view angles did not influence the final outcome of the QV in this study, it helps to determine the optimum QV genome among those 14 models.

The view angles are the same for some models due to the fixed WinW, WinH, and changing WDis and EDis. This change, regardless of the view content, often has an impact on the view to outdoor; 13% of the room in models 2264 and 2552 had more than 90-degree horizontal and vertical view angles. These models were the first and third best ranks of the optimum QV solutions.

In all of these 14 solutions, the WinW, WinHH, and WinSH were fixed, and just WinW, WDis, and EDis changed. The WinSH and WinHH had minimum and maximum possible

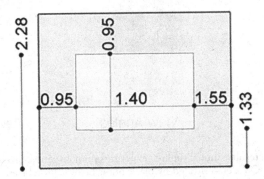

Figure 13.17 Annual sunlight exposure optimum genome elevation and parameter values

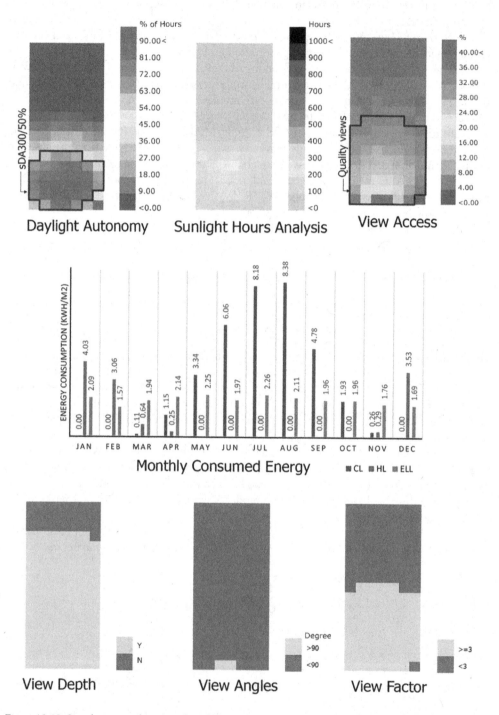

Figure 13.18 Simulation results: annual sunlight exposure optimum genome (n.b. Above: view analysis, areas with QV demonstrated by black lines; Below: daylight and energy analysis, areas with sDA$_{300/50\%}$ demonstrated by black lines; there were no areas in "sunlight hours analysis", which had sunlight more than 250 hours.

Table 13.17 View simulation results with a smaller grid size

Model in the optimisation process	View depth	View angles	VF >= 3	View access	QV
2264	79.41	13.73	80.98	100.00	79.41
2552	79.41	11.76	80.78	100.00	79.41
Base	79.41	15.10	80.78	100.00	79.41
2806	79.41	9.02	80.39	100.00	78.82
1476	79.41	10.59	80.39	100.00	79.02
1060	79.41	7.84	80.20	100.00	78.82

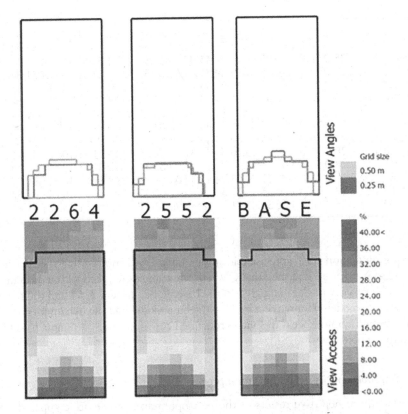

Figure 13.19 View comparison among the optimum view genomes and the base model (n.b. Down: view access, points with quality views are demonstrated by a black line; Middle: view angles analysis with default grid size; Up: view angles analysis with a smaller grid size.

values, respectively; therefore, the WinH had the highest possible value. By decreasing the WinW, the number of points in a room, with view angles of more than 90 degrees, decreased.

Although model 2264 differed in WinW from the model 2552, the number of points obtained with the view angles of more than 90 degrees was the same for both models. To better understand the reasoning behind this, the grid size was reduced to half the size (0.25 m), and the simulation was rerun for these models. With a smaller grid and more precise simulation, model 2264 had approximately 2% more points with view angles of more than 90 degrees, and approximately 1% had a VF value greater than 3. As shown in Figure 13.19, the points at the room end in model 2552 had less view access.

Table 13.18 Parameters and objectives of view optimum genomes

Model in optimisation process	Total optimum genome	Parameters						WWR	Objectives								EUI	Fitness Function
		WinW	WinH	WinHH	WinSH	WDis	EDis		Daylight		Quality View						
									ASE	sDA	View depth = 3X	View angles > 90	View factor > = 3	View access	Majority		
		m	m	m	m	m	m	%	%	%	%	%	%	%	%	kWh/m²	
2264	1	3.40	1.52	2.28	0.76	0.35	0.15	42.75	38.66	52.94	82.35	12.61	81.51	100	81.51	79.52	96.48
2552	3	3.20	1.52	2.28	0.76	0.15	0.55	40.23	36.13	52.94	82.35	12.61	81.51	100	81.51	81.79	85.91
2806	8	2.80	1.52	2.28	0.76	0.15	0.95	35.20	34.45	47.06	82.35	9.24	81.51	100	81.51	82.84	71.28
1476	10	3.00	1.52	2.28	0.76	0.15	0.75	37.72	34.45	49.58	82.35	10.92	81.51	100	81.51	83.70	69.54
1060	17	2.60	1.52	2.28	0.76	0.35	0.95	32.69	34.45	47.06	82.35	8.40	81.51	100	81.51	83.89	63.36
1787	17	2.20	1.52	2.28	0.76	0.75	0.95	27.66	31.93	45.38	82.35	5.88	81.51	100	81.51	84.57	61.62
2859	18	2.20	1.52	2.28	0.76	0.95	0.75	27.66	31.93	45.38	82.35	5.88	81.51	100	81.51	84.57	61.62
1975	25	2.60	1.52	2.28	0.76	0.95	0.35	32.69	34.45	47.06	82.35	8.40	81.51	100	81.51	84.39	59.59
276	27	2.80	1.52	2.28	0.76	0.75	0.35	35.20	34.45	47.06	82.35	9.24	81.51	100	81.51	84.51	58.70
1201	28	2.80	1.52	2.28	0.76	0.35	0.75	35.20	34.45	47.06	82.35	9.24	81.51	100	81.51	84.51	58.70
1422	62	2.40	1.52	2.28	0.76	0.55	0.95	30.17	34.45	47.06	82.35	6.72	81.51	100	81.51	85.87	48.44
591	87	2.60	1.52	2.28	0.76	0.55	0.75	32.69	34.45	47.06	82.35	8.40	81.51	100	81.51	87.21	38.37
1294	6	3.00	1.52	2.28	0.76	0.35	0.55	37.72	34.45	50.42	82.35	10.92	81.51	100	81.51	82.80	77.90
1088	7	3.00	1.52	2.28	0.76	0.55	0.35	37.72	35.29	47.90	82.35	10.92	81.51	100	81.51	81.90	77.72

Results of simulation with a precise grid size in Table 13.18 showed that the two optimum models of 2264 and 2552 had the same QV. The view angles reduced with a drop in WinW, resulting in lower view angles compared to the base model. Model 2264 had the highest VF among the three models. Consequently, since the base model had the maximum possible QV, during the optimisation process this value remained constant and it attempts to stabilise its other improved objectives.

13.14.6 Improvement in the optimisation process

In this section, the objective results of the best optimum genome are compared with the objective results of the base model and the objective average values. This demonstrates the improvement in the optimisation process and the process of meeting the requirement of LEED rating system.

The average QV of the studied models was 24.14% (Table 13.10). This objective was significantly improved in the best optimum genome (81.51%). Thus, the QV in the optimal model was more than 2.3 times higher than the average, which also met the view requirements specified in LEED. The QV in the optimum genome increased by about 1% in comparison to the base model.

Energy consumption of the studied models has the lowest standard deviation among other objectives (Table 13.10). This represents that the difference between the minimum and maximum EUI was 13 kWh/m²/y. Although, with the average EUI of 90.91 kWh/m²/y, these low differences could not make a difference in meeting the LEED requirements in the optimisation process; moreover, the EUI in the optimum genome decreased by about 2% in comparison to the base model.

Although both daylight indices rose significantly compared to an average of themselves, there was no significant change in these indices compared to the base model. sDA in the optimum genome (52.94%) increased more than two times the sDA average value (17.13%); therefore, the chance of getting the minimum score of daylighting in LEED increased. The ASE value in the best optimum genome (38.66%) was equal to the highest found values, more than five times higher than the average.

Given that the ASE average value (5.80%) was less than 10%, it was apparent that the optimisation process reduced the amount of this objective to meet the needs of the LEED; however, since the weight of each objective in the method used was the same for all the objectives of this study, the best optimum genome with the highest fitness function had no acceptable ASE value. Given this, in such projects, in order to find models that can meet LEED requirements in the daylighting alongside the other objectives, the weight of the ASE in Formula 7 should be increased.

Because of the relatively large size of the window used in the base model, the optimisation process did not change the LEED credits of this specific model; however, the model's ability to maintain the positive aspects of a larger window while reducing the negative impacts are clearly observable. With no change in daylighting, the best optimum genome had a better-quality view and consumed less energy than the base model. Certainly, a significant impact on the result of the optimisation could be achieved by choosing a smaller base model.

13.15 Conclusion

The method presented in this chapter addresses the theoretical and methodological gap in configuring window systems for the design of office buildings. The design of window systems directly affects aspects of a building's quality and performance, including building energy performance, daylight gain, and visual comfort. From literature, a reliable method for assessment of the Quality of View (QV) has not yet been introduced. The evidence on the optimisation of window system design presented here focused on energy performance and daylight aspects of windows. The few studies that concentrated on outside view tended to present qualitative methods to maximise WWR. Given this, the development of a framework was presented for the quantitative evaluation of QV, which considered several factors, including view access, view angles, and view factor.

This framework provides a foundation method for evaluating the QV for office environments. A multi-objective optimisation method was introduced in this research design as a decision-making tool to assist building designers and engineers to achieve an optimised window system. This tool considered three optimisation objectives, namely: energy usage, received daylight, and quality of the view. The aim was to minimise energy usage while maximising received daylight and quality of the view. The optimisation framework utilised Rhinoceros software for modelling the building, Grasshopper environment with two plug-ins (Ladybug and Honeybee) to calculate the energy usage and daylight, and a Python-based plug-in to evaluate the QV. The optimisation algorithm (HypE), which is a hypervolume-based evolutionary algorithm, was applied using Octopus (a Grasshopper plug-in).

The optimisation operation was applied to a case study, which was a reference room in an office building based on a room specification defined for standardising dynamic evaluations in office environments. After validating the energy and daylighting simulation results with that specified room (Reinhart reference room), the location used was Tehran (Iran). The results of the proposed optimisation procedure were provided as a set of optimum solutions. Further to the obtained values for the optimisation objectives, a fitness function was introduced to better evaluate the performance of each configuration by weighting different objectives. The solution packages provide decision-makers with potential options based on their performance expectations. The optimisation results showed that the suggested research framework can improve the daylighting and QV results in comparison to these average optimisation values. As for the EUI, this improvement was about 12%. The optimum solutions proved the efficiency of the optimisation framework in finding the best window system to satisfy all studied objectives. It was revealed that it is possible to provide a satisfactory QV performance, for more than 80% of the reference room points, while minimising the energy usage and maximising daylight.

Other findings of note included that in order to meet official standards set for office buildings relating to view, that room geometry must be within the set of optimisation variables. The low complexity of the reference building may introduce other difficulties in satisfying the predefined standards. Similarly, additional building elements such as blinds, shades, and glazing play a crucial role in controlling solar radiation, light amounts, and glare. These must therefore be considered.

Finally, there is an opportunity for further studies to investigate the impact of shading and light control strategies on the studied optimisation objectives. Adding shading devices to façades provides an opportunity for simultaneous reduction in radiation transmission and heat gain energies, as well as a higher capability for controlling day-lighting (Hashemi, 2014); however, the effects of such devices (e.g. blinds, screens, and shutters to the glazed surfaces, as well as implementing control strategies) on the quality of the view to outside have been under-researched. It is proffered that future research should explore variations in the external environment of the building to understand the resulting impact on the triple analysis of daylight, visibility, and energy. The view indices examined in this research design were internal, and external indices such as view content and external distance were not considered in this case study. Analysis of additional contexts are also recommended.

Acknowledgments

The authors acknowledge the contributions of Peiman Pilechiha, Mohammadjavad Mahdavinejad, and Phillippa Carnemolla to the underpinning research published in this chapter.

References

Ahmad, M., Bontemps, A., Sallée, H., & Quenard, D. (2006). Experimental investigation and computer simulation of thermal behaviour of wallboards containing a phase change material. *Energy and Buildings*, 38(4), 357–366. https://doi.org/10.1016/j.enbuild.2005.07.008

Akhmetkal, M., Maulken, A., Roman, P., Irina, S., & Aruzhan, B. (2015). Great silk road on the territory of Kazakhstan: From past to future. *Journal of Resources and Ecology*, 6(2), 114–118. https://doi.org/10.5814/j.issn.1674-764X.2015.02.009

Al Horr, Y., Arif, M., Kaushik, A., Mazroei, A., Katafygiotou, M., & Elsarrag, E. (2016). Occupant productivity and office indoor environment quality: A review of the literature. *Building and Environment*, 105, 369–389. https://doi.org/10.1016/j.buildenv.2016.06.001

Alrubaih, M. S., Zain, M. F. M., Alghoul, M. A., Ibrahim, N. L. N., Shameri, M. A., & Elayeb, O. (2013). Research and development on aspects of daylighting fundamentals. *Renewable and Sustainable Energy Reviews*, 21, 494–505. https://doi.org/10.1016/j.rser.2012.12.057

Altomonte, S. (2008). Daylight for energy savings and psycho-physiological well-being in sustainable built environments. *Journal of Sustainable Development, 1*(3), 3–16. https://doi.org/10.5539/jsd.v1n3p3

Andersen, M. (2015). Unweaving the human response in daylighting design. *Building and Environment, 91*, 101–117. https://doi.org/10.1016/j.buildenv.2015.03.014

Aries, M. B. C. (2005). *Human lighting demands: Healthy lighting in an office environment.* (Doctor of Philosophy, Technische Universiteit Eindhoven, Faculteit Bouwkunde, Eindhoven). Retrieved October 27, 2019, from https://research.tue.nl/files/1883894/200512454.pdf

Aries, M. B. C., Veitch, J. A., & Newsham, G. R. (2007). *Physical and psychological discomfort in the office environment.* Paper presented at the Symposium of the Dutch Light and Health Research Foundation. Retrieved October 20, 2019, from https://research.tue.nl/en/publications/physical-and-psychological-discomfort-in-the-office-environment

Aries, M. B. C., Veitch, J. A., & Newsham, G. R. (2010). Windows, view, and office characteristics predict physical and psychological discomfort. *Journal of Environmental Psychology, 30*(4), 533–541. https://doi.org/10.1016/j.jenvp.2009.12.004

Asadi, E., Da Silva, M. G., Antunes, C. H., & Dias, L. (2012). Multi-objective optimization for building retrofit strategies: A model and an application. *Energy and Buildings, 44*, 81–87. https://doi.org/10.1016/j.enbuild.2011.10.016

ASHRAE. (2007). *Energy standard for buildings except low-rise residential buildings* Vol. 90, pp. 186). ASHRAE. Retrieved October 28, 2019.

ASHRAE. (2011). *ANSI/ASHRAE/USGBC/IES standard 189.1–2011, standard for the design of high-performance green buildings* (p. 114). ASHRAE. Retrieved October 27, 2019, from https://bcgreencare.ca/system/files/resource-files/PREVIEW_Standard%20for%20the%20Design%20of%20High-Performance%20Green%20Buildings.pdf

ASHRAE. (2013). *ANSI/ASHRAE standard 169–2013, climatic data for building design.* Tullie Circle. Retrieved October 28, 2019.

Bader, J., Deb, K., & Zitzler, E. (2010). Faster hypervolume-based search using Monte Carlo sampling. In M. Ehrgott, B. Naujoks, T. J. Stewart, & J. Wallenius (Eds.), *Multiple criteria decision making for sustainable energy and transportation systems* (pp. 313–326). Springer. Retrieved October 28, 2019.

Bader, J., & Zitzler, E. (2011). HypE: An algorithm for fast hypervolume-based many-objective optimization. *Evolutionary Computation, 19*(1), 45–76. https://doi.org/10.1162/EVCO_a_00009

Beume, N., Fonseca, C. M., Lopez-Ibanez, M., Paquete, L., & Vahrenhold, J. (2009). On the complexity of computing the hypervolume indicator. *IEEE Transactions on Evolutionary Computation, 13*(5), 1075–1082. https://doi.org/10.1109/TEVC.2009.2015575

Beute, F., & de Kort, Y. A. W. (2014). Salutogenic effects of the environment: Review of health protective effects of nature and daylight. *Applied Psychology: Health and Well-Being, 6*(1), 67–95. https://doi.org/10.1111/aphw.12019

Boyce, P., Hunter, C., & Howlett, O. (2003). *The benefits of daylight through windows* (p. 88). Rensselaer Polytechnic Institute, Troy. Retrieved October 28, 2019, from www.lrc.rpi.edu/programs/daylight-dividends/pdf/DaylightBenefits.pdf

Carlucci, S., Cattarin, G., Causone, F., & Pagliano, L. (2015a). Multi-objective optimization of a nearly zero-energy building based on thermal and visual discomfort minimization using a non-dominated sorting genetic algorithm (NSGA-II). *Energy and Buildings, 104*, 378–394. https://doi.org/10.1016/j.enbuild.2015.06.064

Carlucci, S., Causone, F., De Rosa, F., & Pagliano, L. (2015b). A review of indices for assessing visual comfort with a view to their use in optimization processes to support building integrated design. *Renewable and Sustainable Energy Reviews, 47*, 1016–1033. https://doi.org/10.1016/j.rser.2015.03.062

Chartered Institution. (2014). *Daylighting – a guide for designers: Lighting for the Built Environment.* Chartered Institution of Building Services Engineers. ISBN: 9781906846480

Crawley, D. B., Lawrie, L. K., Winkelmann, F. C., Buhl, W. F., Huang, Y. J., Pedersen, C. O., Strand, R. K., Liesen, R. J., Fisher, D. E., Witte, M. J., & Glazer, J. (2001). EnergyPlus: Creating a new-generation building energy simulation program. *Energy and Buildings, 33*, 319–331. https://doi.org/10.1016/S0378-7788(00)00114-6

De Rosa, M., Bianco, V., Scarpa, F., & Tagliafico, L. A. (2014). Heating and cooling building energy demand evaluation; a simplified model and a modified degree days approach. *Applied Energy, 128*, 217–229. https://doi.org/10.1016/j.apenergy.2014.04.067

Diakaki, C., Grigoroudis, E., & Kolokotsa, D. (2008). Towards a multi-objective optimization approach for improving energy efficiency in buildings. *Energy and Buildings, 40*(9), 1747–1754. https://doi.org/10.1016/j.enbuild.2008.03.002

Dino, I. (2012). Creative design exploration by parametric generative systems in architecture. *Middle East Technical University Journal of the Faculty of Architecture*(1), 207–224. https://doi.org/10.4305/METU.JFA.2012.1.12

Dino Ipek, G., & Üçoluk, G. (2017). Multiobjective design optimization of building space layout, energy, and daylighting performance. *Journal of Computing in Civil Engineering, 31*(5). https://doi.org/10.1061/(ASCE)CP.1943-5487.0000669

Djongyang, N., Tchinda, R., & Njomo, D. (2010). Thermal comfort: A review paper. *Renewable and Sustainable Energy Reviews, 14*(9), 2626–2640. https://doi.org/10.1016/j.rser.2010.07.040

Domínguez-Amarillo, S., Fernandez-Aguera, J., & Fernandez-Aguera, P. (2018). Teaching innovation and the use of social networks in architecture: Learning building services design for smart and energy efficient buildings. *Archnet-IJAR, 12*(1), 367–375. https://doi.org/10.26687/archnet-ijar.v12i1.1298

Dubois, M.-C. (2003). Shading devices and daylight quality: An evaluation based on simple performance indicators. *Lighting Research & Technology, 35*(1), 61–74. https://doi.org/10.1191/1477153503li062oa

Edwards, L., & Torcellini, P. (2002). *Literature review of the effects of natural light on building occupants* (p. 54). Office of Scientific and Technical Information (OSTI). Retrieved November 8, 2019, from www.osti.gov/servlets/purl/15000841

Energy, U. S. D. O. (2013). *US energy information administration*. US Department of Energy, United States Government Printing Office. Retrieved October 28, 2019.

Engstrom, L. (2016). *Fast style transfer in tensorflow*. https://github.com/lengstrom/fast-style-transfer

Fang, Y., & Cho, S. (2019). Design optimization of building geometry and fenestration for daylighting and energy performance. *Solar Energy, 191*, 7–18. https://doi.org/10.1016/j.solener.2019.08.039

Farley, K. M. J., & Veitch, J. A. (2001). *A room with a view: A review of the effects of windows on work and well-being* (p. 33). National Research Council Canada, Institute for Research in Construction. Retrieved October 27, 2019, from http://irc.nrc-cnrc.gc.ca/fulltext/rr/rr136/rr136.pdf

Futrell, B. J., Ozelkan, E. C., & Brentrup, D. (2015). Bi-objective optimization of building enclosure design for thermal and lighting performance. *Building and Environment, 92*, 591–602. https://doi.org/10.1016/j.buildenv.2015.03.039

Ghaffarianhoseini, A., Berardi, U., & Ghaffarianhoseini, A. (2015). Thermal performance characteristics of unshaded courtyards in hot and humid climates. *Building and Environment, 87*, 154–168. https://doi.org/10.1016/j.buildenv.2015.02.001

Ghaffarianhoseini, A., Rehman, A. U., Naismith, N., & Mehdipoor, A. (2016). *Green Building Assessment Schemes: A critical comparison among BREEAM, LEED, and Green Star NZ*. Paper presented at the International Conference on Sustainable Built Environment (SBE), Seoul, Korea. http://bimdirectory.com/Documents/Publications/2016_SELECTED_PROCEEDING_5_GREENBUILDINGGa.pdf

Halawa, E., Ghaffarianhoseini, A., Ghaffarianhoseini, A., Trombley, J., Hassan, N., Baig, M., Yusoff, S. Y., & Azzam Ismail, M. (2018). A review on energy conscious designs of building façades in hot and humid climates: Lessons for (and from) Kuala Lumpur and Darwin. *Renewable and Sustainable Energy Reviews, 82*, 2147–2161. https://doi.org/10.1016/j.rser.2017.08.061

Hashemi, A. (2014). Daylighting and solar shading performances of an innovative automated reflective louvre system. *Energy and Buildings, 82*, 607–620. https://doi.org/10.1016/j.enbuild.2014.07.086

Hee, W. J., Alghoul, M. A., Bakhtyar, B., Elayeb, O., Shameri, M. A., Alrubaih, M. S., & Sopian, K. (2015). The role of window glazing on daylighting and energy saving in buildings. *Renewable and Sustainable Energy Reviews, 42*, 323–343. https://doi.org/10.1016/j.rser.2014.09.020

Heerwagen, J. H., & Orians, G. H. (1986). Adaptations to windowlessness: A study of the use of visual decor in windowed and windowless offices. *Environment and Behavior, 18*(5), 623–639. https://doi.org/10.1177/0013916586185003

Hensen, J. L. M., & Lamberts, R. (2012). *Building performance simulation for design and operation.* Routledge. ISBN: 1134026358

Heschong, L., & Saxena, M. (2003). *Windows and offices: A study of office worker performance and the indoor environment.* California Energy Commission. Retrieved October 28, 2019, from www.h-m-g.com/downloads/Daylighting/A-9_Windows_Offices_2.6.10.pdf

Heschong, L., Wymelenberg, v. D., Andersen, M., Digert, N., Fernandes, L., Keller, A., Loveland, J., McKay, H., Mistrick, R., & Mosher, B. (2012). *Approved method: IES spatial daylight autonomy (sDA) and annual sunlight exposure (ASE)* (p. 14). IES-Illuminating Engineering Society. Retrieved October 27, 2019, from https://books.google.com/books?id=KtWbmwEACAAJ

Hiyama, K., & Wen, L. (2015). Rapid response surface creation method to optimize window geometry using dynamic daylighting simulation and energy simulation. *Energy and Buildings, 107,* 417–423. https://doi.org/10.1016/j.enbuild.2015.08.035

Iommi, M. (2019). Daylighting performances and visual comfort in Le Corbusier's architecture. The daylighting analysis of seven unrealized residential buildings. *Energy and Buildings, 184,* 242–263. https://doi.org/10.1016/j.enbuild.2018.12.014

Kasinalis, C., Loonen, R. C. G. M., Cóstola, D., & Hensen, J. L. M. (2014). Framework for assessing the performance potential of seasonally adaptable facades using multi-objective optimization. *Energy and Buildings, 79,* 106–113. https://doi.org/10.1016/j.enbuild.2014.04.045

Khatami, N., & Hashemi, A. (2017). Improving thermal comfort and indoor air quality through minimal interventions in office buildings. *Energy Procedia, 111,* 171–180. https://doi.org/10.1016/j.egypro.2017.03.019

Knecht, C. (2004). Urban nature and well-being: Some empirical support and design implications. *Berkeley Planning Journal, 17*(1). https://doi.org/10.5070/BP317111508

Ko, W. H., Brager, G., Schiavon, S., & Selkowitz, S. (2017). *Building envelope impact on human performance and well-being: Experimental study on view clarity.* UC Berkeley, Center for the Built Environment. Retrieved October 27, 2019, from https://escholarship.org/uc/item/0gj8h384

Konis, K., Gamas, A., & Kensek, K. (2016). Passive performance and building form: An optimization framework for early-stage design support. *Solar Energy, 125,* 161–179. https://doi.org/10.1016/j.solener.2015.12.020

Krüger, E. L., & Dorigo, A. L. (2008). Daylighting analysis in a public school in Curitiba, Brazil. *Renewable Energy, 33*(7), 1695–1702. https://doi.org/10.1016/j.renene.2007.09.002

Krüger, E. L., Tamura, C., & Trento, T. W. (2018). Identifying relationships between daylight variables and human preferences in a climate chamber. *Science of The Total Environment, 642,* 1292–1302. https://doi.org/10.1016/j.scitotenv.2018.06.164

Lake, I. R., Lovett, A. A., Bateman, I. J., & Langford, I. H. (1998). Modelling environmental influences on property prices in an urban environment. *Computers, Environment and Urban Systems, 22*(2), 121–136. https://doi.org/10.1016/S0198-9715(98)00012-X

Lee, J. W., Jung, H. J., Park, J. Y., Lee, J. B., & Yoon, Y. (2013). Optimization of building window system in Asian regions by analyzing solar heat gain and daylighting elements. *Renewable Energy, 50,* 522–531. https://doi.org/10.1016/j.renene.2012.07.029

Li, D. H. W., Lau, C. C. S., & Lam, J. C. (2004). Predicting daylight illuminance by computer simulation techniques. *Lighting Research & Technology, 36*(2), 113–128. https://doi.org/10.1191/1365782804li108oa

Lotte, R. G., Haala, N., Karpina, M., de Aragão, L. E. O. e. C., & Shimabukuro, Y. E. (2018). 3D façade labeling over complex scenarios: A case study using convolutional neural network and structure-from-motion. *Remote Sensing, 10,* 1435. https://doi.org/10.3390/rs10091435

Mahdavinejad, M. J., Matoor, S., Feyzmand, N., & Doroodgar, A. (2012). Horizontal distribution of illuminance with reference to window wall ratio (WWR) in office buildings in hot and dry climate, case of Iran, Tehran. *Applied Mechanics and Materials, 110–116,* 72–76. https://doi.org/10.4028/www.scientific.net/AMM.110-116.72

Mangkuto, R. A., Rohmah, M., & Asri, A. D. (2016). Design optimisation for window size, orientation, and wall reflectance with regard to various daylight metrics and lighting energy demand: A case study of buildings in the tropics. *Applied Energy*, 164, 211–219. https://doi.org/10.1016/j.apenergy.2015.11.046

Manzan, M., & Padovan, R. (2015). Multi-criteria energy and daylighting optimization for an office with fixed and moveable shading devices. *Advances in Building Energy Research*, 9(2), 238–252. https://doi.org/10.1080/17512549.2015.1014839

Mardaljevic, J. (2000). Simulation of annual daylighting profiles for internal illuminance. *International Journal of Lighting Research and Technology*, 32(3), 111–118. https://doi.org/10.1177/096032710003200302

Markus, T. A. (1967). The function of windows – A reappraisal. *Building Science*, 2(2), 97–121. https://doi.org/10.1016/0007-3628(67)90012-6

Marler, R. T., & Arora, J. S. (2010). The weighted sum method for multi-objective optimization: New insights. *Structural and Multidisciplinary Optimization*, 41(6), 853–862. https://doi.org/10.1007/s00158-009-0460-7

Mhalas, A., Kassem, M., Crosbie, T., & Dawood, N. (2013). A visual energy performance assessment and decision support tool for dwellings. *Visualization in Engineering*, 1(1), 7. https://doi.org/10.1186/2213-7459-1-7

Mitrache, A. (2012). Ornamental art and architectural decoration. *Procedia – Social and Behavioral Sciences*, 51, 567–572. https://doi.org/10.1016/j.sbspro.2012.08.207

Najafi, M., & Shariff, M. K. B. M. (2011). Factors influencing public attachment to mosques in Malaysia. *Archnet-IJAR*, 5(3), 7–24. www.scopus.com/inward/record.uri?eid=2-s2.0-84859884255&partnerID=40&md5=138738617372e1eb1d98652bd493bd79

Ne'Eman, E. (1974). Visual aspects of sunlight in buildings. *Lighting Research & Technology*, 6(3), 159–164. https://doi.org/10.1177/096032717400600304

Ne'Eman, E., & Hopkinson, R. G. (1970). Critical minimum acceptable window size: A study of window design and provision of a view. *Lighting Research & Technology*, 2(1), 17–27. https://doi.org/10.1177/14771535700020010701

Neira, J. (2019). *Geoship's bioceramic domes for the homeless are designed to last for 500 years*. Designboom.

Neto, A. H., & Fiorelli, F. v. A. S. (2008). Comparison between detailed model simulation and artificial neural network for forecasting building energy consumption. *Energy and Buildings*, 40, 2169–2176. https://doi.org/10.1016/j.enbuild.2008.06.013

Ochoa, C. E., Aries, M. B. C., van Loenen, E. J., & Hensen, J. L. M. (2012). Considerations on design optimization criteria for windows providing low energy consumption and high visual comfort. *Applied Energy*, 95, 238–245. https://doi.org/10.1016/j.apenergy.2012.02.042

Omrany, H., Ghaffarianhoseini, A., Ghaffarianhoseini, A., Raahemifar, K., & Tookey, J. (2016). Application of passive wall systems for improving the energy efficiency in buildings: A comprehensive review. *Renewable and Sustainable Energy Reviews*, 62, 1252–1269. https://doi.org/10.1016/j.rser.2016.04.010

Paule, B., Boutiller, J., Pantet, S., Sutter, Y., & Sutter, Y. (2018). *A lighting simulation tool for the new European daylighting standard*. Paper presented at the Building Simulation and Optimization 2018, Emmanuel College, University of Cambridge. Retrieved October 21, 2019, from www.ibpsa.org/proceedings/BSO2018/1A-5.pdf

Pour Rahimian, F., Chavdarova, V., Oliver, S., Chamo, F., & Potseluyko Amobi, L. (2019). OpenBIM-Tango integrated virtual showroom for offsite manufactured production of self-build housing. *Automation in Construction*, 102, 1–16. https://doi.org/10.1016/j.autcon.2019.02.009

Pour Rahimian, F., Seyedzadeh, S., Oliver, S., Rodriguez, S., & Dawood, N. (2020). On-demand monitoring of construction projects through a game-like hybrid application of BIM and machine learning. *Automation in Construction*, 110. https://doi.org/10.1016/j.autcon.2019.103012

Rahimian, F. P., Seyedzadeh, S., & Glesk, I. (2019). OCDMA-based sensor network for monitoring construction sites affected by vibrations. *Journal of Information Technology in Construction*, 24, 299–317.

Reinhart, C. F., Jakubiec, J. A., & Ibarra, D. (2013). *Definition of a reference office for standardized evaluations of dynamic façade and lighting technologies*. 13th Conference of International Building

Performance Simulation Association, pp. 3645–3652. Retrieved October 28, 2019, from www.ibpsa. org/proceedings/BS2013/p_1029.pdf

Reinhart, C. F., Mardaljevic, J., & Rogers, Z. (2006). Dynamic daylight performance metrics for sustainable building design. *LEUKOS*, *3*(1), 7–31. https://doi.org/10.1582/LEUKOS.2006.03.01.001

Reinhart, C. F., & Walkenhorst, O. (2001). Validation of dynamic RADIANCE-based daylight simulations for a test office with external blinds. *Energy and Buildings*, *33*(7), 683–697. https://doi. org/10.1016/S0378-7788(01)00058-5

Reinhart, C. F., & Weissman, D. A. (2012). The daylit area – correlating architectural student assessments with current and emerging daylight availability metrics. *Building and Environment*, *50*, 155–164. https://doi.org/10.1016/j.buildenv.2011.10.024

Rupp, R. F., Vásquez, N. G., & Lamberts, R. (2015). A review of human thermal comfort in the built environment. *Energy and Buildings*, *105*, 178–205. https://doi.org/10.1016/j.enbuild.2015.07.047

Seyedzadeh, S., Agapiou, A., Moghaddasi, M., Dado, M., & Glesk, I. (2021). WON-OCDMa system based on MW-ZCC codes for applications in optical wireless sensor networks. *Sensors*, *21*. https://doi. org/10.3390/s21020539

Seyedzadeh, S., & Pour Rahimian, F. (2021a). Building energy data-driven model improved by multi-objective optimisation. In *Data-driven modelling of non-domestic buildings energy performance* (pp. 99–109). Springer. https://doi.org/10.1007/978-3-030-64751-3_6

Seyedzadeh, S., & Pour Rahimian, F. (2021b). Building energy performance assessment methods. In *Data-driven modelling of non-domestic buildings energy performance* (pp. 13–30). Springer. https://doi. org/10.1007/978-3-030-64751-3_2

Seyedzadeh, S., & Pour Rahimian, F. (2021c). Machine learning for building energy forecasting. In *Data-driven modelling of non-domestic buildings energy performance* (pp. 41–76). Springer. https://doi. org/10.1007/978-3-030-64751-3_4

Seyedzadeh, S., & Pour Rahimian, F. (2021d). Machine learning models for prediction of building energy performance. In *Data-driven modelling of non-domestic buildings energy performance* (pp. 77–98). Springer. https://doi.org/10.1007/978-3-030-64751-3_5

Seyedzadeh, S., & Pour Rahimian, F. (2021e). Multi-objective optimisation and building retrofit planning. In *Data-driven modelling of non-domestic buildings energy performance* (pp. 31–39). Springer. https://doi.org/10.1007/978-3-030-64751-3_3

Seyedzadeh, S., Pour Rahimian, F., Oliver, S., Rodriguez, S., & Glesk, I. (2020a). Machine learning modelling for predicting non-domestic buildings energy performance: A model to support deep energy retrofit decision-making. *Applied Energy*, *279*, 115908. https://doi.org/10.1016/j.apenergy.2020.115908

Seyedzadeh, S., Pour Rahimian, F., Rastogi, P., & Glesk, I. (2019). Tuning machine learning models for prediction of building energy loads. *Sustainable Cities and Society*, *47*, 101484. https://doi. org/10.1016/j.scs.2019.101484

Seyedzadeh, S., Rahimian, F. P., Glesk, I., & Roper, M. (2018). Machine learning for estimation of building energy consumption and performance: A review. *Visualization in Engineering*, *6*(1), 5. https://doi. org/10.1186/s40327-018-0064-7

Seyedzadeh, S., Rahimian, F. P., Oliver, S., Glesk, I., & Kumar, B. (2020b). Data driven model improved by multi-objective optimisation for prediction of building energy loads. *Automation in Construction*, *116*. https://doi.org/10.1016/j.autcon.2020.103188

Standard, C. E. D. (2018). *EN 17037. Daylight of Buildings*. Retrieved October 27, 2019.

Tagliabue, L. C., Buzzetti, M., & Arosio, B. (2012). Energy saving through the sun: Analysis of visual comfort and energy consumption in office space. *Energy Procedia*, *30*, 693–703. https://doi. org/10.1016/j.egypro.2012.11.079

Toutou, A. M. Y. (2018). *A parametric approach for achieving optimum residential building performance in hot arid Zone* (Master of Science (MS) in Architectural Engineering, Alexandria University). Retrieved October 27, 2019, from www.secheresse.info/spip.php?article84630

Tzempelikos, A., & Shen, H. (2013). Comparative control strategies for roller shades with respect to daylighting and energy performance. *Building and Environment*, *67*, 179–192. https://doi.org/10.1016/j. buildenv.2013.05.016

Uribe, D., Bustamante, W., & Vera, S. (2017). Seasonal optimization of a fixed exterior complex fenestration system considering visual comfort and energy performance criteria. *Energy Procedia, 132*, 490–495. https://doi.org/10.1016/j.egypro.2017.09.676

U.S. Department of Energy. (2001). *Human factors/ergonomics handbook for the design for ease of maintenance*. Springer. ISBN: 978-1-60119-821-1

Vanhoutteghem, L., Skarning, G. C. J., Hviid, C. A., & Svendsen, S. (2015). Impact of façade window design on energy, daylighting and thermal comfort in nearly zero-energy houses. *Energy and Buildings, 102*, 149–156. https://doi.org/10.1016/j.enbuild.2015.05.018

Walsh, J. W. T. (1951). The early years of illuminating engineering in Great Britain. *Transactions of the Illuminating Engineering Society, 16*(3_IEStrans), 49–60. https://doi.org/10.1177/147715355101600301

Wang, W., Zmeureanu, R., & Rivard, H. (2005). Applying multi-objective genetic algorithms in green building design optimization. *Building and Environment, 40*(11), 1512–1525. https://doi.org/10.1016/j.buildenv.2004.11.017

Yang, L., Yan, H., & Lam, J. C. (2014). Thermal comfort and building energy consumption implications – A review. *Applied Energy, 115*, 164–173. https://doi.org/10.1016/j.apenergy.2013.10.062

Yu, S.-M., Han, S.-S., & Chai, C.-H. (2007). Modeling the value of view in high-rise apartments: A 3D GIS approach. *Environment and Planning B: Planning and Design, 34*(1), 139–153. https://doi.org/10.1068/b32116

Yu, X., Su, Y., & Chen, X. (2014). Application of RELUX simulation to investigate energy saving potential from daylighting in a new educational building in UK. *Energy and Buildings, 74*, 191–202. https://doi.org/10.1016/j.enbuild.2014.01.024

Zejnilovic, E., & Husukic, E. (2018). Culture and architecture in distress – Sarajevo experiment. *Archnet-IJAR, 12*(1), 11–36. https://doi.org/10.26687/archnet-ijar.v12i1.1289

Zhai, Y., Wang, Y., Huang, Y., & Meng, X. (2019). A multi-objective optimization methodology for window design considering energy consumption, thermal environment and visual performance. *Renewable Energy, 134*, 1190–1199. https://doi.org/10.1016/j.renene.2018.09.024

Zhang, J., Liu, N., & Wang, S. (2019). A parametric approach for performance optimization of residential building design in Beijing. *Building Simulation*. https://doi.org/10.1007/s12273-019-0571-z

14 Artificial intelligence image processing for on-demand monitoring of construction projects

14.1 Introduction

It has often been argued that inspections, progress monitoring, and comparing 'as planned' with 'as-built' conditions in construction projects do not readily add tangible intrinsic value to end users. In large-scale construction projects, the process of monitoring the implementation of every single part of buildings to reflect them in Building Information Modelling (BIM) models can often become highly labour intensive and error-prone, due to the vast amount of data produced in the form of schedules, reports, photo logs, etc. This is important as the complexity of construction projects and the individualistic approach needed to accommodate all individual aspects of (usually unique) building projects invariably results in delays and errors. In this respect, construction project management and monitoring processes are predominantly still conducted through traditional processes, using 2D drawings, reports, schedules, and photo logs. This system not only makes it rather cumbersome and time-consuming checking documents against each other, but also tends to invite errors to occur; however, considerable improvements have now been made in ICT integration (Kolo et al., 2014; Pour Rahimian et al., 2020; Stewart, 2007). For example, different technologies and systems can be recently implemented on construction sites to improve project communication, coordination, planning, and monitoring, including web-based technologies, cloud computing, BIM, and tracking technologies (Ibem & Laryea, 2014; Adwan & Al-Soufi, 2016). These novel applications are usually used in different technological combinations to improve the construction monitoring and allow for comparison of the as-planned and as-built models (Alsafouri & Ayer, 2018; Seyedzadeh et al., 2021). BIM capabilities are therefore no longer limited to geometry representation (i.e. 3D virtual objects) and can be used to enhance many other aspects of construction projects, such as information management (through semantically rich models), inbuilt intelligence and analysis (via active knowledge-based systems and simulation), as well as collaboration and integration (through digital data exchange) (Gu et al., 2014).

BIM has been widely used in construction projects to improve communication among various parties during different phases of design and project delivery (Sheikhkhoshkar et al., 2019; Svalestuen et al., 2017); however, due to a myriad of issues, such as the unpredictable pace of works, continually changing site environments, and the need for constant synchronisations, BIM use can sometimes hinder the monitoring of construction projects. This is despite the fact that adopting on-demand data acquisition techniques in conjunction with BIM models to compare the state of sites with 'as-planned models' has been supported (Han, Lin, & Golparvar-Fard, 2015; Roh et al., 2011; Teizer, 2015), but the success of this is often influenced by the time-consuming and error-prone nature of these activities. Notwithstanding these

DOI: 10.1201/9781003106944-14

factors, the interoperability between the tools and systems have been identified as the main obstacle, which is presently hard to overcome (Pour Rahimian et al., 2019).

In order to address this methodological and technical challenge, this chapter presents a framework and proof-of-concept prototype for on-demand automated simulation of construction projects, which integrates cutting-edge Information and Communication Technology (ICT) solutions; namely, image processing, machine learning, BIM, and virtual reality. This proof-of-concept prototype facilitates the bilateral coordination of information flow between construction sites and BIM models. Through this framework, computer vision and machine learning (ML) techniques are presented to help prepare and compose site photographs with the as-planned BIM models. The interoperability and integration of these techniques are facilitated by the aid of virtual reality (VR) game engines, such as the Unity engine. The proposed hybrid application of image processing and BIM is expected to enable facilitation of an on-demand as-built model updates for construction progress tracking. The integration performed in the VR environment (VRE) uses a game engine to enable users to actively participate in the progress evaluation, whilst also highlighting and reporting any inconsistencies.

This chapter outlines an overview of existing technologies for automated construction monitoring, focusing on image processing and visualisation methods. The proposed framework for integration of BIM, ML, and VR is then discussed, followed by the details of prototype design, applied techniques and algorithms, development strategies, and system architecture of the game-like VRE. This concludes with a short discussion on improvements and future research opportunities.

14.2 Related studies

14.2.1 Overview of real-time data collection technologies

The data collection technologies for construction project monitoring can be divided into three main categories: (i) enhanced information technologies, such as multimedia, emails, voice, and hand-held computing; (ii) geospatial technologies, such as Geographic Information System (GIS), Global Positioning System (GPS), barcoding and Quick Response (QR) coding, Radio Frequency Identification (RFID), and Ultra-Wide Band (UWB) tags; and (iii) image-based technologies, including photogrammetry, videogrammetry, and laser scanning (Omar & Nehdi, 2016; Rahimian et al., 2019).

GPS and GIS are commonly used automated asset-tracking systems, which can be used for the analysis of construction site equipment operations (Song et al., 2006). Location information of construction elements is used to identify and track equipment activity to compile safety information in order to improve decision-making and site management processes (Pǎtrǎucean et al., 2015). When GIS is integrated with BIM, it can help construction managers identify the best spaces for tower cranes, including material progress in supply chain management (Chen et al., 2015).

Barcoding allows for product identification and is considered one of the most cost-effective construction tool monitoring methods; however, this method can sometimes be time-consuming, due to the required reading of the tag proximity and its limitations of not being able to capture all of the information in the barcode label required by the end user. Barcode technology is also considered somewhat unreliable as the tags are prone to damage in harsh construction environments, and they can also become detached or lost. QR, however, provides much more additional information, and this is commonly exploited due to the increased popularity of mobile phones and tablets being able to use QR code-reading applications.

Moreover, the implementation of QR codes in conjunction with BIM technology has also been suggested as a method of improving communication on site, particularly in relation to health and safety (Lorenzo et al., 2014).

RFID incorporates the use of radiofrequency waves instead of light waves, which thereby overcomes the distance issue (Omar & Nehdi, 2016). This technology has been widely used in conjunction with BIM to leverage control and monitoring of construction projects (Xie et al., 2015), as well as productivity optimisation and health and safety improvements (Jaselskis & El-Misalami, 2003). Ultra-Wideband (UWB) signals are considered reliable, being able to penetrate walls and obstructions, and are generally accepted as being one of the most accurate systems for distance positioning, monitoring, and tracking (Jochen Teizer, 2007).

Laser scanning technology allows for 3D object scanning and data collection of existing environments. Data is generally stored in the form of a 3D point cloud, which can be used to create a digital twin of a building (Lu & Brilakis, 2019; Woodhead et al., 2018). The primary factors that make the scan-to-BIM method less practical include cost of the equipment, time-consuming process for data collection and BIM model creation, and incorporation of errors (Bosché et al., 2015). Although this method is mainly used for documentation and renovation projects, it could be used for automating construction progress monitoring, especially through a combination of 3D point cloud acquisition and three-dimensional object recognition to categorise construction elements and 3D models associated with the construction schedule (Turkan et al., 2012). In addition to laser scanning, light detection and ranging (LiDAR) technology is also an approach for monitoring construction projects by employing robotics and creating 3D modelling (Ladha & Singh, 2018).

All of these cutting-edge technologies offer several benefits, yet equally, pose specific drawbacks, specifically with their respective inconsistences in supporting the monitoring of construction sites (Asgari & Rahimian, 2017); however, capturing site images is still considered one of the easiest and least labour-intensive methods of gathering information from construction sites (Omar & Nehdi, 2016). The advent of new photo capturing technologies, through such applications presented in camera drones and also the incorporation of depth image cameras, has further facilitated and promoted scene recording on construction sites. The following section reviews the various technologies and approaches used for processing construction site photos for monitoring purposes.

14.2.2 Machine learning–based image processing for construction progress monitoring

ML techniques found their applications in the building energy field in the 1990s (Seyedzadeh & Pour Rahimian, 2021c, 2021d; Seyedzadeh et al., 2018), but their use in the field of construction monitoring is very new, and their advantage has not been fully exploited. That being said, automatic image-based modelling techniques for progress monitoring and defects detention has been an area of interest for many researchers, in both construction and computer science domains (Guo et al., 2017; Ham et al., 2016; Seo et al., 2015). Integration of as-built photographs with 4D modelling using time-lapsed photos was proposed for construction progress monitoring (Golparvar-Fard et al., 2009). Later, the same group (Golparvar-Fard et al., 2015) suggested an automated monitoring system for employing daily construction images with an Industry Foundation Classes (IFC) model-based BIM. In their study, visualisation was achieved through the creation of a 4D as-built model from point cloud images, performing an image classification for progress detection and an as-planned model from BIM. This work was then followed by Braun et al. (2015), performing an automated comparison of the actual state and planed state of construction through photometric methods in order to detect discrepancies and adjustment of the construction schedules.

Kim et al. (2013) proposed a methodology based on image processing for automated update of 4D models via incorporating Red-Green-Blue (RGB) colour image acquisition, in accordance with specific instructions. Progress identification was also performed by applying 3D-CAD-based image filters (Wu et al., 2010). Roh et al. (2011) integrated as-planned BIM models with as-built projects data extracted from the site photographs, which was then over-laid in a 3D walkthrough environment to help estimate delays in the construction progress.

Site photography is generally time-consuming and challenging in some aspects, where many researchers have suggested the use of Unmanned Aerial Vehicles (UAVs) for this purpose (Han et al., 2015; Kluckner et al., 2011; Lin et al., 2015). Golparvar-Fard et al. (2015) developed an ML method to detect the ongoing progress of construction. They utilised an image data set, including progress and no-progress predefined photos, and trained a classification model to be used to predict new images (Dawood et al., 2020; Oliver et al., 2020). Image classification was also used to detect construction materials to build a BIM model to support automated progress monitoring (Dimitrov & Golparvar-Fard, 2014).

It has also been suggested that employing image synthesis methods can help improve the accuracy of classifiers. Soltani et al. (2016) demonstrated the efficiency of synthetic images in training vision detectors of construction equipment. Rashidi et al. (2016) compared three different classification models for material detection and managed to automatically detect concrete, oriented strand board, and red brick from the still images. In addition, Kim and Kim (2018) used a histogram of oriented gradient visual descriptors for training a set of synthetically created images to help facilitate site equipment detection.

14.2.3 Virtual and augmented reality for construction project monitoring

Initially used in the gaming industry, VR applications have long since entered the AEC domain, allowing better visualisation and simulation of various scenarios (Whyte, 2002). Kim and Kano (2008) advocated the superiority of related VR images over traditional photographs taken from construction sites for progress monitoring. It has been proffered that VR images can provide a realistic location and condition of structure elements compared to 3D CAD models (Huang et al., 2007). VR technologies have widely been used for design and construction prototyping by modelling and visualising different activities in order to identify potential risks and optimise the construction process (Li et al., 2008). This has also been used to help effectively manage design alterations and improve communication with clients (Li et al., 2008). Retik et al. (2006) developed a hybrid VR interface integrated with telepresence and video communication systems to facilitate remote construction monitoring.

More recently, AR has received additional attention in the construction industry because of its ability to superimpose virtual objects onto real-world scenes. Several research initiatives have used AR to provide more accurate interactive site visualisation (Behzadan & Kamat, 2007), construction worksite planning (Wang, 2007), underground infrastructure planning and maintenance (Schall et al., 2009), comparison of as-built and as-planned images on construction sites (Golparvar-Fard et al., 2009), and visualisation of equipment operation and tasks (Hammad, 2009). The effectiveness of using AR tools in supporting decision-making and conveying the sophisticated knowledge to all parties engaged in construction has also been overtly advocated (Golparvar-Fard et al., 2009).

Kopsida and Brilakis (2016) proposed a method to enhance the inspection and progress monitoring for interior activities using HoloLens (a combination of hardware, mixed reality, and AI). The application projects a 3D as-planned model onto the real-world scene, to identify inconsistencies with actual construction. Ratajczak et al. (2019) integrated AR with BIM and a location-based management system as a mobile application to facilitate progress monitoring and communication of construction project parties. The application was supported by Google Tango–ready smartphones to display a 3D BIM model overlaid onto as-built images. This also delivers information on construction tasks and materials/ technical data.

14.2.4 Summary of progress monitoring methods

The various methods used for construction progress monitoring are summarised in Table 14.1, highlighting the limitations and advantages of each.

Table 14.1 Data acquisition and site visualisation construction progress monitoring

Type	Method	Advantage	Limitation	References
Geospatial techniques	GPS & GIS	Identification of the optimal location for the equipment	Limited to outdoor operation	(Chen et al., 2015; Pətrəucean et al., 2015; Song et al., 2006)
	Barcode & QS	Cost-effective, no extra device to read the codes	Time consuming, requires line-of-sight	(Lorenzo et al., 2014; Tserng et al., 2005)
	RFID	Do not require line-of-sight, close proximity, individual reading and direct contact	Prone to error in presence of metals and liquids, time-consuming	(Jaselskis & El-Misalami, 2003; Jochen Teizer, 2007; Xie et al., 2015)
	UWB	Do not require line-of-sight, close proximity, individual reading and direct contact	High cost and no mini device or daily necessity-embedded tool	(Cho et al., 2010; Jochen Teizer, 2007)
Imaging techniques	Laser scanning	Automated data collection, high accuracy	Expensive equipment, laborious	(Lu & Brilakis, 2019; Turkan et al., 2012; Woodhead et al., 2018)
	Photogrammetry	Calculation of completion percentage and measurement of the project progress	Sensitivity to lighting conditions, high time complexity	(Dai & Lu, 2010; Memon et al., 2005)
	Videogrammetry	Low time complexity, detection of moving equipment	Lower accuracy than photogrammetry	(Brilakis et al., 2011; Dai et al., 2013)

(*Continued*)

Table 14.1 (Continued)

Type	Method	Advantage	Limitation	References
VR/AR techniques	VR from site photos	Provide realistic location and condition of structure element, remote construction monitoring	Inability to check with as-planned model	(Kim & Kano, 2008; Li et al., 2008; Retik et al., 2006)
	AR	Superimpose as-planned onto as-built image	No support for remote monitoring, and automated actions, depends on the device technologies	(Behzadan & Kamat, 2007; Golparvar-Fard et al., 2009; Hammad, 2009)

14.2.5 Challenges in system integration

Although there have been tools, including geospatial and imaging techniques, to enhance progress tracking of construction projects, these applications are not yet able to effectively identify the inconsistencies between as-built and as-planned models. Moreover, there is a need for a decision support system for project monitoring to support effective communication among the involved parties. The reviewed literature shows that there has been significant success in employing VR and AR applications to support decision-making at design stages and handling complicated tasks in facilities management; however, this potential has not been fully utilised in the construction phase to enhance project progress monitoring and reporting, mainly due to specific conditions of construction sites that make them different from any other environment. It was mentioned that previous studies have been successful in integrating AR and BIM models to provide a tool for construction progress inspection; however, these tools rely heavily on the accuracy of utilised AR tools, and are thereby prone to error. On the other hand, the use of such solutions for proper communications among construction parties requires the presence of contributors on construction sites. This issue is one of the main raising challenges in construction management. The use of VR to tackle this problem has been very limited as it requires multidisciplinary cooperation to realise the interoperability of it with 3D information modelling and advanced AI methods.

The framework and prototype presented in this chapter presents an approach for integrating ML, artificial intelligence, image processing, VR, and BIM technologies aligned with gamification approaches in order to address this particular research gap. Unlike previous attempts that utilised ML and BIM directly as a means for identifying work progress or diagnosing particular problems, this framework benefits from those technologies for the preparation of an interactive virtual environment to be manipulated by a game engine. Therefore, this tool allows effective utilisation of these technologies in order to support selective examination of various building characteristics at different times and by different people, hence making the system more usable for all professionals involved in the project.

14.3 Project framework

The core component of the hybrid system described in this chapter is a game-like VRE, providing integration between ML-based image processing and BIM. As such, the developed system architecture consists of four major elements: (i) image capturing, (ii) image processing, (iii) BIM authoring, and (iv) VR game authoring and object linking. This research proved the concept that the integration of computer vision, image processing, BIM, and VR can facilitate the automatic update of a digital model, store the data in a standard file format, and display project progress information in a structured manner. This also demonstrates this approach can be used as an effective tool for communicating with different parties involved in construction projects to aid decision-making.

However, despite the fact that laser scanning and photogrammetry are leading methods for collecting special and geometric information, the still image photography method was adopted in this study, as this is considered an inexpensive and hassle-free method that has become the industry standard for gathering construction site progress information.

Figure 14.1 presents a schematic diagram of the proposed system architecture. Interoperability and flawless information exchange were seen as a significant driver in this research.

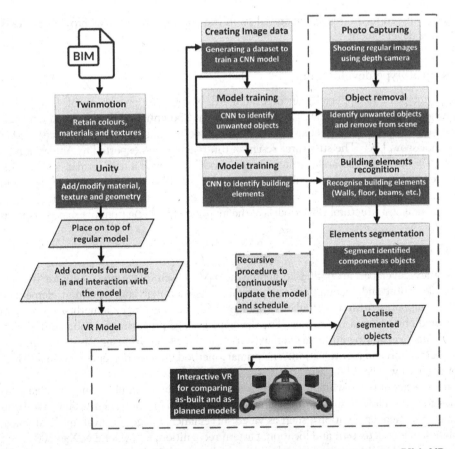

Figure 14.1 Proposed framework flowchart: construction progress monitoring using BIM, VR, and image processing technologies

Therefore, Autodesk Naviswork (NWD) file format and IFC as a standard file format were used throughout this research. First, entities from the BIM IFC model links to the Unity to create the VR model. From this model, a dataset of synthetic RGB-D images is generated. Then, neural networks are trained using the created data enriched with real-world depth images to create classifier models.

Construction site images were captured every day using a depth camera, then regenerated by copying the same camera settings and location (localisation) within the BIM model. These images were stored in cloud-based storage along with the BIM model. An ML classification technique, Convolutional Neural Network (CNN), was applied to these photographs to detect and identify different objects and building components. Image processing was then utilised to remove unwanted objects from the scenes for further clarity.

Recognised and classified elements were linked to the related actionable tasks from the time schedules linked to the BIM model. The extracted details were then overlaid onto the as-planned BIM models for comparison. The superimposed construction status data was then transferred to the game engine through scripting or exporting IFC into an Autodesk FilmBoX (FBX) format, then integrated with the virtual environment. This provided the system with a higher level of immersion, improved visualisation, and enhanced interaction. The VRE contained both as-planned and as-built models for comparison and identification of potential discrepancies.

The proposed technique is elaborated on in the following sections by providing details of prototype development.

14.4 Prototype development

The prototype of the application incorporating ML-based image processing, BIM, and VR was implemented in order to showcase the feasibility and potential of the proposed conceptual framework. The research selected a new leisure and sports complex at the University of Strathclyde (Glasgow, UK). The structural design of this building was exported to Autodesk Naviswork (NWD) file format, and the architectural model was saved in IFC format, which made it suitable for unobstructed data exchange between different BIM authoring software applications used in the building industry by various parties. Figure 14.2 (a) and (b), respectively, present architectural and structural BIM models of the mentioned building from the same perspective.

14.4.1 Image processing

This study developed an optimised approach to ML-based image processing for the automatic detection and recognition of the main constructional and structural elements. In order to support a 'neater' virtual environment, the study developed a method for removing unwanted objects from the scenes. The developed image processing method started with a depth camera acquiring multiple overlap RGB-D (colour+depth) images as system inputs for object detection and identification. The primary method used for object detection and image segmentation was based on ML.

Image segmentation, which partitions a set of pixels into multiple groups, was employed to designate building elements in the image coordinates. This method is widely investigated in many practical applications, such as video surveillance, image retrieval, medical imaging analytics, object detection and location, pattern recognition, etc. (Dong & Xie, 2005).

By using a 2D BIM image/video, physical coordinates can be estimated and point cloud or depth image can be generated via structure from motion (SfM) technique (Schonberger &

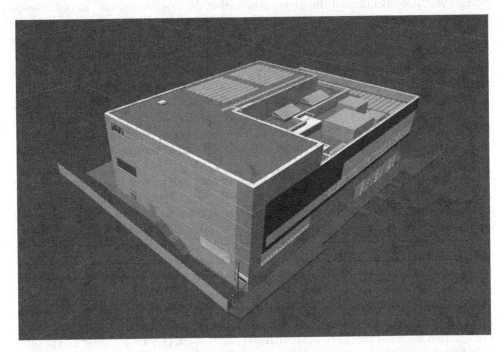

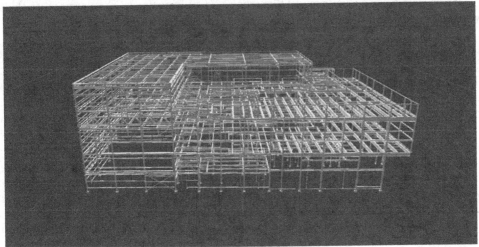

Figure 14.2 (a) Architectural and (b) structural BIM models

Frahm, 2016; Schönberger et al., 2016). In the BIM model, pre-defined 3D data was provided as the prior knowledge, which can help the machine recognise BIM objects via ML algorithms, such as support vector machine (SVM), random forests (RF), and CNN (Seyedzadeh & Pour Rahimian, 2021a, 2021b, 2021e).

A carefully designed dataset was required for training a robust classifier. Image processing for pose estimation (Shotton et al., 2013) and semantic segmentation of the scene

(Song et al., 2006) included labelling each pixel within the depth image with an object-associated category to supervise the ML algorithm and learning the voxel features of the object. In BIM projects, it is common to use a pre-defined 3D model for simulating depth images in various viewpoints, thus generating as many labelled depth images as possible for building a dataset prior to commencing the construction work. The details of generating data for training the network using virtual photogrammetry are presented in the next section.

Taking daily construction activity planning into consideration, a deep neural network was considered an option, as this can automatically tune the parameters (Seyedzadeh et al., 2019a, 2019b), based on the new given data. Therefore, it can effectively learn a new representation of an object along with the progress of a BIM project instead of retraining a new classifier on a daily basis.

Due to the diverse nature of predictions for semantic segmentation and recognition of unwanted objects in the construction scenery, training a single model to perform both tasks accurately is rather difficult. Moreover, the utilised dataset for training a model for image segmentation that includes synthetic images does not include those intended objects, such as humans or machinery. As such, this study first applied a separate network for recognition of these objects. This procedure guaranteed the precision of both models by the utilisation of individual training sets. It was therefore possible to train one model for both purposes, using a comprehensive dataset; however, preparation of such a dataset is often laborious as it requires a graphical mixture of environment and objects, both in the form of synthetic and real images. Moreover, obtaining adequate negative data is another hurdle for creating a recognition model. It should be noted that the negative samples are as valuable as positive records in training an accurate model that requires identifying the desired objects precisely as well as rejecting false detections.

Object removal

Taking photos from a construction site was the first step in creating an as-built model to compare between the current state and the as-planned model. Construction scenes consisted of many tools and materials, which were considered as unwanted objects in the construction progress monitoring application. Furthermore, the presence of these objects can lead to faulty detection of construction elements related to the BIM model. Generally, it is not practical to move all of these objects while taking photos. This study proposed an automated two-stage object removal method in order to address this challenge. The first step was to use a supervised object recognition technique for identifying unwanted objects, and the second step was to fill the area of the detected object in a visually plausible way.

Many approaches for filling a part of an image (inpainting) have been developed by previous studies; however, as the images used in this study contained depth information, it was posited that a suitable region-filling method should be able to estimate the depth filling pixels. This study employed an exemplar-based method developed by (Atapour-Abarghouei & Breckon, 2018) as a suitable means to address this requirement. First, these elements were recognised and segmented using the pre-trained CNN, then the boundary of the target region was identified, where a patch was chosen to be inpainted, and the source area was queried to find the best-matching spot via an appropriate error metric. This study noted that the main advantage of this method over the other inpainting methods was its ability to propagate the texture into the target region.

Gupta et al. (2014) trained a large CNN on RGB-D images to recognise objects and applied a semantic segmentation to infer object masks. For the purpose of demonstration, this study trained CNN to detect humans in the construction scene. For training data, the study selected 3200 positive images (Silberman et al., 2012) and 7500 negative images. Figure 14.3 shows the detail of the method for recognition of objects using RGB-D images and presents how RGB-D images make it possible to calculate the depth and average gradients. This detail was then combined with a fast edge-detection approach to generate enhanced outlines. The contours are used to create 2.5D region candidates through processing characteristics on the depth and colour image. The depth-map is encoded with various channels at each pixel, namely horizontal disparity, height over ground, and the angle of the pixel's local surface. Then the CNNs are trained on RGB-D images to detect intended objects in 2.5D region candidates. Each CNN begins with a set of region proposals and calculate features on them. The box proposals are then classified using a linear support vector machine.

Figure 14.4 shows the original image and with a depth-map, taken from the construction site. Figure 14.5 presents the output of applying object recognition with the trained CNN.

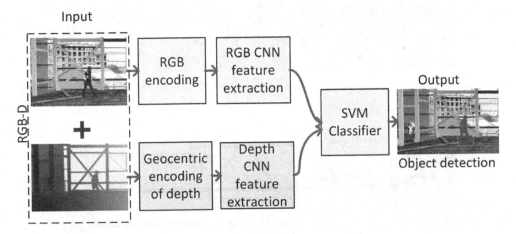

Figure 14.3 Object recognition using colour and depth images

Figure 14.4 (a) Photograph taken from construction site and (b) its depth-map

Figure 14.4 (Continued)

Figure 14.5 Creation of an object mask with human recognition using CNN

When the unwanted object is recognised, the filtered image is passed to the object removal procedure to eliminate it from the scene and fill the gap in both RGB and depth space. The outcome of applying this algorithm in Figure 14.4 is demonstrated in Figure 14.6. The depth-map of the generated image is illustrated in Figure 14.6 (b).

Figure 14.6 (a) Photograph taken from construction site and (b) its depth-map

Building element identification

The next step in the preparation of the taken depth images for use in the VRE was to apply a semantic segmentation method to identify the various building elements. In this step, the pixels were directly labelled to create the segments using a trained CNN. The training data for this network was generated from the BIM model, adding depth information and pixel labels. The use of synthetic images with the aim of semantic segmentation has been widely reported (McCormac et al., 2017; Panagiotakis et al., 2013; Ros et al., 2016); however, due to perfection of those images, the networks can sometimes fail to learn all characterisation of the noisy real photos. Moreover, as previously mentioned, there was already a need to train another network to detect unwanted objects.

For building scene segmentation, this study adopted the FuseNet algorithm (Hazirbas et al., 2017), as a fully fusion-based CNN developed for semantic segmentation of RGB-D images. After training the ML model with the generated dataset, the model was applied to the image that resulted from object removal. Figure 14.7 demonstrates the architecture of the

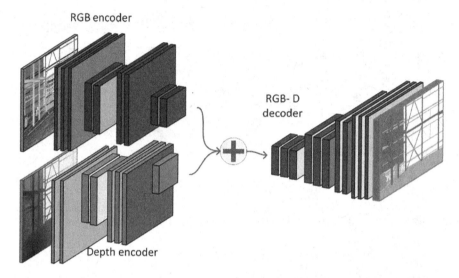

Figure 14.7 Semantic segmentation incorporating depth into colour image

Figure 14.8 Output of semantic segmentation for identifying structural building parts

network for semantic segmentation. The network contains encoders for obtaining features from RGB and depth images and a decoder that maps the feature into the original input resolution. Afterwards, the elements from the depth encoders are fused into the feature-maps of the RGB part.

Figure 14.8 presents the outcome of the segmentation, with the coloured areas indicating different building elements.

14.4.2 Linking the IFC model and the Unity game engine

The Unity game engine was used as the main platform to integrate various technologies, i.e. BIM, ML-based image processing, and VR. For succinctness, hereafter, the Unity game engine will be referred to as Unity. This study adopted a previously developed method of linking a digital model (Pour Rahimian et al., 2019) to support integration of the information contained in the BIM model. The method was established based on the use of IFC file format and the incorporation of programming with C#.

Entities from IFC models were linked to Unity through procedurally generated regular expressions, with no families. They were, however, combined with spatial localisation, where families were present. The primary challenge in this process was due to the geometric representation characteristics and Unity equivalents. This required a supporting planar geometry intermediary library linking Unity's plain ignorant, mixed Cartesian, and barycentric coordinate systems with coplanar, relative Cartesian coordinate systems. To tackle this issue, the system engaged mixed ambivalent interactions with geometry and took advantage of the game engine (in order to optimise interactions with the environment).

Naming conventions between IFC model, FBX, and Unity environment models were found to differ unpredictably depending on the applied export to-import processes and the source application. Several characteristics of the naming convention in Unity, typically swapped delimiters, replaced, removed, and amended characters, and varied in case sensitivity retention, which prevented direct one-to-one linking or effective character-by-character comparison. This issue was resolved using procedurally generated regular expressions and iterative searching over the entity record set. The model was extended to include a top-level record array with records either without parents, omitting representation layers, or entities presumed to be physical entities, which significantly reduced the processing search set. This subsequently reduced the processing time.

Families presented two additional challenges for linking. This was partially due to categorical naming returning a set of candidates, whereas family-less comparisons returned single associations. It was also because of the need to name family-named entities hierarchically, which required stepped iteration through nested game objects for full name generation. Nevertheless, due to failing to identify a non-spatial attribute to filter candidates, identification of the appropriate representation relied on localisation. This task was achieved by using bounding box centre points, which were generated by Unity during mesh construction and lazily evaluated by the IFC Library.

Figure 14.9 illustrates the procedure of integrating BIM to VR and linking segmented photos and the flow of information between each step.

14.4.3 Virtual photogrammetry for NN training

Point cloud generation in the real world has several notable forms, including LiDAR, laser scanning, and Structure-from-Motion (SfM). Each method produces a vector map of points representing the physical elements and pixel information from the device's point of capture.

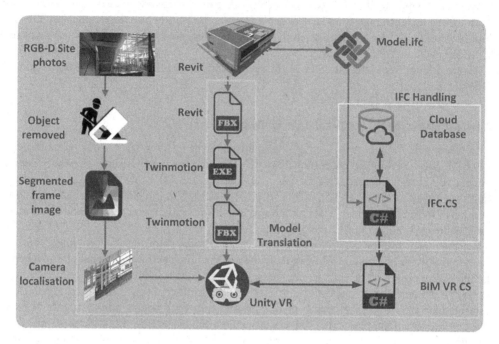

Figure 14.9 VRE from BIM model flowchart overlying real-world images

Virtual worlds in terms of physical representations, discrete interval scale, and rendering definition are not meaningfully different from the real world. This enables the application of many traditional and contemporary photogrammetry methods with varying levels of algorithmic complexity; however, while many methods would be transferable to the virtual world, real-world techniques often attract imperfections in the collected images. For example, in a large zone in a virtual world rendered with acrylic paint, there may be few if any identifying features, instead of a featureless image with constant colour pixels. Although this is rarely the case in virtual models, a method needs to work with everything or nothing. This reduces the potential for SfM. Similarly, implementations of real-world tools in a virtual world are often significantly different from how they function in reality. LiDAR, for example, uses omnidirectional pulses, where, for example, if this were to be implemented literally in the virtual world, a countable but impractical number of ray casts would be required, which would place significant demands on the engine.

Instead, at the point of triggering a function in the virtual world, the triggering objects that reside and interrogate a physical object database would step out of the world. For instance, in the case of a game with human agents, rather than raycast in every direction, the script would locate all players in the database and then separate the (x, y) and (z) dimensions. Using the (x, y) boundaries, a simple rectangle would be created around the agent, and an intersection calculation would be carried out for each line until a horizontal intersection occurs; otherwise, play would be ignored. Assuming that it was identified, in the next stage, a vertical intersect assertion would be made. The simplest solution at this point could raycast only within the boundaries of the box. A more complex solution may use relative coordinate systems and spatial caching to reduce the number of cases required. For example, for any given intersect, the boundary box method may be applied to infer points that do not merit raycasting.

In order to avoid a complicated algorithm based on existing techniques used in the real world, this study facilitated virtual photogrammetry using a simple application of the game engine's physics component and its camera class functionality. Using the camera class's camera point to virtual world coordinate translation function, each pixel's relative location in the virtual world is therefore identified. A raycast is sent from that location, matching the base orientation of the camera, which then attempts to intersect with any object with a MeshCollider. Upon successfully hitting an object, a struct is created containing the 2D and 3D coordinate system locations, including an extension to the default Mesh class, which accommodates the binding of external data, such as IFC and the triangle on the mesh that was hit – the latter implicitly linking a barycentric coordinate system to the struct. As mentioned in the planar geometry system, triangular meshes are translated into planar surface sets. This set is already bound with the mesh, and each plane constituents mesh triangles. Each point struct is then added to the point cloud dictionary, which may then be exported for producing composite images or helping scalable discrete reconstruction.

However, the library is susceptible to two issues common with other forms of photogrammetry, i.e. nonuniform intervals and point density. This is in contrast to the real world, where the process is reproducible and does not require an additional journey through the model. At any point, where the cloud lacks integrity, a camera can be spawned to collect additional data with partially controlled precision. Partial is used here as a caveat for the inherent limitations of casting a finite number of times on varying distance surfaces.

Figures 14.10 (a) and (b) show the camera view and spatial linking between image pixels and mesh continuous spatial information from the localised camera and BIM model. Every pixel represents an object containing 3D coordinates and existing and mapped colouring.

Figure 14.11 demonstrates applying discrete spatial constraints on a virtual environment translation. Spatial data was decoupled into continuous, discrete, and photogrammetric vectors, representing the real position of the object in the virtual environment, interval equivalent of that object, and its position as it appeared in the primitive reconstruction.

Figure 14.10 Camera (a) and spatial distance heat-map depth view (b)

Figure 14.10 (Continued)

Figure 14.11 Discrete distance heat-map view

Synthetic image data was collected primarily using the virtual photogrammetry methods demonstrated in Figure 14.10 and Figure 14.11, and procedurally generating random camera locations in spaces or near the objects of interest. Using the floor perimeter, the camera location boundaries were obtained and the number of floor elements used as a proxy. Cameras were first spawned around human height, enabling initial data collection regarding the current space, primarily making ceiling height identification and partially identifying zone boundaries. Once these had been identified, a camera can store an image and use the learned spatial boundaries to choose a new camera location within the space. The virtual photogrammetry and camera movement in the space can then be iteratively applied to the partial mapping, ceiling height, and farthest surface distance to constrain the camera angle.

Virtual photogrammetry is then used to bind pixels to their constituent entities in the virtual world, and if bound, to update IFC representations. Their material, entity type, and

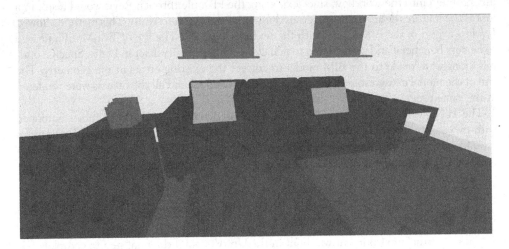

Figure 14.12 Category-mapped scene view

where relevant, the families are extracted from the IFC schema. This is bound to the raw image data included in UV and ARGB. Pixels are grouped by the entity with which they are associated. Entities with pixel adjacencies can have surface poly-boundary points to check for coplanarity with surfaces on the other entity. The relationships are tested in each imaging set, ensuring most are captured. As shown in Figure 14.12, entities can be turned off such that the surrounding hidden (or partially hidden) entity surfaces can be mapped. If necessary, this could also be used to identify which entities are present in the space, but occluded.

Data was then split into its constituent sets and inserted into a database or tracked in an image relations file. The former was interrogated via SQL (Structured Query Language) and the latter parsed into the PixelRelation class of the VP library. Splicing in a localised real-world image now had a truth network for real-world training.

Classifiers and segmentation models can be used where the data in the virtual world is appropriate, mostly for producing inference networks. The significance of the former is that projects like ImageNet are significant human endeavours, as manual segmentation of images is needed to classify objects. Since the virtual world has already segmented virtual representations, these can be used to estimate confidence in classifications. Using depth from the real world and virtual world, relative scale can be used to refine the expectations. If something seems right, but not linked between both, other profiles from virtual images can be scaled and compared. If an object is still not identified, a case for propagating a specialist classifier for that type of object exists. The family or entity's related images could then be used to train a model to look for those exclusively.

14.4.4 Integration with the game engine

Autodesk Revit viewed and amended the original BIM model of the Strathclyde Sports Centre. The digital model was then imported into the Unity game engine to allow for the development of VR simulation. The Unity game creation engine was the primary platform for creating the application, using C# as the main programming language. The IFC BIM model was exported into the FBX file through a TwinMotion plug-in for Revit. The plug-in was incorporated into the workflow, since exporting the FBX file through Revit would result in a loss of the textures defining the materials during the transition to Unity. This method allowed flawless export of the BIM model with the textures assigned. The use of TwinMotion proved to be very beneficial and contributed to achieving a high-quality asset in Unity. Specific optimisations were made to the BIM model to correct the existing errors in the geometry. The size of the model caused some issues since the entire model and all the objects were rendered at the same time.

The HTC Vive was the primary VR headset used for this work. Vive comes equipped with two controllers and two wall or tripod mountable scanners that allow the user to move around extensively within their personal real-world space. A Windows gaming laptop with an Intel i7 processor and NVidia GTX 1070 graphics card was the machine used to create this application.

14.4.5 Overlaying the segmented images

When the virtual environment was built in the Unity engine, the final step to complete the monitoring tool was to overlay as-built images over the as-planned model. The elaboration of the method as a set of rules was as follows:

*   Where no high-precision camera location information was known, but the entities of interest were identified, they may have been highlighted entirely via the Colour property of the primary material. Changing this property would highlight the desired objects, so that when the user brought them into view, they would have been able to identify whether or not they were of interest.
*   Where camera location was known, but orientation and field of view was not, faces could be identified via generation of a FaceSet through the UnityIFC library, which could create the planes to define the entity. In this case, the script needed to effectively raycast this to the centre of the closest triangle on the largest two faces. Once the face was identified, the barycentric coordinate systems needed to be used to determine which triangle is associated with that specific raycast. The triangle point list direction informed the user as to whether the triangle was visible or not rendered. If the triangle was rendered, then it was the face that the user should see; otherwise, the other largest face was of interest. The identified face's primary material colour would then need to be changed to the highlighting colour.
*   Where all of the above (including the field of view) were known, the system would provide an indication of entities that were not entirely in place. That is not to say the system has definitely identified erroneously positioned columns; however, they might have been highlighted, where a re-measure with a laser distance meter (such as Leica Disto) would have been worthwhile. There were two options for achieving this goal: (i) by using superimposing the segmented image over an n-dimensional image created with the library designed for this project, or (ii) by applying a similar process for direct raycasting from the superimposed segments. In the case of the former, the screen or secondary camera resolution needed to be adjusted to match the metadata on the picture; or if not possible, up to a proportional scale. It should be noted, however, that in this case, additional control for pixel scale would have been required, as well as including buffering the AND failure area to accommodate the larger pixels. Pixels from each would then need to be compared with an AND operation to identify where expected pixels overlie with the segmented sections from the image processor.

- Where the AND failed, there was a potential misalignment of the entities, where each should have been highlighted appropriately using two distinctive colours to demonstrate where the segment and virtual face were out with the boundary of the other.

Figure 14.13 illustrates the proposed procedure for overlaying the segmented images on the BIM model in the VRE.

Camera localisation has several meaningful opportunities depending on the information of entities captured within the view and segmentation. The methods appropriate for this project primarily relied on segmentation and BIM localisation. Depth segmentation is mostly excluded here, but confidence in its results can surpass the random sampling type method described. The following describes an idealised (two-sample) instance, but the principle of RANSAC could be applied in conjunction with it to refine and confirm.

IPS has been reported to be accurate within 300 mm, which is a significant improvement on GPS, reported to be accurate within 5,000 mm. In recent years, research has refined the use of fiducial markers, demonstrating accuracy within 80 mm. In each case, the reported accuracy is without the benefit of verification with segmentation or virtual photogrammetry from BIM models. With the assumption that a floor is segmented, choosing three points on the surface, with two creating a line perpendicular to the camera, forms an isosceles or equilateral triangular hyperplane. Splitting this into a triangle and trapezoid and injecting the z dimension from segmentation should provide enough data to estimate the shear coefficient.

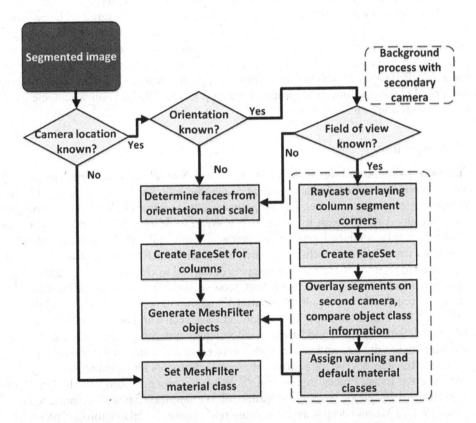

Figure 14.13 Flowchart: overlaying segmented images over the BIM model in VRE

Repeating this process on a second perpendicular triangle will only be consistent if it falls within a reasonable tolerance and if the surface is horizontal. Otherwise, it should be possible to estimate the second shear coefficient and then rearrange shear functions to determine perpendicular horizontal axis orientation. With the orientation of the surface and shear coefficient, each camera's height and orientation can be determined. The results may be confirmed by repeating the process on the virtual planes associated with the segmented surfaces by using the pixel reference to identify the surface.

With or without the previous calibration for height and orientation, classified segmentation may aid location inference. Segmented entities and IPS virtual photogrammetry can be used to search for similar or expected profiles. Considering a beam, for example, the orientation can be determined from another hyperplane, this time between the floor and vertical centre point of the beam, shear coefficient, and taking the ratio of the full height and the second height of another point along the floor/beam, not the entire length. By repeating the process, the results may be refined. The final stage for this method is subsampling segmented and virtual properties and applying RANSAC to make an educated guess from what appears to be in both from segmentation in both virtual and real worlds.

Another option was using fiducial or natural markers, specifically for localisation of the camera. While natural markers are not necessarily present, anyone taking pictures may throw temporary markers to the floor. Segmentation-aided localisation is a matter of confidence in what has been identified. Any markers known to be on the floor can facilitate the previous method. In either case, if a floor is known, IPS accuracy is enough to start using surface-to-surface edge detection on connected, non-coplanar surfaces. Localisation is not an exact science and, under normal circumstances, this is not always guaranteed to work; however, with BIM overlaying, the virtual test frame can be moved and the VP can be regenerated. An educated guess may not always solve the problem, but it does not matter whether the process is achieved by a single process or through a thousand. Localisation doesn't need to happen in seconds or minutes, it just has to fall within an acceptable tolerance and not require human input.

14.5 Discussion

Technology-Supported Integration of Real and Virtual Worlds: Bridging the gap between real and virtual worlds is no longer a matter of technological barriers, but rather a resource management issue. The tools that are necessary to link, translate, and fuse mixed-reality data now exist in isolation. It is down to development teams to figure out how they may be combined to produce a practical utility under the constraints of funding and team capacity. Some techniques, such as Drone LiDAR scanning, have long since been used to produce discrete spatial mapping of the real world, and this can be translated into virtual components. To a lesser extent, tools such as Tango-enabled mobiles can produce and translate point-clouds to coloured 3D meshes in near real time. Pour Rahimian et al. (2019) demonstrated the translation between vector and discrete virtual environments by linking cartesian and barycentric coordinate systems. In this study, bidirectional linking between discrete and vector worlds were extended to incorporate raster representations, ultimately developing a virtual photogrammetry tool for mixed reality.

ML-Assisted Image Processing: Python's SciKit-learn contains many flexible deep learning utilities designed for image classification, object identification, and semantic segmentation. Whilst ML and deep learning tools are not without their flaws, through progressive interactive training with reinforcement, transfer learning, and input homogenisation, they can perform well beyond expectations. Linking virtual and real worlds, however, requires testing, tweaking, and reinforcement, which under normal circumstances would require significant team involvement (which cannot always be undertaken in parallel) and must be carefully managed.

Paradigm Shift in Project Monitoring: The process of developing and proving solutions for research objectives similar to this study do not need to rely on the real world initially. Through abstracting the process, there is no reason why individual components from virtual and real-world utilities or data sources cannot be interchangeable. For example, a virtual photogrammetry module for the Unity game engine that was developed in this study can differ from real photogrammetry only by imperfections in the surfaces they scan. Although virtual models can produce ideal surface maps, there is a challenge in introducing imperfections. The idea behind converging realities is that tasks that are constrained by serial time, equipment allocation, and team capacity are not subject to the same constraints. If a team has one drone in the real world, they can generate data specific to the target building(s) at the rate at which the camera is capable of capturing UV and point cloud data. The properties of the building and environment are immutable, the weather and lighting are situational, and the rendering is not specifically controllable by the pilot.

Expedited Data Collection: In contrast, virtual worlds are constrained only by the amount of processing time and devices that can be afforded to the project. Data collection can be automatic and in parallel, and the environmental and target features are mutable. The virtual world can be spawned randomly with rendering material types, resolution, and imperfections unique to a given instance; even the target can be sampled from a repository of buildings. Drones in the virtual world do not need pilots, nor do they have physical representations, meaning more than one can collect data during a single collection. Unity can be compiled for Linux machines, which enable parallel processing. In short, by the time the real-world pilot had travelled to the site and generated data for a single target, the virtual world could produce thousands of datasets from any number of targets with a flexible selection of the characteristics that were otherwise immutable in the real world.

Converging Realities: The aim of attempting to converge realities was to take actions that are applicable to virtual world data applicable to data from the real world by gradually blurring the lines between the two. It is acknowledged that real and virtual world renderings are never going to be identical, and therefore, training models on purely virtual data alone would unlikely be fruitful; however, between readily available rendering material images and Fast Style Transfer (FST), CNNs creating a progressive interactive training ensemble can be made easier. FSTs, as demonstrated by Engstrom (2016), learn the common characteristics of a given training image's artistic style by comparing it with thousands of images with distinct artistic styles, including works from historically famous artists. Once a model has been trained, an image can be converted from its raw state to a style-transferred equivalent.

The Virtual Photogrammetry Technique: In this study, rather than attempting to create a complicated algorithm based on existing techniques used in the real world, virtual photogrammetry was facilitated using a simple application of the game engine's physics component. Currently constrained by the screen resolution, not a physical constraint, any given camera is sent a request to capture what it sees at the time step it receives the request. At the end of the time, step a call back to capture the spatial and physical object data. This process is by no means perfect and not implicitly transferable to the real world, but it lays the foundation of progressing to a practical tool.

Role of Open Standards, Interoperability, and Social Psychology: This research suggested that the next disruptive innovation in AEC software will not be in the form of cutting-edge design functionality, but rather, greater consideration for social psychology and its role in effective computer-mediated communication. Pour Rahimian et al. (2019), in contrast, argued that focus should be on interoperability with an emphasis on accommodating immersive virtual environments. This chapter presented an intermediate stage that demonstrated a collection of tools that partially bridge the gap between these two suggestions. Open standards can facilitate interoperability not only between traditional vector CAD vendors but also packages that are not directly linked to AEC. Pour Rahimian et al. (2019), for example, discussed the potential application of a BIM library to reside entirely outside of virtual environments. The library does not presuppose that the interface has any graphical interface to the extent that the model may be interrogated without an interfacing script. This project tackles the opposite side of communication problems in accordance to a converging realities approach, in which linking real and virtual worlds has many objective and subjective benefits. Through the ability to mix worlds, the subjective perception of proximity between involved parties can be heightened with an inherent increase in media richness while reducing the risk of communication breakdown from initially unverifiable conflicts.

Application, Functionality, Impacts, and Contributions of the Study: This application of the proposed method in data science is yet to be established; however, it has the potential to be its most significant contribution to scientific knowledge. The framework and prototype presented in this study formed methodological and technical foundations for a converging realities approach to self-propagating hierarchical deep learning ensembles, creating an AI network that aims to progressively introduce real-world images to virtual world datasets, via homogenisation of real and virtual world images. A weakness of this kind of process, however, is its initial attempts to 'rationalise' the real world. This project is well suited to mitigating this traction problem. Linking both the real and virtual world enables deterministic assertion of the presence of equipment and construction features through spatial comparison of real and virtual photogrammetry data. This serves to solve two problems. First, where an entity is proven to be present in both worlds (but not identified by an image classifier), additional images from either world may be introduced to its training data (or can be used to produce a child branch in the network). Second, the information in the virtual world is explicitly linked to the virtual entity; therefore, it can be bound to pixels in a deterministic manner. In short, this library's functionality may reduce the need for difficult segmentation while providing a mechanism for reinforcement and transfer learning.

14.6 Conclusion

This chapter presented the theoretical and technical challenge, the methodological and technical gap in the emerging digital analytical tools of machine learning and computer vision, embracing advanced visualisation media including BIM and interactive game-like immersive VR interfaces, in order to leverage the automation of construction progress monitoring. To achieve this goal, the works presented here proposed a framework and proof-of-concept prototype of a hybrid system that was capable of importing and processing construction site images and integrating them with the nD building information models within a game-like immersive VRE. This was discussed against reviewed literature that, despite the wide adoption of modern technologies in construction progress monitoring, the use of BIM, as an as-planned constructional model, has not been fully exploited – predominantly due to interoperability issues between different technologies.

This study responded to this challenge by presenting a platform that can support continuous system updates to enable construction managers, clients, and other relevant stakeholders to effectively compare building construction (as-built) with an as-planned BIM model for the purpose of deficiency detection. The resulting prototype, based around the principles of remote construction project monitoring, took advantage of ML and image processing to remove unwanted objects, recognising and extracting main building characteristics, and overlaying these on the corresponding as-planned nD components. VR technologies, including Unity and HTC Vive, provided a virtual environment to allow better user interaction with these elements.

It is therefore argued that this proposed platform could help construction companies more effectively follow work as it progresses, specifically to diagnose discrepancies and deviations without interrupting on-site operations. This remote managing tool can also save a significant amount of time by providing a more accurate comparison of constructed parts with the as-planned BIM model. This powerful virtual environment can also be used to present stage updates of building progress development to clients. The facilitated automatic update of the model makes it possible to have an on-demand schedule method without the need for sophisticated wireless-enabled devices. One of the main contributions of this project was providing a technical exemplar for the integration of ML and image processing approaches with immersive and interactive BIM interfaces – the algorithms and programme codes of which can help improve the replicability of these approaches by other scholars.

The methods of object recognition, image processing, and overlying real images over an as-planned BIM model were suggested for completion and demonstration of the presented prototype; however, different approaches might be adopted in future to achieve better identification and segmentation of the building elements, considering the studied construction site situation and demands. Moreover, advanced positioning systems can be employed for easier localisation of such components through extracting camera location. Furthermore, the prospective photo-shooting procedures can be further enhanced by means of contemporary UAVs, which are capable of pre-programmed and autonomous aviation and actions. Hence, the utilisation of advanced small drones can automate image acquisition. This is possible by defining a home point and a route in the virtual environment, then adapting it to the real-world location, and finally transferring the flight plan to the drone via a waypoint path. As most advanced UAVs encompass 360-degree sensors, they are able to perform safe indoor manoeuvres. The use of these UAVs can also help reduce health and safety risks, allowing image capture even when works are in progress.

Acknowledgements

The authors would like to gratefully acknowledge the generous funding received from Advanced Forming Research Centre (AFRC), under the Route to Impacts Scheme (grant numbers: R900014–107 & R900021–101), which enabled the team to conduct this study, which was nominated for the buildingSMART International Award 2019. This work would also not be feasible without the generous PhD funding for the second author, which was co-funded by the Engineering the Future scheme from the University of Strathclyde and the Industry Funded Studentship Agreement with arbnco Ltd. (Studentship Agreement Number: S170392–101).

The authors also would like to acknowledge the contributions of Stephen Oliver, Sergio Rodriguez, Nashwan Dawood, Andrew Agapiou, Vladimir Stankovic, Frahad Chamo, Minxiang (Jeremy) Ye, Andrew Graham, and Kasia Kozlowska to the underpinning research presented in this chapter. Finally, valuable advice received from various managers in AFRC, especially David Grant, is highly appreciated.

References

Alsafouri, S., & Ayer, S. K. (2018). Review of ICT implementations for facilitating information flow between virtual models and construction project sites. *Automation in Construction*, 86, 176–189. https://doi.org/10.1016/j.autcon.2017.10.005

Asgari, Z., & Rahimian, F. P. (2017). Advanced virtual reality applications and intelligent agents for construction process optimisation and defect prevention. *Procedia Engineering*, 196, 1130–1137. https://doi.org/10.1016/j.proeng.2017.08.070

Atapour-Abarghouei, A., & Breckon, T. P. (2018). *Extended patch prioritization for depth filling within constrained exemplar-Based RGB-D image completion*. Lecture Notes in Computer Science (including subseries Lecture Notes in Artificial Intelligence and Lecture Notes in Bioinformatics), Springer.

Behzadan, A. H., & Kamat, V. R. (2007). Georeferenced registration of construction graphics in mobile outdoor augmented reality. *Journal of Computing in Civil Engineering*, 21, 247–258. https://doi.org/10.1061/(ASCE)0887-3801(2007)21:4(247)

Bosché, F., Ahmed, M., Turkan, Y., Haas, C. T., & Haas, R. (2015). The value of integrating Scan-to-BIM and Scan-vs-BIM techniques for construction monitoring using laser scanning and BIM: The case of cylindrical MEP components. *Automation in Construction*, 49, 201–213. https://doi.org/10.1016/j.autcon.2014.05.014

Braun, A., Tuttas, S., Borrmann, A., & Stilla, U. (2015). A concept for automated construction progress monitoring using BIM-based geometric constraints and photogrammetric point clouds. *Journal of Information Technology in Construction*, 20, 68–79.

Brilakis, I., Fathi, H., & Rashidi, A. (2011). Progressive 3D reconstruction of infrastructure with videogrammetry. *Automation in Construction*, 20, 884–895. https://doi.org/10.1016/j.autcon.2011.03.005

Chen, K., Lu, W., Peng, Y., Rowlinson, S., & Huang, G. Q. (2015). Bridging BIM and building: From a literature review to an integrated conceptual framework. *International Journal of Project Management*, 33, 1405–1416. https://doi.org/10.1016/j.ijproman.2015.03.006

Cho, Y. K., Youn, J. H., & Martinez, D. (2010). Error modeling for an untethered ultra-wideband system for construction indoor asset tracking. *Automation in Construction*, 19, 43–54. https://doi.org/10.1016/j.autcon.2009.08.001

Dai, F., & Lu, M. (2010). Assessing the accuracy of applying photogrammetry to take geometric measurements on building products. *Journal of Construction Engineering and Management*, 136, 242–250. https://doi.org/10.1061/(ASCE)CO.1943-7862.0000114

Dai, F., Rashidi, A., Brilakis, I., & Vela, P. (2013). Comparison of image-based and time-of-flight-based technologies for three-dimensional reconstruction of infrastructure. *Journal of Construction Engineering and Management*, 139, 69–79. https://doi.org/10.1061/(ASCE)CO.1943-7862.0000565

Dawood, N., Rahimian, F., Seyedzadeh, S., & Sheikhkhoshkar, M. (2020). *Enabling the development and implementation of digital twins*. Proceedings of the 20th International Conference on Construction Applications of Virtual Reality, Teesside University. ISBN: 9780992716127

Dimitrov, A., & Golparvar-Fard, M. (2014). Vision-based material recognition for automated monitoring of construction progress and generating building information modeling from unordered site image collections. *Advanced Engineering Informatics*, 28, 37–49. https://doi.org/10.1016/j.aei.2013.11.002

Dong, G., & Xie, M. (2005). Color clustering and learning for image segmentation based on neural networks. *IEEE Transactions on Neural Networks*, 16, 925–936. https://doi.org/10.1109/TNN.2005.849822

Engstrom, L. (2016). *Fast style transfer in tensorflow*. https://github.com/lengstrom/fast-style-transfer

Golparvar-Fard, M., Peña-Mora, F., Arboleda, C. A., & Lee, S. (2009). Visualization of construction progress monitoring with 4D simulation model overlaid on time-lapsed photographs. *Journal of Computing in Civil Engineering.* https://journals.vgtu.lt/index.php/JCEM/article/view/11803?toggle_hypothesis=off

Golparvar-Fard, M., Peña-Mora, F., & Savarese, S. (2015). Automated progress monitoring using unordered daily construction photographs and IFC-based building information models. *Journal of Computing in Civil Engineering, 29*, 04014025. https://doi.org/10.1061/(ASCE)CP.1943-5487.0000205

Gu, N., Singh, V., & London, K. (2014). *BIM ecosystem: The coevolution of products, processes, and people* (pp. 197–210). John Wiley & Sons.

Guo, H., Yu, Y., & Skitmore, M. (2017). Visualization technology-based construction safety management: A review. *Automation in Construction, 73*, 135–144. https://doi.org/10.1016/j.autcon.2016.10.004

Gupta, S., Girshick, R., Arbelaez, P., & Malik, J. (2014). *Learning rich features from RGB-D images for object detection and segmentation supplementary material.* Computer Vision – Eccv 2014, Pt Vii. Springer.

Ham, Y., Han, K. K., Lin, J. J., & Golparvar-Fard, M. (2016). Visual monitoring of civil infrastructure systems via camera-equipped unmanned aerial vehicles (UAVs): A review of related works. *Visualization in Engineering, 4*, 1. https://doi.org/10.1186/s40327-015-0029-z

Hammad, A. (2009). Distributed augmented reality for visualising collaborative construction tasks. *Mixed Reality In Architecture, Design And Construction, 23*, 171–183. https://doi.org/10.1007/978-1-4020-9088-2_11

Han, K. K., & Golparvar-Fard, M. (2015). Appearance-based material classification for monitoring of operation-level construction progress using 4D BIM and site photologs. *Automation in Construction, 53*, 44–57. https://doi.org/10.1016/j.autcon.2015.02.007

Han, K. K., Lin, J. J., & Golparvar-Fard, M. (2015). *A formalism for utilization of autonomous vision-based systems and integrated project models for construction progress monitoring.* 2015 Conference on Autonomous and Robotic Construction of Infrastructure.

Hazirbas, C., Ma, L., Domokos, C., & Cremers, D. (2017). *FuseNet: Incorporating depth into semantic segmentation via fusion-based CNN architecture.* Lecture Notes in Computer Science (including subseries Lecture Notes in Artificial Intelligence and Lecture Notes in Bioinformatics), Springer.

Huang, T., Kong, C. W., Guo, H. L., Baldwin, A., & Li, H. (2007). A virtual prototyping system for simulating construction processes. *Automation in Construction, 16*, 1–21. https://doi.org/10.1016/j.autcon.2006.09.007

Ibem, E. O., & Laryea, S. (2014). Survey of digital technologies in procurement of construction projects. *Automation in Construction, 46*, 11–21. https://doi.org/10.1016/j.autcon.2014.07.003

Adwan, E. J., & Al-Soufi, A. (2016). A review of ICT technology in construction. *International Journal of Managing Information Technology, 8*, 01–21. https://doi.org/10.5121/ijmit.2016.8401

Jaselskis, E. J., & El-Misalami, T. (2003). Implementing radio frequency identification in the construction process. *Journal of Construction Engineering and Management, 129*, 680–688. https://doi.org/10.1061/(ASCE)0733-9364(2003)129:6(680)

Jochen Teizer, D. L. A. M. S. (2007). *Rapid automated monitoring of construction site activities using ultra-wideband.* 24th International Symposium on Automation and Robotics in Construction (ISARC).

Kim, C., Kim, B., & Kim, H. (2013). 4D CAD model updating using image processing-based construction progress monitoring. *Automation in Construction, 35*, 44–52. https://doi.org/10.1016/j.autcon.2013.03.005

Kim, H., & Kano, N. (2008). Comparison of construction photograph and VR image in construction progress. *Automation in Construction, 17*, 137–143. https://doi.org/10.1016/j.autcon.2006.12.005

Kim, H., & Kim, H. (2018). 3D reconstruction of a concrete mixer truck for training object detectors. *Automation in Construction, 88*, 23–30. https://doi.org/10.1016/j.autcon.2017.12.034

Kluckner, S., Birchbauer, J. A., Windisch, C., Hoppe, C., Irschara, A., Wendel, A., Zollmann, S., Reitmayr, G., & Bischof, H. (2011). *AVSS 2011 demo session: Construction site monitoring from highly-overlapping MAV images.* 2011 8th IEEE International Conference on Advanced Video and Signal Based Surveillance, AVSS 2011, IEEE.

Kolo, S. J., Rahimian, F. P., & Goulding, J. S. (2014). Offsite manufacturing construction: A big opportunity for housing delivery in Nigeria. *Procedia Engineering*, 85, 319–327. https://doi.org/10.1016/j.proeng.2014.10.557

Kopsida, M., & Brilakis, I. (2016). *BIM registration methods for mobile augmented reality-based inspection* (pp. 201–208). CRC Press.

Ladha, S., & Singh, R. (2018). *Monitoring construction of a structure.* https://patents.google.com/patent/US20180012125A1/en

Li, H., Huang, T., Kong, C. W., Guo, H. L., Baldwin, A., Chan, N., & Wong, J. (2008). Integrating design and construction through virtual prototyping. *Automation in Construction*, 17, 915–922. https://doi.org/10.1016/j.autcon.2008.02.016

Lin, J. J., Han, K. K., & Golparvar-Fard, M. (2015). *A framework for model-driven acquisition and analytics of visual data using UAVs for automated construction progress monitoring (ASCE).* American Society of Computing in Civil Engineering.

Lorenzo, T. M., Bossi, B., Cassano, M., & Todaro, D. (2014). BIM and QR code: A synergic application in construction site management. *Procedia Engineering*, 85, 520–528. https://doi.org/10.1016/j.proeng.2014.10.579

Lu, R., & Brilakis, I. (2019). Digital twinning of existing reinforced concrete bridges from labelled point clusters. *Automation in Construction*, 105, 102837. https://doi.org/10.1016/J.AUTCON.2019.102837

McCormac, J., Handa, A., Leutenegger, S., & Davison, A. J. (2017). *SceneNet RGB-D: Can 5M synthetic images beat generic imagenet pre-training on indoor segmentation?* Proceedings of the IEEE International Conference on Computer Vision.

Memon, ZA., Majid, M. Z., & Mustaffar, M. (2005). *An automatic project progress monintoring model by integrating auto CAD and digital photos 1605–1617.* https://ascelibrary.org/doi/abs/10.1061/40794%2819%29151

Oliver, S., Seyedzadeh, S., Rahimian, F., Dawood, N., & Rodriguez, S. (2020). Cost-effective as-built BIM modelling using 3D point-clouds and photogrammetry. *Current Trends in Civil & Structural Engineering-CTCSE*, 4(5), 000599. https://doi.org/10.33552/CTCSE.2020.04.000599

Omar, T., & Nehdi, M. L. (2016). Data acquisition technologies for construction progress tracking. *Automation in Construction*, 70, 143–155. https://doi.org/10.1016/j.autcon.2016.06.016

Panagiotakis, C., Papadakis, H., Grinias, E., Komodakis, N., Fragopoulou, P., & Tziritas, G. (2013). Interactive image segmentation based on synthetic graph coordinates. *Pattern Recognition*, 46, 2940–2952. https://doi.org/10.1016/j.patcog.2013.04.004

Pătrăucean, V., Armeni, I., Nahangi, M., Yeung, J., Brilakis, I., & Haas, C. (2015). State of research in automatic as-built modelling. *Advanced Engineering Informatics*, 29, 162–171. https://doi.org/10.1016/j.aei.2015.01.001

Pour Rahimian, F., Chavdarova, V., Oliver, S., & Chamo, F. (2019). OpenBIM-Tango integrated virtual showroom for offsite manufactured production of self-build housing. *Automation in Construction*, 102, 1–16. https://doi.org/10.1016/j.autcon.2019.02.009

Pour Rahimian, F., Seyedzadeh, S., Oliver, S., Rodriguez, S., & Dawood, N. (2020). On-demand monitoring of construction projects through a game-like hybrid application of BIM and machine learning. *Automation in Construction*, 110, 103012. https://doi.org/10.1016/j.autcon.2019.103012

Rahimian, F. P., Seyedzadeh, S., & Glesk, I. (2019). OCDMA-based sensor network for monitoring construction sites affected by vibrations. *Journal of Information Technology in Construction*, 24, 299–317.

Rashidi, A., Sigari, M. H., Maghiar, M., & Citrin, D. (2016). An analogy between various machine-learning techniques for detecting construction materials in digital images. *KSCE Journal of Civil Engineering*, 20, 1178–1188. https://doi.org/10.1007/s12205-015-0726-0

Ratajczak, J., Schweigkofler, A., Riedl, M., & Matt, D. T. (2019). *Augmented reality combined with location-based management system to improve the construction process, quality control and information flow* (pp. 289–296). Springer.

Retik, A., Mair, G. M., & Fryer, R. (2006). *Integrating virtual reality and telepresence to remotely monitor construction sites: A ViRTUE project* (pp. 459–463). Springer.

Roh, S., Aziz, Z., & Peña-Mora, F. (2011). An object-based 3D walk-through model for interior construction progress monitoring. *Automation in Construction, 20,* 66–75. https://doi.org/10.1016/j.autcon.2010.07.003

Ros, G., Sellart, L., Materzynska, J., Vazquez, D., & Lopez, A. M. (2016). *The synthia dataset: A large collection of synthetic images for semantic segmentation of urban scenes.* Proceedings of the IEEE Computer Society Conference on Computer Vision and Pattern Recognition.

Schall, G., Mendez, E., Kruijff, E., Veas, E., Junghanns, S., Reitinger, B., & Schmalstieg, D. (2009). Handheld augmented reality for underground infrastructure visualization. *Personal and Ubiquitous Computing, 13,* 281–291. https://doi.org/10.1007/s00779-008-0204-5

Schonberger, J. L., & Frahm, J.-M. (2016). *Structure-from-motion revisited.* The IEEE Conference on Computer Vision and Pattern Recognition (CVPR), IEEE.

Schönberger, J. L., Zheng, E., Pollefeys, M., & Frahm, J.-M. (2016). *Pixelwise view selection for unstructured multi-view stereo.* European Conference on Computer Vision (ECCV).

Seo, J., Han, S., Lee, S., & Kim, H. (2015). Computer vision techniques for construction safety and health monitoring. *Advanced Engienering Informatics, 29,* 239–251. https://doi.org/10.1016/j.aei.2015.02.001

Seyedzadeh, S., Agapiou, A., Moghaddasi, M., Dado, M., & Glesk, I. (2021). WON-OCDMA system based on MW-ZCC codes for applications in optical wireless sensor networks. *Sensors, 21.* https://doi.org/10.3390/s21020539

Seyedzadeh, S., & Pour Rahimian, F. (2021a). Building energy data-driven model improved by multi-objective optimisation. In *Data-driven modelling of non-domestic buildings energy performance* (pp. 99–109). Springer. https://doi.org/10.1007/978-3-030-64751-3_6

Seyedzadeh, S., & Pour Rahimian, F. (2021b). Building energy performance assessment methods. In *Data-driven modelling of non-domestic buildings energy performance* (pp. 13–30). Springer. https://doi.org/10.1007/978-3-030-64751-3_2

Seyedzadeh, S., & Pour Rahimian, F. (2021c). Machine learning for building energy forecasting. In *Data-driven modelling of non-domestic buildings energy performance* (pp. 41–76). Springer. https://doi.org/10.1007/978-3-030-64751-3_4

Seyedzadeh, S., & Pour Rahimian, F. (2021d). Machine learning models for prediction of building energy performance. In *Data-driven modelling of non-domestic buildings energy performance* (pp. 77–98). Springer. https://doi.org/10.1007/978-3-030-64751-3_5

Seyedzadeh, S., & Pour Rahimian, F. (2021e). Multi-objective optimisation and building retrofit planning. In *Data-driven modelling of non-domestic buildings energy performance* (pp. 31–39). Springer. https://doi.org/10.1007/978-3-030-64751-3_3

Seyedzadeh, S., Pour Rahimian, F., Rastogi, P., & Glesk, I. (2019a). Tuning machine learning models for prediction of building energy loads. *Sustainable Cities and Society, 47.* https://doi.org/10.1016/j.scs.2019.101484

Seyedzadeh, S., Rahimian, F. P., Glesk, I., & Roper, M. (2018). Machine learning for estimation of building energy consumption and performance: A review. *Visualization in Engineering, 6,* 5. https://doi.org/10.1186/s40327-018-0064-7

Seyedzadeh, S., Rastogi, P., Pour Rahimian, F., Oliver, S., Glesk, I., & Kumar, B. (2019b). *Multi-objective optimisation for tuning building heating and cooling loads forecasting models.* 36th CIB W78 2019 Conference.

Sheikhkhoshkar, M., Pour Rahimian, F., Kaveh, M. H., Hosseini, M. R., & Edwards, D. J. (2019). Automated planning of concrete joint layouts with 4D-BIM. *Automation in Construction, 107,* 102943. https://doi.org/10.1016/j.autcon.2019.102943

Shotton, J., Girshick, R., Fitzgibbon, A., Sharp, T., Cook, M., Finocchio, M., Moore, R., Kohli, P., Criminisi, A., Kipman, A., & Blake, A. (2013). Efficient human pose estimation from single depth images. *IEEE Transactions on Pattern Analysis and Machine Intelligence, 35,* 2821–2840. https://doi.org/10.1109/TPAMI.2012.241

Silberman, N., Hoiem, D., Kohli, P., & Fergus, R. (2012). *Indoor segmentation and support inference from RGBD images lecture notes in computer science.* ECCV'12: Proceedings of the 12th European conference on Computer Vision – Volume Part V, Springer.

Soltani, M. M., Zhu, Z., & Hammad, A. (2016). Automated annotation for visual recognition of construction resources using synthetic images. *Automation in Construction, 62,* 14–23. https://doi.org/10.1016/j.autcon.2015.10.002

Song, J., Haas, C. T., & Caldas, C. H. (2006). Tracking the location of materials on construction job sites. *Journal of Construction Engineering and Management, 132,* 911–918. https://doi.org/10.1061/(ASCE)0733-9364(2006)132:9(911)

Stewart, R. A. (2007). IT enhanced project information management in construction: Pathways to improved performance and strategic competitiveness. *Automation in Construction, 16,* 511–517. https://doi.org/10.1016/j.autcon.2006.09.001

Svalestuen, F., Knotten, V., Lædre, O., Drevland, F., & Lohne, J. (2017). Using building information model (BIM) devices to improve information flow and collaboration on construction sites. *Journal of Information Technology in Construction, 22,* 204–219.

Teizer, J. (2015). Status quo and open challenges in vision-based sensing and tracking of temporary resources on infrastructure construction sites. *Advanced Engineering Informatics, 29,* 225–238. https://doi.org/10.1016/j.aei.2015.03.006

Tserng, H. P., Dzeng, R. J., Lin, Y. C., & Lin, S. T. (2005). Mobile construction supply chain management using PDA and bar codes. *Computer-Aided Civil and Infrastructure Engineering, 20,* 242–264. https://doi.org/10.1111/j.1467-8667.2005.00391

Turkan, Y., Bosché, F., Haas, C. T., & Haas, R. (2012). Automated progress tracking using 4D schedule and 3D sensing technologies. *Automation in Construction, 22,* 414–421. https://doi.org/10.1016/j.autcon.2011.10.003

Wang, X. (2007). Using augmented reality to plan virtual construction worksite. *International Journal of Advanced Robotic Systems, 4,* 501–512. https://doi.org/10.5772/5677

Whyte, J. (2002). *Virtual reality and the built environment.* https://doi.org/10.4324/9780080520667

Woodhead, R., Stephenson, P., & Morrey, D. (2018). Digital construction: From point solutions to IoT ecosystem. *Automation in Construction, 93,* 35–46. https://doi.org/10.1016/J.AUTCON.2018.05.004

Wu, Y., Kim, H., Kim, C., & Han, S. H. (2010). Object recognition in construction-site images using 3D CAD-based filtering. *Journal of Computing in Civil Engineering, 24,* 56–64. https://doi.org/10.1061/(ASCE)0887-3801(2010)24:1(56)

Xie, H., Shi, W., & Issa, R. R. A. (2015). Using RFID and real-time virtual reality simulation for optimization in steel construction. *Journal of Information Technology, 16,* 291–308.

15 Digitalisation of Architecture, Engineering, and Construction

Immersive technologies and unmanned aerial vehicles

15.1. Introduction

Digital transformation in Architecture, Engineering, and Construction (AEC) has witnessed an increase in the uptake and adoption of a wide range of technologies to support business operations. Part of this research includes the adoption of technologies relating to the use and application of Unmanned Aerial Vehicles (UAVs), also known as drones (Dawood et al., 2020; Oliver et al., 2020). This is seen as a valuable addition to the immersive technologies already used in the industry. Given this, the purpose of this chapter is to present the current status of employing UAVs and immersive technologies in AEC. In particular, how these support digitalisation and the specific ways these technologies support applications, either individually or collectively, as part of wider integration measures. These issues are explored through a literature review using meta-synthesis in order to evaluate and integrate findings into a single context.

As a precursor to this discussion, the rapid development of information technology and systems have played a pivotal role in expediting the implementation of digitalisation in industry (Elghaish et al., 2020). For example, increased labour costs have been countered by decreasing costs of technologies, which in turn has encouraged the advancement of new digitisation processes in the quest to move towards Industry 4.0 (Golizadeh et al., 2018; Newman et al., 2020). Part of the movement includes Building Information Modelling (BIM), which is seen as one of the main embodiments of digitalisation (Howard & Björk, 2008; Pour Rahimian et al., 2020). More recently, the use of UAVs and immersive technologies have become very popular. UAVs, also known as drones, flying robots, or Unmanned Aerial Systems (UAS), have experienced an unusual growth as they have become much more affordable (Martinez et al., 2020). This made many UAV-based applications more widely available for the automation of construction processes, including gathering information (e.g. taking photos), surveying construction sites (Bang et al., 2017), monitoring construction progress (Liu et al., 2014), inspecting built infrastructure (Kim et al., 2015), and assessing safety of construction sites (Gonçalves et al., 2017).

AEC has also started to more readily embrace immersive technologies, especially Virtual Reality (VR), Augmented Reality (AR), and Mixed Reality (MR). That being said, the concept of VR was established many years ago as it 'replaced' users' perception of the surrounding environment with a computer-generated artificial 3D environment (Zhou et al., 2012). AR technology integrates objects in both the real and virtual worlds, allowing real-time interaction. This has been seen to enhance users' perception, where real and virtual entities overlap (Hou et al., 2013). In essence, VR and AR are based on the engagement and involvement of visual sensations from the real world regardless of the establishment of immersion or the

DOI: 10.1201/9781003106944-15

mechanisms of the display (Wolfartsberger, 2019). This has now encompassed MR, a combination of reality and virtual, which blends real and virtual worlds to generate a brand-new environment – where physical and digital objects simultaneously interact (Chi et al., 2013). These approaches offer AEC significant potential, transforming the ways in which organisations consume and interact with information (Boton, 2018).

The adoption of the aforementioned technologies is expected to lead to: improved project delivery (Ammar et al., 2018), enhanced communication among stakeholders, thereby supporting informed decision-making (Elghaish et al., 2019), reduced on-site injuries and fatalities (Aghimien et al., 2019), and improved productivity (Leviäkangas et al., 2017); however, research exploring the potential applications of using UAVs and immersive technologies (either individually or integrated with other technologies) is rather limited. Given the paucity of research in this area, this chapter presents and discusses the latest developments and applications of these technologies in AEC, including trends and ways through which these technologies can be integrated and exploited for market uptake. This discussion includes: (i) an overview of technologies, (ii) applications available in AEC, and (iii) key findings and observations.

15.2. Methodology

The content and structure of this chapter is presented on findings generated through a literature review, where according to Webster and Watson (2002), the purpose of which is to address the research gap by identifying, evaluating, and integrating the previous findings from relevant and similar studies. Following this approach, the literature review presented here aimed to: (i) understand the progress achieved by other researchers to build the research base that can be used as a point of departure (Cohen et al., 2013); (ii) build an integrated context that respects the arrangement between the different ideas, while showing contradictions between theories in order to build a reliable argument (Saunders et al., 2016); and (iii) articulate specific statements and build arguments around each statement, using different views to link them together in a single context. This naturally includes epistemological underpinnings. Epistemology is a philosophical term that discusses the nature, structure, and scope of knowledge, and epistemology explores the sources that shape knowledge, such as perception, memory, and different types of reasoning (Goldman, 2004). Whilst several positions exist [*cf* empiricism, rationalism, skepticism, pragmatism, etc.], one epistemological position is termed 'interpretivist' or 'interpretivism', which refers to discovering the reality from human views (human experience and perceptions) (Cohen et al., 1994), and is often adopted to understand the socially constructed contexts and beliefs (Willis et al., 2007). This approach is therefore followed in this chapter, as it more naturally resonates and relates to this enquiry approach.

In summary, the critical literature review presented here provides a cognate-specific review of research in this field. This aimed to explore the specific issue and gap in knowledge (Arbnor & Bjerke, 2008), where publications were sourced using an online search, using prominent cognate domain-specific journals and conference articles, targeted according to specific criteria, namely: the category of journals (Q1 and Q2) using Scopus classification – with the date of publication from 2008 onwards – and the inclusion of major online databases for conference papers: IEEE, Thomson Reuters' Web of Science, ProQuest (ABI/INFORM), ScienceDirect (Elsevier), and Xplore. Through these searches, different keywords were used, such as 'virtual reality in construction', 'augmented and mixed reality in construction', 'drones applications in construction', and 'health and safety using drones in construction'. These papers were categorised into different themes and sub-themes regarding immersive technologies and UAVs. From this, interpretive analytical techniques (critical and meta-synthesis techniques), were then used to analyse findings. The interpretive analytical techniques aimed to evaluate and integrate findings into a single context to identify the key elements (Leary & Walker, 2018). Accordingly, these were listed in relation to each theme, summarised, and contextually interpreted (Brinkmann, 2013, p. 62).

15.3 Unmanned aerial vehicles: technologies and applications

15.3.1. *Applications and benefits*

Over recent years, UAVs have been used across a wide range of construction activities, especially for their ability to capture data (Asadi et al., 2020). One of the main reasons for this increased use acknowledges that UAV-based application can often surpass conventional methods on construction sites in terms of accuracy, efficiency, and cost effectiveness (Green-wood et al., 2019); however, whilst acknowledging that literature in this area is limited, a number of publications have explored the benefits of UAVs in AEC, albeit largely explora-tory. For example, Irizarry and Costa (2016) explored the potential benefits of a small-scale aerial drone on construction sites for safety managers, leading to recommendations support-ing the need for safety inspection drones. Other work by Liu et al. (2014) overviewed UAV developments and their possible applications in civil engineering. This presented a summary of potential UAV applications in seismic risk assessment, transportation, disaster response, construction management, surveying and mapping, and flood monitoring and assessment.

From a construction management perspective, Irizarry and Costa (2016) presented an exploratory case study that identified potential applications of visual assets obtained from UASs, particularly for construction management tasks. Results from this study revealed sev-eral potential applications, mainly for project progress monitoring, job-site logistics, evalu-ation of safety, quality inspections, and general management tasks. Work by Goessens et al. (2018) examined the feasibility of building real-scale structures (masonry) with custom-built drones, highlighting issues of precision, behaviour, etc., including the need to establish the first guidelines for the design of 'drone-compatible' construction elements. This also high-lighted that the use of UAVs in construction was very promising, not just in terms of the drones per se, but also regarding the transition from the laboratory stage to the construction of real structures with complex geometry, connections, and finishings. On this theme, Li et al. (2018) examined 3D path-planning algorithms for drones in the indoor environment. This study highlighted a novel approach for planning universal paths for drones using a voxel model, by computing a 3D buffer around obstacles using approximate Euclidean distance transformation. On the mapping theme, Álvares et al. (2018) presented an exploratory study for 3D mapping assessment of buildings and construction sites using UAS imagery to support construction-management tasks. Findings from two case studies showed that 3D mapping from UAS imagery offered views from different perspectives, which supported greater inter-activity and manipulation of 3D models.

Similar construction applications included the work by Bogue (2018), which discussed the role of drones and autonomous ground vehicles regarding their use in the construction industry – from site surveying to the monitoring of project progress, inventory management, transportation and logistics, health and safety, etc. For example, Goessens et al. (2018) cap-tured the merits of adopting UAS technology throughout the lifecycle of a project, which also proposed a multidimensional framework that focused on four dimensions: lifecycle, managed object, potential role, and stakeholder engagement. This resonates with a study by Greenwood et al. (2019) regarding a review of UAV developments in civil-infrastructure applications, including post-disaster reconnaissance, infrastructure-component monitoring, geotechnical engineering, and construction management. That being said, whilst using drones in the construction industry has many significant advantages (Li & Liu, 2019), it is equally important to acknowledge that barriers to wider uptake also exist. For example, Golizadeh et al. (2019) investigated barriers to adoption (albeit from an Australian context). In summary, a number of applications are now starting to permeate the market. These are significant and offer AEC several opportunities for exploitation. This work also includes the wider issue of human performance and the relationship between performance and UAS experience Kim et al. (2019a).

15.3.2. Automated surveying

One of the typical operations undertaken on construction sites is that of inspections, where progress is visually assessed against pre-determined performance. In this respect, a traditional method of completing this task is to assign staff to go on 'inspection rounds' with printed checklists, taking photographs as needed to document progress (Frank, 2012). From this inspection, all gathered information is then placed into files or models, which is then transferred to contractors and other relevant staff (Kim et al., 2014); however, this process is often labour-intensive, where the efforts of maintaining updated models and schedules become increasingly complex, especially with large construction projects. Surveying applications have been used to alleviate this challenge to some extent (Barbarella et al., 2017), but their selection often depends on the surrounding terrain and size of the area that needs to be surveyed (Oskouie et al., 2016). They are also limited in range, labour-intensive, costly, prone to measurement errors, and consume a considerable amount of time (Tang et al., 2010); however, UAVs have now been promoted as a potential solution to these concerns. For example, UAV technology has been shown to be accurate and reliable (Siebert & Teizer, 2014).

Siebert and Teizer (2014) developed a novel approach for evaluating the performance of a newly designed as-built UAV system in test-bed and field-realistic environments. This evaluation focused on the magnitude of errors associated with UAV-based photogrammetric approaches compared with conventional surveying techniques used for ground truth measurements. Similar studies by Shazali and Tahar (2019) assessed the geometric accuracy of a 3D model using UAV images, noting that errors between the actual measurement and the generated 3D model were less than 4 cm. Work by Moon et al. (2019) evaluated the potential and usability of data integration by comparing the data processed through photogrammetry (based on laser scanning), focusing on earthworks. This study proposed a method for generating and merging hybrid point-cloud data acquired from laser scanning and UAV-based image processing.

From a data acquisition perspective, Freimuth and König (2018) focused on improving as-built data generation using a framework employing autonomous UAVs to fully automate the acquisition of structural objects. Results highlighted that the prototype framework was capable of guiding a UAV around building structures effectively, enabling it to capture as-built information at an object level. Work by Marmo et al. (2019) compared the original blueprints of the Basento River bridge with a photogrammetric survey carried out by UAV, noting a good agreement between the blueprints and the surveyed geometry. Finally, Park et al. (2019) proposed a framework for the automated registration of UAV and point clouds using 2D local feature points in images taken from UAVs and unmanned ground vehicles

(UGVs). This study identified the optimal angle at which to detect sufficient points matching the images taken by the point clouds, confirming that full automation of spatial data collection and registration from a scattered environment (e.g. construction or disaster sites) by UAVs was feasible without human intervention.

15.3.3. Information management and visualisation

The use and application of UAVs are becoming an essential part of virtual design and construction (VDC), providing architects and engineers with new and efficient ways of visualising and analysing structural elements. This includes increased emphasis on the issue of integration among UAV technology, information management, and visualisation. For example, Lu and Davis (2018) presented a framework for integrating unordered images, geometric models, and the surrounding environment on Google Earth. This used two major components: UAV-centric image alignment and processing, and a Keyhole Markup Language (KML)-based image and 3D model-management system. This system provided construction engineers with a low-cost and low-technology barrier solution to capture and represent a dynamic construction site. Similar studies by Puppala et al. (2018) developed 3D models using UAV-based photogrammetry to assess the health conditions of structures, focusing on material performance. This engaged different image datasets on the condition of infrastructure assets, where 3D models were then developed and assessed.

Research by Ajibola et al. (2019) developed a model that integrated weighted averaging and additive median filtering algorithms to improve the accuracy and quality of a Digital Elevation Model (DEM) produced by UAV. Results identified an increase of 88% in the accuracy of the fused DEM. On this theme, Li and Liu (2019) presented an approach using neural networks to extract road information from remote-sensing images using a camera sensor equipped with UAV (Seyedzadeh et al., 2021). Similarly, Kim et al. (2019b) proposed a UAV-assisted robotic approach for data collection and processing, noting that technologies could be deployed in cluttered environments to support timely decision-making. This resonates with the work by Ham and Kamari (2019), who presented an automated method for enabling practitioners to assess the status of construction sites using selective visual data to support data-driven decision-making.

15.3.4. Monitoring, inspection, and safety

Construction managers, job superintendents, and safety coordinators often have to undertake periodic inspections for job-site progress monitoring, safety assessment, etc. (Ajibola et al., 2019). Drones and UAVs are now increasingly being used to do these tasks. That being said, whilst commercial drones can track and inspect sites faster, better, and more accurately, surprisingly, many construction professionals still inspect and monitor their sites in the traditional way (Bosche & Haas, 2008).

From a progress monitoring and inspection perspective, Bang et al. (2017) proposed a method of generating a construction site panorama using an image stitching technique, which focused on pre-processing to help managers identify various construction site conditions more readily. Similarly, Vick and Brilakis (2018) presented a novel model for automatically detecting layered road design surfaces through as-built point-cloud data using UAS. Other applications include the work by Kim and Kim (2018), who used First Personal View of a quadcopter drone to monitor on-site status, enabling findings to be communicated to various construction professionals. Work by Kim et al. (2017) presented a crack identification strategy that combined hybrid image-processing with UAV technology. This system was able to successfully measure cracks thicker than 0.1 mm with a maximum length-estimation error of 7.3%. Additionally, Freimuth and König (2018) developed an application that integrated UAVs and BIM for visual inspection tasks. This application enabled operators to plan inspections in a 3D environment based on BIM data – triggering inspections automatically. A recent study conducted by Ficapal and Mutis (2019) presented a framework for detecting, diagnosing, and evaluating thermal bridges in façade systems using infrared thermography and UAVs. This focused on the general performance of the building envelope, including its state of deterioration, obsolescence, energy consumption, and functionality (Seyedzadeh & Pour Rahimian, 2021a, 2021b). Findings from this study have been used to facilitate remedial actions.

From a safety perspective, the work by Gheisari and Esmaeili (2019) developed a usability study and heuristic evaluation of a small-scale quadcopter equipped with a camera (as a safety inspection tool on construction sites). This concluded that UASs could be used as a safety-inspection assistant, providing safety managers with voice interaction with construction workers and real-time access to videos and images from around the job site. This resonates with the work by de Melo and Costa (2019), who used two case studies for assessing UASs in the collection of visual assets regarding compliance with safety regulations. A conceptual framework was presented to support safety planning and control, noting that this integration helped support decision-making, especially in tasks that involved a high risk of accidents. Other studies focusing on construction safety include Kim et al. (2019a), who developed a UAV-assisted visual monitoring method for automatically measuring proximities among construction entities. This was proffered as a safety intervention method, thereby promoting a safer working environment for construction workers. On this issue, Liu et al. (2019) proposed a safety-inspection method that integrated UAV with BIM. This used a dynamic BIM model, which was created by aggregating safety information with a BIM model, facilitating synchronous navigation of UAV video with dynamic BIM to support safety evaluations and safety training. Finally, the work by Gheisari and Esmaeili (2019) conducted a study that investigated the effectiveness of using UASs to improve safety operations. Findings indicated that safety activities could be improved using UASs, particularly through monitoring boom vehicles such as cranes, especially in the proximity of overhead power lines, unprotected edges, or openings.

15.4. Immersive technologies

15.4.1. *Project monitoring and control*

The use of immersive technologies in AEC continues to provide a number of significant benefits, particularly for monitoring project performance. For example, these technologies can now provide users with a number of visualisation opportunities to see building progress against schedules. This is especially useful for monitoring and tracking performance on personal computers, tablets, and smartphones. Examples of these types of applications include the work by Park and Kim (2013), who complemented existing research in this area to support scheduling and monitoring by connecting AR material tracking on-site. Similar work in construction site monitoring includes Ratajczak et al. (2019), who noted that context-specific information on progress and key performance indicators could also be captured.

As previously mentioned, other studies have also focused on safety monitoring. This has not started to include immersivity. For example, systems combining immersive technologies and real-time tracking can be used to improve the situational awareness of construction workers to avoid hazardous situations (Cheng & Teizer, 2013; Kim et al., 2017). Some studies are also starting to focus on project scheduling, where Kim et al. (2018) developed an AR-based 4D CAD system that connected 4D and 5D objects with real field images and AR objects against schedule information. Similar work on scheduling includes research by Ratajczak et al. (2019), who developed a field application that integrated a location-based management system into BIM along with an AR platform. This was able to detect scheduling deviations, with users being able to visualise construction progress in AR on issues such as daily progress, performance, and context-specific information/documents on scheduled tasks.

15.4.2. Facilities management

Facilities Management (FM) can loosely be described as the process of managing all assets and services supporting the product, environment infrastructure, or servitisation components underpinning the business entity. In this respect, FM is considered to be one of the growth areas for immersive technologies; however, to date, no commonly accepted integrated data model has been proposed in this area to support FM. That being said, BIM models have now gained increased acceptance in AEC, but these are not yet robust or mature enough to adequately support data exchange in the post-handover (operation) phase. Some studies have started to investigate maintenance inspections; for example, Sampaio et al. (2012) presented two prototype applications to support the performance of periodic inspections and the monitoring of interior and exterior wall maintenance, using VR technology. Other work in this area includes Irizarry et al. (2013), who developed a system called InfoS-POT (a mobile AR tool) to support facilities managers' information on the assets they maintain. Similarly, the work by Williams et al. (2015) developed a fully automated process to help facilities managers access real-time information using an AR environment linked to BIM models in order to help them make better FM decisions. Finally, other FM research includes that of Paulo Carreira et al. (2018), who evaluated the implementation of a building management system using game engine technologies (in a VR environment) to integrate with FM activities.

15.4.3. Provision of project information on-site

Given the need to manage a vast array of project information within AEC, it has been acknowledged that immersive technologies could be used to pool digital data and documentation into a central prism for greater interaction. This is seen as a game-changer. Examples include Yeh et al. (2012), who presented a wearable device that could project construction drawings and related information, which not only helped engineers minimise the need to search for the latest documentation, but also provided time savings associated with decision-making. On a similar theme, Kim et al. (2018) developed a system using AR to facilitate the accurate exchange of project information and 4D objects with personnel. On-site construction management using mobile computing technology was also demonstrated by Kim et al. (2013), which engaged mobile computing technology to deliver project information in an AR environment. BIM and AR was also evaluated by Chu et al. (2018), particularly for its effectiveness and integration. This highlighted that task efficiency could be enhanced, along with an improved ability to retrieve information, especially with BIM/AR systems that supported cloud-based storage capabilities.

15.4.4. Team collaboration and communication

Construction projects often involve significant collaboration among project disciplines, including contractors, designers, managers, engineers, subcontractors, etc. In this respect, successful partnerships are predominantly those that are able to maximise collaboration and communication; however, more often than not, it is not always possible to have all project team members available at the same time, yet alone be available on-site. This can sometimes be challenging, especially where errors or design challenges need to be resolved. This is now starting to change, as the use of immersive technologies are now enabling stakeholders to 'virtually' visit a site, to see these challenges, take notes, share video views, and make collective decisions in real time.

Examples of this type of immersive technology include Pejoska et al. (2016), who highlighted that immersive technologies could improve on-site project information accessibility, whilst also supporting more effective communication. The use of immersive technologies can therefore be seen as a useful mechanism for facilitating collaboration and communication between the design team. For example, Goulding et al. (2012) demonstrated an approach for integrating collaborative design teams to facilitate project integration. This used a game environment supported by a web-based VR cloud platform to facilitate collaboration and decision-making. Similar work by Chalhoub and Ayer (2018) examined the application of MR technologies for electrical construction design communication. This compared the performance of 18 electrical construction personnel who were tasked with building similar conduit assemblies against a traditional paper-based approach. BIM data and VR have also been examined for collaborative decision-making, where Du et al. (2018) developed a real-time synchronisation system. This used an innovative cloud-based BIM metadata interpretation and communication method that allowed users to visualise BIM model updates/changes in VR headsets automatically.

Other studies have also focused on facilitating project communication between. For example, Lin et al. (2015) proposed a visualised environment that facilitated construction discussion using AR in a multiscreen environment. Communication was also explored by Zaker and Coloma (2018), who investigated the application of a VR-based workflow system. A VR integrated collaboration workflow case study was used to demonstrate how AEC firms could overcome the challenge of collaboration. Improved communication and interactivity was also demonstrated by Du et al. (2018), who developed a cloud-based multi-user VR headset system that facilitated interpersonal project communication within an interactive VR environment.

15.4.5. Training and education

VR-based training simulators have been successfully employed in many industries (such as aviation) to help train operators and professionals in a safe environment (Sacks et al., 2013). In AEC, this approach has now started to occur more frequently, especially given its effectiveness; arguably, as the industry tends to use a lot of machinery, there are significant savings to be made (yet alone the benefits associated with health and safety). Acknowledging this potential, immersive technologies can be used to support construction workers, especially to support decision-making (Zhao & Lucas, 2015). Additionally, immersive technologies also provide a much safer training environment, with a corresponding reduced risk of injury (Goulding et al., 2012).

Several studies have focused on the effectiveness of using immersive technologies to avoid exposure to hazardous job sites whilst also improving the overall training provision and experience (Albert et al., 2014; Shi et al., 2019). For example, the use and application of immersive technology-based systems have proven to be particularly successful in AEC, providing personnel with real practical and safety experience (Le et al., 2015; Pedro et al., 2016; Sacks et al., 2013). The use of VR technologies have also started to permeate the training market. For example, 360-degree VR has been used to simulate construction safety challenges. This 360-degree VR is able to deliver a panoramic view of a real-world environment with a high sense of presence (Portalés et al., 2018). In contrast to traditional VR, 360-degree VR offers fast digital job site generation, easy-to-produce simulations, and high levels of realism due to the inherent photography techniques used with this technology.

Whilst immersive technologies are improving the speed and quality of training, it is also important to acknowledge allowing distant experts to be present (Sacks et al., 2013). Moreover, these technologies have demonstrated their potential in education and training. These technologies can also be used to simulate accessibility design reviews using VR and MR mockups, allowing participation of many stakeholders – both novice and professional (Wu et al., 2019). This transcendence can be used to bridge the gap between experts and new employees, using AR technology and simulated visualisations to enhance new employees' spatial and temporal understanding of complex construction processes (Costa et al. (2019). These approaches can also be used to deliver teaching programmes to support construction courses (Bashabsheh et al., 2019; Bosché et al., 2016; González, 2018; Hou et al., 2017; Meža et al., 2014; Shanbari et al., 2016; Shirazi & Behzadan, 2015; Wang et al., 2018; Zhao et al., 2016).

15.4.6. *Quality management*

From a quality management perspective, immersive technology solutions are continuing to play a key role. For example, the work by Zhao and Lucas (2015) demonstrated how quality management in construction projects could be enhanced using defect detection, highlighting that this can be achieved much earlier in the project lifecycle by coordinating multiple 3D models. This was considered to be much better than using a 2D screen, where issues can be more easily overlooked. It has also been advocated that the application of immersive technologies can save costs and time, especially by resolving potential issues in the virtual world before something is built (Portalés et al., 2018).

Examples of studies that have focussed on quality management in construction using immersive technologies include (but are not limited to): Hou et al. (2015), who developed an AR system to improve product-assembly productivity and performance by lowering cognitive workload via AR; Fazel and Izadi (2018), who developed an affordable interactive multi-marker AR tool for constructing free-form modular surfaces to enhance the accuracy and quality of construction; Ahn et al. (2019), who proposed an approach for visualising vital information within a user's field of view during manufacturing processes – offering considerations for implementing projection-based AR in practice; Mirshokraei et al. (2019), who presented a web-based system for enhancing quality management during the execution phase of structural elements by integrating BIM with AR technology; and Costa et al. (2019), who proposed a spatial AR system for leveraging BIM standards to automatically orchestrate the necessary tasks to help operators tack weld beam attachments to ensure maximum flexibility during the beam-assembly stage.

Other studies in quality management have focused on a wide range of subjects, from defect management to the development of inspection tools to control construction quality. These studies include (but are not limited to): Kwon et al. (2014), who developed a system that combined BIM, image-matching, and AR technologies to improve construction defect management and quality management; Hernández et al. (2018), who developed building self-inspection techniques using AR and BIM technologies to support construction workers in the inspection process – to control quality and ensure specifications are implemented in accordance with the design; and Portalés et al. (2018), who developed an AR-based interactive tool to help support prefabricated building inspection, allowing inspectors to document the process digitally with images and 3D data, along with traditional measurements and annotations.

15.5. Discussion and key findings

Given the complexity and overlap of some of the subject matter presented in this chapter, the discussion and findings are presented through two core themes, namely: 'Immersive Technologies' and 'Unmanned Vehicles Technologies'. This representation can be seen in Figure 15.1.

15.5.1. Immersive technologies

The following section outlines the main areas through which immersive technologies are now starting to have an impact on AEC. Whilst not exhaustive, these are grouped under four application areas as follows:

- **Construction safety training and construction education** was seen as the most popular research area for immersive technologies (covering 35% of the articles reviewed). This is probably due to a number of factors, not least of which is the importance of this area in resolving construction accidents and increased demand for new and innovative approaches to safety. The corollary of these studies seem to additionally: (i) call for more safety interventions using immersive technologies, (ii) highlight the risks beforehand (prior to construction), to reduce or mitigate these before site-work begins, and (iii) to evaluate and quantify the long-term impact of using immersive technology-driven training platforms to support accident reduction.

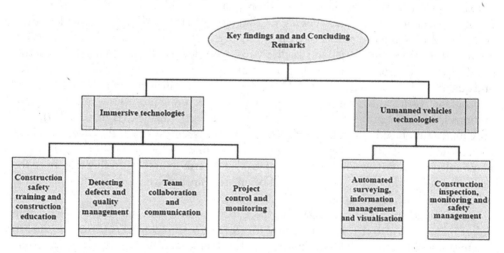

Figure 15.1 Immersive technologies and unmanned vehicles technologies applications

- **Defect and quality management** was given high importance from a research impact perspective. Core issues raised covered the financial implications of rework in construction caused by quality issues and the high cost of on-site inspections. In this respect, calls for new technologies to enhance construction quality and facilitate inspection processes were high on the agenda. The main directions for this work call for: (i) greater emphasis of technologies to support prefabrication and modular construction; (ii) the need to enhance quality of the finished product, whilst also minimising the number of defects; and (iii) the wider use of immersive technologies to facilitate quality inspections, particularly on materials (which often requires architects and engineers to verify that the correct products have been installed in accordance with requirements).

- **Team collaboration and communication** was then seen as the next most important issue, noting the importance of this in project collaboration. This was seen as being pivotal in securing project success. Work in this area now includes active participation in the design process – not just for clients, but for contractors and all other vested parties. For example, a VR headset can facilitate everything from proposed layouts to the visualisation of design choices. This, therefore, enables and facilitates more meaningful and valuable discussions among project participants. That being said, not many articles seemed to cover communication and collaboration during the pre-planning stages. It is therefore acknowledged that this underrepresented area needs some attention, particularly on the use of immersive technologies to facilitate collaboration. This collaboration and communication would allow rich engagement of discourse, especially in relation to the proposed concept. This would also help avoid (or at least minimise) any misunderstandings when discussing complex details – arguably better than either 2D drawings or even 3D BIM models.

- **Project control and monitoring** was seen as the next most important category for AEC engagement with immersive technologies (covering 16% of the articles reviewed). In this remit, AR was seen as a core enabler, particularly its ability in project scheduling. This also includes being able to show 'as-planned vs. as-built' as part of AR's visualisation of progress. This was also shown to be particularly suitable for project scheduling and progress tracking; however, AR did not seem to be used to solve other on-site construction challenges, such as differing site conditions, managing design defects, or applications covering error monitoring. Perhaps this will change in the future? Another much needed area of research concerns FM, where it is also proffered that additional research is needed. Specifically, the use of AR and immersive technologies to support FM and servitisation would be especially beneficial.

15.5.2. Unmanned aerial vehicle technologies

The following section outlines the main areas through which UAV technologies are now starting to have an impact on AEC. Of particular note here, the propensity and use of UAVs and corresponding applications in AEC is a relatively new innovation, where initial interest has been mainly exploratory – examining potential application areas. Early work started to permeate the market around 2012, with a significant increase in publications post 2018. This impetus is expected to rise exponentially in the future (Gheisari & Esmaeili, 2019). To date, UAV-based applications have focused on addressing several niche themes. Whilst not exhaustive, these are broadly grouped under two application areas, these being:

- **Existing applications of UAVs** have mainly been developed to address core AEC challenges, the remits of which include: automated surveying, information management, visualisation, construction inspection and monitoring, and health- and safety-related matters; however, the review noted a paucity of work in the areas surrounding process. More specifically, the integration of process with the technical capabilities of UAVs. In some respects, this may be due in part to the lack of technical skills and experience of AEC practitioners in terms of using UAVs for construction activities. In essence, studies have focused on the technical aspects of UAVs' applications rather than on the UAV implementation process for wider use in the industry. This has arguably somewhat stifled or limited progress. Golizadeh et al. (2019) observed that research was needed to address this gap. In summary, whilst the existing applications of UAVs have started to attract considerable attention, it is expected that a number of new areas will start to emerge over the next few years in line with technological developments and evidence indicators arising from existing application areas.

- **Emerging issues** in the use of UAVs is continuing to grow. That being said, extant literature has not yet fully reported these developments sufficiently in order to predict trends. Further studies are therefore needed to provide in-depth exploration of UAVs within AEC. It is advocated that additional studies are needed on the ability of UAVs to potentially disrupt the industry, insofar as their ability to remove many labour-intensive tasks, or indeed areas of high health and safety risk. Whilst legislation, cost, and the limited number of applications available on the market may be an issue for some applications, these alone are likely to dissipate once the wider benefits are extolled though integration with other technologies such as BIM, AR, and predictive simulation. Some of these issues are now starting to emerge, with applications now embracing FM, servitisation, demolition, civil engineering infrastructure projects, maintenance, structural assessment, surveying, project monitoring and control, data mining, applications supporting the use of digital twins, etc.

15.6. Conclusion

This chapter presented a critical evaluation of digitalisation tools employed in AEC. Two main technologies (UAV technologies and immersive technologies) were explored and critically discussed to evaluate: (i) progress made, (ii) how well these have supported the industry, (iii) how these are helping AEC move towards Industry 4.0, and (iv) future trends and opportunities arising from these two technologies. In this respect, each topic was analysed individually to assess its strengths, weaknesses, capabilities, integration with other technologies, and the viability of the solutions emanating from each of these topics.

This analysis revealed a number of important findings. For example, it was encouraging to note that AEC has really started to delve into these technologies in order to derive value across a number of disciplines. Of particular interest was the need to bring technologies together in order to enhance synergy. In this respect, some progress has been made in the integration and implementation of MR with BIM. Application areas included work on project tracking and progress monitoring, drones for scanning construction sites and assets-producing models, and comparing captured images with 4D BIM to measure progress, etc. Most of these developments were embryonic (at the conceptual stage), but some had developed into highly advanced commercial products for market uptake. Interestingly, much of these developments are also available as outsourced services through proprietary systems and bespoke companies. The issue here seems to be investment vs. return (cf. early adopters). Perhaps maturity and scalability will provide additional solutions over time?

Statistical analysis of existing research in immersive technologies rested in four main areas of applications (defect and quality management; team collaboration and communication; project control and monitoring; and UAV technologies). These application areas included a number of streams, from health and safety to defect management and quality control. In addition, work with UAVs provided encouraging findings, especially in automated surveying, information management, and visualisation. That being said, it is also important to acknowledge that these developments are (arguably) still in their infancy, which in some areas will need additional work to support the veracity of findings. Finally, with the advent of real-time data capture and monitoring, the increased use of autonomous vehicles, improved sensing capabilities of drones (GPS, thermal, acoustic etc), and the increased prevalence of AR, MR, and BIM (including support services such as blockchain and distributed ledgers), it is contended here that AEC is well on the way to meeting Industry 4.0.

References

Aghimien, D. O., Aigbavboa, C. O., Oke, A. E., & Thwala, W. D. (2019). Mapping out research focus for robotics and automation research in construction-related studies. *Journal of Engineering, Design and Technology*. https://doi.org/10.1108/JEDT-09-2019-0237

Ahn, S., Han, S., & Al-Hussein, M. (2019). 2D drawing visualization framework for applying projection-based augmented reality in a panelized construction manufacturing facility: Proof of concept. *Journal of Computing in Civil Engineering, 33*(5). https://doi.org/10.1061/(ASCE)CP.1943-5487.0000843

Ajibola, I. I., Mansor, S., Pradhan, B., & Shafri, H. Z. M. (2019). Fusion of UAV-based DEMs for vertical component accuracy improvement. *Measurement, 147*, 106795. https://doi.org/10.1016/j.measurement.2019.07.023

Albert, A., Hallowell, M. R., Kleiner, B., Chen, A., & Golparvar-Fard, M. (2014). Enhancing construction hazard recognition with high-fidelity augmented virtuality. *Journal of Construction Engineering and Management, 140*(7). https://doi.org/10.1061/(ASCE)CO.1943-7862.0000860

Álvares, J. S., Costa, D. B., & de Melo, R. R. S. (2018). Exploratory study of using unmanned aerial system imagery for construction site 3D mapping. *Construction Innovation, 18*(3), 301–320. https://doi.org/10.1108/CI-05-2017-0049

Ammar, M., Russello, G., & Crispo, B. (2018). Internet of things: A survey on the security of IoT frameworks. *Journal of Information Security and Applications, 38*, 8–27. https://doi.org/10.1016/j.jisa.2017.11.002

Arbnor, I., & Bjerke, B. (2008). *Methodology for creating business knowledge* (3rd ed.). Sage. ISBN: 1446202526

Asadi, K., Suresh, A. K., Ender, A., Gotad, S., Maniyar, S., Anand, S., Noghabaei, M., Han, K., Lobaton, E., & Wu, T. (2020). An integrated UGV-UAV system for construction site data collection. *Automation in Construction, 112*, 103068. https://doi.org/10.1016/j.autcon.2019.103068

Bang, S., Kim, H., & Kim, H. (2017). UAV-based automatic generation of high-resolution panorama at a construction site with a focus on preprocessing for image stitching. *Automation in Construction, 84*, 70–80. https://doi.org/10.1016/j.autcon.2017.08.031

Barbarella, M., De Blasiis, M. R., & Fiani, M. (2017). Terrestrial laser scanner for the analysis of airport pavement geometry. *International Journal of Pavement Engineering, 20*(4), 1–15. https://doi.org/10.1080/10298436.2017.1309194

Bashabsheh, A. K., Alzoubi, H. H., & Ali, M. Z. (2019). The application of virtual reality technology in architectural pedagogy for building constructions. *Alexandria Engineering Journal, 58*(2), 713–723. https://doi.org/10.1016/j.aej.2019.06.002

Bogue, R. (2018). What are the prospects for robots in the construction industry? *Industrial Robot: An International Journal, 45*(1). https://doi.org/10.1108/IR-11-2017-0194

Bosché, F., Abdel-Wahab, M., & Carozza, L. (2016). Towards a mixed reality system for construction trade training. *Journal of Computing in Civil Engineering*, 30(2), 04015016. https://doi.org/10.1061/ (ASCE)CP.1943-5487.0000479

Bosche, F., & Haas, C. T. (2008). Automated retrieval of 3D CAD model objects in construction range images. Automation in Construction, 17(4), 499–512. https://doi.org/10.1016/j.autcon.2007.09.001

Boton, C. (2018). Supporting constructability analysis meetings with immersive virtual reality-based collaborative BIM 4D simulation. *Automation in Construction*, 96, 1–15. https://doi.org/10.1016/j. autcon.2018.08.020

Brinkmann, S. (2013). *Qualitative interviewing: Understanding qualitative research*. Oxford University Press. ISBN: 0199861390

Chalhoub, J., & Ayer, S. K. (2018). Using mixed reality for electrical construction design communication. *Automation in Construction*, 86, 1–10. https://doi.org/10.1016/j.autcon.2017.10.028

Cheng, T., & Teizer, J. (2013). Real-time resource location data collection and visualization technology for construction safety and activity monitoring applications. *Automation in Construction*, 34, 3–15. https://doi.org/10.1016/j.autcon.2012.10.017

Chi, H. L., Kang, S. C., & Wang, X. (2013). Research trends and opportunities of augmented reality applications in architecture, engineering, and construction. *Automation in Construction*, 33, 116–122. https://doi.org/10.1016/j.autcon.2012.12.017

Chu, M., Matthews, J., & Love, P. E. (2018). Integrating mobile building information modelling and augmented reality systems: An experimental study. *Automation in Construction*, 85, 305–316. https:// doi.org/10.1016/j.autcon.2017.10.032

Cohen, L., Manion, L., & Morrison, K. (1994). *Educational research methodology*. Metaixmio.

Cohen, L., Manion, L., & Morrison, K. (2013). *Research methods in education* (7th ed.). Routledge. ISBN: 113572203X

Costa, C. M., Rocha, L. F., Malaca, P., Costa, P. G., Moreira, A. P., Tavares, P., Sousa, A., & Veiga, G. (2019). Collaborative welding system using BIM for robotic reprogramming and spatial augmented reality. *Automation in Construction*, 106, 1–12. https://doi.org/10.1016/j.autcon.2019.04.020

Dawood, N., Rahimian, F., Seyedzadeh, S., & Sheikhkhoshkar, M. (2020). *Enabling the development and implementation of digital twins*. Proceedings of the 20th International Conference on Construction Applications of Virtual Reality, Teesside University. ISBN: 9780992716127

de Melo, R. R. S., & Costa, D. B. (2019). Integrating resilience engineering and UAS technology into construction safety planning and control. *Engineering, Construction and Architectural Management*, 26(11). https://doi.org/10.1108/ECAM-12-2018-0541

Du, J., Shi, Y., Zou, Z., & Zhao, D. (2018). CoVR: Cloud-based multiuser virtual reality headset system for project communication of remote users. *Journal of Construction Engineering and Management*, 144(2), 040171091-0401710919. https://doi.org/10.1061/(ASCE)CO.1943-7862.0001426

Elghaish, F., Abrishami, S., & Hosseini, M. R. (2020). Integrated project delivery with blockchain: An automated financial system. *Automation in Construction*, 114, 1–16. https://doi.org/10.1016/j. autcon.2020.103182

Elghaish, F., Abrishami, S., Hosseini, M. R., Abu-Samra, S., & Gaterell, M. (2019). Integrated project delivery with BIM: An automated EVM-based approach. *Automation in Construction*, 106. https:// doi.org/10.1016/j.autcon.2019.102907

Fazel, A., & Izadi, A. (2018). An interactive augmented reality tool for constructing free-form modular surfaces. *Automation in Construction*, 85, 135–145. https://doi.org/10.1016/j.autcon.2017.10.015

Ficapal, A., & Mutis, I. (2019). Framework for the detection, diagnosis, and evaluation of thermal bridges using infrared thermography and unmanned aerial vehicles. *Buildings*, 9(8), 179. https://doi. org/10.3390/buildings9080179

Frank, G. C. (2012). *Construction quality: Do it right or pay the price*. Pearson Higher Education, Inc. ISBN: 0133002837

Freimuth, H., & König, M. (2018). Planning and executing construction inspections with unmanned aerial vehicles. *Automation in Construction*, 96, 540–553. 0.1016/j.autcon.2018.10.016

Gheisari, M., & Esmaeili, B. (2019). Applications and requirements of unmanned aerial systems (UASs) for construction safety. *Safety Science*, 118, 230–240. https://doi.org/10.1016/j.ssci.2019.05.015

Goessens, S., Mueller, C., & Latteur, P. (2018). Feasibility study for drone-based masonry construction of real-scale structures. *Automation in Construction, 94,* 458–480. https://doi.org/10.1016/j.autcon.2018.06.015

Goldman, A. I. (2004). Group knowledge versus group rationality: Two approaches to social epistemology. *Episteme, 1*(1), 11–22.

Golizadeh, H., Hon, C. K., Drogemuller, R., & Hosseini, M. R. (2018). Digital engineering potential in addressing causes of construction accidents. *Automation in Construction, 95,* 284–295. https://doi.org/10.1016/j.autcon.2018.08.013

Golizadeh, H., Hosseini, M. R., Edwards, D. J., Abrishami, S., Taghavi, N., & Banihashemi, S. (2019). Barriers to adoption of RPAs on construction projects: A task – technology fit perspective. *Construction Innovation.* https://doi.org/10.1108/CI-09-2018-0074

Gonçalves, P., Sobral, J., & Ferreira, L. A. (2017). Unmanned aerial vehicle safety assessment modelling through petri nets. *Reliability Engineering & System Safety, 167,* 383–393. https://doi.org/10.1016/j.ress.2017.06.021

González, N. A. A. (2018). Development of spatial skills with virtual reality and augmented reality. *International Journal on Interactive Design and Manufacturing (IJIDeM), 12*(1), 133–144. https://doi.org/10.1007/s12008-017-0388-x

Goulding, J., Nadim, W., Petridis, P., & Alshawi, M. (2012). Construction industry offsite production: A virtual reality interactive training environment prototype. *Advanced Engineering Informatics, 26*(1), 103–116. https://doi.org/10.1016/j.aei.2011.09.004

Greenwood, W. W., Lynch, J. P., & Zekkos, D. (2019). Applications of UAVs in civil infrastructure. *Journal of Infrastructure Systems, 25*(2), 04019002. https://doi.org/10.1061/(ASCE)IS.1943-555X.0000464

Ham, Y., & Kamari, M. (2019). Automated content-based filtering for enhanced vision-based documentation in construction toward exploiting big visual data from drones. *Automation in Construction, 105.* https://doi.org/10.1016/j.autcon.2019.102831

Hernández, J. L., Martín Lerones, P., Bonsma, P., Van Delft, A., Deighton, R., & Braun, J.-D. (2018). An IFC interoperability framework for self-inspection process in buildings. *Buildings, 8*(2), 32. https://doi.org/10.3390/buildings8020032

Hou, L., Chi, H. L., Tarng, W., Chai, J., Panuwatwanich, K., & Wang, X. (2017). A framework of innovative learning for skill development in complex operational tasks. *Automation in Construction, 83,* 29–40. https://doi.org/10.1016/j.autcon.2017.07.001

Hou, L., Wang, X., Bernold, L., & Love, P. E. (2013). Using animated augmented reality to cognitively guide assembly. *Journal of Computing in Civil Engineering, 27*(5), 439–451. https://doi.org/10.1061/(ASCE)CP.1943-5487.0000184

Hou, L., Wang, X., & Truijens, M. (2015). Using augmented reality to facilitate piping assembly: An experiment-based evaluation. *Journal of Computing in Civil Engineering, 29*(1). https://doi.org/10.1061/(ASCE)CP.1943-5487.0000344

Howard, R., & Björk, B.-C. (2008). Building information modelling – experts' views on standardisation and industry deployment. *Advanced Engineering Informatics, 22*(2), 271–280. https://doi.org/10.1016/j.aei.2007.03.001

Irizarry, J., & Costa, D. B. (2016). Exploratory study of potential applications of unmanned aerial systems for construction management tasks. *Journal of Management in Engineering, 32*(3), 05016001.

Irizarry, J., Gheisari, M., Williams, G., & Walker, B. N. (2013). InfoSPOT: A mobile augmented reality method for accessing building information through a situation awareness approach. *Automation in Construction, 33,* 11–23. https://doi.org/10.1016/j.autcon.2012.09.002

Kim, C., Park, T., Lim, H., & Kim, H. (2013). On-site construction management using mobile computing technology. *Automation in Construction, 35,* 415–423. https://doi.org/10.1016/j.autcon.2013.05.027

Kim, D., Liu, M., Lee, S., & Kamat, V. R. (2019a). Remote proximity monitoring between mobile construction resources using camera-mounted UAVs. *Automation in Construction, 99,* 168–182. https://doi.org/10.1016/j.autcon.2018.12.014

Kim, H., Lee, J., Ahn, E., Cho, S., Shin, M., & Sim, S.-H. (2017). Concrete crack identification using a UAV incorporating hybrid image processing. *Sensors, 17*(9), 2052. https://doi.org/10.3390/s17092052

Kim, H. S., Kim, S.-K., Borrmann, A., & Kang, L. S. (2018). Improvement of realism of 4D objects using augmented reality objects and actual images of a construction site. KSCE *Journal of Civil Engineering*, 22(8), 2735–2746. https://doi.org/10.1007/s12205-017-0734-3

Kim, J. W., Kim, S. B., Park, J. C., & Nam, J. W. (2015). Development of crack detection system with unmanned aerial vehicles and digital image processing. *Advances in structural engineering and mechanics (ASEM15)*, 25–29. https://doi.org/10.1061/(ASCE)CF.1943-5509.0001185

Kim, M. K., Sohn, H., & Chang, C. C. (2014). Automated dimensional quality assessment of precast concrete panels using terrestrial laser scanning. *Automation in Construction*, 45, 163–177.

Kim, P., Park, J., Cho, Y. K., & Kang, J. (2019b). UAV-assisted autonomous mobile robot navigation for as-is 3D data collection and registration in cluttered environments. *Automation in Construction*, 106, 102918. https://doi.org/10.1016/j.autcon.2019.102918

Kim, S., & Kim, S. (2018). Opportunities for construction site monitoring by adopting first personal view (FPV) of a drone. *Smart Structures and Systems*, 21(2), 139–149. https://doi.org/10.12989/sss.2018.21.2.139

Kwon, O.-S., Park, C.-S., & Lim, C.-R. (2014). A defect management system for reinforced concrete work utilizing BIM, image-matching and augmented reality. *Automation in Construction*, 46, 74–81. https://doi.org/10.1016/j.autcon.2014.05.005

Le, Q. T., Pedro, A., & Park, C. S. (2015). A social virtual reality based construction safety education system for experiential learning. *Journal of Intelligent & Robotic Systems*, 79(3–4), 487–506. https://doi.org/10.1007/s10846-014-0112-z

Leary, H., & Walker, A. (2018). Meta-analysis and meta-synthesis methodologies: Rigorously piecing together research. *TechTrends*, 62(5), 525–534. https://doi.org/10.1007/s11528-018-0312-7

Leviäkangas, P., Paik, S. M., & Moon, S. (2017). Keeping up with the pace of digitization: The case of the Australian construction industry. *Technology in Society*, 50, 33–43. https://doi.org/10.1016/j.techsoc.2017.04.003

Li, F., Zlatanova, S., Koopman, M., Bai, X., & Diakité, A. (2018). Universal path planning for an indoor drone. *Automation in Construction*, 95, 275–283. https://doi.org/10.1016/j.autcon.2018.07.025

Li, Y., & Liu, C. (2019). Applications of multirotor drone technologies in construction management. International *Journal of Construction Management*, 19(5), 401–412. https://doi.org/10.1080/15623599.2018.1452101

Lin, T. H., Liu, C. H., Tsai, M. H., & Kang, S. C. (2015). Using augmented reality in a multiscreen environment for construction discussion. *Journal of Computing in Civil Engineering*, 29(6), 04014088. https://doi.org/10.1061/(ASCE)CP.1943-5487.0000420

Liu, D., Chen, J., Hu, D., & Zhang, Z. (2019). Dynamic BIM-augmented UAV safety inspection for water diversion project. *Computers in Industry*, 108, 163–177. https://doi.org/10.1016/j.compind.2019.03.004

Liu, P., Chen, A. Y., Huang, Y.-N., Han, J.-Y., Lai, J.-S., Kang, S.-C., Wu, T.-H., Wen, M.-C., & Tsai, M.-H. (2014). A review of rotorcraft unmanned aerial vehicle (UAV) developments and applications in civil engineering. *Smart Struct Syst*, 13(6), 1065–1094. https://doi.org/10.12989/sss.2014.13.6.1065

Lu, X., & Davis, S. (2018). Priming effects on safety decisions in a virtual construction simulator. *Engineering, Construction and Architectural Management*, 25(2), 273–294. https://doi.org/10.1108/ECAM-05-2016-0114

Marmo, F., Demartino, C., Candela, G., Sulpizio, C., Briseghella, B., Spagnuolo, R., Xiao, Y., Vanzi, I., & Rosati, L. (2019). On the form of the Musmeci's bridge over the Basento river. *Engineering Structures*, 191, 658–673. https://doi.org/10.1016/j.engstruct.2019.04.069

Martinez, J. G., Gheisari, M., & Alarcón, L. F. (2020). UAV integration in current construction safety planning and monitoring processes: Case study of a high-rise building construction project in Chile. *Journal of Management in Engineering*, 36(3). https://doi.org/10.1061/%28ASCE%29ME.1943-5479.0000761

Meža, S., Turk, Ž., & Dolenc, M. (2014). Component based engineering of a mobile BIM-based augmented reality system. Automation in onstruction, 42, 1–12. https://doi.org/10.1016/j.autcon.2014.02.011

Mirshokraei, M., De Gaetani, C. I., & Migliaccio, F. (2019). A web-based BIM – AR auality management system for structural elements. *Applied Sciences*, 9(19). https://doi.org/10.3390/app9193984

Moon, D., Chung, S., Kwon, S., Seo, J., & Shin, J. (2019). Comparison and utilization of point cloud generated from photogrammetry and laser scanning: 3D world model for smart heavy equipment planning. *Automation in Construction*, 98, 322–331. https://doi.org/10.1016/j.autcon.2018.07.020

Newman, C., Edwards, D., Martek, I., Lai, J., Thwala, W. D., & Rillie, I. (2020). Industry 4.0 deployment in the construction industry: A bibliometric literature review and UK-based case study. *Smart and Sustainable Built Environment*. https://doi.org/10.1108/SASBE-02-2020-0016

Oliver, S., Seyedzadeh, S., Rahimian, F., Dawood, N., & Rodriguez, S. (2020). Cost-effective as-built BIM modelling using 3D point-clouds and photogrammetry. *Current Trends in Civil & Structural Engineering-CTCSE*, 4(5), 000599. https://doi.org/10.33552/CTCSE.2020.04.000599

Oskouie, P., Becerik-Gerber, B., & Soibelman, L. (2016). Automated measurement of highway retaining wall displacements using terrestrial laser scanners. *Automation in Construction*, 65, 86–101. https://doi.org/10.1016/j.autcon.2015.12.023

Park, C.-S., & Kim, H.-J. (2013). A framework for construction safety management and visualization system. *Automation in Construction*, 33, 95–103. https://doi.org/10.1016/j.autcon.2012.09.012

Park, J., Kim, P., Cho, Y. K., & Kang, J. (2019). Framework for automated registration of UAV and UGV point clouds using local features in images. *Automation in Construction*, 98, 175–182. https://doi.org/10.1016/j.autcon.2018.11.024

Paulo Carreira, Castelo, T., Gomes, C. C., Ferreira, A., Ribeiro, C., & Costa, A. A. (2018). Virtual reality as integration environments for facilities management: Application and users perception. *Engineering, Construction and Architectural Management*, 25(1). https://doi.org/10.1108/ECAM-09-2016-0198

Pedro, A., Le, Q. T., & Park, C. S. (2016). Framework for integrating safety into construction methods education through interactive virtual reality. *Journal of Professional Issues in Engineering Education and Practice*, 142(2). https://doi.org/10.1061/(ASCE)EI.1943-5541.0000261

Pejoska, J., Bauters, M., Purma, J., & Leinonen, T. (2016). Social augmented reality: Enhancing context-dependent communication and informal learning at work. *British Journal of Educational Technology*, 47(3), 474–483. https://doi.org/10.1111/bjet.12442

Portalés, C., Casas, S., Gimeno, J., Fernández, M., & Poza, M. (2018). From the paper to the tablet: On the design of an AR-based tool for the inspection of pre-fab buildings. Preliminary results of the SIRAE project. *Sensors*, 18(4), 1262. https://doi.org/10.3390/s18041262

Pour Rahimian, F., Seyedzadeh, S., Oliver, S., Rodriguez, S., & Dawood, N. (2020). On-demand monitoring of construction projects through a game-like hybrid application of BIM and machine learning. *Automation in Construction*, 110, 103012. https://doi.org/10.1016/j.autcon.2019.103012

Puppala, A. J., Congress, S. S., Bheemasetti, T. V., & Caballero, S. R. (2018). Visualization of civil infrastructure emphasizing geomaterial characterization and performance. *Journal of Materials in Civil Engineering*, 30(10), 04018236. https://doi.org/10.1061/(ASCE)MT.1943-5533.0002434

Ratajczak, J., Riedl, M., & Matt, D. T. (2019). BIM-based and AR Application combined with location-based management system for the improvement of the construction performance. *Buildings*, 9(5), 118. https://doi.org/10.3390/buildings9050118

Sacks, R., Perlman, A., & Barak, R. (2013). Construction safety training using immersive virtual reality. *Construction Management and Economics*, 31(9), 1005–1017. https://doi.org/10.1080/01446193.2013.828844

Sampaio, A., Rosário, D., & Gomes, A. (2012). Monitoring interior and exterior wall inspections within a virtual environment. *Advances in Civil Engineering*. https://doi.org/10.1155/2012/780379

Saunders, M., Lewis, P., & Thornhill, A. (2016). *Research methods for business students* (7th ed.). Pearson Education. ISBN: 9780273716860

Seyedzadeh, S., Agapiou, A., Moghaddasi, M., Dado, M., & Glesk, I. (2021). WON-OCDMA system based on MW-ZCC codes for applications in optical wireless sensor networks. *Sensors*, 21. https://doi.org/10.3390/s21020539

Seyedzadeh, S., & Pour Rahimian, F. (2021a). Building energy performance assessment methods. In *Data-driven modelling of non-domestic buildings energy performance* (pp. 13–30). Springer. https://doi.org/10.1007/978-3-030-64751-3_2

Seyedzadeh, S., & Pour Rahimian, F. (2021b). Machine learning for building energy forecasting. In *Data-driven modelling of non-domestic buildings energy performance* (pp. 41–76). Springer. https://doi.org/10.1007/978-3-030-64751-3_4

Shanbari, H., Blinn, N., & Issa, R. R. (2016). Using augmented reality video in enhancing masonry and roof component comprehension for construction management students. *Engineering, Construction and Architectural Management, 23*(6), 765–781. https://doi.org/10.1108/ECAM-01-2016-0028

Shazali, A. S. A., & Tahar, K. N. (2019). Virtual 3D model of Canseleri building via close-range photogrammetry implementation. *International Journal of Building Pathology and Adaptation, 38*(1), 217–227. https://doi.org/10.1108/IJBPA-02-2018-0016

Shi, Y., Du, J., Ahn, C. R., & Ragan, E. (2019). Impact assessment of reinforced learning methods on construction workers' fall risk behavior using virtual reality. *Automation in Construction, 104*, 197–214. https://doi.org/10.1016/j.autcon.2019.04.015

Shirazi, A., & Behzadan, A. H. (2015). Content delivery using Augmented Reality to enhance students' performance in a building design and assembly project. *Advances in Engineering Education, 4*(3). https://eric.ed.gov/?id=EJ1076141

Siebert, S., & Teizer, J. (2014). Mobile 3D mapping for surveying earthwork projects using an unmanned aerial vehicle (UAV) system. *Automation in Construction, 41*, 1–14. https://doi.org/10.1016/j.autcon.2014.01.004

Tang, P., Huber, D., & Akinci, B. (2010). Characterization of laser scanners and algorithms for detecting flatness defects on concrete surfaces. *Journal of Computing in Civil Engineering, 25*(1), 31–42. https://doi.org/10.1061/ ASCECP.1943-5487.0000073

Vick, S., & Brilakis, I. (2018). Road design layer detection in point cloud data for construction progress monitoring. *Journal of Computing in Civil Engineering, 32*(5). https://doi.org/10.1061/(ASCE)CP.1943-5487.0000772

Wang, C., Li, H., & Kho, S. Y. (2018). VR-embedded BIM immersive system for QS engineering education. *Computer Applications in Engineering Education, 26*(3), 626–641. https://doi.org/10.1002/cae.21915

Webster, J., & Watson, R. T. (2002). Analyzing the past to prepare for the future: Writing a literature review. *MIS Quarterly*, xiii–xxiii.

Williams, G., Gheisari, M., Chen, P.-J., & Irizarry, J. (2015). BIM2MAR: An efficient BIM translation to mobile augmented reality applications. *Journal of Management in Engineering, 31*(1). https://doi.org/10.1061/(ASCE)ME.1943-5479.0000315

Willis, J. W., Jost, M., & Nilakanta, R. (2007). *Foundations of qualitative research: Interpretive and critical approaches*. Sage.

Wolfartsberger, J. (2019). Analyzing the potential of Virtual Reality for engineering design review. *Automation in Construction, 104*, 27–37. https://doi.org/10.1016/j.autcon.2019.03.018

Wu, W., Hartless, J., Tesei, A., Gunji, V., Ayer, S., & London, J. (2019). Design assessment in virtual and mixed reality environments: Comparison of novices and experts. *Journal of Construction Engineering and Management, 145*(9). https://doi.org/10.1061/(ASCE)CO.1943-7862.0001683

Yeh, K.-C., Tsai, M.-H., & Kang, S.-C. (2012). On-site building information retrieval by using projection-based augmented reality. *Journal of Computing in Civil Engineering, 26*(3), 342–355. https://doi.org/10.1061/(ASCE)CP.1943-5487.0000156

Zaker, R., & Coloma, E. (2018). Virtual reality-integrated workflow in BIM-enabled projects collaboration and design review: A case study. *Visualization in Engineering, 6*(1). https://doi.org/10.1186/s40327-018-0065-6

Zhao, D., & Lucas, J. (2015). Virtual reality simulation for construction safety promotion. *International Journal of Injury Control and Safety Promotion, 22*(1), 57–67. https://doi.org/10.1080/17457300.2013.861853

Zhao, D., McCoy, A., Kleiner, B., & Feng, Y. (2016). Integrating safety culture into OSH risk mitigation: A pilot study on the electrical safety. *Journal of Civil Engineering and Management, 22*(6), 800–807. https://doi.org/10.3846/13923730.2014.914099

Zhou, J., Lee, I., Thomas, B., Menassa, R., Farrant, A., & Sansome, A. (2012). In-situ support for automotive manufacturing using spatial augmented reality. *International Journal of Virtual Reality, 11*(1), 33–41. https://doi.org/10.20870/IJVR.2012.11.1.2835

16 Optical code division multiple access–based sensor network for monitoring construction sites affected by vibrations

16.1. Introduction

Due to the progressive demand for more accurate structural health monitoring of large-scale facilities (e.g. modern high-speed railways and bridges), there has been a significant uptake in the development of optical sensor networks (OSNs). These can help mitigate the issues of conventional electric sensors, especially their sensitivity to electromagnetic interferences and larger sizes. Anecdotally, existing fibre-optic infrastructures are not widely used by OSNs, due in part to the lack of appropriate multiplexing techniques. In order to address the implementation issues of optical sensors in urban areas, this chapter proposes an efficient and cost-effective system for supporting the vibration sensing of unequally distributed points. The proposed system takes advantage of the Spectral Amplitude Coding (SAC) Optical Code Division Multiple Access (SAC-OCDMA) technique, to provide differentiated services in the physical layer with varying code weights. This system utilises more wavelengths (i.e. higher power) to the farthest sensing points in order to retrieve vibration signals more effectively. The mechanism of SAC for OSN is elaborated using simulation results, including the impact of transmission distance and the procedure of allocating codes to different zones. These results indicate the suitability of the proposed system for implementation in existing fibre-optic infrastructures. Numerical analysis shows a high capacity of the sensor network to deploy SAC. Moreover, the proposed system addresses the implementation issues in optically sensing of structures distributed in urban areas. The development approach of this research is presented through the following sections.

16.2. Core literature and knowledge gap

Literature highlights a myriad of advantages of using sensors for a range of applications. In many respects, this is because of a number of factors, not least are sensor size, accuracy, and immunity to electromagnetic noises, where optical sensors are now also being advocated as solutions for structural health monitoring (SHM) (Kakaee et al., 2014; Rahimian et al., 2019). The use of optical multiplexing techniques has also improved the efficiency of optical sensing for distributed sensors' network concerning cost and system complexity (Li et al., 2004). In such systems, fibre-optic sensing is either implemented as intrinsic (in which the fibre is used as the sensing element) or extrinsic (where fibre is a medium to communicate the sensed signal). Seminal literature advocates the success of optical sensing in the construction industry when used for various monitoring purposes, including gas leaks (Shabaneh et al., 2014), temperature (Woyessa et al., 2016), strain (Li et al., 2008), reinforced concrete beams (Lu & Xie, 2007), and building cladding systems (Unzu et al., 2013).

DOI: 10.1201/9781003106944-16

Accurate monitoring of vibration is essential for assuring the structural health of large facilities in order to evaluate the structural condition and to identify (and mitigate) risks of internal damage at early stages before they actually develop further (Alavi et al., 2016; Chae et al., 2012; Dawood et al., 2020). Since the use of traditional electric sensors in measuring vibration is noticeably exposed to electromagnetic interference, the practicality of these systems has always been limited to single-point simplex measuring only. Researchers have taken the advantages of optical vibration sensors through their capability of eliminating electromagnetic interference. In addition, optical sensors provide the characteristic of light signal transmission, which enables multiplex measurements from multiple points over a shared media (i.e. one optical fibre) to form a distributed OSN. The advantages of optical sensor networks make them more economical and allow the implementation of a broad range of essential applications such as real-time monitoring of large civil engineering structures.

Taiwo et al. (2013) employed Fibre Bragg Gratings (FBG) with a minimum bandwidth for a narrowband reflection of the spectrum coming from a broadband light source. The vibration was then detected based on the intensity modulation produced from the vibration of the FBG; however, one disadvantage of FBG-based sensors is that they are temperature sensitive, which can hinder the vibration detection. Later, a temperature compensation technique for these sensors was introduced (Woyessa et al., 2016). Other research includes Zhang and Bao (2008), who assessed the practicability of a fully distributed vibration sensor based on a fibre diversity detection sensor. Similarly, Talebinejad et al. (2009) proposed an accelerometer based on FBG using the stiffness of the optical fibre. The sensor was assessed for monitoring vibration for an actual bridge. Thakur et al. (2011) applied FBG and photonic crystal fibre sensors for SHM of composite and found that the crystal fibre sensor was a better alternative as it was less sensitive to temperature. Zhang et al. (2017) also monitored vibration and deformation distribution identification of a long-span rigid-frame bridge using sensors. Ge et al. (2013) developed a particular intensity-modulated fibre-optic accelerometer for vibration monitoring of wind turbine blades. Brillouin scattering-based sensors have also been introduced to examine distributed vibration along the optical fibre. The systems operate based on the backscattered light from the sensing fibre to reflect the vibration. Each sensor point has to be accessed at a time slot, which might not be the best option for real-time vibration sensing (Hotate & Ong, 2003). Klar et al. (2014) evaluated a Brillouin scattering optical time domain analysis for monitoring the tunnelling process with ground displacements. Li et al. (2018) used a distributed fibre vibration sensor for monitoring pipeline in China. They tested the system over an area of 131 km, where thousands of vibrations were detected daily.

These multiplexing techniques include time division multiplexing (TDM), wavelength division multiplexing (WDM), frequency modulated carrier wave (Li et al., 2004), and optical code division multiple access (OCDMA) (Seyedzadeh et al., 2021; Taiwo et al., 2014).

Various multiplexing techniques to collect data from sensors have already found application in building monitoring. WDM has commonly been used in static strain sensing and has also been integrated with each other techniques, resulting in hybrid multiplexing systems (Noura et al., 2013). In WDM, each sensor is assigned a given slice of the input spectrum, which is provided by a coherent optical source. WDM is mostly suitable for sensing a few points only, since by increasing the number of sensors the system cost increases significantly due to the high cost of fabricating multi-wavelength lasers (Cheng et al., 2011). In the TDM approach, sensors transmit signals at different time slots, which demand precise network synchronisation. TDM can significantly expand the number of sensors in the time domain; however, it suffers from transmission loss and is also restrained by light source power. Therefore, only a few sensors (i.e. a maximum of 10 sensors) can be supported by this system (Dai et al.,

2009). Wang et al. (2011) introduced a serial TDM-OSN based on FBG. It was concluded that achieving high signal quality was difficult due to the presence of 12 FBGs.

Although these multiplexing methods facilitate sensing different structures using one shared medium, they generally fail in the utilisation of the ever-growing fibre-optic infrastructures in urban areas. The main reason is that OSN with current technologies are developed to monitor nodes with almost the same distances, so that they require a separate fibre wiring. A solution must receive signals from different distances with almost the same level. Figure 16.1 illustrates an example of how sensing data from a structure can be transmitted over an urban fibre-optic infrastructure.

To address these issues, several studies advocated the adoption of OCDMA sensor networks due to their random-access capabilities (Ko et al., 2010; Kwong, 2002; Tseng et al., 2013; Yang & Kwong, 2002). In a sensor network based on the OCDMA system approach, each sensor is assigned a unique address called a signature code. This unique feature enables individual sensor identification and provides its location in the building (Seyedzadeh et al., 2019). Yen and Chen (2016) proposed a hybrid wavelength/time/spatial optical CDMA system to increase the capacity of OSN.

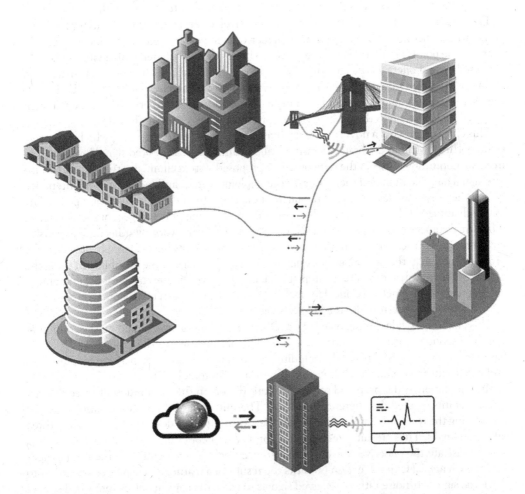

Figure 16.1 Schematic of existing urban fibre-optic infrastructure for transmitting sensor data

Implementation can use either optical or electric sensors. In this respect, several studies have focussed on sensors' placement in different structures, including tower buildings (Yi et al., 2014, 2015), bridges (Chen et al., 2014; Wan et al., 2013), railways (Filograno et al., 2012), tunnels (Li et al., 2008), and complex structures (Li et al., 2016). When it comes to sensor selection for vibration monitoring, FBGs have been the preferred choice when it came to a sensor type in most OSNs, due to their low price and high performance. When it comes to a sensor placement topology, the focus has been on equally distributed sensor points; however, in SHM of structures having different distances from each other (i.e. near-far problem), it is not always possible to set up the sensor network base with equal distances from all the sensing points. The near-far problem is the inability of a receiver to receive a weak signal from farther distances in the presence of a strong signal that is transmitted from a nearer source. A distributed sensor network consists of varying fibre lengths amongst the sensing points, which results in uneven power attenuation, making it hard to detect the transmitting signals from points with longer fibre lengths. To address these challenges and research gap, this study proposes an optical sensing system based on the variable weight spectral amplitude coding (VW-SAC) network approach. In this system, higher weight codes (providing a better signal quality) are assigned to the nodes requiring higher quality (Seyedzadeh Kharazi et al., 2011). The weight of the code is the number of wavelengths assigned to the user or sensing point, where the more wavelengths carry higher optical power. Hence, the signals from farther sensing points are carried by more power to better distinguish detectable vibrations. The proposed system offers a cost-effective method for vibration monitoring of urban structures, including tower buildings, bridges, tunnels, etc. Moreover, because the SAC-OCDMA system is somewhat immune to most fibre effects (Seyedzadeh Kharazi et al., 2012; Seyedzadeh et al., 2016a), this solves the problems of WDM and TDM regarding these issues.

This chapter presents a novel collimator-based VW-SAC sensor network, which is capable of supporting distributed vibration sensing while maintaining the desired signal level delivery from all monitoring nodes in the network. The proposed system eliminates the demand for in-line signal amplification and thereby eases the implementation and reduces the maintenance cost of the sensing nodes located at the farthest distances from the data collection point. The main advantage of this approach is that vibration monitoring of heterogeneous structures at different distances can be performed using a single sensor network by utilising inexpensive components. Furthermore, by noting the fact that optical fibres have a large bandwidth, the proposed system in this study can be designed to take advantage of the possibility of using the existing fibre infrastructure already in place. Hence, this system presents a robust and practical vibration sensing network, ideal for monitoring constructions in metropolitan areas where optical fibre infrastructures are readily available.

This chapter provides an overview of OCDMA and SAC systems, including the configuration of intended distributed sensor networks. The architecture of the proposed VW-SAC-based sensor system is explained, including the parameters used for performance analysis and the results of conducted OptiSystem (Optiwave Photonics Software, 2021), in addition to the simulation of the proposed sensor network (based on the assumption of three distinct sensing points located at different distances). The outcome of the system is also discussed concerning the received signals of sensors in the radio frequency domain by considering three vibration levels. These results are expanded for a system containing nine monitoring points to demonstrate the effective transmission distance, including allocation of the right weight set for each area. The final section presents the results of a mathematical model developed to drive the signal-to-noise ratio of received signals to show the capacity of the proposed system in supporting simultaneous active sensors. This mathematical model includes assumptions set to worst cases for calculating an upper boundary for the noise level to 'guarantee' simulation for real case scenarios. Section 16.5 presents the expansion of a mathematical model and the numerical result. The chapter concludes with a discussion on findings and relevance to practice and contemporary discourse.

16.3. VW-SAC code and architecture

In the proposed system, all sensing points were grouped based on the distance from the monitoring unit, where a matrix with a suitable code weight was assigned for each. This means construction sites that are located approximately the same distance from the control unit are clustered in one group or zone. For example, assuming we have nine sensors, of which four are located in 14–15 km, three in 11–12 km, and two 7–8 km from the control unit, then three zones exist. The next stage is to assign the highest weight to the farthest zone, a lower one to the zone with the medium distance, and the lowest weight to the nearest zone. In the proposed VW-SAC system, sensing points with higher weights carry vibration signals using a greater number of wavelength or chips, and hence are higher. In this approach, the attenuation caused by optical fibres is compensated by increasing the transmission power by adding additional wavelengths instead of signal amplification. Therefore, the system can be utilised in any existing infrastructure without the need for electric power in any sensing location. The design of the system and choice of the codes are a way to retrieve almost the same power from all distances to overcome the near-far problem.

Chips are designated for different nodes with codes developed for the SAC-OCDMA system. These codes are differentiated by their characteristics, such as maximum cross-correlation (maximum number of wavelengths shared between two pairs of codes), length, and supportable weight. Cross-correlation is the maximum number of overlapping wavelengths between codewords, which determines the interference level of a system. A code family with a high cross-correlation provides higher information security (but lower performance). The importance of using codes for sensing applications rather than assigning consecutive wavelengths to the nodes is to provide security in the transmission layer while maintaining maximum bandwidth utilisation. The code length defines the total wavelength required for supporting the sensing points in different zones. Generally, higher weights increase the overall code length, where higher code weights intensify the quality of one group of sensing signals; however, due to the higher length, the overall performance of the system is decreased. Therefore, to construct a system with higher capacity (whilst still maintaining minimum security), it is essential to select an appropriate code.

Among several VW-code families (Anas et al., 2016; Kakaee et al., 2015; Liang et al., 2008; Seyedzadeh et al., 2017), this study uses VW Khazani-Syed (VW-KS) code, which was initially developed based on the single-weight KS code (Seyedzadeh et al., 2016b). The KS code is based on matrix construction, where the two sub-codes A = [110] and B = [011] are used to construct the basic matrix. The structure of this code is arranged in such a way that cross-correlation (the number of the overlapping chip(s) between two different users' codes), R, is either zero or one, which results in the reduction of the multiple access interference (MAI) effect. Figure 16.2 shows the basic KS-code matrix with a weight of 4. The overlapping chips between each pair of codes are depicted by dotted lines.

The number of rows K_B, also known as the basic number of users and number of columns, N_B, or basic code length, are calculated by following equations:

$$K_B = \frac{W}{2} + 1 \tag{1}$$

$$R_{1,2}=1$$

$$C_{B_4} = \begin{bmatrix} 1 & 1 & 0 & 1 & 1 & 0 & 0 & 0 & 0 \\ 0 & 1 & 1 & 0 & 0 & 0 & 1 & 1 & 0 \\ 0 & 0 & 0 & 0 & 1 & 1 & 0 & 1 & 1 \end{bmatrix}$$

$$R_{1,3}=1 \qquad\qquad R_{2,3}=1$$

Figure 16.2 Construction of the basic matrix for KS code with weight of 4

And

$$N_B = 3\sum_{i=1}^{W/2} i \qquad\qquad\qquad (2)$$

For the detail of code constructing VW codes for a large number of nodes, refer to Section 16.3.

16.3.1. Recovering signals from multiple distances

When each sensing node's wavelengths are identified, upon a vibration exposure these coded signals are modulated by collimators installed on-site. As signals from all nodes travel over a shared medium, the signal must be optically decoded. This means that chips or wavelengths dedicated to each node should be retrieved while cancelling the effect of signals from other nodes. Optical signals are then converted into the electrical domain to obtain the vibration frequency. A high-intensity oscillation will generally result in retrieving a higher frequency. This procedure is discussed more in the results section. Here, the approach for receiving the correct spectrum and cancelling noise is elaborated.

Among several detection techniques developed for the SAC system to cancel MAI and decode the desired signals, three widely used ones are complementary subtraction (CS) detection, AND subtraction detection, and direct decoding (DD) (Seyedzadeh et al., 2016b). Both CS and AND utilise balanced detection in which two decoders (upper and lower) are required in a single receiver to eliminate the effect of MAI. The upper decoder detects the desired code, while the lower decoder retrieves the binary logical AND of desired and interfering code (the interferer signal of other nodes having an overlapping chip with the desired code) for AND detection and the complement of the upper decoder in the case of CS. DD only deploys one decoder unlike the others, which reduces the number of filters and receiver complexity by detecting the non-overlapping code of the desired signal.

Generally, FBGs are employed to filter the coveted wavelengths within both the encoder and decoder. After forming the code words for each sensor point, the signals that have been modulated by vibration sensors travel over the fibre. In order to avoid power splitting across the whole wavelengths, FBGs are consecutively structured; however, such an arrangement can cause unequal losses in different chips (due to the FBGs insertion loss affecting passing signals). Therefore, to prevent power reduction of various chips, the FBGs are arranged in the opposite order of the encoder in the receiver part. This technique also eliminates the delay imposed by FBGs at the decoder.

The choice of deployed detection technique is dependent on the utilised code family (Seyedzadeh & Pour Rahimian, 2021a, 2021b; Seyedzadeh et al., 2013). It has been shown that codes with high cross-correlation values (≥ 3), such as integer lattice optical orthogonal code (Djordjevic & Vasic, 2004), are best decoded via balanced detection-based methods, while DD is more appropriate for lower cross-correlation codes (e.g. VW-KS and VW-MS (Seyedzadeh et al., 2017)). In this respect, this study uses the DD detection technique for retrieving vibration modulated optical signals; for example, the hardware experiment of a VW-OCDMA system using DD transmitted over optical fibre has been successfully demonstrated (Seyedzadeh et al., 2014).

16.4. Expanding OCDMA code for larger sensors

The process of VW-KS code construction is a combination of two algorithms, which are the fixed mapping technique and dynamic weight assignment. The first technique is a simple mapping of codes with same and different weights that have been initially constructed using the single weight KS technique. The second technique, dynamic weight assignment, uses the value of [1 2 1] combination to arrange sub-codes A and B, accordingly (Anas et al., 2016). The main aim of developing dynamic weight assignment was to support few numbers of differentiated services without wasting extra bandwidth.

16.4.1. Mapping technique

First, the number of sensor points is increased by mapping the basic matrix, C_B, for the required weight. Based on the assumption that the total number of requisite sensors in one region (which determines the allocated weight) is N, the basic matrix repeated by $M = N / N_B c$ times, where N_B is the number of nodes in C_B as the following matrix:

$$C(M) = \begin{bmatrix} C_B(1) & 0 & 0 & 0 & 0 \\ 0 & C_B(2) & 0 & 0 & 0 \\ 0 & 0 & C_B(3) & 0 & 0 \\ 0 & 0 & 0 & 0 & \\ 0 & 0 & 0 & 0 & C_B(M) \end{bmatrix} \qquad (3)$$

Here $C_B(m)$ is the mth mapping sequence, where $m = 1, 2, \ldots, M$. Each '0' in a mapping matrix is a sequence of zeros with same size of C_B. The maximum cross-correlation λ_c between codes within the same matrix is one and from different mappings is zero. Hence, using the mapping technique a maximum cross-correlation of 1 is obtained for all users.

Sensor points are grouped based on the distance from the monitoring unit where a matrix with suitable code weight is assigned for them. This means that construction sites that are located at approximately the same distance from the control unit are clustered into one zone. For example, if we had nine sensors in total, of which four were located at 18–20 km, three in 12–5 km, and two 8–10 km from the control unit, then we will have three zones. The next process is to assign the highest weight to the farthest zone, a lower one to the zone with the medium distance, and the lowest weight to the nearest zone.

If the number of points is higher than one basic matrix, an additional number of mapped matrices are assigned to the group to support all users. Then, matrices with different code weights are joined together using mapping methods to construct the overall matrix. The structure of VW-code construction to support Q number of multiple distances (weights) using KS code is presented in equation 4. $C_{BW_q}(m)$ shows the mth mapping basic matrix for the qth region, which is appointed with weight of W_q.

$$
\begin{bmatrix}
\begin{bmatrix} C_{Bw_1}(1) & 0 & 0 \\ 0 & 0 & 0 \\ 0 & 0 & C_{Bw_1}(M_1) \end{bmatrix} & 0 & 0 & 0 \\[2em]
0 & \begin{bmatrix} C_{Bw_2}(1) & 0 & 0 \\ 0 & 0 & 0 \\ 0 & 0 & C_{Bw_2}(M_2) \end{bmatrix} & 0 & 0 \\[2em]
0 & 0 & 0 & 0 \\[1em]
0 & 0 & 0 & \begin{bmatrix} C_{Bw_Q}(1) & 0 & 0 \\ 0 & 0 & 0 \\ 0 & 0 & C_{Bw_Q}(M_{jQ}) \end{bmatrix}
\end{bmatrix}
\tag{4}
$$

The total code length of VW-MS code, Lt, with mapping technique is expressed as:

$$
L_t = \sum_{q=1}^{Q} \left(\left(M_q \times 3 \sum_{i=1}^{W_q/2} i \right) \right)
\tag{5}
$$

16.4.2. *Dynamic weight assignment and hybrid method*

Dynamic weight assignment reduces the code length by considering two important parameters to determine the position of sub-codes insertion. They are the value of three-column combination (3CC) and cross-correlation between two different users, x and y, i.e. $R_{x,y}$. The procedure of dynamic weight assignment is elaborated by Anas et al. (2016).

When the number of requesting nodes with the same weight (sensors in the same region) is equal to the maximum number of users for that particular mapping, $K_W = K_{max}(M)$, a fixed mapping technique is the best choice for VW-KS codes. In the case of $K_W < K_{max}(M)$, a dynamic weight assignment technique can result in a reduced code length and utilised bandwidth, provided that the number of mappings is one, i.e. $M = 1$; however, for $M > 1$, a combination of both techniques results in better bandwidth utilisation. In this method, it is important to determine the number of maximum nodes in the previous mapping, $K_{max}(M-1)$. For a large number of requesting users, K_W, where the number of mapping is greater than one ($M > 1$), fixed mapping technique is the best choice for nodes up to $K_{max}(M-1)$. While for the remaining users, $K_r = K_W - K_{max}(M-1)$, their codes can be constructed using

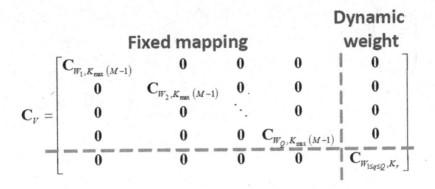

Figure 16.3 General form of hybrid fixed-dynamic weight assignment technique

dynamic weight assignment. The codes of different weights, generated using fixed mapping for $1 \leq K_W \leq K_{max}(M - 1)$, are appended diagonally, and later combined with codes for remaining users, K_r generated using dynamic weight assignment technique. The general form of the hybrid fixed-dynamic weight assignment technique is depicted in Figure 16.3, and its code length is given by:

$$N_V = \left[\sum_{q=1}^{Q} \sum_{k=1}^{K_{max}(M-1)} N_{r,q} + \frac{3}{4} \sum_{q=1}^{Q} K_{r,q} W_q \right] \tag{6}$$

16.5. System description

This study assumed that the sensing points were distributed in different zones in which the nodes inside the same zone had almost the same distance from the base control unit, and that different zones were located unequally from the base unit. Figure 16.4 shows three zones with different sensing nodes and the base control unit. The main idea of this work was to assign a specific code weight for each zone, where higher weights were allocated for zones with a farther distance from the base. In this case, nodes were categorised into zones as equivalents to services in communication systems according to their distances from the control unit. It should be noted that dividing nodes into the zones does not mean they are near each other, but rather are logically distributed with similar spans.

Based on the assumed three sensing zones and nine sensing points, the weight assignment technique was used to construct the desired code with the weight of 6, 4, and 2. The structure of the code and the wavelengths assigned for chips (each one in the code representing the presence of the correspondent wavelength) can be seen in Figure 16.5. Nine codewords were generated using VW-KS code with chip spacing of 0.4 nm and a code length of 30. These codes support four sensing points in zone 3 (with the longest distance), three in zone 2 (with the medium distance), and two points in zone 1 (the nearest to the control unit).

Figure 16.6 presents the architecture of a VW-SAC OCDMA system designed for multiple sensor nodes. Amplified spontaneous emission broadband source (ASE-BBS) was

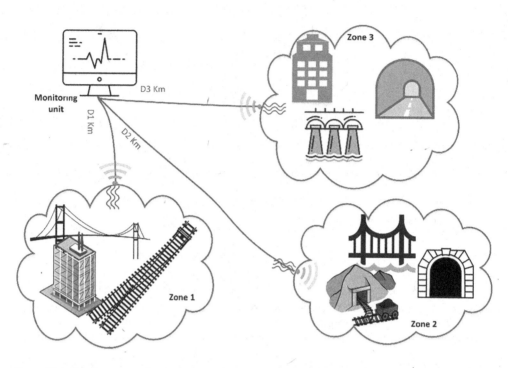

Figure 16.4 Configuration of optical sensors for monitoring structures distributed over three zones with different distances from the control unit

Seq. #	Code #	W = 6																		W = 4									W = 2		
		192.2	192.25	192.3	192.35	192.4	192.45	192.5	192.55	192.6	192.65	192.7	192.75	192.8	192.85	192.9	192.95	193	193.05	193.1	193.15	193.2	193.25	193.3	193.35	193.4	193.45	193.5	193.55	193.6	193.65
1	C_1	1	1	0	1	1	0	1	1	0	0	0	0	0	0	0	0	0	0	0	0	0	0	0	0	0	0	0	0	0	0
	C_2	0	1	1	0	0	0	0	0	0	1	1	0	1	1	0	0	0	0	0	0	0	0	0	0	0	0	0	0	0	0
	C_3	0	0	0	0	1	1	0	0	0	0	1	1	0	0	0	1	1	0	0	0	0	0	0	0	0	0	0	0	0	0
	C_4	0	0	0	0	0	0	0	0	1	1	0	0	0	1	1	0	1	1	0	0	0	0	0	0	0	0	0	0	0	0
2	C_5	0	0	0	0	0	0	0	0	0	0	0	0	0	0	0	0	0	0	1	1	0	1	1	0	0	0	0	0	0	0
	C_6	0	0	0	0	0	0	0	0	0	0	0	0	0	0	0	0	0	0	0	1	1	0	0	0	1	1	0	0	0	0
	C_7	0	0	0	0	0	0	0	0	0	0	0	0	0	0	0	0	0	0	0	0	0	0	1	1	0	1	1	0	0	0
3	C_8	0	0	0	0	0	0	0	0	0	0	0	0	0	0	0	0	0	0	0	0	0	0	0	0	0	0	0	1	1	0
	C_9	0	0	0	0	0	0	0	0	0	0	0	0	0	0	0	0	0	0	0	0	0	0	0	0	0	0	0	0	1	1

Figure 16.5 Code construction for vibration system and wavelengths for each chip in an OCDMA system

used as the optical source. The spectrum of BBS is sent through an optical circulator and a 1×3 optical coupler (for supporting three zones). Each port of the coupler is then connected to a $1 \times N$ coupler (N is the number of sensing points in each zone) after transmission over a single mode fibre (SMF) with different lengths of 8, 12, and 15 km. Then the output port of the secondary couplers are connected to the sensing points to collect the vibration signals. The encoder of VW-SAC comprises a collimator, vibration box, and

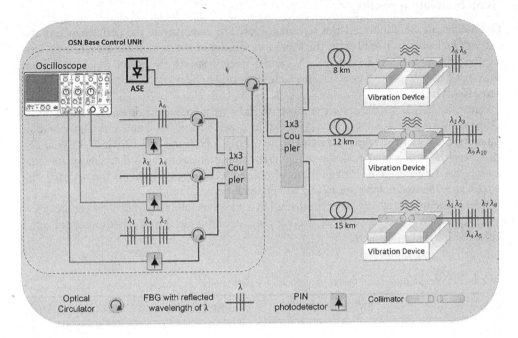

Figure 16.6 VW-SAC OCDMA system structure of fibre vibration sensor for monitoring three nodes in different areas

a series of FBGs. Due to the property of VW-KS code being double weighted, two wavelengths can be reflected using one FBG with a bandwidth twice the chip spacing. The generated vibration causes a modulation that can be detected by decoders. In the conducted simulation set-up, the study used a Mach Zehnder modulator to mimic the behaviour of the collimator in sensing vibration. Three different frequencies of 70, 140, and 210 MHz were used to represent low, medium, and high vibration, respectively. It should be noted that these values are examples from previous experimental research and were selected to demonstrate the operation of the proposed system. In a real-world implementation, the intensity of vibration is not discrete; in addition, the range is determined by the type of structure and variation of the quiver.

The modulated signals travel back to the couplers, are combined, and then guided to the detection section by a circulator. Here, the combined signals are split among decoders using a coupler. FBGs are also used to filter the desired wavelengths at the receiver. As mentioned, DD can be used for recovery of sensor signals in which only the non-overlapping wavelengths (the chips without cross-correlation with other codes) are detected at the receiver for the desired users. FBGs with a bandwidth of 0.4 nm were arranged in the opposite order of the encoder. As the node with weight 2 used only one filter in both the encoder and decoder, the order was not important, and the decoder was composed of one wavelength.

The decoded optical signals are finally converted into an electrical domain employing PIN photodetectors and are then sent to a three-channel oscilloscope to be translated using fast Fourier transform.

The proposed system was simulated using OptiSystem version 12 software, with default values used for optical and electrical components.

16.6. Simulation results

Three different vibration scenarios were considered to demonstrate the performance of the system (Table 16.1). To illustrate the vibration sensing in the system from each zone, one sensing point was selected. Nodes N_1, N_2, and N_3 were respectively located 8, 12, and 15 km away from the base control unit. The frequencies of other nodes were randomly assigned from 0 to 210 MHz.

Figure 16.7(a) to (c) illustrate the power in radio frequency (RF) domain of three nodes for experiments 1 to 3, respectively. As it can be observed, low, medium, and high vibration signal were constantly retrieved at frequencies of 66.3, 139.8, and 207.1 MHz at the encoder. Moreover, the signals of all nodes in different configurations were revived at almost the same peak power (± 1 dB). The configuration of utilised weights was based on the selected distances, and different combinations can be applied for desired results.

Hence, the weight of each zone is selected in a way that sufficient optical power for delivery of a noiseless signal could be received at the base station where all signals were collected and vibration was monitored.

Table 16.1 Configuration of vibration frequencies for three nodes

Node	EXP 1 (MHz)	EXP 2 (MHz)	EXP 2 (MHz)
N_1 (W =2)	70	0	210
N_2 (W =4)	140	70	70
N_3 (W =5)	210	210	70

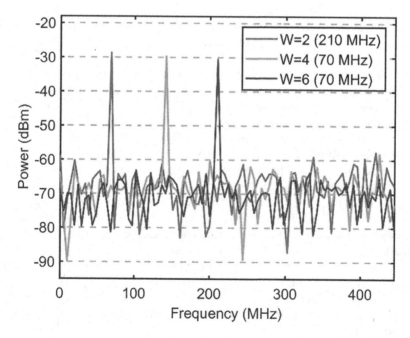

Figure 16.7 RF signal power received for three nodes for experiments (a)–(c) (Rahimian et al., 2019)

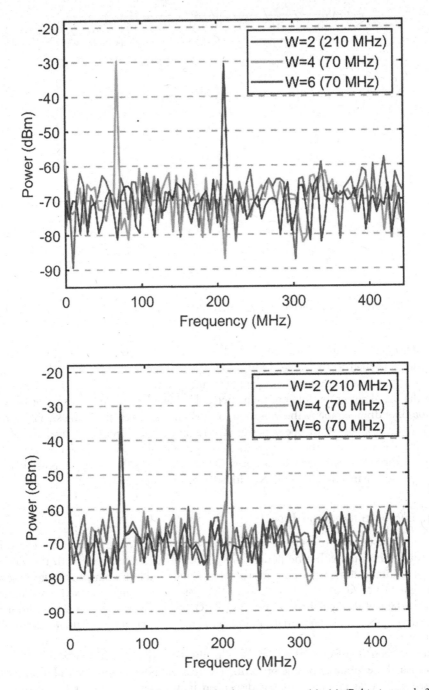

Figure 16.7 RF signal power received for three nodes for experiments (a)–(c) (Rahimian et al., 2019)

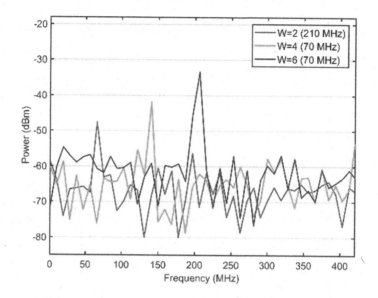

Figure 16.8 RF signal power received for a distance of 20 km for all nodes (Rahimian et al., 2019)

An additional experiment was conducted to check the scenario in which the different weights were applied to the same distance. Here, the fibre length for all nodes was set to 20 km and the vibration frequencies adopted from experiment one in Table 16.1. Figure 16.8 depicts the signal power for different frequencies. The signal for node N_3 was received with a power of -33.4 dBm, enabling the vibration to be distinctly detectable. The signals of nodes N_1 and N_2 were obtained with peak power of -47.6 and -41.8 dBm, respectively. This lower power retrieval was due to the fact that the received optical power of these nodes was lower than N_3, as they were assigned with lower weights. This shows that it was impossible to recognise vibrations happening in node N_1 with such a configuration.

16.6.1. Effect of transmission distance

In order to evaluate the performance of the proposed system under the effect of the transmission medium, a set-up with nine nodes (two points in the nearest zone with the allocated weight of 2, three points in zone 2 with the weight of 4, and four points at the farthest zone with weight 6) was investigated. The codes were generated using the fixed mapping technique. The plot of the signal-to-noise ratio (SNR) against average transmission distance is presented in Figure 16.9. The length of optical fibres for each zone is demonstrated as the separate x-axis.

It can be seen that below the transmission distance equal to 56 km, the SNRs of all nodes satisfied the minimum value of 15 dB. The reason SNRs of nodes with higher weights decreased with a faster rate was that dispersion had a more significant impact on them (Seyedzadeh Kharazi et al., 2011); however, all worst-case scenarios were selected for investigation to provide more robust results. In reality, the vibration frequencies were far lower than test case findings of this study, and the chirping effect was not considered a real problem. Therefore, in-line boosting was not required, nor dispersion compensation.

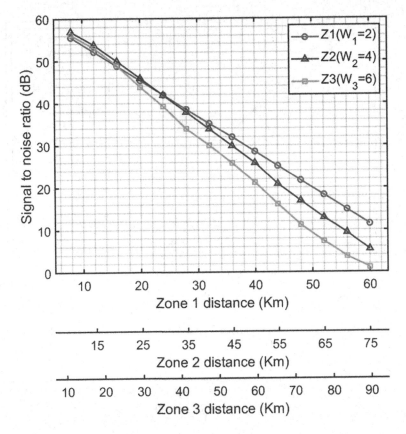

Figure 16.9 SNR of received signals from three zones vs. transmission distance (Rahimian et al., 2019)

To elaborate the determination of the right weight for different distances, this study assumed that zones 3 and 2 were located 2 and 2.5 times, respectively, of zone 1 from the control unit. The aim was to assign weights for these three zones located at 36, 72, and 90 km from the data collection point. Figure 16.10 illustrates two sets of weights allocated for nine sensor points distributed in different zones. The first set (i.e. 6, 4, and 2) could not deliver sufficient signals for nodes located in zones 2 and 3. Utilising the second set including weights 10, 8, and 2, signals from all nodes received sufficient power with acceptable SNRs. That means the second set of weights allowed for the monitoring of vibrations of structures located at the target distances. Hence, the selection of weights for each zone depends on the distances of all zones from the signal collection point and the number of sensing points located in each area.

16.7. VW-OCDMA sensor system capacity

In order to determine the approximate capacity of the VW-OCDMA sensor system, a mathematical model based on the upper bound calculation (Seyedzadeh et al., 2013) was developed. In this model, shot and thermal noises were taken into account to derive the SNR of

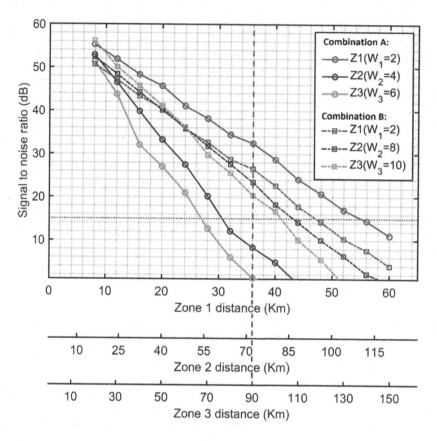

Figure 16.10 SNR for different code weights vs. transmission distance (Rahimian et al., 2019)

the received signals. It should be noted that as DD detection was used, the phase-induced intensity noise was avoided in the proposed system.

The noise variance of a photocurrent emitted from optical source as a result of the detection of an ideally unpolarised thermal light was calculated as:

$$\langle I^2 \rangle = \langle I_{shot}^2 \rangle + \langle I_{thermal}^2 \rangle \tag{3}$$

Where I_{shot} and $I_{thermal}$ represent the shot noise and thermal noise, respectively. The coherence time of the thermal source, τc, is given by (Smith et al., 1998):

$$Tc = \frac{\int_0^\infty G^2(V)dv}{\left[\int_0^\infty G^2(v)dv\right]^2} \tag{4}$$

Where $G(v)$ is the source power spectral density (PSD).

This mathematical model considers the following assumptions:

a) Each power spectral component has identical spectral width.
b) Each node receives equal total power.
c) Each vibration is considered as a modulated signal at frequency of 210 MHz.
d) The number of nodes in every zone is almost the same.

The PSD of the received signals can be expressed as (Prucnal, 2006):

$$r(v) = \frac{P_{sr}}{\Delta u} \sum_{k=1}^{K} d_k \sum_{i=1}^{N} ck(i) \Pi(i) \tag{5}$$

Here P_{sr} is the effective power of source at the receiver, Δv is the bandwidth of optical source, K and N are the number of sensor nodes and total code length, respectively, d_k is the modulated information of kth vibrated nodes, which is either "1" or "0", $(d_k \in 0,1), c_k(i)$ is the ith element of the kth KS-code sequence, and $\Pi(i)$ is a function written as:

$$\Pi(i) = u\left[v - v_0 - \frac{\Delta v}{2N}(-N + 2i - 2)\right] - u\left[v - v_0 - \frac{\Delta v}{2N}(-N + 2i)\right] = u\left[\frac{\Delta u}{2N}\right] \tag{6}$$

and $u[v]$ is the unit step function expressed as:

$$u[v] = \begin{cases} 1, & v \geq 0 \\ 0, & v < 0 \end{cases} \tag{7}$$

The code properties using DD is expressed as:

$$\sum_{i=1}^{N} ck(i)cl(i) = \begin{cases} W_k / 2, & k = l \\ 0 & k \neq l \end{cases} \tag{8}$$

The photocurrent of the desired node's signal is therefore:

$$I_R = \frac{RP_{sr} W_k}{2N_u} \tag{9}$$

Where $\mathfrak{R}$ represents receiver responsivity, and P_{sr} is the received power. It was assumed that the configuration of the VW-OCDMA sensor system was structured in a way that signals from nodes in different distances were received with almost the same power; therefore, $P_{sr} \times W_k$ remained constant for all nodes and could be replaced by P_c. Indeed, this was the main idea of varying weights for unequally distributed sensors. For longer distances, P_{sr} decreased; however, different higher weights, W_k, ensured that the same received power would be attained from these nodes. It is true that in reality the noise of these signals with different wavelengths and same total power is not the same, as this model gives an upper-bound approximation of SNR; therefore, this issue can be ignored.

The variance of shot noise in the photocurrent can be calculated as:

$$\left\langle I^2_{shoot} \right\rangle = \frac{eBRP_c}{N\upsilon} \tag{10}$$

The thermal noise is given as:

$$\left\langle I^2_{thermal} \right\rangle = \frac{4K_b T_n B}{R_L} \tag{11}$$

where B is the electrical bandwidth, K_b is Boltzmann's constant, T_n is received noise temperature, and RL represents the receiver load resistor.

Hence, SNR of VW-SAC-OCDMA can be written as:

$$SNR = \frac{I_R}{\left\langle I^2_{shot} \right\rangle + \left\langle I^2_{thermal} \right\rangle} = \frac{\dfrac{R^2 p_c^2}{2N_\upsilon^2}}{\dfrac{eBRP_c}{N_\upsilon} + \dfrac{4K_b T_n B}{R_L}} \tag{12}$$

The parameters for mathematical analysis are listed in Table 16.2.

Figure 16.11 presents the plot of SNR versus the number of simultaneous sensor nodes with different configurations. The three-zones configuration was simulated using (6, 4, 2), (8, 6, 4), and (10, 6, 4) code combinations. The system capacity is illustrated for four zones with assigned code weights of 8, 6, 4, and 2. It can be seen that increasing the code weight can expand system capacity.

In conclusion, different configurations can be designed for each specific optical sensor. In this analysis, it was assumed that each node group (zone) with different weight had the same share of total number of nodes; however, it should be noted that if the number of nodes with lower weights are dominated, then the system capacity would increase due to the reduction

Table 16.2 Mathematical parameters adopted from Seyedzadeh et al. (2017) to evaluate system capacity

Symbol	Parameter	Value
η	Photodetector quantum efficiency	0.6
Δv	Linewidth of broadband source	3.75 THz
λ0	Operating wavelength	1550 nm
Pc	Received optical power	−10 dBm
Tn	Receiver noise temperature	300 K
R₁	Receiver load resistor	1030 Ω
e	Electron charge	1.6×10^{-19} C
h	Planck's constant	6.66×10^{-34} Js
K_b	Boltzmann's constant	1.38×10^{-23} J/K

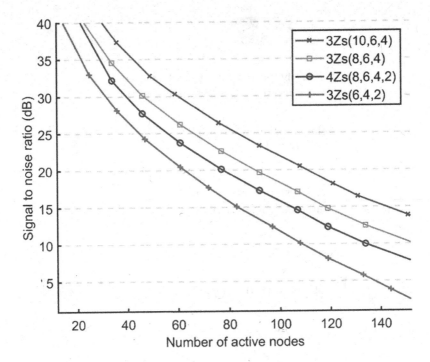

Figure 16.11 SNR of received signal vs. active nodes (Rahimian et al., 2019)

of total code length, N_v, and vice versa. Figure 16.12 illustrates a rule of thumb approach to select the code weights for the required number of zones and sensing points.

To support a large number of nodes distributed in several zones, more flexible OCDMA codes such as VW-MS (Seyedzadeh et al., 2017) can often result in better SNR and facilitated configuration.

16.8. Conclusion

This research presented the functionality gap between the emerging optical sensor networks and urban SHM in order to leverage more efficient remote vibration sensing on construction sites. From extant literature, despite the broad adoption of optical sensing in SHM, there are several challenges in monitoring various structures located in urban areas, including high cost of implementation of WDM systems, demand for synchronisation in TDM network,; and inability of sensing support of various zones distributed in an urban area without the need for independent wiring. This study presented a response to these challenges, in particular, the need for cost-effective monitoring using existing infrastructure and fibre links connecting these dispersed construction areas. Moreover, the need to minimise possible detrimental effects arising from non-linearities due to signal amplification, dispersion, and time jitter (which otherwise would need to be carefully dealt with in case of employing competing approaches including WDM or TDM systems). From this, the proposed VW-SAC OCDMA solution does not require traffic management or system synchronisation, and thanks to the

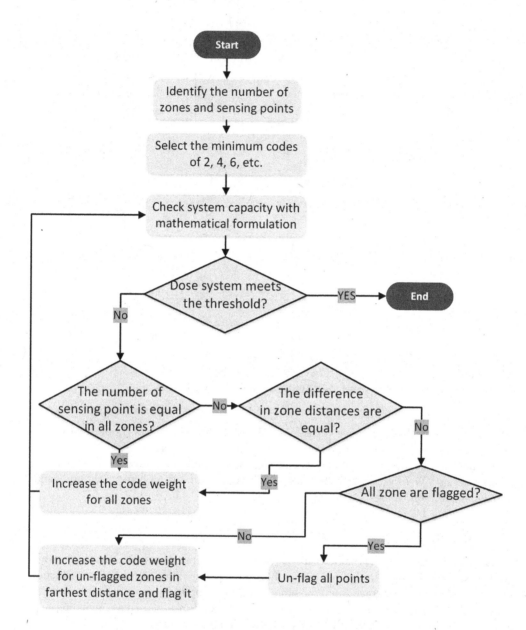

Figure 16.12 Code weight for sensor configurations flowchart

nature of the utilised optical source, this approach is resilient to performance degradation caused by fibre non-linearities.

An optical multiplexing system for monitoring vibration in unequally distributed nodes was proposed based on a VW-SAC OCDMA system. The proposed system engaging nine nodes with different distances from the control unit were investigated, using optical simulation software. Low, medium, and high vibrations were considered as sinusoidal modulation with different frequencies, and the corresponding received signals were presented in the radio frequency domain for the three nodes in different zones. The results indicated that the applied vibration frequencies were explicitly obtained for all weights in various scenarios. Furthermore, the conducted simulation showed how by increasing the fibre length (i.e. the distance of nodes from the base) that the nodes with lower weight experienced signal power degradation, making the vibration detection more difficult. The effect of transmission fibre and selection of appropriate code combination for different zones were numerically analysed by considering nine sensing points distributed in three locations. Through mathematical approximation, the capacity of VW-OCDMA for vibration sensing was demonstrated as SNR against the number of nodes, indicating that the proposed system had the potential to support a high number of nodes distributed with uneven distances. The performance of the system may be further improved using more flexible OCDMA code families. The simulation and mathematical results highlight the suitability of this system to support vibration sensing with simple implementation and high accuracy. If the security of monitoring data is not an issue, then the use of zero cross-correlation codes are recommended as they provide better codeword cardinality using the same bandwidth.

Acknowledgements

The authors would like to thank Professor Glesk at the University of Strathclyde (UK) for his professional expertise, contribution, insight, and support of this chapter.

References

Alavi, A. H., Hasni, H., Lajnef, N., Chatti, K., & Faridazar, F. (2016). An intelligent structural damage detection approach based on self-powered wireless sensor data. *Automation in Construction*, 62, 24–44. https://doi.org/10.1016/j.autcon.2015.10.001

Anas, S. B. A., Seyedzadeh, S., Mokhtar, M., & Sahbudin, R. K. Z. (2016). Variable weight Khazani-Syed code using hybrid fixed-dynamic technique for optical code division multiple access system. *Optical Engineering*, 55, 106101. https://doi.org/10.1117/1.OE.55.10.106101

Chae, M. J., Yoo, H. S., Kim, J. Y., & Cho, M. Y. (2012). Development of a wireless sensor network system for suspension bridge health monitoring. *Automation in Construction*, 21, 237–252. https://doi.org/10.1016/j.autcon.2011.06.008

Chen, B., Wang, X., Sun, D., & Xie, X. (2014). Integrated system of structural health monitoring and intelligent management for a cable-stayed bridge. *The Scientific World JournalWorldJournal*, 2014, 689471. https://doi.org/10.1155/2014/689471

Cheng, H. C., Wu, C. H., Yang, C. C., & Chang, Y. T. (2011). Wavelength division multiplexing/spectral amplitude coding applications in fiber vibration sensor systems. *IEEE Sensors Journal*, 11, 2518–2526. https://doi.org/10.1109/JSEN.2011.2128308

Dai, Y., Liu, Y., Leng, J., Deng, G., & Asundi, A. (2009). A novel time-division multiplexing fiber Bragg grating sensor interrogator for structural health monitoring. *Optics and Lasers in Engineering*, 47(10), 1028–1033. https://doi.org/10.1016/j.optlaseng.2009.05.012

Dawood, N., Rahimian, F., Seyedzadeh, S., & Sheikhkhoshkar, M. (2020). *Enabling the development and implementation of digital twins*. Proceedings of the 20th International Conference on Construction Applications of Virtual Reality, Teesside University. ISBN: 9780992716127

Djordjevic, I. B., & Vasic, B. (2004). Combinatorial constructions of optical orthogonal codes for OCDMA systems. *Communications Letters, IEEE*, 8, 391–393. https://doi.org/10.1109/LCOMM.2004.831331

Filograno, M. L., Guillen, P. C., Rodriguez-Barrios, A., Martin-Lopez, S., Rodriguez-Plaza, M., Á, A.-A., & Gonzalez-Herraez, M. (2012). Real-time monitoring of railway traffic using Fiber Bragg grating sensors. *IEEE Sensors Journal*, 12(1), 85–92. https://doi.org/10.1109/JSEN.2011.2135848

Ge, Y., Kuang, K. S., & Quek, S. T. (2013). Development of a low-cost bi-axial intensity-based optical fibre accelerometer for wind turbine blades. *Sensors and Actuators, A: Physical, 197*, 126–135. https://doi.org/10.1016/j.sna.2013.03.016

Hotate, K., & Ong, S. S. L. (2003). Distributed dynamic strain measurement using a correlation-based Brillouin sensing system. *IEEE Photonics Technology Letters, 15*(2), 272–274. https://doi.org/10.1109/LPT.2002.806107

Kakaee, M., Essa, S., Abd, T., & Seyedzadeh, S. (2015). Dynamic quality of service differentiation using fixed code weight in optical CDMA networks. *Optics Communications, 355*, 342–351. https://doi.org/10.1016/j.optcom.2015.03.046

Kakaee, M. H., Seyedzadeh, S., Adnan Fadhil, H., Barirah Ahmad Anas, S., & Mokhtar, M. (2014). Development of Multi-Service (MS) for SAC-OCDMA systems. *Optics & Laser Technology, 60*, 49–55. https://doi.org/10.1016/j.optlastec.2014.01.002

Klar, A., Dromy, I., & Linker, R. (2014). Monitoring tunneling induced ground displacements using distributed fiber-optic sensing. *Tunnelling and Underground Space Technology, 40*, 141–150. https://doi.org/10.1016/j.tust.2013.09.011

Ko, J., Kim, Y., & Park, C.-S. (2010). Fiber Bragg grating sensor network based on code division multiple access using a reflective semiconductor optical amplifier. *Microwave and Optical Technology Letters, 52*, 378–381. https://doi.org/10.1002/mop.24913

Kwong, W. C. (2002). Design of multilength optical orthogonal codes for optical CDMA multimedia networks. *IEEE Transactions on Communications, 50*, 1258–1265. https://doi.org/10.1109/TCOMM.2002.801499

Li, C., Zhao, Y. G., Liu, H., Wan, Z., Zhang, C., & Rong, N. (2008). Monitoring second lining of tunnel with mounted fiber Bragg grating strain sensors. *Automation in Construction, 17*, 641–644. https://doi.org/10.1016/j.autcon.2007.11.001

Li, G., Zhu, J., Sun, R., Lin, X., Yang, W., Zeng, K., Wang, F., Wang, C., & Zhou, B. (2018). *Pipe line safety monitoring using distributed optical fiber vibration sensor in the china west-east gas pipeline project*. Asia Communications and Photonics Conference, ACP. https://doi.org/10.1109/ACP.2018.8596159

Li, H.-N., Ren, L., Jia, Z.-G., Yi, T.-H., & Li, D.-S. (2016). State-of-the-art in structural health monitoring of large and complex civil infrastructures. *Journal of Civil Structural Health Monitoring, 6*, 3–16. https://doi.org/10.1007/s13349-015-0108-9

Li, H. N., Li, D. S., & Song, G. B. (2004). Recent applications of fiber optic sensors to health monitoring in civil engineering. *Engineering Structures, 26*, 1647–1657. https://doi.org/10.1016/j.engstruct.2004.05.018

Liang, W., Yin, H., Qin, L., Wang, Z., & Xu, A. (2008). A new family of 2D variable-weight optical orthogonal codes for OCDMA systems supporting multiple QoS and analysis of its performance. *Photonic Network Communications, 16*, 53–60. https://doi.org/10.1007/s11107-008-0117-2

Lu, S.-w., & Xie, H.-q. (2007). Strengthen and real-time monitoring of RC beam using "intelligent" CFRP with embedded FBG sensors. *Construction and Building Materials, 21*, 1839–1845. https://doi.org/10.1016/j.conbuildmat.2006.05.062

Noura, A., Seyedzadeh, S., & Anas, S. B. A. (2013). *Simultaneous vibration and humidity measurement using a hybrid WDM/OCDMA sensor network*. Paper presented at the 4th International Conference on Photonics, ICP 2013 – Conference Proceeding. https://doi.org/10.1109/ICP.2013.6687101

Optiwave Systems Inc. (2021). C2MI. Retrieved August 17, 2021, from https://www.c2mi.ca/en/partenaire/optiwave-systems-inc/

Prucnal, P. R. (2006). *OPTICAL CODE DIVISION MULTIPLE ACCESS Fundamentals and Applications* (Vol. OCDMA-based sensor network for monitoring construction sites affected by vibrations. Journal of Information Technology in Construction). CRC, Taylor & Francis. ISBN:9780849336836

Rahimian, F. P., Seyedzadeh, S., & Glesk, I. (2019). OCDMA-based sensor network for monitoring construction sites affected by vibrations. *Journal of Information Technology in Construction, 24*, 299–317.

Seyedzadeh Kharazi, S., Amouzad Mahdiraji, G., Sahbudin, R. K. Z., Mokhtar, M., Abas, A. F., & Anas, S. B. A. (2012). Effects of fiber dispersion on the performance of optical CDMA systems. *Optical Communications, 33*, 311–320. https://doi.org/10.1515/joc-2012-0057

Seyedzadeh Kharazi, S., Amouzad Mahdiraji, G., Sahbudin, R. K. Z. K. Z., Mokhtar, M., Abas, A. F. F., & Anas, S. B. A. B. A. (2011). *Performance analysis of a variable-weight OCDMA system under the impact of fiber impairments.* 2nd International Conference on Photonics, IEEE.

Seyedzadeh, S., Agapiou, A., Moghaddasi, M., Dado, M., & Glesk, I. (2021). WON-OCDMA system based on MW-ZCC codes for applications in optical wireless sensor networks. *Sensors, 21.* https://doi.org/10.3390/s21020539

Seyedzadeh, S., Glesk, I., Pour Rahimian, F., & Kwong, W. C. (2019). *Variable weight code division multiple access system for monitoring vibration of unequally distributed points.* International Conference on Transparent Optical Networks. https://doi.org/10.1109/ICTON.2019.8840283

Seyedzadeh, S., Mahdiraji, G. A., Sahbudin, R. K. Z., Abas, A. F., & Anas, S. B. A. (2014). Experimental demonstration of variable weight SAC-OCDMA system for QoS differentiation. *Optical Fiber Technology, 20,* 495–500. https://doi.org/10.1016/j.yofte.2014.05.013

Seyedzadeh, S., Moghaddasi, M., & Anas, S. B. A. (2016a). Effects of fibre impairments in variable weight optical code division multiple access system. *IET Optoelectronics, 10,* 221–226. https://doi.org/10.1049/iet-opt.2016.0011

Seyedzadeh, S., Moghaddasi, M., & Anas, S. B. A. (2016b). Variable-weight optical code division multiple access system using different detection schemes. *Journal of Telecommunications and Information Technology, 2016,* 50–59. Retrieved October 16, 2019, from www.infona.pl/resource/bwmeta1.element.baztech-fc836baf-c704-4b10-a947-a1ab1588cd5c/content/partDownload/75ac68d0-1229-3fe9-8815-7a127b2c2eb9

Seyedzadeh, S., & Pour Rahimian, F. (2021a). Building energy performance assessment methods. In *Data-driven modelling of non-domestic buildings energy performance* (pp. 13–30). Springer. https://doi.org/10.1007/978-3-030-64751-3_2

Seyedzadeh, S., & Pour Rahimian, F. (2021b). Modelling energy performance of non-domestic buildings. In *Data-driven modelling of non-domestic buildings energy performance* (pp. 111–133). Springer. https://doi.org/10.1007/978-3-030-64751-3_7

Seyedzadeh, S., Rahimian, F. P., Glesk, I., & Kakaee, M. H. (2017). Variable weight spectral amplitude coding for multiservice OCDMA networks. *Optical Fiber Technology, 37,* 53–60. https://doi.org/10.1016/j.yofte.2017.07.002

Seyedzadeh, S., Sahbudin, R. K. Z., Abas, A. F., & Anas, S. B. A. (2013). *Weight optimization of variable weight OCDMA for triple-play services.* Paper presented at the 4th International Conference on Photonics, ICP 2013 – Conference Proceeding, Melaka. https://doi.org/10.1109/ICP.2013.6687080

Shabaneh, A. A., Girei, S. H., Arasu, P. T., Rahman, W. B. W. A., Bakar, A. A. A., Sadek, A. Z., Lim, H. N., Huang, N. M., & Yaacob, M. H. (2014). Reflectance response of tapered optical fiber coated with graphene oxide nanostructured thin film for aqueous ethanol sensing. *Optics Communications, 331,* 320–324. https://doi.org/10.1016/j.optcom.2014.06.035

Smith, E. D. J., Blaikie, R. J., & Taylor, D. P. (1998). Performance enhancement of spectral-amplitude-coding optical CDMA using pulse-position modulation. *IEEE Transactions on Communications, 46,* 1176–1185. https://doi.org/10.1109/26.718559

Taiwo, A., Seyedzadeh, S., Sahbudin, R. K. Z., Yaacob, M. H., Mokhtar, M., & Taiwo, S. (2013). *Performance comparison of OCDMA codes for quasi-distributed fiber vibration sensing.* Paper presented at the 2013 IEEE 4th International Conference on Photonics (ICP). https://doi.org/10.1109/ICP.2013.6687084

Taiwo, A., Seyedzadeh, S., Taiwo, S., Sahbudin, R. K. Z., Yaacob, M. H., & Mokhtar, M. (2014). Performance and comparison of fiber vibration sensing using SAC-OCDMA with direct decoding techniques. *Optik – International Journal for Light and Electron Optics, 125,* 4803–4806. https://doi.org/10.1016/j.ijleo.2014.04.055

Talebinejad, I., Fischer, C., & Ansari, F. (2009). Serially multiplexed FBG accelerometer for structural health monitoring of bridges. *Smart Structures and Systems, 5*(4), 345–355. https://doi.org/10.12989/sss.2009.5.4.345

Thakur, H. V., Nalawade, S. M., Saxena, Y., & Grattan, K. T. V. (2011). All-fiber embedded PM-PCF vibration sensor for structural health monitoring of composite. *Sensors and Actuators, A: Physical, 167*(2), 204–212. https://doi.org/10.1016/j.sna.2011.02.008

Tseng, S.-P., Yen, C.-T., Syu, R.-S., & Cheng, H.-C. (2013). Employing optical code division multiple access technology in the all fiber loop vibration sensor system. *Optical Fiber Technology, 19,* 627–637. https://doi.org/10.1016/j.yofte.2013.10.001

Unzu, R., Nazabal, J. A., Vargas, G., Hernández, R. J., Fernández-Valdivielso, C., Urriza, N., Galarza, M., & Lopez-Amo, M. (2013). Fiber optic and KNX sensors network for remote monitoring a new building cladding system. *Automation in Construction, 30,* 9–14. https://doi.org/10.1016/j.autcon.2012.11.005

Wan, C., Hong, W., Liu, J., Wu, Z., Xu, Z., & Li, S. (2013). Bridge assessment and health monitoring with distributed long-gauge FBG sensors. *International Journal of Distributed Sensor Networks, 2013,* 494260. https://doi.org/10.1155/2013/494260

Wang, Y., Gong, J., Wang, D. Y., Dong, B., Bi, W., & Wang, A. (2011). A quasi-distributed sensing network with time-division-multiplexed fiber bragg gratings. *IEEE Photonics Technology Letters, 23*(2), 70–72. https://doi.org/10.1109/LPT.2010.2089676

Woyessa, G., Nielsen, K., Stefani, A., Markos, C., & Bang, O. (2016). Temperature insensitive hysteresis free highly sensitive polymer optical fiber Bragg grating humidity sensor. *Optics Express, 24,* 1206. https://doi.org/10.1364/OE.24.001206

Yang, G. C., & Kwong, W. C. (2002). *Prime codes with applications to CDMA optical and wireless networks.* Artech House. ISBN: 1580530737

Yen, C.-T., & Chen, C.-M. (2016). A study of three-dimensional optical code-division multiple-access for optical fiber sensor networks. *Computers & Electrical Engineering, 49,* 136–145. https://doi.org/10.1016/j.compeleceng.2015.02.016

Yi, T.-H., Li, H.-N., & Zhang, X.-D. (2014). Health monitoring sensor placement optimization for Canton Tower using immune monkey algorithm. *Structural Control and Health Monitoring, 22,* 123–138. https://doi.org/10.1002/stc.1664

Yi, T. H., Li, H. N., & Zhang, X. D. (2015). Health monitoring sensor placement optimization for Canton Tower using virus monkey algorithm. *Smart Structures and Systems, 15,* 1373–1392. https://doi.org/10.12989/sss.2015.15.5.1373

Zhang, J., Tian, Y., Yang, C., Wu, B., Wu, Z., Wu, G., Zhang, X., & Zhou, L. (2017). Vibration and deformation monitoring of a long-span rigid-frame bridge with distributed long-gauge sensors. *Journal of Aerospace Engineering, 30*(2). https://doi.org/10.1061/(ASCE)AS.1943-5525.0000678

Zhang, Z., & Bao, X. (2008). Continuous and damped vibration detection based on fiber diversity detection sensor by Rayleigh backscattering. *Journal of Lightwave Technology, 26*(7), 832–838. https://doi.org/10.1109/JLT.2008.919446

17 Blockchain integrated project delivery

An automated financial system

17.1. Introduction

Integrated Project Delivery (IPD) is a delivery approach that is characterised by (i) early involvement of project participants (AIA, 2007; Allison et al., 2018), (ii) sharing risk/rewards (Allison et al., 2018; Ballard et al., 2015), (iii) replacing the tender stage by a buyout stage without traditional bidding (AIA, 2007), and (iv) deferring paying profits until all project works are completed (Roy et al., 2018). Given these characteristics, it has been proffered that IPD requires a distinguished financial management approach, as well as a collaboration platform (Allison et al., 2018; Roy et al., 2018). A review of literature indicates that the required financial system and the collaboration platform for IPD projects must satisfy several requirements (Durdyev et al., 2019). These included: (i) the establishment of a readable/consistent accounting system (Roy et al., 2018), (ii) an ability for all project participants to be able to check cost records for each other (Allison et al., 2018), (iii) that all recorded data should be immutable, to achieve the desired trust environment (Allison et al., 2018; Roy et al., 2018), and (iv) that the collaboration platform should be inaccessible to third parties (Ma et al., 2018; Parn & Edwards, 2019).

Given these issues, IPD requires high levels of information and communication technology adoption in order to enable all vested parties to be able to interact and share sensitive data (Ahmad et al., 2018). It has been advocated, however, that blockchain could be an ideal solution to this challenge, because: (i) it is defined as a distributed ledger that is advantaged by decentralising the operation across the network through a specific consensus mechanism (i.e. peer to peer) (Abrishami & Elghaish, 2019), and (ii) all data are presented as blocks, which are immutable once joined the chain, thereby facilitating self-authentication for all new recorded data (Kinnaird et al., 2018; Turk & Klinc, 2017).

Recently, the research and practitioner communities have been exploring various forms of using blockchain in Architecture, Engineering, and Construction (AEC) (Dawood et al., 2020). One particular area of interest is that of construction management – using smart contracts to automate payments without appointing a third party – sharing data through a decentralised platform (Abrishami & Elghaish, 2019; ICE, 2018; Turk & Klinc, 2017). As such, the AEC industry has now started to explore blockchain opportunities in creating immutable financial systems (Lamb, 2018; Turk & Klinc, 2017). This includes sharing information in secured platforms (ICE, 2018; Lamb, 2018), using smart contracts to automate payments (Lamb, 2018; Turk & Klinc, 2017).

Another advanced methodology used in AEC is that of Building Information Modelling (BIM), which is designed to enhance project delivery (Azhar et al., 2012). That being said, a number of deficiencies still exist, most notably the lack of integration methods that foster

DOI: 10.1201/9781003106944-17

BIM adoption (Nawi et al., 2014). The most advanced form of BIM implementation used today is BIM level 3, where this relies on a delivery approach that facilitates collaboration and sharing risk/rewards among project parties (Wickersham, 2009). In this respect, AEC is now investigating the feasibility of integrating blockchain into construction processes in order to accelerate collaboration, maximise trust, and reduce costs by minimising third-party involvement in legal/financial tasks (Li et al., 2019b; Tozzi, 2018). With this in mind, taking advantage of the interrelationships between BIM and IPD has been recommended by a number of sources (Allison et al., 2018; Ma & Ma, 2017; Nawi et al., 2014). Moreover, integrating BIM with blockchain has also been encouraged (ICE, 2018; Lamb, 2018; Mason & Escott, 2018). This includes the capabilities of blockchain to offer solutions for meeting deficiencies experienced in existing financial systems. Among these, Abrishami and Elghaish (2019) presented some generic cases of the application of blockchain in AEC industry. Similarly, Turk and Klinc (2017) and Wang et al. (2017) discussed the potential of blockchain for enhancing construction management processes and tools. Other studies have also investigated the integration of blockchain and BIM, such as the work by Mathews et al. (2017), which demonstrated how blockchain could enhance collaboration through BIM.

Acknowledging these benefits, it seems that the solutions presented to date only embrace the theoretical realm, where (predominantly) their contribution remains confined to conceptual frameworks or theoretical models. The work presented in this chapter aims to provide readers with an insight into the contextual issues surrounding the practical capabilities of blockchain in AEC (and construction management in particular). In doing so, it extends existing research studies by moving beyond theoretical models to a fully developed working solution. This is presented as follows: (i) presenting the precepts for developing a framework to build an automated financial system using blockchain (hyperledger fabric), considering BIM throughout the process; (ii) building the blockchain network components and smart contract – including all IPD transactions such as reimbursed costs, profit, and cost saving – for an IPD project; and (iii) testing the proposed IPD-based blockchain framework through developing a proof-of-concept, using the IBM blockchain platform Beta 2.

In summary, this chapter is structured into several sections. It commences with the theoretical background in order to provide insight into the essence of the topic and gap in literature. This is followed by the development of the framework that describes the validation process – the case project. It concludes with setting out several recommendations for practitioners, clarifying the implications from this study and future opportunities.

17.2. Theoretical background

17.2.1. *Distributed ledger technology*

From a distributed ledger perspective, Tapscott and Tapscott (2016) defined blockchain as a distributed ledger that records all shared data among different members in a network. Each transaction represents a block in the network, and subsequently, new blocks are linked to previous blocks, in order to create a chain (Li et al., 2018a). From this, the interrelationships among all blocks maximise the opportunity of security (Liang et al., 2017). That is, each block carries data and hash for previous blocks in order to reduce the chance of hacking or interference (Nofer et al., 2017). Li et al. (2018a) defined two categories of blockchain networks (BCNs): specifically, *public BCN* and *private BCN*. The *public BCN* can be accessed publicly under the generic consensus mechanism; however, this remains secure due to its level of cryptography – similar to Bitcoin (Andoni et al., 2019). The second category was termed *private BCN*, characterised by having pre-identified users, for which specific consensus must be identified clearly (Li et al., 2018b). The *private BCN* represents a single BCN platform for an organisation, where data are centralised in this organisation, albeit this is decentralised between network users (Andoni et al., 2019).

Kumar and Mallick (2018) defined BCN as a tamper-proof technology that was fit for multifunction, whilst also presenting a promising technology to avoid a range of poor practice across various industries. BCN has also been acknowledged as being able to provide high-level security, as the block recorder is able to check all recorded data in terms of sequence, including the interrelationship of data in the network (Banafa, 2017). This ability prevents the likelihood of tampering with data in BCN (Kumar & Mallick, 2018). As such, BCN is seen as being efficient in supporting computing solutions (ICE, 2018; Lamb, 2018; Seyedzadeh & Pour Rahimian, 2021a, 2021b; Turk & Klinc, 2017). Given this, implementing blockchain cost has been justified, especially compared with the cost of using third parties to implement financial tasks (Ammous, 2018), which, since the price of making a transaction relies on the size of the blockchain network and its load, these costs can be optimised by adding specific provisions through a smart contract (Fröwis & Böhme, 2019).

Finally, BCNs also include nodes. These are divided into two categories: the network member nodes and the orderer peer nodes (that direct information inside the BCN). This distinction is important, as smart contracts tend to include a set of functions for sending any new data to a BCN. These functionalities and relationships are explained in the following section.

17.2.2. Smart contracts

The development of smart contracts dates backs to around 1994, defined as an automated system for performing contract terms such as payment transactions – using an automated/agreed protocol (Christidis & Devetsikiotis, 2016; Tapscott & Tapscott, 2016). In doing so, a traditional trusted third party is not needed, due to contract terms being executed based on pre-identified consensus mechanisms (Mason, 2017). In this respect, Peters and Panayi (2016) proposed a comprehensive definition for a smart contract: a platform for enforcing and monitoring the data entered by trusted sources, to be stored in BCN, based on pre-identified contract terms, where these pre-identified terms should be coded/written using a program language like Go (see Donovan and Kernighan (2015) for further details). Blockchain features and BCNs are still evolving, particularly the ability to transfer cryptocurrency/data over blockchain (Christidis & Devetsikiotis, 2016). Moreover, Andoni et al. (2019) noted that smart contracts tend to use peer-to-peer (PTP) networks that enable multi-trusted parties to manage data simultaneously, where each chain in the BCN carries its own data and is subsequently stored in a ledger, according to agreed-upon consensus mechanisms (Watanabe et al., 2016). These developments in smart contracts therefore reduce the dependency on lawyers/third parties to execute and monitor contract terms such as financial transactions, but equally, improve the accuracy and transparency of data (Mason & Escott, 2018). Smart contracts have also been seen to provide additional benefits, including an automatic audit for all transferred data (Christidis & Devetsikiotis, 2016). This enables greater validity, as the data is immutable, transparent, and secure.

From a development perspective, smart contracts have a close affinity with the chaincode in the hyperledger fabric, where chaincode ensures that all transactions are linked and sequenced properly. So, in order to provide additional clarity on the origin of smart contracts (and how they work), it is important to explain the structure of the hyperledger fabric as a blockchain platform and the way in which chaincode operates.

Klaokliang et al. (2018) described the structure of hyperledger as follows:

- **Ledger:** a set of blocks that records multiple transactions;
- **Peer:** a pool that contains ledgers and smart contracts;
- **Chaincode:** the smart contract to perform transactions according to the hyperledger concept;
- **Channel:** the path which the transaction and blocks take – to be allocated among different peers;
- **Endorsement policy:** a set of instructions that provide specific metrics to the peer, to decide if the received transaction was valid – or invalid (Hyperledger, 2018);
- **Ordering service:** a node, Ordering Service Node (OSN), exploited to order transactions and blocks based on an agreed-upon consensus mechanism such Kafka (see Javaid et al., 2019). This node should include specific information regarding the size of blocks, maximum time, and number of allowed transactions for each block, before assigning it to the peer through the channel (Androulaki et al., 2018; Hyperledger, 2018).
- **Consensus mechanism:** a set of protocols designed to ensure that all networks' nodes work according to the agreed-upon conditions and defined steps to endorse transactions (Andoni et al., 2019; Cachin & Vukolić, 2017; Kasireddy, 2017).

17.2.3. Blockchain and smart contracts in construction

Within the construction sector, it appears that blockchain has not been widely adopted; however, several attempts have been made to use this in business models (Tozzi, 2018). For example, 'Bimchain' presents a concept for integrating BIM into blockchain in the form of a plug-in for BIM platforms (Bimchain, 2018; Lamb, 2018). Similarly, from a smart contracts perspective, Fox (2019) noted several cases of smart contract adoption in the construction industry. These are seen as especially beneficial for delivering agreed contracts automatically, enabling parties to update any variations, enhancing copyright for project documentation, making automated payments among project parties, or acting as a claim submission platform (Lamb, 2018; Tozzi, 2018). As such, smart contracts are now considered invaluable, especially in terms of the automation of some construction processes that traditionally rely on multi-interactions – and that require contribution from project participants to make decisions (Mason, 2017; Mason & Escott, 2018).

Cardeira (2015) highlighted that late payments and insolvencies in the construction industry often led to several claims, for which adopting smart contracts could be used to significantly reduce the negative consequences (Fox, 2019). Similarly, ICE (2018) and Lamb (2018) contended that a smart contract was a simple and quick executable solution, which made it promising for business developments. In fact, complex transactions are relatively expensive, where adopting smart contracts could reduce such accumulative costs (Seetharaman, 2018).

Another challenge facing the industry is that of uncertainty. In this respect, uncertainties in construction payments are a challenge for establishing reliable cash flows, which subsequently often leads to claims, which in turn then affects business growth (Carmichael & Balatbat, 2010; Elghaish et al., 2019a). To mitigate this problem, a construction trust account has been recommended (Cardeira, 2015), where smart contracts work as trust accounts that hold money to be transferred automatically to the party who deserves it (Cardeira, 2015).

Project participants can engage with trust in smart contracts outputs, given that all the embedded data are immutable and decentralised (Lamb, 2018; Mason & Escott, 2018).

Finally, Koutsogiannis and Berntsen (2019) argued that digital construction could be seen as an integrated process. With the growth of digitalisation across AEC, smart contracts are now being implemented for a wide range of activities. The utilisation of smart contracts with cryptocurrencies can also provide a contract draft, where specific funds can be kept to avoid the common insolvency issues or late payments (Cardeira, 2015). In addition, the cross-verification approach used from several references helps create an efficient, robust, secure, and reliable system, which also helps build trust among project parties and environment (Mason, 2017; Mason & Escott, 2018).

17.2.4. BIM, IPD, and smart contracts integration

There are significant advantages to integrating smart contracts with BIM and IPD. Turk and Klinc (2017) stated that blockchain platforms (i.e. Ethereum, hyperledger) could be integrated with BIM to add new features. These features include an ability to record all changes in 3D BIM models throughout the design and construction stages, which subsequently enable stakeholders to track these changes easily (Lohry, 2015). In addition, Mason and Escott (2018) asserted that BIM integration with smart contracts was attainable, especially with the foreseeable increase in the number of sensors in devices, up to almost 25 billion. Moreover, BIM level 2 attainment was minimising paper-based communications and exchange (Gibbs et al., 2015). Acknowledging this, a platform that shares information among project parties with high levels of transparency and tracking functionality is much needed (Mosey, 2014). This approach was endorsed by Parn and Edwards (2019), who recommended the utilisation of blockchain with Common Data Environment (CDE), in order to enable tracking recorded data with recorders – as data is stored as a set of nodes.

From a contract perspective, Cousins (2018) argued that BIM processes required a 3D contractual model that included data for validation and authorisation of all possible tasks. In this respect, Bimchain could be used to minimise the gap between 3D BIM models and paper-based legal documentations (Bimchain, 2018). Moreover, this is also seen as a viable attempt of managing BIM using smart contracts that enable automated payments, insurance, and project information tracking (Bimchain, 2018; Lamb, 2018). As such, smart contracts can be coded for integration into BIM process/platforms, specifically to enable and automate traditional provisions. In doing so, this facilitates stakeholders' access to all the data available in a secure way, where project funds can be managed and released based on a set of agreed-upon rules (Cardeira, 2015; Fox, 2019). Additionally, blockchain can provide a secure and collaborative environment for the BIM process (Abrishami & Elghaish, 2019; Ahmad et al., 2018), where project parties secure the same benefits and access to all information. This in turn also allows stakeholders to have control of project changes, due to the main principle of blockchain regarding neutrality (Abrishami & Elghaish, 2019).

From an IPD perspective, Mathews et al. (2017) contended that IPD required a high level of trust and a collaboration network among core team members. This resonates with the notion that all IPD members should be "all for one, and one for all" (Ashcraft, 2012). On this theme, blockchain's capabilities, transparency, immutability, and automated data validation is seen as an opportunity to create new propositions (Abrishami & Elghaish, 2019; Watanabe et al., 2016). These benefits and rewards (tangible or intangible) can be extracted (Elghaish et al., 2019a; Pishdad-Bozorgi et al., 2013). Additionally, blockchain allows several participants to work collaboratively in a single project, which supports a data-driven digital environment for better project delivery (Abrishami & Elghaish, 2019; Koutsogiannis & Bern-tsen, 2019). This is particularly advantageous, as the combination of BIM and blockchain can provide incorruptible, reliable, and transparent systems for recording, updating, and maintaining project databases Bimchain (2018), Cousins (2018).

In summary, blockchain and smart contracts can be used to further enhance collaboration in the construction industry (Oraee et al., 2019). This openness and transparency helps keep participants informed not only of project status, but also changes in 3D BIM design, site pro-cedures, and the supply/flow of materials (Mathews et al., 2017).

17.2.5. Decision criteria for selecting the blockchain platform for IPD

One of the major hallmarks of IPD is its compensation system for allocating gain and pain ratios among project participants (Ashcraft, 2014; Fischer et al., 2017); however, this arrangement necessitates a cooperative contracting relationship that ties the individual success of participants to the overall success of achieving project objectives (Ahmad et al., 2019; AIA, 2007; Pishdad-Bozorgi, 2017). In this respect, project participants must agree upon a suitable compensation scheme (Pishdad-Bozorgi et al., 2013), with this scheme determining the proportions of cost overruns, cost underruns, and any other fees within the total budget – under the agreed cost (Fischer et al., 2017; Pishdad-Bozo-rgi et al., 2013). The cost scheme typically comprises direct, indirect, and overhead costs, and also captures the risk/reward proportions based on the degree of achievement during project delivery (Love et al., 2011; Pishdad-Bozorgi et al., 2013; Zhang & Li, 2014). In IPD, three components (or limbs) are defined, where Limb 1 represents the reimburse-ment of project costs and captures all project implementation costs (guaranteed), Limb 2 refers to the overhead costs for all participants, in addition to the profit (at-risk), and Limb 3 covers the pain or gain ratios (the contractual agreement) (Raisbeck et al., 2010; Zhang & Li, 2014).

Table 17.1 shows the characteristics of IPD in terms of financial processes. This presents five common *permissioned blockchain* platforms. The suitable platform is the one chosen where its characteristics match the corresponding IPD characteristics. The five platforms are sum-marised as follows:

- **Hyperledger fabric** (as previously discussed);
- **Ethereum**, an open and programmable blockchain platform that: (i) enables anyone to sign up and create an Ethereum account; (ii) allows users to build decentralised applica-tions and deploy smart contracts; (iii) uses cryptocurrency called Ether, and the consen-sus mechanism is not fabricated (Bahga & Madisetti, 2016);
- **R3 Corda** is designed as a specialised distributed ledger platform for the financial indus-try; it is classified as a permissioned blockchain platform, where a token can be sent using smart contract (Sandner, 2017);
- **Ripple** is an open payment system (as well as a digital currency), which is called 'XRP', engaging a consensus mechanism called Ripple Consensus Algorithm (RPCA), which is not fabricated. This has an open-source project for smart contracts (Armknecht et al., 2015);
- **Quorum** is designed to provide security whilst also maintaining a desired level of privacy for financial and banking services. For further insight into this area, please refer to Baliga et al. (2018) for additional details.

Table 17.1 Permissioned blockchain platforms and IPD financial characteristics

No	IPD characteristics	References	Blockchain platforms					References for blockchain platforms
			Hyperledger fabric	*Ethereum*	*R3 Corda*	*Ripple*	*Quorum*	
1	IPD's core team members are pre-identified entities, and all members should acquire the same information at the same time of releasing it.	(AIA, 2007; Allison et al., 2018; Rowlinson, 2017; Zhang & Li, 2014)	✓	✗	✓	✓	✓	(Fersht, 2018; Prerna, 2018)
2	Risk/rewards are shared among parties; this requires all parties to be able to track project progress (cost and schedule) by the same degree of interest, regardless of their location.	(Ballard et al., 2015; Pishdad-Bozorgi & Srivastava, 2018; Zhang & Chen, 2010; Zhang & Li, 2014)	✓	✗	✓	✓	✓	(Fersht, 2018; Team, 2019)
3	A new party can join any time after the core team members are formulated.	(Ashcraft, 2012; Elghaish et al., 2019a)	✓	✗	✓	✓	✓	(Androulaki et al., 2018; Androulaki et al., 2017; Fersht, 2018; Hirai, 2017)
4	Three financial transactions should be invoked at each payment milestone (reimbursed cost, profit, and cost saving).	(Ballard et al., 2015; Roy et al., 2018; Thomsen et al., 2009)	✓	✓	✓	✗	✓	(Cachin, 2016; cointelegraph, 2019; Dhillon et al., 2017; Fersht, 2018)
5	The consensus mechanism needs to be flexible, as it can be changed based on agreed-upon conditions.	(Ahmad et al., 2018)	✓	✗	✓	✗	✗	(Baliga, 2017; Cachin, 2016; Cachin and Vukolić, 2017; Wang et al., 2018)
6	IPD's core team members come from different backgrounds. Therefore, the financial system should be friendly for various users, understandable and, flexible; a platform with more commercial packages is preferred.	(Allison et al., 2018; Ma & Ma, 2017)	✓	✗	✗	✗	✗	(Ranade et al., 2018; Van Mölken, 2018; Woodside and Amiri, 2018)
7	Financial transactions should be invoked and recorded in specific tokens (fiat currencies such as dollars).	(Allison et al., 2018; Roy et al., 2018)	✓	✗	✓	✗	✓	(Dhillon et al., 2017; Fersht, 2018; Hyperledger, 2018; Kiviat, 2015; Prerna, 2018)

From this list, it is therefore important to choose a mechanism that is modular and flexible. This will enable IPD parties to develop the mechanism according to the team and project environment. In this respect, Ethereum, Ripple, and Quorum would not be particularly suitable for IPD; however, consensus mechanism, privacy, sending transactions (as fiat currency or tokens), and functionality of smart contracts are the main distinctions among the listed platforms. Of these, the Ethereum platform is a private blockchain, hence any interested entity can join based on agreed-upon algorithms (Prerna, 2018; Valenta & Sandner, 2017), though it was not designed for business networks.

Regarding the consensus mechanism, Ripple and Quorum use probabilistic and major voting techniques, respectively (Fersht, 2018). Accordingly, these two platforms are not sufficiently flexible for designing a consensus mechanism based on agreement among IPD's core team members. That being said, the R3 Corda permissioned blockchain platform enables network participants to modularise the consensus mechanism, where transactions can be sent and recorded as fiat currencies (Sandner, 2017; Valenta & Sandner, 2017).

The hyperledger fabric consensus mechanism is modular and can be fabricated according to the agreed-upon terms among network (project) participants (Androulaki et al., 2018; Androulaki et al., 2017; Cachin, 2016; Valenta & Sandner, 2017). Regarding the applicability of permissioned blockchain platforms, several commercial packages (in cooperation with hyperledger) can facilitate its implementation. For example, IBM blockchain cloud, Oracle blockchain platform, and SAP cloud are available packages, among others (Adler et al., 2018; Van Mölken, 2018).

From Table 17.1, the characteristics of IPD are evaluated in terms of financial processes through five common *permissioned blockchain* platforms. From this, it can be seen that the hyperledger fabric platform is more appropriate to the demands and needs of IPD projects.

17.3. Research gap

Research on blockchain has received considerable attention over recent years (Ahmad et al., 2018; Turk & Klinc, 2017), where research has demonstrated the importance of implementing specific features of blockchain (like smart contracts) in the automation of payments in the construction industry (Mason, 2017; Mason & Escott, 2018), with additional evidence to acknowledge the wide applicability of blockchain and smart contracts. As an example, Mathews et al. (2017) argued that integrating BIM and blockchain could maximise trust among project participants in AEC, and Abrishami and Elghaish (2019) proposed that blockchain could be useful for enhancing supply chain management; however, to date, available research on these topics has not gone beyond the proposal stage, *ergo* conceptual proposals and recommendations. As such, research covering these topic areas has been limited to 'theoretical conceptualisation' or possible applications of AEC blockchain.

This is a significant gap for AEC and one that requires attention. This is particularly so, especially with the growth of adopting BIM level 2 and subsequent transition to full integration through BIM level 3 (Succar & Kassem, 2015). In addition, there is also an increasing need for project participants to use IPD (Pishdad-Bozorgi et al., 2013; Rowlinson, 2017). This extends to embrace integration measures for BIM and IPD (Wickersham, 2009). Whilst there have been some barriers to this integration (particularly with financial management), consensus remains that these barriers are not untenable. It is proffered here that there is a need to embrace these issues in order to share risks and rewards; however, this requires an automated/immutable system that can record financial matters such as profit, cost savings, etc., where monetary values through IPD can be distributed to core team members when all project works have been delivered (Ballard et al., 2015; Pishdad-Bozorgi & Srivastava, 2018; Roy et al., 2018).

By comparing the requirements of an efficient IPD financial system and blockchain capabilities (specifically hyperledger fabric), it was determined that using hyperledger fabric for building an IPD financial system was a viable option. This builds upon the proven capabilities highlighted by previous studies; see Abrishami and Elghaish (2019), Li et al. (2019a), and Nawari and Ravindran (2019). These solutions provide a response to widespread recommendations and, in particular, the need to exploit blockchain integration with BIM. In doing so, outcomes are expected to enhance AEC financial processes, particularly through IPD projects (Bimchain, 2018; Lamb, 2018; Mathews et al., 2017).

17.4. Research methodology and design

In the process of developing this suggestion further, the critical challenge was to ensure the final outcome (workable solution) would be 'fit for purpose'. In doing so, this would also need to be tested and validated (as a proof of concept), using the permissioned blockchain (hyperledger fabric platform). Acknowledging this, the blockchain approach was used, as this was recognised as being able to provide a secure platform to transfer data of a sensitive nature (as previously discussed).

The work presented in this chapter uses an experiment as the principal method for testing all assumptions, particularly on the effectiveness and workability of the proposed automated financial system for IPD. The experiment approach was adopted as this can often reveal whether 'real data' supports or refutes conceptualisations or projections. For example, according to Zellmer-Bruhn et al. (2016, p. 400), "Controlled experiments isolate causal variables and enable a strong test of the robustness of a theory: they provide convincing evidence for theories, especially when followed by field studies." Given this, there was a need to countercheck the appropriateness and validity of all assumptions based on 'cause and effect', particularly to determine and observe a match between data and theory through the experiment approach (Shadish et al., 2002; Yin, 1981). The research design layout, stages, and logic can be seen in Figure 17.1.

Figure 17.1 presents the approach adopted to develop the proposed solution to address the deficiencies of IPD financial management. This integrates blockchain and smart contract requirements (hyperledger fabric platform). This approach was used to develop the 'proof-of-concept' so that this could be tested using the following tools:

- **IBM blockchain cloud beta 2 platform** – chosen for functionality and usability (Abrishami & Elghaish, 2019). This does not require skilled operators or high levels of competence. It is an easy-to-use tool and applicable for general practitioners across AEC, even junior and novice users (Hyperledger, 2018).
- **IBM VSCode blockchain extension** – chosen for its ability to write the smart contract. Provides templates to help novices write functions correctly in line with requirements.

17.5. Framework development

Framework development is divided into three main sections (following the three main phases of IPD discussed earlier). The first section focuses on preparing the BCN before its deployment. This should be implemented throughout the IPD pre-construction stage. The second section highlights the process of developing a mechanism for managing all IPD transactions

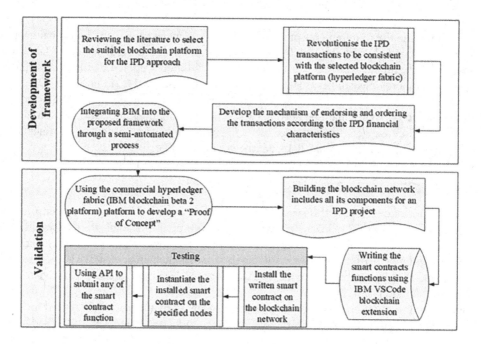

Figure 17.1 Research design and tools

within the IPD construction stage. This includes enabling parties who finish their agreed-upon works at early stages of the project timeline (to follow other contractors without needing to attend all meetings) – see Figure 17.1. The third section presents the close-out stage, which is different from traditional approaches in terms of determining the owed profit proportions for owners and non-owner parties.

The framework presented was specifically designed to integrate the triple processes of IPD, blockchain, and BIM to visualise the flow of information. The framework also provides details on the BIM dimensions and parameters. This information feeds the proposed IPD-blockchain system, using 4D to inform the payment schedule for all IPD core team members, where 5D provides the cost data (see Figure 17.2).

Figure 17.2 shows the entire process of implementing the permissioned blockchain using the hyperledger fabric across the IPD stages. Every stage of IPD includes different types of information and different tasks. Given this, the framework was designed to facilitate easy implementation. For example, it can inform users of inputs and outcomes of each IPD stage, along with the progress of developing or utilising the blockchain network. At this juncture it is important to note that each set of IPD stages has different levels of information and distinct characteristics. Acknowledging this, it was important that the BCN was developed and used according to the characteristics of IPD stages, as discussed next.

17.5.1. Conceptualisation to buyout

There are three major sections in this stage. The first concerns building the network components, where each party in the IPD core team represents a peer node in the blockchain network. This peer node carries its own ledgers in the deployment stage, as well as two peers: one to

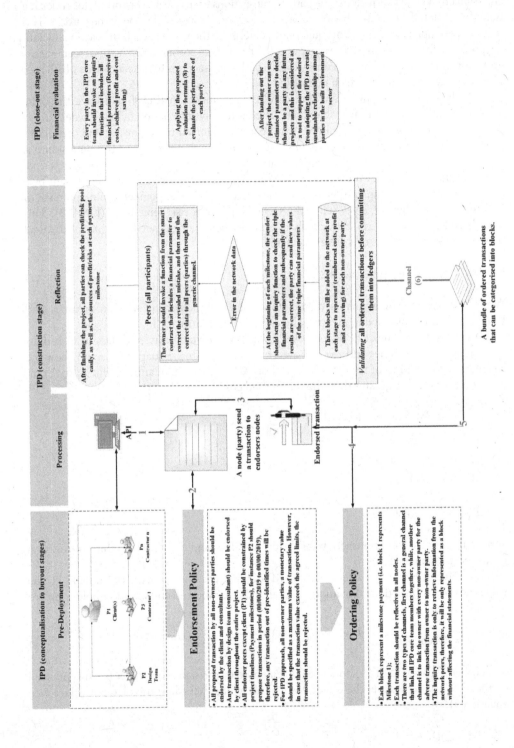

Figure 17.2 The framework: IPD-based hyperledger fabric

represent the ordering peer, and the peer is called the orderer peer. The second section covers the endorsement policy. This includes the path of transactions from one party to others for endorsing transactions and defines who should endorse transactions proposed (by one of the parties). This requires developing mathematical equations to determine the value of each transaction, including proposals for new terms in consistency for the blockchain technology. The third section covers the ordering policies. This concerns the path of the transactions to be recorded in which peer (project party) and through which nominated channel.

From the outset, it is important that core team members in IPD should have the same level of information/details available to all. Therefore, any transaction by non-owner parties (including contractors and consultant team) can be endorsed by the owner and consultant peers. Given that not all contractors finish their tasks at the same time, the time stamp is a part of the endorsement policy. Each contractor was limited to act in a specific period, extracted from the project timeline (4D BIM); however, any proposed transaction sent by a contractor beyond the ranges specified was deemed invalid. The compensation approach of IPD relies on reimbursing costs below the specified profit at risk percentage (LIMB-3). This value is coded for each party individually to prevent these being exceeded. Equation 1 shows how the IPD reimbursed costs were calculated:

$$RMVoT_i = \sum Limbs(1, 2 \, and \, 3) \leq PMVoLimbs \tag{1}$$

Where $RMVoT_i$ is the reimbursed monetary value of transaction for contractor I, and PMVoLimbs is the planned monetary value of limbs for contractor I.

Other transactions must be also invoked by any non-owner party, and these transactions were the profit/risk values and the achieved cost-saving value. Equations 2 and 3 show the calculation mechanism of these two transactions when the total planned value of the compensation structure is greater than the reimbursed costs:

$$T_{2p} = PMVoLimbs - RMVoT1 =$$
$$\begin{cases} (+)Profit \\ (+)(Profit + Cost \, saving) \\ (-)Monetary \, value \, of \, risks \end{cases} \tag{2}$$

$$T_{3CS} = T_{2p} - Limb3 \tag{3}$$

Where T_{2p} is the second transaction for the profit values and the T_{3cs} is the third transaction for the cost-saving values.

If the value of $RMVoT_i$ exceeds that of PMVoLimbs, the non-owner party should split the value into two transactions. Equation 4 presents the reimbursed costs as the whole compensation structure:

$$T_{1R} = PMVoLimbs \tag{4}$$

Another transaction (T_{2R}) should be implemented by the same contractor (i) and endorsed by the client; it represents the direct costs of all works exceeding the planned values (see equation 5).

$$T_{2R} = \sum DCALimb3 \tag{5}$$

The value of transaction 2R should be assigned to all other peer nodes that carry the stamp (which identifies the trigger of the transaction and the time of this).

It is important to appreciate that the interrelationships among project parties on the blockchain network should be identified on the endorsement path. The proposed framework assumes that the owner is committed to endorse transactions invoked by any non-owner party; however, in the case of mistakes (in previous transactions made by the client), the client has an ability to invoke a retrieved payment to receive money back, which should be endorsed only by the payer non-owner party.

The IPD smart contract should include specific functions to record the proposed financial transactions. In this respect, there are three main IPD financial functions to be coded: (1) reimbursed costs pool, (2) profit pool, and (3) cost-saving pool. Each function should include identifier parameters such as sender, value, milestone, and the trade package. Given that the IPD agreement can accept the inclusion of new members anytime during the project stages, the smart contract should also include a function for this purpose with specific parameters such as the name, trade package, and contact details. In order to maximise transparency and security for IPD parties, the profit pool can also be capped. For example, by a certain monetary value for each milestone, accumulatively, etc., where profits can be checked/endorsed automatically for any new transaction.

The ordering process presents the main part of the hyperledger fabric network component. From the IPD context, the ordering policy refers to managing and controlling relationships among project parties. That is, the movement of endorsed transactions should be pre-identified by nominating the channel for transferring the transaction data. In this respect, the ordering process presented in this study was specifically designed to follow the sequence of the project timeline, including the distinguished relationships among IPD project team members. To extend IPD characteristics in sharing all acquired data to all participants, the genotype of each transaction follows: (1) the transaction number (i.e. 1, 2, etc.); (2) the respondent (owner and non-owner parties); and (3) the endorsement status (which peer has accepted the transaction), based on the invoked party, where the endorsement policy defines which peers should endorse the transaction (Figure 17.2).

For each payment milestone, all non-owner parties implementing works based on the project timeline (4D BIM) should invoke three transactions according to the agreed-upon endorsement policy. Once all of the invoked transactions have been endorsed, the total reimbursed cost transaction is gathered in a block (i.e. block 1 for the May payment milestone). Accordingly, this block is then shared with all parties' peers through a channel. Subsequently, the other two transactions that carry the profit and cost saving are then transferred to all parties' peers, to make sure all IPD core team members have the same amount of information to enable them to make decisions (see Figure 17.2). In summary, the IPD project requires two main channels: the main channel to transfer the transactions among all parties and another individual channel among all non-owner parties and the owner (in case of errors), so that erroneous transactions can be revoked by the owner.

17.5.2. Construction stage (processing and reflection)

The processing of a transaction in hyperledger fabric includes four major stages. These stages are tailored to fit into the BIM and IPD contexts. This requires information from the BIM models to be uniquely identified in line with the characteristics of IPD. In addition, the related tasks with hyperledger fabric are also presented. These four stages are described as follows and highlighted in Figure 17.2, using numerical indications from 1 to 5:

- **Sending a transaction proposal to specific peer nodes**: according to the project timeline (4D BIM), non-owner parties who implement works should initiate request transactions using API to invoke the chaincode function. The framework relies on IBM blockchain, where the IBM cloud offers an Application Programming Interface (API) screen that can manage the blockchain network nodes, channels, and peers. Every network member can use this API screen to log in and invoke any function to record new data in the hyperledger. As stated in the endorsement policy, the transaction should be sent for endorsement to other pre-identified peers (see Figure 17.2, processing and reflection sections).
- **Endorsing proposed transactions:** all transactions should meet the mentioned endorsement policies, including the maximum value of each transaction and the planned time to invoke the transaction (see Figure 17.2, processing and reflection sections). Once a transaction has been endorsed, it returns to the transaction sender to begin the ordering process.
- **Ordering endorsed transaction:** all endorsed transactions should be transferred to the ordering peer node for signatures to be double-checked. Subsequently, transactions are then ordered chronologically (as there is an interrelationship between the transactions and a precedence for each transaction as planned in the 4D BIM), based on the agreed-upon ordering policy in the pre-deployment stage. Hence, the architecture of chaincode represents the number of transactions, respondent, and endorser (i.e. T1, consultant, owner) (see Figure 17.2, processing and reflection sections). Accordingly, based on the timestamp, the transactions are packaged into a block, to be sent to peers for commitment. The architecture of chaincode for the proposed three transactions (reimbursed costs, profit, and cost saving) are arranged as illustrated in Figure 17.3 and coded as function parameters in the IPD smart contract.
- **Committing transaction:** all ordered and packaged transactions are broadcasted to pre-identified peer nodes through the ordering policy, as identified in Figure 17.2 (processing and reflection sections). For illustration, all ordered transactions proposed by non-owner parties should be broadcasted to all peer nodes through a channel using API. Additionally, transactions originating from the owner party should be transferred to all peers (project parties), to make them aware of changes in the final statements of the three main IPD transactions.

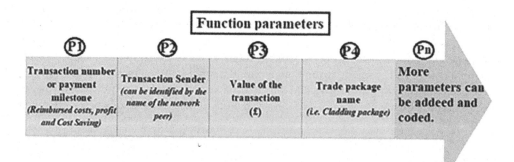

Figure 17.3 IPD-based smart contract transaction architecture

17.5.3. Close-out stage

At each milestone, the same process should be repeated; however, the accumulative value of project profit should be checked throughout the ordering service. All profit transactions for each milestone are gathered in a ledger, hence the profit node (profit pool) includes a bundle of ledgers. The summation of the profit requested by all non-owner parties is also presented in the ledger, where each profit ledger is linked to the previous one to achieve the conditions, as formulated in equation 7:

$$VAPT = \Sigma\, AVoPT_{(Ln,Pn)} \leq PLimb3_{(Mn,Ln)} \tag{7}$$

Where VAP is the valid accumulative profit transaction, $AVoPT_{(Ln, Pn)}$ is the accumulative value of profit transactions, stated in Ledger (n) for Party (p), and $PLimb3_{(Mn, Ln)}$ is the planned monetary value of Limb 3 for payment milestone (n), stated in ledger (n). As discussed, IPD supports sustainable relationships among owner and non-owner parties. Accordingly, a financial evaluation for all parties can be retrieved from the hyperledger fabric network, with evaluation parameters as presented in equation 8:

$$f\left(AFP_{ij}\right) = \begin{cases} C = ARC_{ij} - Planned\,Limbs\,(1\,\&\,2) & (-) = C \geq 0 \\ P = APP_{ij} - Planned\,Limb\,3 & (-)\,P > 0 \\ CS = ACS_{ij}\,/\,Planned\,Limbs\,(1\,\&\,2) & CA \geq 0 \end{cases} \tag{8}$$

Where C represents the paid cost, AFP_{ij} is the accumulative financial parameters for party (i) that is appointed to implement trade package (j) (£), P represents the profit parameter, ARC_{ij} is the accumulative reimbursed cost (£), APP_{ij} is the accumulative planned profit (£), CS represents the cost saving, and ACS_{ij} is the accumulative cost saving (£).

As previously mentioned, the three parameters can articulate a performance indicator for the entire IPD financial progress. In this respect, Table 17.2 illustrates how these parameters are applied.

In summary, the evaluation of the parameters provides parties with conditions applied through IPD. In this respect, an inquiry function should be coded into the IPD smart contract in order to support the collection of all information needed to undertake the proposed financial evaluation.

Table 17.2 Evaluation of financial parameters at the IPD close-out stage

Parameter	Values	Indication
C	Zero	The package has been implemented as planned and there is no achieved cost saving.
	(+)	There is a cost overrun and a part of the profit proportion is consumed.
	(-)	There is cost saving equal to the estimated value from this parameter.
P	Zero	The estimated profit is achieved.
	(-)	There is a cost overrun and a proportion of the profit is consumed as a cost.
CA	Zero	There is no achieved cost saving. This case is accompanied with the C equals zero parameter.
	>Zero	There is achieved cost underrun and the profit percentage is completely achieved.

17.5.4. Interoperability: BIM, IPD, and blockchain

Figure 17.4 illustrates the interrelationships between BIM tools and the chaincode hyperledger fabric within the IPD implementation stages. During the IPD pre-construction stages (particulary documentation and buyout stages), BIM dimensions – 3D, 4D, (Scheduling) and 5D (Cost) – provide the information needed to develop the chaincode system. Information from BIM includes the dates (start and end dates) for each trade package, which is coded into the endorsement and ordering policies; this also includes the total cost for each package and maximum estimated profit for each non-owner party, which is used for validiting profit transactions per payment milestones and accumulatively at further milestones (see Figure 17.2, the endorsement policy section). In addition, the chaincode hyperledger fabric was designed using the data from the BIM model, which includes issues such as defining the number of peers (peer per party) and the required functions to be written in IPD's smart contract format.

As shown in Figure 17.4, once the construction stage commences, non-owner parties who have implemented works are able to invoke smart contract functions by retrieving values from 5D BIM. This includes the financial resources for implementing the agreed-upon works, counting the remaining profit-at-risk percentage based on agreed-upon values in the IPD buyout stage, and determining whether cost savings were achieved through API. This process is replicated throughout the close-out stage. Since all risks/rewards are shared during the close-out stage, all parties can request the net amount of total profit, cost saving, and the reimbursed costs. Subsequently, based on the agreed-upon risk/rewards proportions during the buyout stages, each party can retrieve the owed proportion in each term: profit, cost saving, and risks (Ashcraft,

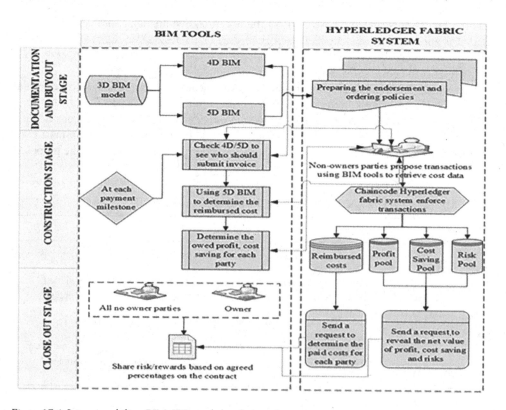

Figure 17.4 Interoperability: BIM, IPD, and the chaincode system

2012; Teng et al., 2017). This involves assessing the performance of each party, according to the achieved profit, compared to the planned profit, using 5D BIM.

In contrast to traditional accounting systems that record owed profit, cost saving, and profits for each party, the chaincode hyperledger fabric prevents parties from amending achieved percentages. This can be particulalry advantageous for endendering trust, especially where parties leave at different stages.

17.6. Proof-of-concept

17.6.1. *The blockchain (permissioned) web-based IPD*

In the process of developing this proof-of-concept, ten main steps were used to create the blockchain network using IBM blockchain platform Beta 2 (Figure 17.5). This IBM Beta platform enables enterprises to develop and extend their networks. The IPD-based blockchain proof-of-concept was developed based on the hyperledger fabric (as discussed). The hyperledger fabric

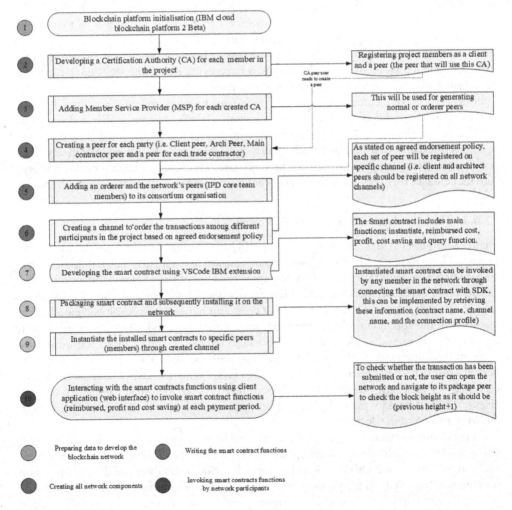

Figure 17.5 Proof-of-concept blockchain logic underpinning the IPD framework

includes specific components such as Certificate Authorisation (CA), Member Service Provider (MSP), peers, and channels, where each peer (project party) needs to have a CA as well as MSP to identify its presence in the network. The channel role is to move the information (transaction) to a set of peers (project parties) according to an agreed-upon endorsement policy. For instance, a client should have all information regarding reimbursed cost, profit, and cost saving for all participants. Moreover, the client peer should be selected when instantiating the smart contract. Similarly, the design team (architect) is also responsible for developing the network so that other participants (i.e. contractors and trade contractors) can join. These interactions are crucial. Figure 17.5 presents the blockchain network logic for automating financial transactions in the construction industry. This identifies all processes, accompanied by detailed requirements to clarify the nature of each step and responsible parties for each one.

From a development perspective, there was a need to embed smart contracts with specific algorithms. For this study, the IBM VSCode extension for blockchain was used to write these functions.

17.6.2. The case project

The case study presented here engaged a property development company wishing to build a compound of 100 identical houses. The specification for each house was as follows: (i) the gross floor area was approximately 192 m^2; (ii) the house had a single floor; (iii) from the Revit architectural plan, spaces included a master bedroom (with its own facilities of a bathroom and a robe room), three bedrooms, a large living room, kitchen, dining room, an additional bathroom, family room, and one utility room.

Project works were categorised into five trade packages (general works, ceiling, lighting fixture, finishing, and doors and windows packages). The client engaged IPD to deliver this project. Acknowledging these requirements, a core project team was formed. This consisted of an architectural firm and five trade contractors (project core group), which involved trade contractors as appropriate during kick-off meetings. The blockchain network therefore included all IPD core team members (client, five contractors, and one consultant).

17.6.3. The blockchain network: IBM blockchain beta 2

Figure 17.6 presents the seven participants (IPD core team members) used in this case project. These were classified as network members, where each party was represented by a peer. In order to create a peer, two main components were created beforehand, notably: the Certification Authority (CA) and Member Service Provider (MSP). Figure 17.6 identifies the CA for each party and one for the orderer peer. This network was developed for all seven members of the core team (client, architect, main contractor, and four trade contractors), covering the doors and windows, finishing works, ceiling works, and lighting fixture works.

Figure 17.7 illustrates the IPD core team's organisations, where each participant is identified by a distinct MSP. This approach was used to validate the identity of network members, especially for data sent between parties, where receivers are identified through their MSP (Figure 17.7). Where for example, the orderer peer operates as a node in the network, with the MSP presented in the organisation list.

As previously discussed, the channel is a main constituent part of the blockchain network and is used to move the data between network parties. Figure 17.8 shows the channel for this IPD project case. This is called 'ipdchannel', which includes members and data paths (when functions are invoked) to record new data on the network, including which parties should receive the data. In IPD projects, all core team members should receive the same amount of data in the same sequence.

Figure 17.6 Developed blockchain network based on IPD

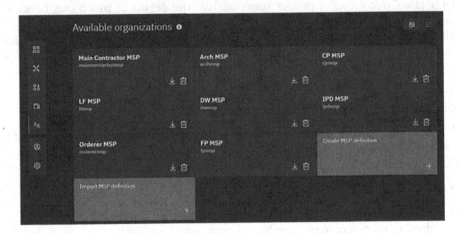

Figure 17.7 MSP for organisational members

17.6.4. *Smart contract alignment to IPD financial terms*

The IBM VSCode extension was used to build the smart contract (chaincode) functions, where it was packaged, installed, and instantiated to the specific channel and peer. From the framework presented earlier (Figure 17.4), the chaincode included a number of substantial functions such as instantiate and query function; however, users are able to add various additional functions to govern the purpose of the chaincode.

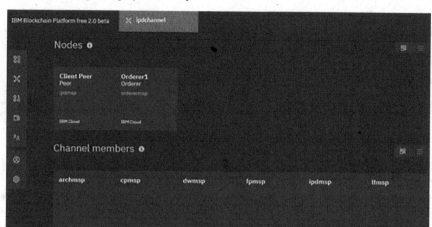

Figure 17.8 Channel assignment: data and network participants

In the prototype presented here, four functions were added to perform the proposed purpose of the framework: (1) add participants; (2) cost saving; (3) reimbursed costs; and (4) profit (recording of project transactions), the details of which can be seen in Figure 17.9.

All financial transactions were defined through specific parameters. For example, the sender, trade package, payment milestone, and the value of this transaction (see Figure 17.9).

17.6.5. Smart contract on the blockchain network

After the chaincode was developed, parties are able to subsequently invoke one of the transactions in accordance with the agreed-upon endorsement policy (see Figure 17.2). In order to provide additional clarity here, Figure 17.10 presents the installed smart contract and associated functions, where the smart contract is uploaded to the smart contract panel in the network (which contain the endorsement policies). The smart contract is then installed and instantiated in all peers (project parties), allowing project parties to invoke the four main functions (reimbursed costs, profit, saving, and query) at each payment milestone.

Upon completion of the IPD project, all project parties are able to use this system to invoke the query function in order to estimate the recorded amount of money available in each pool (reimbursed cost, profit, and cost saving). This type of inclusive engagement is one of that factors that can help and facilitate a wider rate of uptake with IPD adoption, especially given that the main barriers to conventional approaches is the lack of trust. In this respect, blockchain for business networks allows all participants an equal opportunity of tracking financial transactions.

```
1   'use strict';
2
3   const { Contract } = require('fabric-contract-api');
4
5   class MyContract extends Contract {
6
7     //update ledger with a greeting
8     async instantiate(ctx) {
9       let greeting = { text: 'Instantiate was called!' };
10      await ctx.stub.putState('GREETING', Buffer.from(JSON.stringify(greeting)));
11    }
12
13    //add a member along with their email, name, address, and number
14    async addMember(ctx, email, name, address, phoneNumber) {
15      let member = {
16        name: name,
17        address: address,
18        number: phoneNumber,
19        email: email
20      };
21      await ctx.stub.putState(email, Buffer.from(JSON.stringify(member)));
22      return JSON.stringify(member);
23    }
24
25    async Reimbuirisiedcosts(ctx, sender, milestone, value, package) {
26      let member = {
27        sender: sender,
28        milestone: milestone,
29        value: value,
30        package: package
31      };
32      await ctx.stub.putState(sender, Buffer.from(JSON.stringify(member)));
33      return JSON.stringify(member);
34    }
35
36    async profit(ctx, sender, milestone, value, package) {
37      let member = {
38        sender: sender,
39        milestone: milestone,
40        value: value,
41        package: package
42      };
43      await ctx.stub.putState(sender, Buffer.from(JSON.stringify(member)));
44      return JSON.stringify(member);
45    }
46    // look up data by key
47    async costsaving(ctx, sender, milestone, value, package) {
48      let member = {
49        sender: sender,
50        milestone: milestone,
51        value: value,
52        package: package
53      };
54      await ctx.stub.putState(sender, Buffer.from(JSON.stringify(member)));
55      return JSON.stringify(member);
56    }
57
58    // look up data by key
59    async query(ctx, key) {
60      console.info('querying for key: ' + key );
61      let returnAsBytes = await ctx.stub.getState(key);
62      let result = JSON.parse(returnAsBytes);
63      return JSON.stringify(result);
64    }
65  }
66
67  module.exports = MyContract;
```

Figure 17.9 Snapshot of chaincode based on IPD financial transactions

17.7. Discussion

The proof-of-concept presented in this chapter highlighted a workable procedure for addressing the challenges and knowledge gap identified at the outset, where it was acknowledged that a major barrier to the adoption of IPD in AEC was managing financial transactions among project parties (Durdyev et al., 2019; Elghaish et al., 2019b). In order to systematically address these issues, a methodology was presented, along with a working

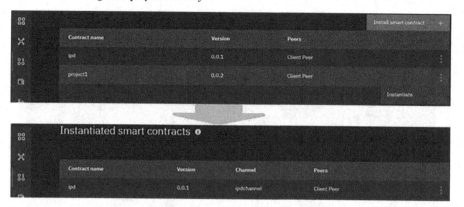

Figure 17.10 Snapshot of instantiated smart contract on the blockchain IPD network

prototype. Some of the main issues arising from this solution are discussed through the following points:

- **The profit pooling** – paying profits after all project works have been completed, regardless of the trade packages timeline (Roy et al., 2018), was an issue that needed to be addressed. This was solved through this system, as all profit transactions are received by the profit pool after passing the automated endorsement and the validation processes. Subsequently, all recorded values are immutable, negating potential amendment and subsequent problems in the entire network.
- **Inconsistencies** – one of the initial concerns noted inconsistencies in financial matters relating to IPD, specifically accounting between owner and non-owner parties (Ashcraft, 2011; Kent & Becerik-Gerber, 2010; Lichtig, 2006), the result of which caused misunderstanding. This challenge contradicts the main purpose of IPD, which is to create sustainable relationships. Therefore, in order to address this issue, the hyperledger fabric was used, which has a single consistent electronic format for recording data, where all parties receive data in the same sequence, amounts, and tokens (i.e. currencies).
- **Management** – the concern of decentralised teams through IPD (as there is no dominant party), requiring intensive meetings to make decisions (Ashcraft, 2012; Roy et al., 2018). The solution presented here addresses this issue through the utilisation of IPD-based blockchain, which reduces the need for such intensive meetings, particularly as all financial issues are managed through the hyperledger fabric network.
- **Decision-making** – one particular advantage of using a system such as this is its ability to facilitate decision-making among IPD core team members. This is enhanced through the prescribed endorsement policies and underpinning rigour associated with the algorithms. These not only define the paths of decisions in terms of identity of decision-makers and the effectiveness of previous decisions but also the validation of financial data across the decision paths.
- **Targets** – IPD targets can be seen to create sustainable relationships among AEC parties (Pishdad-Bozorgi & Beliveau, 2016); however, this requires performance evaluation at the end of the project, especially in terms of achieved rewards against risks for each party. In this respect, the hyperledger fabric network deployed in this prototype is able to engage the inquiry function to provide all recorded accumulative risk/reward values

for each node (party) in the network. As such, the owner can determine parties who achieved their targets for informed decision-making.

- **Transactions** – the need to secure transaction payments. In this respect, network blockchain uses cryptocurrencies instead of fiat currencies, where the contradiction between the private ledger at the bank and the distributed ledger in the blockchain is directly managed through the hyperledger fabric. This depends on tokens for transactions – to build the network (and link with the smart contract). The IPD approach requires merely recording three main transactions (profit, cost saving, and the reimbursed costs), where actual money can then be sent to traditional bank accounts.

- **Legal issues** – the need to embrace legal concepts and issues. These issues have been acknowledged in literature (Elghaish et al., 2019b). In this respect, contractual challenges of coding legal concepts were reflected in this study. With this in mind, this prototype developed a smart contract that included certain functions, where non-coded expressions and elements were prevented from influencing the efficiency of the entire financial process.

- **Methodology** – One of the challenges of proposing a new methodology to address challenges highlighted at the beginning of this chapter is to defend approaches based on 'need'. This should not only be supported by empirical evidence per se, but also follow structured scientific principles and logic-driven methodologies. IPD adopters were therefore used as the primary focus (or research lens), particularly given IPD's goals, merits, and solutions. The solution presented here combines (and aligns) IPD, BIM, and blockchain together through dynamic processes to exploit all available capabilities.

17.8. Conclusion

This chapter presented readers with an overview of the challenges facing AEC, and in particular, the need to move away from conventional approaches that have stifled innovation. In this respect, a methodology was presented that opened up new discourse horizons on promoting the adoption of IPD, BIM, and blockchain in AEC, which goes beyond current studies that have only concentrated on conceptual issues. The premise of this chapter was to demonstrate a real-life application of this synergy, which it was hoped would stimulate further work and development in future studies.

Whilst literature on IPD highlighted a number of challenges and opportunities, the combination of IPD, BIM, and blockchain provides a solution for many of the IPD challenges, particularly on the financial aspects. In this regard, this study is unique insofar as it is the first of its kind to showcase blockchain and smart contract technologies to address the financial management deficiencies of IPD – in particular, the capabilities of hyperledger fabric and alignment with IPD features.

On reflection, whilst this study presented a viable working platform for dealing with the financial challenges of implementing IPD projects, the development team acknowledged that there were significantly more benefits to be realised beyond the 'proof-of-concept' stage. For example, given that this prototype was fully integrated and validated (to retrieve data from the BIM model to the blockchain network), this opens another avenue of exploration, specifically relating this to develop the IPD endorsement policy. Additionally, there is also much more to be done to improve its functionality (vertically and horizontally). A case in point here is that of contingency costs – to record incurred values – where unneeded proportions and legal terms can be coded for automation within the smart contract. In this respect, this methodology could also be extended to embrace other platforms such as Oracle (following the same methodology of the IBM blockchain platform). Other areas of note include the need to embrace different procurement approaches, including Design-Build. This would, however, need to mirror the needs of these different procurement approaches, as terms and conditions vary considerably (including contractual relationships).

In broader terms, the work presented in this chapter forms a point of departure to move beyond IPD financial management systems *per se*, as the capabilities of blockchain technology have wider applicability to other areas of IPD projects. Operational flexibility is of keen interest to AEC. This was acknowledged from the outset when developing this project, particularly regarding scalability. It was therefore important that this flexibility was embedded into the system in order to accommodate multiple parties, security, privacy measures, etc. Since the hyperledger fabric is permissioned blockchain, specific details such as Certificate Authorisation, Member Service Provider, and peers/channels all need to be considered.

Finally, whilst the functionality and applicability of this framework was validated using one single case study, it is recommended that further testing and validation is undertaken with projects that vary in type, scope, and scale. This additional data would help strengthen the veracity of processes, dependencies, and functionality. Future work might also wish to reflect on the need to observe the attitudes of project stakeholders regarding future application areas (*cf.* emerging need/priorities).

References

Abrishami, S., & Elghaish, F. A. K. (2019). *Revolutionising AEC financial system within project delivery stages: A permissioned blockchain digitalised framework*. 36th CIB W78 2019 Conference: ICT in Design, Construction and Management in Architecture, Engineering, Construction and Operations, 2019. International Council for Research and Innovation in Building and Construction, pp. 199–210.

Adler, J., Berryhill, R., Veneris, A., Poulos, Z., Veira, N., & Kastania, A. (2018). Astraea: A decentralized blockchain oracle. *IEEE Blockchain Technical Briefs*. https://doi.org/10.1109/Cybermatics_2018.2018.00207

Ahmad, I., Azhar, N., & Chowdhury, A. (2018). Enhancement of IPD characteristics as impelled by information and communication technology. *Journal of Management in Engineering, 35*(1), 04018055. https://doi.org/10.1061/(ASCE)ME.1943-5479.0000670

Ahmad, I., Azhar, N., & Chowdhury, A. (2019). Enhancement of IPD characteristics as impelled by information and communication technology. *Journal of Management in Engineering, 35*(1), 04018055. https://doi.org/10.1061/(ASCE)ME.1943-5479.0000670

AIA. (2007). *Integrated project delivery: A guide*. Retrieved May 10, 2019, from www.aia.org/resources/64146-integrated-project-delivery-a-guide

Allison, M., Ashcraft, H., Cheng, R., Klawens, S., & Pease, J. (2018). *Integrated project delivery: An action guide for leaders*. Retrieved August 15, 2019, from https://leanipd.com/integrated-project-delivery-an-action-guide-for-leaders/

Ammous, S. (2018). *The bitcoin standard: The decentralized alternative to central banking*. John Wiley & Sons. ISBN: 1119473861

Andoni, M., Robu, V., Flynn, D., Abram, S., Geach, D., Jenkins, D., McCallum, P., & Peacock, A. (2019). Blockchain technology in the energy sector: A systematic review of challenges and opportunities. *Renewable and Sustainable Energy Reviews, 100*, 143–174. https://doi.org/10.1016/j.rser.2018.10.014

Androulaki, E., Barger, A., Bortnikov, V., Cachin, C., Christidis, K., De Caro, A., Enyeart, D., Ferris, C., Laventman, G., & Manevich, Y. (2018). *Hyperledger fabric: A distributed operating system for permissioned blockchains*. Proceedings of the Thirteenth EuroSys Conference, 2018. ACM, 30. https://.doi.org/10.1145/3190508.3190538

Androulaki, E., Cachin, C., De Caro, A., Kind, A., & Osborne, M. (2017). *Cryptography and protocols in hyperledger fabric*. Paper presented at the Real-World Cryptography Conference, Columbia University. https://scholar.googleusercontent.com/scholar?q=cache:3qWjsamMzR8J:scholar.google.com/&hl=en&as_sdt=0,5

Armknecht, F., Karame, G. O., Mandal, A., Youssef, F., & Zenner, E. (2015). *Ripple: Overview and outlook*. Springer. ISBN: 978-3-319-22845-7

Ashcraft, H. W. (2011). *Negotiating an integrated project delivery agreement*. Retrieved September 15, 2019, from www.hansonbridgett.com/-/media/Files/Publications/NegotiatingIntegratedProjectDeliveryAgreement.pdf

Ashcraft, H. W. (2012). *IPD framework*. Retrieved September 10, 2019, from www.hansonbridgett.com/Publications/pdf/ipd-framework.aspx

Ashcraft, H. W. (2014). Integrated project delivery: A prescription for an ailing industry. *Construction Law International*, 9, 21. Retrieved September 15, 2019, from www.hansonbridgett.com/-/media/Files/Publications/CLInt_9_4_December_2014_Ashcraft.pdf

Azhar, S., Khalfan, M., & Maqsood, T. (2012). Building information modelling (BIM): Now and beyond. *Construction Economics and Building*, 12(4), 15–28.

Bahga, A., & Madisetti, V. K. (2016). Blockchain platform for industrial internet of things. *Journal of Software Engineering and Applications*, 9(10), 533. https://.doi.org/10.4236/jsea.2016.910036

Baliga, A. (2017). *Understanding blockchain consensus models*. Retrieved September 20, 2019, from https://pdfs.semanticscholar.org/da8a/37b10bc1521a4d3de925d7ebc44bb606d740.pdf?_ga=2.21200635.1919538867.1522092864-1798624458.1520283070&source=post_page

Baliga, A., Subhod, I., Kamat, P., & Chatterjee, S. (2018). *Performance evaluation of the quorum blockchain platform*. arXiv preprint arXiv:1809.03421. https://arxiv.org/abs/1809.03421

Ballard, G., Dilsworth, B., Do, D., Low, W., Mobley, J., Phillips, P., Reed, D., Sargent, Z., Tillmann, P., & Wood10, N. (2015). How to make shared risk and reward sustainable. *Lean Construction Journal*, 2015, 257–266. Retrieved April 25, 2019, from www.leanconstruction.org/media/docs/lcj/2015/LCJ_15_003.pdf

Banafa, A. (2017). IoT and blockchain convergence: Benefits and challenges. *IEEE Internet of Things*. Retrieved September 25, 2019, from https://iot.ieee.org/newsletter/january-2017/iot-and-blockchain-convergence-benefits-and-challenges.html

Bimchain. (2018). *Accelerating BIM through the blockchain*. Retrieved April 14, 2019, from https://bimchain.io/

Cachin, C. (2018). *Architecture of the hyperledger blockchain fabric*. Workshop on Distributed Cryptocurrencies and Consensus Ledgers. Retrieved September 30, 2019, from www.zurich.ibm.com/dccl/papers/cachin_dccl.pdf

Cachin, C., & Vukolić, M. (2017). *Blockchain consensus protocols in the wild*. arXiv preprint arXiv:1707.01873. Retrieved October 6, 2019, from https://arxiv.org/abs/1707.01873

Cardeira, H. (2015). *Smart contracts and their applications in the construction industry*. Paper presented at the New Perspectives in Construction Law Conference. Retrieved October 15, 2019, from https://leanipd.com/integrated-project-delivery-an-action-guide-for-leaders/

Carmichael, D. G., & Balatbat, M. C. (2010). A contractor's analysis of the likelihood of payment of claims. *Journal of Financial Management of Property and Construction*, 15(2), 102–117. https://.doi.org/10.1108/13664381011063412

Christidis, K., & Devetsikiotis, M. (2016). Blockchains and smart contracts for the internet of things. *Ieee Access*, 4, 2292–2303. https://.doi.org/10.1109/ACCESS.2016.2566339

Cointelegraph. (2019). *What is ripple: Everything you need to know*. Retrieved April 19, 2019, from https://cointelegraph.com/ripple-101/what-is-ripple

Cousins, S. (2018). *Blockchain could hold the key to unlocking BIM level 3*. Retrieved May 5, 2019, from www.constructionmanagermagazine.com/technology/blockchain-could-hold-key-unlocking-bim-level-3/

Dawood, N., Rahimian, F., Seyedzadeh, S., & Sheikhkhoshkar, M. (2020). *Enabling the development and implementation of digital twins*. Proceedings of the 20th International Conference on Construction Applications of Virtual Reality, Teesside University. ISBN: 9780992716127

Dhillon, V., Metcalf, D., & Hooper, M. (2017). The hyperledger project. In *Blockchain enabled applications* (pp. 139–149). Springer. https://doi.org/10.1007/978-1-4842-3081-7_10

Donovan, A. A., & Kernighan, B. W. (2015). *The go programming language* (Vol. 1). Addison-Wesley Professional. ISBN: 0134190564

Durdyev, S., Hosseini, M. R., Martek, I., Ismail, S., & Arashpour, M. (2019). Barriers to the use of integrated project delivery (IPD): A quantified model for Malaysia. *Engineering, Construction and Architectural Management*. https://.doi.org/10.1108/ECAM-12-2018-0535

Elghaish, F., Abrishami, S., Abu Samra, S., Gaterell, M., Hosseini, M. R., & Wise, R. (2019a). Cash flow system development framework within integrated project delivery (IPD) using BIM tools. *International Journal of Construction Management*, 1–16. https://.doi.org/10.1080/15623599.2019.1573477

Elghaish, F., Abrishami, S., Hosseini, M. R., Abu-Samra, S., & Gaterell, M. (2019b). Integrated project delivery with BIM: An automated EVM-based approach. *Automation in Construction*, 106, 102907. https://doi.org/10.1016/j.autcon.2019.102907

Fersht, P. (2018). *The top 5 enterprise blockchain platforms you need to know about*. Retrieved April 19, 2019, from www.horsesforsources.com/top-5-blockchain-platforms_031618

Fischer, M., Khanzode, A., Reed, D., & Ashcraft, H. W. (2017). *Integrated project delivery*. John Wiley & Sons. ISBN: 9781118415382

Fox, S. (2019). *Why construction needs smart contracts*. Retrieved April 13, 2019, from www.thenbs.com/knowledge/why-construction-needs-smart-contracts

Fröwis, M., & Böhme, R. (2019). *The operational cost of ethereum airdrops*. Paper presented at the ESORICS 2019 International Workshops, DPM 2019 and CBT 2019, Luxembourg. https://doi.org/10.1007/978-3-030-31500-9_17

Gibbs, D.-J., Emmitt, S., Lord, W., & Ruikar, K. (2015). BIM and construction contracts – CPC 2013's approach. *Proceedings of the Institution of Civil Engineers – Management, Procurement and Law*, 168(6), 285–293. https://doi.org/10.1680/jmapl.14.00045

Hirai, Y. (2017). *Defining the ethereum virtual machine for interactive theorem provers*. International Conference on Financial Cryptography and Data Security, 2017 Sliema, Malta. Springer, pp. 520–535. https://doi.org/10.1007/978-3-319-70278-0_33

Hyperledger. (2018). *A blockchain platform for the enterprise (transaction flow)*. Retrieved September 10, 2019, from https://hyperledger-fabric.readthedocs.io/en/release-1.3/txflow.html

ICE. (2018). *Blockchain technology in the construction industry digital transformation for high productivity*. Retrieved August 10, 2019, from www.ice.org.uk/ICEDevelopmentWebPortal/media/Documents/News/Blog/Blockchain-technology-in-Construction-2018-12-17.pdf

Javaid, H., Hu, C., & Brebner, G. (2019). *Optimizing validation phase of hyperledger fabric*. 2019 IEEE 27th International Symposium on Modeling, Analysis, and Simulation of Computer and Telecommunication Systems (MASCOTS), IEEE, pp. 269–275. https://doi.org/10.1109/MASCOTS.2019.00038

Kasireddy, P. (2017). *Fundamental challenges with public blockchains*. Retrieved April 14, 2019, from https://medium.com/@preethikasireddy/fundamental-challenges-with-public-blockchains-253c800e9428

Kent, D. C., & Becerik-Gerber, B. (2010). Understanding construction industry experience and attitudes toward integrated project delivery. *Journal of Construction Engineering and Management*, 136(8), 815–825. https://.doi.org/10.1061/(ASCE)CO.1943-7862.0000188

Kinnaird, C., Geipel, M., & Bew, M. (2018). *Blockchain technology: How the inventions behind bitcoin are enabling a network of trust for the built environment*. Retrieved January 18, 2020, from www.arup.com/perspectives/publications/research/section/blockchain-and-the-built-environment

Kiviat, T. I. (2015). Beyond bitcoin: Issues in regulating blockchain tranactions. *DUKE LAW JOURNAL*, 65, 569. https://scholarship.law.duke.edu/dlj/vol65/iss3/4

Klaokliang, N., Teawtim, P., Aimtongkham, P., So-In, C., & Niruntasukrat, A. (2018). *A novel IoT authorization architecture on hyperledger fabric with optimal consensus using genetic algorithm*. 2018

Seventh ICT International Student Project Conference (ICT-ISPC), IEEE, pp. 1–5. https://doi.org/10.1109/ICT-ISPC.2018.8523942

Koutsogiannis, A., & Berntsen, N. (2019). *Blockchain and construction: The how, why and when.* Retrieved July 16, 2019, from www.bimplus.co.uk/people/blockchain-and-construction-how-why-and-when/

Kumar, N. M., & Mallick, P. K. (2018). Blockchain technology for security issues and challenges in IoT. *Procedia Computer Science, 132*, 1815–1823. https://doi.org/10.1016/j.procs.2018.05.140

Lamb, K. (2018). *Blockchain and smart contracts: What the AEC sector needs to know.* CDBB. Retrieved September 8, 2019.

Li, J., Greenwood, D., & Kassem, M. (2019a). Blockchain in the built environment and construction industry: A systematic review, conceptual models and practical use cases. *Automation in Construction, 102*, 288–307. https://doi.org/10.1016/j.autcon.2019.02.005

Li, J., Greenwood, D., & Kassem, M. (2019b). Blockchain in the construction sector: A socio-technical systems framework for the construction industry. In *Advances in informatics and computing in civil and construction engineering* (Vol. 1, pp. 51–57). Springer. https://doi.org/10.1007/978-3-030-00220-6_7

Li, Z., Barenji, A. V., & Huang, G. Q. (2018a). Toward a blockchain cloud manufacturing system as a peer to peer distributed network platform. *Robotics and Computer-Integrated Manufacturing, 54*, 133–144. https://doi.org/10.1016/j.rcim.2018.05.011

Li, Z., Kang, J., Yu, R., Ye, D., Deng, Q., & Zhang, Y. (2018b). Consortium blockchain for secure energy trading in industrial internet of things. *IEEE Transactions on Industrial Informatics, 14*(8), 3690–3700. https://.doi.org/10.1109/TII.2017.2786307

Liang, X., Shetty, S., Tosh, D., Kamhoua, C., Kwiat, K., & Njilla, L. (2017). *Provchain: A blockchain-based data provenance architecture in cloud environment with enhanced privacy and availability.* Proceedings of the 17th IEEE/ACM International Symposium on Cluster, Cloud and Grid Computing, IEEE Press, pp. 468–477.

Lichtig, W. A. (2006). *The integrated agreement for lean project delivery* (Vol. 26). Blackwell Publishing Ltd. ISBN: 9781444319675

Lohry, M. (2015). *Blockchain enabled co-housing.* Retrieved August 16, 2019, from https://medium.com/@MatthewLohry/blockchain-enabled-co-housing-de48e4f2b441

Love, P. E., Davis, P. R., Chevis, R., & Edwards, D. J. (2011). Risk/reward compensation model for civil engineering infrastructure alliance projects. *Journal of Construction Engineering and Management, 137*(2), 127–136. https://doi.org/10.1061/(ASCE)CO.1943-7862.0000263

Ma, Z., & Ma, J. (2017). Formulating the application functional requirements of a BIM-based collaboration platform to support IPD projects. *KSCE Journal of Civil Engineering, 21*(6), 2011–2026. https://doi.org/10.1007/s12205-017-0875-4

Ma, Z., Zhang, D., & Li, J. (2018). A dedicated collaboration platform for integrated project delivery. *Automation in Construction, 86*, 199–209. https://doi.org/10.1016/j.autcon.2017.10.024

Mason, J. (2017). Intelligent contracts and the construction industry. *Journal of Legal Affairs and Dispute Resolution in Engineering and Construction, 9*(3), 04517012. https://doi.org/10.1061 . . . %23sthash.wMgQa56Q.dpuf

Mason, J., & Escott, H. (2018, May). *Smart contracts in construction: Views and perceptions of stakeholders.* Proceedings of FIG Conference, FIG. Retrieved July 7, 2019, from https://uwe-repository.worktribe.com/output/868722/smart-contracts-in-construction-views-and-perceptions-of-stakeholders

Mathews, M., Robles, D., & Bowe, B. (2017). *BIM+ blockchain: A solution to the trust problem in collaboration?* Paper presented at the CITA BIM Gathering 2017, Croke Park. https://doi.org/10.21427/D73N5K

Mosey, D. (2014). BIM and related revolutions: A review of the Cookham Wood trial project. *Society of Construction Law.* www.scl.org.uk/papers/bim-and-related-revolutions-review-cookham-wood-trial-project

Nawari, N. O., & Ravindran, S. (2019). Blockchain technology and BIM process: Review and potential applications. *Journal of Information Technology in Construction (ITcon), 24*(12), 209–238. www.itcon.org/2019/12

Nawi, M. N. M., Haron, A. T., Hamid, Z. A., Kamar, K. A. M., & Baharuddin, Y. (2014). Improving integrated practice through building information modeling-integrated project delivery (BIM-IPD) for

Malaysian industrialised building system (IBS) construction projects. *Malaysia Construction Research Journal (MCRJ)*, *15*(2), 29–38. www.cream.my/publication/download-mcrj.html

Nofer, M., Gomber, P., Hinz, O., & Schiereck, D. (2017). Blockchain. *Business & Information Systems Engineering*, *59*(3), 183–187. https://doi.org/10.1007/s12599-017-0467-3

Oraee, M., Hosseini, M. R., Edwards, D. J., Li, H., Papadonikolaki, E., & Cao, D. (2019). Collaboration barriers in BIM-based construction networks: A conceptual model. *International Journal of Project Management*, *37*(6), 839–854. https://.doi.org/10.1016/j.ijproman.2019.05.004

Parn, E. A., & Edwards, D. (2019). Cyber threats confronting the digital built environment: Common data environment vulnerabilities and block chain deterrence. *Engineering, Construction and Architectural Management*, *26*(2), 245–266. https://.doi.org/10.1108/ECAM-03-2018-0101

Peters, G. W., & Panayi, E. (2016). Understanding modern banking ledgers through blockchain technologies: Future of transaction processing and smart contracts on the internet of money. In *Banking beyond banks and money* (pp. 239–278). Springer. https://doi.org/10.1007/978-3-319-42448-4

Pishdad-Bozorgi, P. (2017). Case studies on the role of integrated project delivery (IPD) approach on the establishment and promotion of trust. *International Journal of Construction Education and Research*, *13*(2), 102–124. https://doi.org/10.1080/15578771.2016.1226213

Pishdad-Bozorgi, P., & Beliveau, Y. J. (2016). Symbiotic relationships between integrated project delivery (IPD) and trust. *International Journal of Construction Education and Research*, *12*(3), 179–192. https://doi.org/10.1080/15578771.2015.1118170

Pishdad-Bozorgi, P., Moghaddam, E. H., & Karasulu, Y. (2013). *Advancing target price and target value design process in IPD using BIM and risk-sharing approaches.* Paper presented at the the 49th ASC Annual International Conference California Polytechnic State University. Retrieved January 15, 2019, from http://ascpro0.ascweb.org/archives/cd/2013/paper/CPRT115002013.pdf

Pishdad-Bozorgi, P., & Srivastava, D. (2018). *Assessment of Integrated Project Delivery (IPD) Risk and Reward Sharing Strategies from the Standpoint of Collaboration: A Game Theory Approach.* Paper presented at the Construction Research Congress 2018. https://doi.org/10.1061/9780784481271.020

Prerna. (2018). *Hyperledger vs ethereum – which blockchain platform will benefit your business?* Retrieved May 1, 2019, from www.edureka.co/blog/hyperledger-vs-ethereum/

Raisbeck, P., Millie, R., & Maher, A. (2010). *Assessing integrated project delivery: A comparative analysis of IPD and alliance contracting procurement routes.* Paper presented at the 26th Annual ARCOM Conference.

Ranade, V., Shrivastava, S., & Sharma, S. (2018). Generalised design of efficient supply chain management system and enterprise resource planning [ERP] system, using two layer blockchain setup on hyperledger fabric and ethereum. *International Journal of Innovative Research in Computer and Communication Engineering*, *6*(7), 6713–6721. https://.doi.org/10.15680/IJIRCCE.2018.0607022

Rowlinson, S. (2017). Building information modelling, integrated project delivery and all that. *Construction Innovation*, *17*(1), 45–49. https://doi.org/10.1108/CI-05-2016-0025

Roy, D., Malsane, S., & Samanta, P. K. (2018). Identification of Critical Challenges For Adoption Of Integrated Project Delivery. *Lean Construction Journal*, 32–52. www.leanconstruction.org/media/docs/ktll-add-read/Transitioning_to_Integrated_Project_Delivery_Potential_barriers_and_lessons_learned.pdf

Sandner, P. (2017). *Comparison of ethereum, hyperledger fabric and corda.* Retrieved April 19, 2019, from https://medium.com/@philippsandner/comparison-of-ethereum-hyperledger-fabric-and-corda-21c1bb9442f6

Seetharaman. (2018). *The cost-cutting potential of blockchain.* Retrieved June 9, 2019, from www.wsj.com/articles/the-cost-cutting-potential-of-blockchain-1520993100

Seyedzadeh, S., & Pour Rahimian, F. (2021a). Building energy data-driven model improved by multi-objective optimisation. In *Data-driven modelling of non-domestic buildings energy performance* (pp. 99–109). Springer. https://doi.org/10.1007/978-3-030-64751-3_6

Seyedzadeh, S., & Pour Rahimian, F. (2021b). Building energy performance assessment methods. In *Data-driven modelling of non-domestic buildings energy performance* (pp. 13–30). Springer. https://doi.org/10.1007/978-3-030-64751-3_2

Shadish, W. R., Cook, T. D., & Campbell, D. T. (2002). *Experimental and quasi-experimental designs for generalized causal inference.* Wadsworth Publishing. ISBN: 0395615569

Succar, B., & Kassem, M. (2015). Macro-BIM adoption: Conceptual structures. *Automation in Construction, 57,* 64–79. https://doi.org/10.1016/j.autcon.2015.04.018

Tapscott, D., & Tapscott, A. (2016). *Blockchain revolution: How the technology behind bitcoin is changing money, business, and the world.* Penguin Random House. ISBN: 0143196871

Team, S. (2019). *The 5 best blockchain platforms for enterprises and what makes them a good fit.* Retrieved October 25, 2019, from https://medium.com/swishlabs/the-5-best-blockchain-platforms-for-enterprises-and-what-makes-them-a-good-fit-1b44a9be59d4

Teng, Y., Li, X., Wu, P., & Wang, X. (2017). Using cooperative game theory to determine profit distribution in IPD projects. *International Journal of Construction Management,* 1–14. https://.doi.org/10.1080/15623599.2017.1358075

Thomsen, C., Darrington, J., Dunne, D., & Lichtig, W. (2009). *Managing integrated project delivery.* Retrieved July 3, 2019, from www.leanconstruction.org/wp-content/uploads/2016/02/CMAA_Managing_Integrated_Project_Delivery_1.pdf

Tozzi, C. (2018). *How blockchain innovation can help cost-efficiency in the construction industry.* Retrieved October 3, 2019, from www.nasdaq.com/article/how-blockchain-innovation-can-help-cost-efficiency-in-the-construction-industry-cm956525

Turk, Ž., & Klinc, R. (2017). Potentials of blockchain technology for construction management. *Procedia Engineering, 196,* 638–645. https://doi.org/10.1016/j.proeng.2017.08.052

Valenta, M., & Sandner, P. (2017). *Comparison of ethereum, hyperledger fabric and corda.* Frankfurt School, Blockchain Center. http://explore-ip.com/2017_Comparison-of-Ethereum-Hyperledger-Corda.pdf

Van Mölken, R. (2018). *Blockchain across Oracle: Understand the details and implications of the blockchain for Oracle developers and customers* (1st ed.). Packt Publishing Ltd. ISBN: 1788472160

Wang, J., Wu, P., Wang, X., & Shou, W. (2017). The outlook of blockchain technology for construction engineering management. *Frontiers of engineering management, 4*(1), 67–75. https://doi.org/10.15302/J-FEM-2017006

Wang, W., Hoang, D. T., Xiong, Z., Niyato, D., Wang, P., Hu, P., & Wen, Y. (2018). A survey on consensus mechanisms and mining management in blockchain networks. *Ieee Access, 7,* 22328–22370. https://doi.org/10.1109/ACCESS.2019.2896108

Watanabe, H., Fujimura, S., Nakadaira, A., Miyazaki, Y., Akutsu, A., & Kishigami, J. (2016). *Blockchain contract: Securing a blockchain applied to smart contracts.* 2016 IEEE International Conference on Consumer Electronics (ICCE), IEEE, pp. 467–468. Retrieved September 7, 2019, from https://ieeexplore.ieee.org/stamp/stamp.jsp?arnumber=7430693

Wickersham, J. (2009). Legal and business implications of building information modeling (BIM) and integrated project delivery (IPD). *Rocket: BIM-IPD Legal and Business Issues.* https://docplayer.net/23232445-Legal-and-business-implications-of-building-information-modeling-bim-and-integrated-project-delivery-ipd.html

Woodside, J. M., & Amiri, S. (2018). *Healthcare hyperchain: Digital transformation in the healthcare value chain.* Paper presented at the Twenty-fourth Americas Conference on Information Systems. Retrieved September 15, 2019, from https://aisel.aisnet.org/amcis2018/Health/Presentations/10/

Yin, R. K. (1981). The case study crisis: Some answers. *Administrative Science Quarterly, 26*(1), 58–65. https://doi.org/10.2307/2392599

Zellmer-Bruhn, M., Caligiuri, P., & Thomas, D. C. (2016). From the Editors: Experimental designs in international business research. *Journal of International Business Studies, 47*(4), 399–407. https://doi.org/10.1057/jibs.2016.12

Zhang, L., & Chen, W. (2010). *The analysis of liability risk allocation for integrated project delivery.* Paper presented at the the 2nd International Conference on Information Science and Engineering. https://doi.org/10.1109/ICISE.2010.5689527

Zhang, L., & Li, F. (2014). Risk/reward compensation model for integrated project delivery. *Engineering Economics, 25*(5), 558–567. https://doi.org/10.5755/j01.ee.25.5.3733

Index

Printed in the United States
by Baker & Taylor Publisher Services